2nd International Conference on Environment, Resources and Earth Sciences (ICERES 2021)

IOP Conference Series: Earth and Environmental Science
Volume 964

Ho Chi Minh City, Vietnam
29 October 2021

Le Van Trung
Vo Le Phu
Armando A. Apan
Marc Descloitres
Tanita Suepa
Nguyen Nhat Huy
Ho Tran Duc
Ha Quang Khai
Pham Tan Thi

ISBN: 978-1-7138-4946-9
ISSN: 1755-1307

Printed from e-media with permission by:

Curran Associates, Inc.
57 Morehouse Lane
Red Hook, NY 12571

Some format issues inherent in the e-media version may also appear in this print version.

Institute of Physics
Dirac House, Temple Back
Bristol BS1 6BE UK

Phone: 44 1 17 929 7481
Fax: 44 1 17 920 0979

techtracking@iop.org

Additional copies of this publication are available from:

Curran Associates, Inc.
57 Morehouse Lane
Red Hook, NY 12571 USA
Phone: 845-758-0400
Fax: 845-758-2633
Email: curran@proceedings.com
Web: www.proceedings.com

TABLE OF CONTENTS

IOP Conf. Series: Earth and Environmental Science **964** (2022) 011001 doi:10.1088/1755-1315/964/1/011001

Le Van Trung, Vo Le Phu, Armando A. Apan, Marc Descloitres, Tanita Suepa, Nguyen Nhat Huy, Ho Tuan Duc, Ha Quang Khai, Pham Tan Thi (Editors)

Environment, Resources, and Earth Sciences

Proceedings of the International Conference on Environment, Resources, and Earth Sciences 2021 (ICERES 2021)

IOP Conf. Series: Earth and Environmental Science **964** (2022) 011001 doi:10.1088/1755-1315/964/1/011001

Editors

Prof. Armando A. Apan
School of Civil Engineering and Surveying & Institute for Life Sciences and the Environment
University of Southern Queensland, Toowoomba 4350, QLD, Australia

Associate Prof. Le Van Trung
Department of Resources and Environmental Informatics
Faculty of Environment & Natural Resources
Ho Chi Minh City University of Technology (HCMUT), VNU HCM
Bldg. B9, 268 Ly Thuong Kiet street, Dist. 10
Ho Chi Minh City, Vietnam

Prof. Marc Descloitres
Senior Geophysicist at French Institute for Sustainable Development (IRD)
Scientific Director of CARE-Rescif Laboratory
Ho Chi Minh University of Technology (HCMUT), VNU HCM
Bldg. B7, 268 Ly Thuong Kiet street, Dist. 10
Ho Chi Minh City, Vietnam

Associate Prof. Vo Le Phu
Department of Environmental Management
Faculty of Environment & Natural Resources
Ho Chi Minh City University of Technology (HCMUT), VNU HCM
Bldg. B9, 268 Ly Thuong Kiet street, Dist. 10
Ho Chi Minh City, Vietnam

Dr. Tanita Suepa
Geo-informatics and Space Technology Development Agency (GISTDA)
88 Moo 9 Tambon Thung Sukala, Amphoe Siracha, Chonburi 20230, Thailand

Associate Prof. Nguyen Nhat Huy
Department of Environmental Engineering
Faculty of Environment & Natural Resources
Ho Chi Minh City University of Technology (HCMUT), VNU HCM
Bldg. B9, 268 Ly Thuong Kiet street, Dist. 10
Ho Chi Minh City, Vietnam

Dr. Ho Tuan Duc
Centre Asiatique de Recherche sur l'Eau, CARE
Ho Chi Minh City University of Technology (HCMUT), VNU HCM
Bldg. B7, 268 Ly Thuong Kiet street, Dist. 10
Ho Chi Minh City, Vietnam

Dr. Pham Tan Thi
Faculty of Applied Sciences
Ho Chi Minh City University of Technology (HCMUT), VNU HCM
Bldg. B4, 268 Ly Thuong Kiet street, Dist. 10
Ho Chi Minh City, Vietnam

Dr. Ha Quang Khai
Department of Environmental Management
Faculty of Environment & Natural Resources
Ho Chi Minh City University of Technology (HCMUT), VNU HCM
Bldg. B9, 268 Ly Thuong Kiet street, Dist. 10, Ho Chi Minh City, Vietnam

IOP Conf. Series: Earth and Environmental Science **964** (2022) 011001 doi:10.1088/1755-1315/964/1/011001

Preface

In parallel with a remarkable economic growth over the past several decades, the consequences of environmental degradation and resources depletion have placed a paramount challenge on the sustainable development track in Vietnam. Accordingly, in the course of the national rapid socio-economic development, issues of environmental pollution, natural resources management coupled with stresses in the age of global change, have increasingly becoming a vital of concern. Therefore, both engineering and management techniques are necessary for environmental pollution control, resources management, and resilience that draw much attention from various stakeholders, including government managers, academia, and industry.

The *International Conference on Environment, Resources and Earth Sciences* (ICERES) is the annual event which was initiated and organised by the Faculty of Environment and Natural Resources (FENR), Ho Chi Minh City University of Technology (HCMUT). Following the success of ICERES 2020, we welcome you to the ICERES 2021 of this year with the theme "Integrated Environmental Technology and Management Solutions for Sustainable and Resilient Development" which was organised in associated with the 17th **Conference on Science and Technology** on 29 October 2021 at HCMUT, Vietnam National University Ho Chi Minh City. The **ICERES 2021** is co-organised by Faculty of Environment and Natural Resources (FENR) and the Asian Center for Water Research (CARE), Ho Chi Minh City University of Technology (HCMUT). The Conference is financially supported by Ho Chi Minh City University of Technology, Vietnam National University Ho Chi Minh City.

The main objective of the Conference is to provide a platform for researchers, academicians, engineers, managers, and postgraduate students for presentation, exhibition, and discussion about the nexus of environmental solutions at both international, regional, and national levels. Further, the annual ICERES Conference is also a venue for presenting their recent studies and latest research results in the field of environmental engineering, resources management, environmental informatics, geospatial technologies, and earth sciences. Most importantly, the ICERES Conference aims to provide a setting environment for scientists and managers to exchange new ideas, innovative thinking, and to explore applicable experiences via face-to-face discussions, as well as establishing collaborative research projects or business relations, and to enable partners for their future collaborations.

The conference program was organised into three (03) sessions on the three main domains, including: *Environmental Engineering; Natural Resources and Environmental Management; and Resources and Environmental Informatics* that cover various topics on (i) Sustainable treatment technologies for water, wastewater, air-borne, and solid wastes; (ii) Policy, approaches, and technologies for the sustainable management practices of natural resources and environment, natural disasters reduction, climate change adaptation and mitigation; (iii) Applications of GIS, GNSS, UAV, remote sensing, and modeling in the field of environment and natural resources, and possibility of utilizing new technologies in managing, protecting and conserving the environments and resources for sustainable development.

Overall, the conference has attracted more than 127 papers from scientists, researchers, government managers, and graduate students. Among them, 70 high-quality papers were recommended to submit for peer-reviewing. Then, the double-blind review process was carried out for the 70 papers, in which each paper has been reviewed for its merit and novelty by at least two reviewers and one editor by matching the content areas. As a result, a total of 39 papers have been finally selected for this proceedings book. We believe that this proceedings book provides a broad overview of recent advance studies in the fields of Environment, Resources and Earth Sciences for readers.

Finally, we would like to express our sincere thanks to (i) Division of Science and Technology – Vietnam National University Ho Chi Minh City (**Prof. Lam Quang Vinh**); (ii) Rector of HCMUT (**Prof. Mai Thanh Phong**); (iii) Asian Center for Water Research – CARE (**Prof. Marc Descloitres** and **Dr. Ho Tuan Duc**) for their help and assistance in administrative works and other supports. Our special thanks go to *Mr. Feichi Gao* and *Ms. Kayleigh Parsons* for their great assistance and have always responded promptly. We would like to thank all the reviewers for their timely and rigorous reviews of the papers and all authors for their interests in the ICERES 2021.

Ho Chi Minh City, November 2021

Le Van Trung
Vo Le Phu
Armando A. Apan
Marc Descloitres
Tanita Suepa
Nguyen Nhat Huy
Ho Tuan Duc
Ha Quang Khai
Pham Tan Thi

IOP Conf. Series: Earth and Environmental Science **964** (2022) 011001 doi:10.1088/1755-1315/964/1/011001

List of Scientific Committees

Andrew Kun-Yi Lin	Department of Environmental Engineering College of Engineering National Chung Hsing University, Taichung, Taiwan
Armando A. Apan	School of Civil Engineering and Surveying & Institute for Life Sciences and the Environment University of Southern Queensland, Toowoomba 4350, Queensland, Australia
Bui Ta Long	Lab of Environmental Modelling Faculty of Environment and Natural Resources Ho Chi Minh City University of Technology (HCMUT) Vietnam National University Ho Chi Minh City
Dang Viet Hung	Department of Environmental Engineering Faculty of Environment and Natural Resources Ho Chi Minh City University of Technology (HCMUT) Vietnam National University Ho Chi Minh City
Dang Vu Bich Hanh	Department of Environmental Engineering Faculty of Environment and Natural Resources Ho Chi Minh City University of Technology (HCMUT) Vietnam National University Ho Chi Minh City
Dao Minh Trung	Thu Dau Mot University (TDMU) Thu Dau Mot city, Binh Duong province, Vietnam
Dao Thanh Son	Department of Environmental Management Faculty of Environment and Natural Resources Ho Chi Minh City University of Technology (HCMUT) Vietnam National University Ho Chi Minh City
Dat Doan	Built Environment Engineering Department Auckland University of Technology WZ 816, 34 St Paul Street

	Auckland 1010, New Zealand
Emilie Strady	Asian Centre for Water Research Ho Chi Minh City University of Technology, Vietnam National University Ho Chi Minh City
Ervin Nurhayati	Department of Environmental Engineering, Faculty of Civil, Planning, and Geo Engineering Institut Teknologi Sepuluh Nopember Surabaya, Indonesia
Idaa Warmadewanthi	Department of Environmental Engineering, Faculty of Civil, Planning, and Geo Engineering Institut Teknologi Sepuluh Nopember Surabaya, Indonesia
Julien Nemery	Faculty of Hydraulics, Civil and Environmental Engineering The Grenoble Institute of Technology, France
Lam Dao Nguyen	Ho Chi Minh City Space Technology Application Center Ho Chi Minh City, Vietnam
Le Meur Mathieu	University of Bordeaux at Montaigne (UBM) – Bordeaux INP, Carnot ISIFoR, France
Le Thi Minh Tam	Asian Centre for Water Research Ho Chi Minh City University of Technology, Vietnam National University Ho Chi Minh City
Le Van Khoa	Department of Environmental Management Faculty of Environment and Natural Resources Ho Chi Minh City University of Technology (HCMUT) Vietnam National University Ho Chi Minh City
Marc Descloitres	Institute for Sustainable Development, France
Nguyen Cong Nguyen	Faculty of Chemistry and Environment

	Ho Chi Minh City University of Technology (HCMUT) Vietnam National University Ho Chi Minh City
Tran Nguyen Hai	Faculty of Environmental and Chemical Engineering Institute of Fundamental and Applied Sciences Duy Tan University, Da Nang, Vietnam
Tran Le Luu	Master Program in Water Technology, Reuse, and Management Vietnamese German University Binh Duong Province, Vietnam
Tran Thi Van	Faculty of Environment and Natural Resources, Ho Chi Minh City University of Technology (HCMUT), Vietnam National University Ho Chi Minh City
Tsung-Yi Lin	Department of Geography National Taiwan Normal University (NTNU) Taipei, Taiwan
Vilas Nitivattananon	Department of Development and Sustainability (DDS) School of Environment, Resources and Development (SERD), Asian Institute of Technology (AIT), Thailand

IOP Conf. Series: Earth and Environmental Science **964** (2022) 011002 doi:10.1088/1755-1315/964/1/011002

Peer Review Declaration

All papers published in this volume of *IOP Conference Series: Earth and Environmental Science* have been peer reviewed through processes administered by the Editors. Reviews were conducted by expert referees to the professional and scientific standards expected of a proceedings journal published by IOP Publishing.

- **Type of peer review: Double-blind**
  The reviewers accept or decline the manuscript based on the scope of the conference, the originality of idea, the scientific content of the work, and the English writing of the manuscript. The editorial board based on the recommendation of the reviewers, including *Publishable without revision (No Revision), Publishable after a few revision (Minor Revision), Publishable only after correcting as recommended, HUGE Revision must be done (Major revision), and REJECT)* to decide that the manuscript should be revised or rejected. Once rejected, there has not been any changed for resubmission.

- **Conference submission management system:**
  https://conferences.hcmut.edu.vn/ICERES2021

- **Number of submissions received: 127** (for the whole Conference), 70 (selected and invited for IOP submission).

- **Number of submissions sent for review: 70**

- **Number of submissions accepted: 33**

- **Acceptance Rate (Number of Submissions Accepted / Number of Submissions Received X 100): 47.1%**

- **Average number of reviews per paper: 2.5**

- **Total number of reviewers involved:** 71 (60 responded)

- **Any additional info on review process:** None

- **Contact person for queries:** Assoc. Prof. Dr. Vo Le Phu (volephu@hcmut.edu.vn), Ho Chi Minh City University of Technology (HCMUT) – Vietnam National University Ho Chi Minh City.

IOP Conf. Series: Earth and Environmental Science **964** (2022) 012001 doi:10.1088/1755-1315/964/1/012001

Numerical study of sediment transport in Thu-Bon estuary and coastal areas of Hoi-An City

Nguyen Thong[1,2], Ho Tuan Duc[2,3], Phan Quang Hung[1,2] and Tran Hai Yen[1,2]

[1] Faculty of Civil Engineering, Ho Chi Minh City University of Technology (HCMUT), 268 Ly Thuong Kiet Street, District 10, Ho Chi Minh City, Vietnam
[2] Vietnam National University Ho Chi Minh City, Linh Trung Ward, Thu Duc District, Ho Chi Minh City, Vietnam
[3] Asian Centre for Water Research, HCMUT, 268 Ly Thuong Kiet Street, District 10, Ho Chi Minh City, Vietnam

nguyenthong@hcmut.edu.vn

Abstract. The area of Cua-Dai estuary and the coastal areas of Hoi-An City have experienced complicated erosion and sedimentation in recent years. Along the coast of Hoi-An, erosion often occurs, whereas in the area of Cua-Dai River, there is an accretion phenomenon that obstructs the waterway navigation from Cua-Dai to Cu-Lao-Cham. Occurrence of sand dunes in the offshore location of Cua-Dai has been recorded at a number of times in recent years. Studying the process of bed morphological change due to the sediment transport in the Thu-Bon river and the influence of monsoons in the area allows to explain the above phenomenon thus an in-depth study to propose appropriate solutions. This study used the numerical model Telemac which combines the hydro-morphodynamic and wave modules. The simulation results show that the main trend of coastal currents caused by tides and waves tends to go southward, leading to coastal erosion especially in the northeast monsoon season as well as sedimentation in the estuarine area. In addition, the model also shows the crucial role of waves in shoreline erosion, with the degree of erosion in the north coast near Cua-Dai being more severe than the southern coast, through the formation of local eddy flow on the north coast.

1. Introduction

Thu-Bon estuary (Cua-Dai) is the outlet of the Vu-Gia – Thu-Bon river system in Quang-Nam province and Da-Nang city. This is a type of estuary that opens and closes seasonally and moves cyclically under the combined effects of river and sea, thus the morphological evolution is quite complicated. Fluctuation processes are controlled by wave characteristics, beach form, sediment characteristics, and resultant near-shore circulation. In terms of morphology, the estuary forms mainly with loose materials consisting of medium to coarse sand and is prone to the water currents. In terms of hydrodynamics, the river is under the combined influence of both marine and river hydrological regimes, but has very distinct seasonal characteristics: opening and eroding river banks during the flood season with high discharge rate; and accreting quickly in the dry season with a duration of about eight months a year, causing difficulties for navigation and economic activities when the locality has to continuously dredge to ensure the depth of navigation.

The interaction between sediment, ocean waves and currents over a range of spatial and temporal scales generate a variety of morphological features observed in the coastal areas. These range from wave ripples [1], sandbars [2] and large scale bedforms [3] [4]. Various projects at different geographic scales have been undertaken to study Cua-Dai, with an aim of better understanding the effects of ocean impacts and sediment transport trends to propose appropriate measures for the local economic development. Some highlight international studies are Vietnamese – Indian project (2002–2003) "Study on erosion/deposition phenomenon in Hoi An coast", Vietnamese – Holland project (2005–2007) "Study on the environmental impact assessment from opening/closing of Tu Hien Inlet phenomenon in Tam Giang-Cau Hai Lagoon", Vietnamese – Sweden project (2005–2011) "Study on the coastal evolution and sustainable management in Vietnam" or recently AFD (France) and GIZ (Germany) projects that focused on coastal erosion in Hoi-An and proposed some hard and soft solutions to mitigate this phenomenon. Concerning the national projects, there are KC.09.24/06–10 "Study on erosion/deposition processes along South Central Vietnamese coast" or Quang Nam Province project (2013–2015) "Study on coastal erosion/deposition processes in Quang Nam province and proposal of the necessary measures for coastline protection". However, access to the full report of these projects is still limited. For published documents in the last decade, many studies monitored the shoreline dynamics of the Cua Dai estuary by using field survey data, geographic information systems techniques, and multi-temporal satellite remote sensing images [5] [6] [7]. Besides carrying out field investigations, several numerical models for explanation and prediction of erosion/deposition processes were applied [8] [9]. However, these studies have not focused on the mechanism of formation and loss of sand dunes in the estuary.

Since waves and nearshore currents cause sedimentation and morphological changes at the estuaries, this research aims to understand the mechanism of the erosion process in Hoi-An beach. The formation and loss of sand dunes in the offshore of Cua-Dai is a physical phenomenon that occurs over a long period (with a cycle of years). Therefore, in order to study this phenomenon, numerical modeling has many advantages over the laboratory scale models. This study uses numerical modeling methods to simultaneously describe 3 related physical phenomena: hydrodynamic model in river and coastal areas, wave model and morphological model. The study outcomes can provide essential information for establishing the operation institution of upstream reservoirs and sand mining in Vu Gia - Thu Bon, as well as recommendations of soft / hard measures to protect Cua-Dai beach but not leaving the bad effects on neighborhoods.

2. Study area and methods

2.1. Study area

The study area includes the downstream network of Thu-Bon river, a part of the north coast bordering Son-Tra peninsula, with a length of about 45 km along the coast and an average width of about 40 km from the shore. The area is about 2594 km^2, including 20 km of downstream river of Vu Gia-Thu Bon system, 5 km along the coastline to the north and the south of the estuary, and 4 km away from the coast. Cua-Dai River mouth is a large estuary area where the Thu Bon River flows into the East Sea. In regard to climate, this site belongs to the tropical climate zone, which is affected by the cold northern wind and has only two main seasons: rainy and dry. The average rainfall ranges from 2000 to 2500 mm, but it is unevenly distributed over time and space. The rainy season mainly occurs between September and December (representing about 80% of the total annual rainfall) and coincides with the tropical typhoon season.

This zone is described by 64571 unstructured triangular elements with the largest mesh of 2000 m (offshore element) whereas the smallest is 50m for the elements in Thu-Bon river (see figure 1). A larger scale of Cua-Dai River mouth can be seen in the figure 2 in which the bed level of this area is also represented.

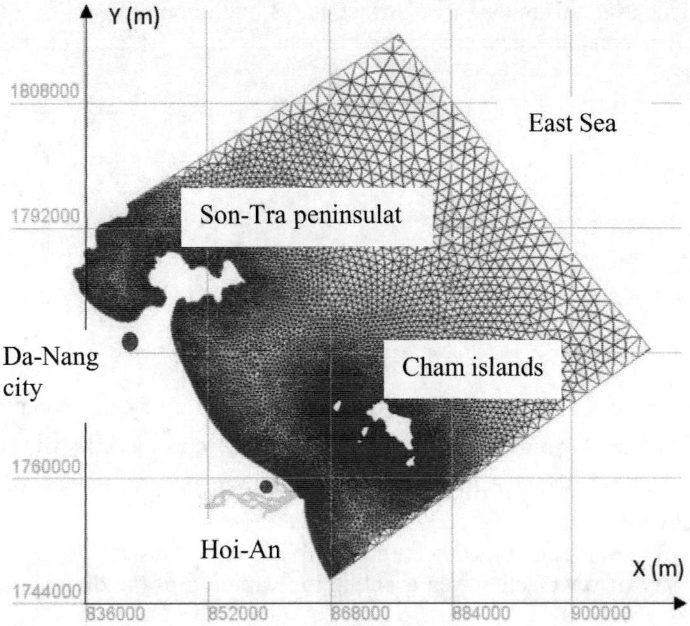

Figure 1. Numerical mesh in the study area.

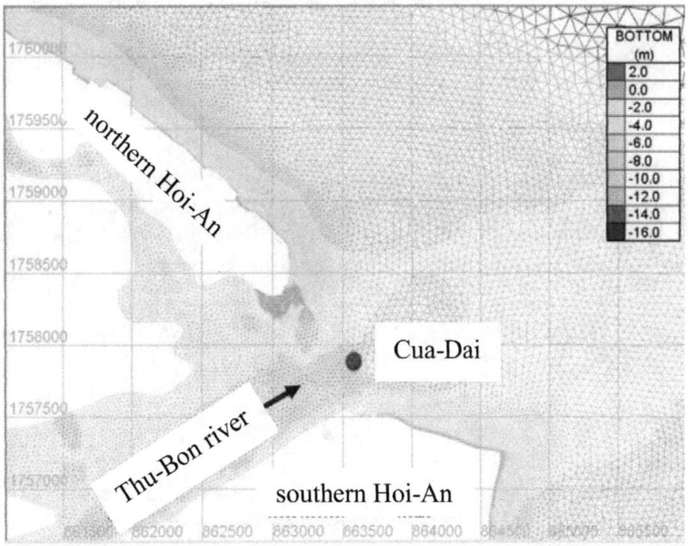

Figure 2. Cua-Dai River mouth.

2.2. Method

In this study, we used the open-source hydrodynamics software TELEMAC 2D, coupling with two-dimensional morphologic module called "SISYPHE" and spectral wave propagation module called "TOMAWAC" to evaluate the erosion mechanism in Cua-Dai estuary area under the combined and complex effects of waves and tides. Users can interfere with these modules using Fortran algorithms to achieve specific computational purposes. Telemac model was developed in 1987 by the Electricity of France (EDF) Company, with the participation of many research organizations around the world. The

Telemac-2D module is used to simulate free surface flows in 2 dimensions of horizontal space [10]. It solves the Saint-Venant equations using the finite-element or finite-volume method as follows:

Continuity equation:

$$\frac{\partial h}{\partial t} + div\left(h\vec{U}\right) = q \tag{1}$$

Momentum equation along x axis:

$$\frac{\partial u}{\partial t} + u\frac{\partial u}{\partial x} + v\frac{\partial u}{\partial y} = -g\frac{\partial Z_s}{\partial x} + F_x + \frac{1}{h}div\left[h\nu_e grad(u)\right] \tag{2}$$

Momentum equation along y axis:

$$\frac{\partial v}{\partial t} + u\frac{\partial v}{\partial x} + v\frac{\partial v}{\partial y} = -g\frac{\partial Z_s}{\partial y} + F_y + \frac{1}{h}div\left[h\nu_e grad(v)\right] \tag{3}$$

where h (m) –depth of the flow, u & v (m/s) – depth-averaged velocity components along x & y axis of velocity $\vec{U}$, q (m/s) – discharge, Z_s(m) – free surface elevation, $F_{x,y}$(m/s^2) – external forces (except gravity, for example, the Coriolis force, wind ...) per mass unit along x & y axis, ν_e (m^2/s) – coefficient of velocity diffusion.

By means of a finite-element type method, TOMAWAC solves a simplified equation for the spectro-angular density of wave action, representing the variation of the directional spectrum density by the following equations:

$$\frac{\partial N}{\partial t} + \frac{\partial \dot{x}N}{\partial x} + \frac{\partial \dot{y}N}{\partial y} + \frac{\partial \dot{k}_x N}{\partial x} + \frac{\partial \dot{k}_y N}{\partial y} = Q(k_x, k_y, x, y, t) \tag{4}$$

with $N(\vec{x}, \vec{y}, \vec{k}, t) = N(x, y, k_x, k_y, t)$ indicates spectrum density of wave.

$\vec{x} = (x, y)$ indicates position in Cartesien cordinates.

$\vec{k} = (k_x, k_y) = (k\sin\theta, k\cos\theta)$ is wave number vector, θ denote the wave propagation direction

$$\dot{g} = \frac{dg}{dt} = \frac{\partial g}{\partial t} + \frac{\partial g}{\partial x}\cdot\frac{\partial x}{\partial t} + \frac{\partial g}{\partial y}\cdot\frac{\partial y}{\partial t} \tag{5}$$

Sediment transport and bed change module SISYPHE can be used to model complex morphodynamic processes for different flow conditions, sediment size classes and sediment transport modes. SISYPHE is applicable to non-cohesive sediments that can be uniform or non-uniform, as well as cohesive sediments and sand mud mixtures. The sediment transport in 2D model is described by the following equation:

$$\frac{\partial hC}{\partial t} + \frac{\partial(hUC)}{\partial x} + \frac{\partial(hVC)}{\partial y} = \frac{\partial}{\partial x}\left(h\varepsilon_s\frac{\partial C}{\partial x}\right) + \frac{\partial}{\partial y}\left(h\varepsilon_s\frac{\partial C}{\partial y}\right) + E - D \tag{6}$$

$$(E - D)_{Zref} = \omega_s\left(C_{eq} - C_{ref}\right) \tag{7}$$

where C = C(x,y,t) is the concentration of sediment in vertical direction

$h = Z_s - Z_f \approx Z_s - Z_{ref}$ is the river stage; E: erosion rate; D: accretion rate; (U,V): average velocity in x, y direction; C_{eq} is equilibrium sediment concentration close to bed; C_{ref} is sediment concentration close to bed; ; ω_s is settling velocity.

2.3. Boundary conditions

The study area includes 2 discharge boundaries on two major tributaries located at about 12km towards the upstream of Cua-Dai estuary: a large branch near Cau-Lau bridge and a small branch in the upstream of Cua-Dai bridge. Each branch is assigned with two discharge rates respectively: the wet period (October - December) with $Q_1 = 700m^3/s$ and $Q_2 = 100m^3/s$, and the dry period (March - July) with $Q'_1 = 250m^3/s$ and $Q'_2 = 50m^3/s$. The river stage on the open boundary in the East Sea is from the astronomical tides with the 9 main waves (S_2, N_2, K_2, M_2, K_1, O_1, P_1, Q_1 and M_4) extracted from the OTIS database with a resolution of $1/30^0$.

The spatial-temporal variations of wave and wind data during the simulation period are from the NOAA's database. Typical wave height in the offshore area ranges from 1.5m to 2.5m and gradually decreases to about 1m in the estuary area. At the location near the North and South shores of Cua-Dai, wave values are below 0.5m (see figure 3).

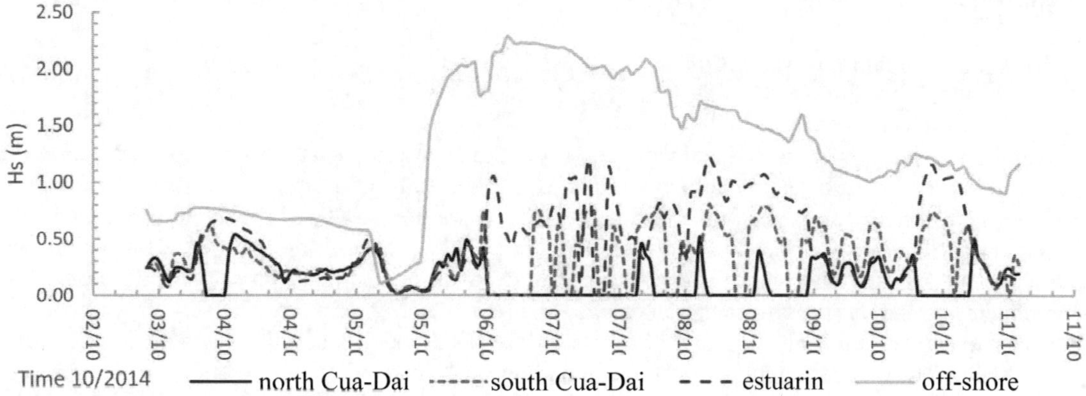

Figure 3. Typical wave height evolution at different locations during the simulation period from October 1, 2014 to October 11, 2014.

During the period between winter and early summer, mostly the Southeast wind blows to the Northwest. The maximum wind speed up to 12.7m/s occurs during the Northeast monsoon (blowing from the sea in winter). In the summer, the area is influenced by the Southwest monsoon with hot and dry westerly wind (from Laos to Central Vietnam) (see figure 4).

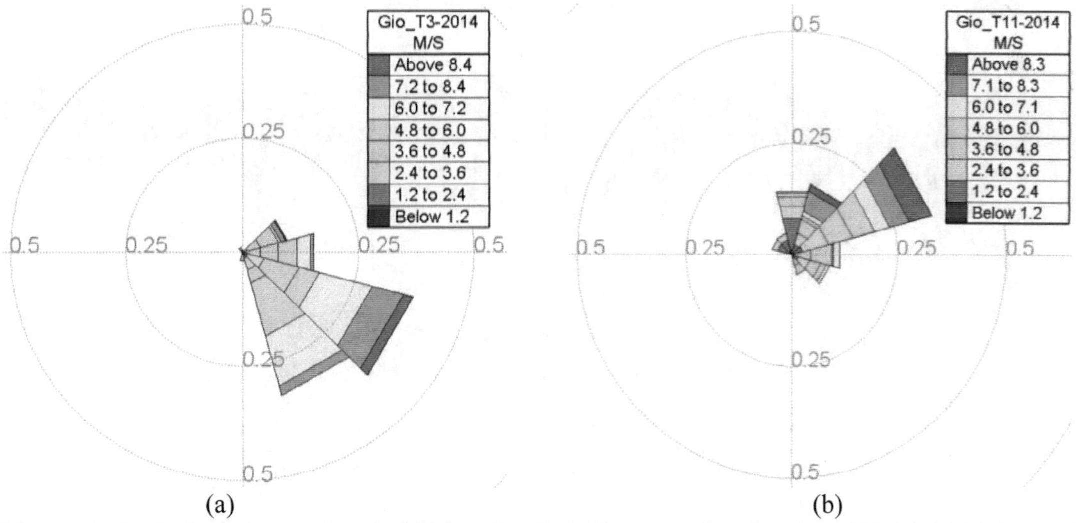

Figure 4. Typical wind rose chart in 2014 at Cua-Dai River mouth, taken from NOOA 2014 data: (a) March 2014, (b) November 2014.

The sediment is considered as non-cohesive sand and the distribution is assumed to be uniform throughout the study area. According to the sediment monitoring, particle sizes were divided into three typical classes: 0.125mm, 0.6 mm, and 1.0 mm; with a ratio of 40%, 30%, 30%, respectively (from AFD project). For sediment flux at the inflow boundary, unit suspended-sediment concentration (m^2/s) of three particle classes at the two tributaries are shown in table 1.

Table 1. Sediment flux at the inflow boundary (m^2/s).

Tributary	$Q_{0.125mm}$	$Q_{0.6mm}$	Q_{1mm}
Large branch (behind Cau-Lau bridge)	$2.644 \cdot 10^{-10}$	$3.526 \cdot 10^{-10}$	$2.644 \cdot 10^{-10}$
Small branch (upstream of Cua-Dai bridge)	$6.545 \cdot 10^{-11}$	$8.727 \cdot 10^{-11}$	$6.545 \cdot 10^{-11}$

3. Model calibration and validation

3.1. Hydrodynamic

The hydraulic model is calibrated and validated using the monitoring data at some typical locations in the wet and dry seasons. The model is sensitive to friction, eddy viscosity, Coriolis as well as space discretization. These parameters have to be adjusted to give the best fit of the model results to field observations for the same location and period of time. The simulation duration is taken within one month to analyze the short-wave components, based on the insignificant impact of the long-wave components on the short-term simulation results. Figure 5 and 6 represent respectively the comparison between observed and simulated stages in October 2014 (model calibration) and in March 2014 (model validation). Accordingly, with the one month water levels, there are 12 tidal wave components. The results from tidal harmonic analysis using the T_tide code of Pawlowicz et al. [11] showed that there is high consistency (> 90%) in amplitude and phase between the simulated and observed data (see table 2 and table 3). Tidal waves in both the calibrated (October 2014) and validated period (March 2014) showed that the principal lunar semidiurnal constituent (M_2), lunisolar diurnal constituent (K_1) and the lunar diurnal constituent (O_1) are dominant with an amplitude of 0.15m - 0.2m but there is a variation in phase between seasons. The wave height Hm0 during wet season can reach 2.5m at offshore locations, higher than in dry season. This result is consistent with the previous measurement data.

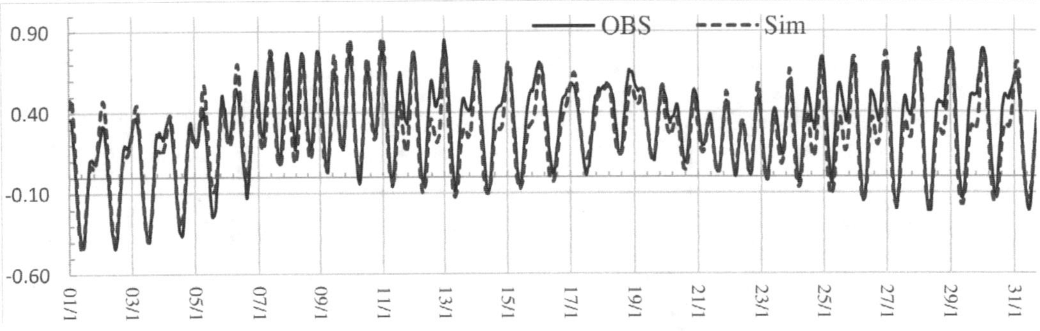

Figure 5. Comparison between observed and simulated stages in October 2014 (Model calibration).

Table 2. Harmonic analysis results for stages in October 2014 (Model calibration).

	Observed data		Simulated results	
Tide	Amplitude (m)	Phase (degree)	Amplitude (m)	Phase (degree)
K_1	0.1908	71.2	0.1844	78.1
M_2	0.1838	264.6	0.1903	252.4
O_1	0.1543	34.6	0.1491	50.5
S_2	0.0747	335.2	0.0928	334.1
N_2	0.0417	311.1	0.0458	313.7
Q_1	0.0298	95.4	0.0317	113.5
NO_1	0.0206	141.3	0.0162	201.0

Tide	Observed data		Simulated results	
	Amplitude (m)	Phase (degree)	Amplitude (m)	Phase (degree)
MO_3	0.0117	202.7	0.0182	305.1
J_1	0.0072	206.7	0.0149	208.1
MK_3	0.0057	231.4	0.0186	312.8
M_4	0.0041	117.1	0.0159	157.6

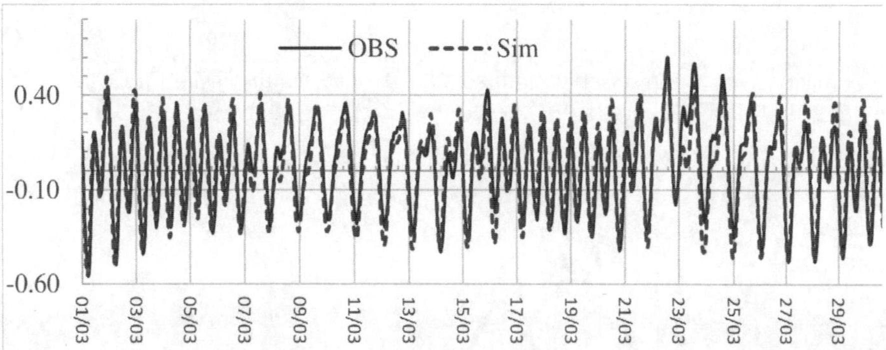

Figure 6. Comparison between observed and simulated stages in March 2014 (Model validation).

Table 3. Harmonic analysis results for stages in March 2014 (Model validation).

Tide	Observed data		Simulated results	
	Amplitude (m)	Phase (degree)	Amplitude (m)	Phase (degree)
M_2	0.1838	84.2	0.2118	69.6
O_1	0.1612	10.4	0.1642	10.7
K_1	0.1500	302.7	0.1654	304.8
S_2	0.0822	354.9	0.0907	335.9
N_2	0.0416	219.7	0.0433	213.2
NO_1	0.0279	15.1	0.0144	348.6
Q_1	0.0236	150.0	0.0299	165.9
MO_3	0.0127	5.1	0.0021	289.1
M_4	0.0081	118.3	0.0021	160.9
MS_4	0.0059	39.1	0.0019	247.8

3.2. Wave

The wave model is also calibrated using the monitoring data at some typical locations in October – November 2014 and then validated in October 2016. Figure 7 and figure 8 show respectively the comparison between observed and simulated wave height HM_0 as the calibration and validation of the wave model.

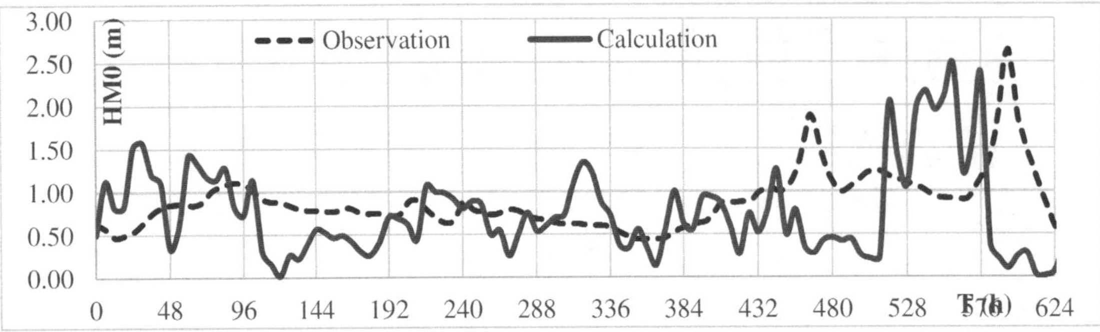

Figure 7. Comparison between observed and simulated wave height HM_0 from October 16, 2014 to November 12, 2014 (Model calibration).

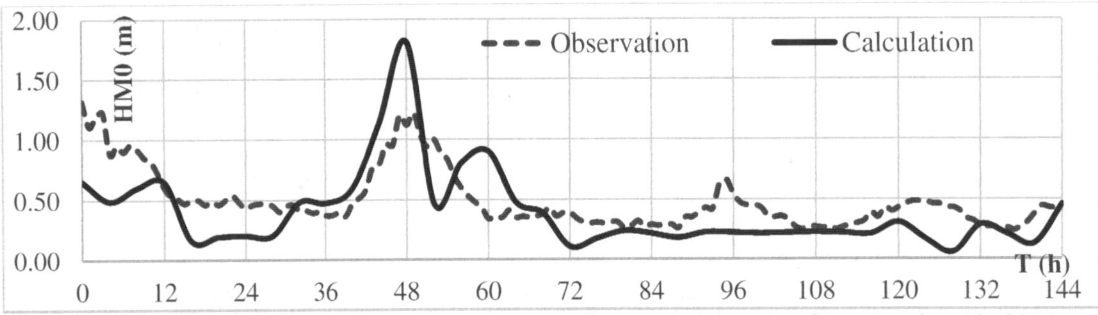

Figure 8. Comparison between observed and simulated wave height HM_0 from October 19, 2016 to October 25, 2016 (Model validation).

4. Study of sediment transport

4.1. Sediment transport

Simulation results reveal that the sediment transport occurred in all directions with the highest amount at the estuary. However, we noted a more significant sediment rate in the North whereas the sediment is shifted very far to the South coast of Cua-Dai although with a low amount (see figure 9 and 10).

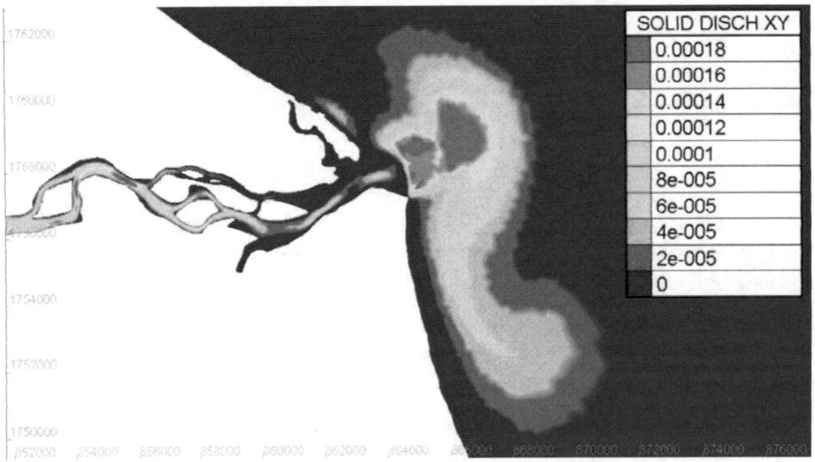

Figure 9. Sediment transport (m^2/s) in the rainy season.

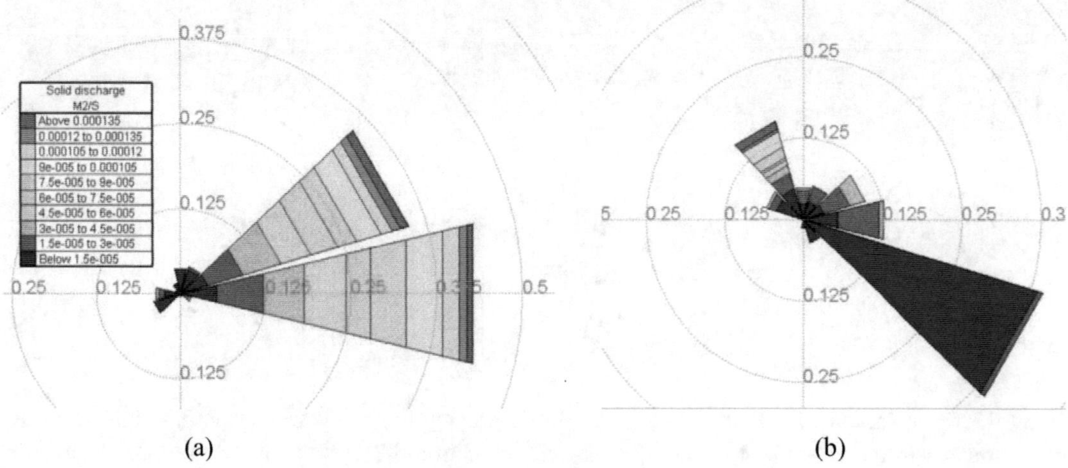

Figure 10. Distribution of sediment transport in the rainy season (m^2/s), at different locations: (a) at the estuary, (b) 1000 m from thc estuary along the river axis.

4.2. Bed morphology evolution

In figure 11, the simulation of sediment transport for 2 different periods (rainy season versus dry season) are presented. In addition, to investigate the morphological change of Thu-Bon estuary, three study sections have been applied (1-1, 2-2 and 3-3). The results show that erosion occurs mostly in the north coast of Cua-Dai estuary (about 4.5km) and happens more seriously in the Northeast monsoon period (October 2014), with the maximum erosion depth of about 0.85m (at section 1-1). In addition, sedimentation is formed in the offshore of Cua-Dai towards Cu Lao Cham, which is consistent with the observed situation in recent years.

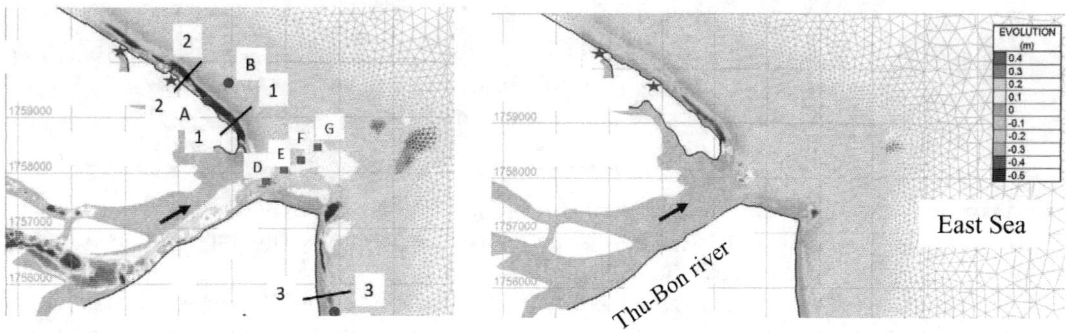

Figure 11. Simulation results of erosion (accretion) (a) from October 1, 2014 to November 15, 2014; (b) from March 1, 2014 to April 15, 2014.

4.3. Impact of wave

One of the factors that can be detrimental to the phenomenon of coastal erosion in Hoi An comes from the action of waves. To elucidate the influence of this factor, we tested a scenario during rainy season in which the wave effect is ignored. The simulation results of the change of bottom elevation are presented in figure 13. Comparison of the results between two scenarios (with and without waves) shows the crucial role of waves in coastal erosion phenomenon in Hoi-An city. In the case of ignoring the influence of waves on the Hoi An coast, especially on the northern shore, the erosion is greatly reduced and the beach is almost stable.

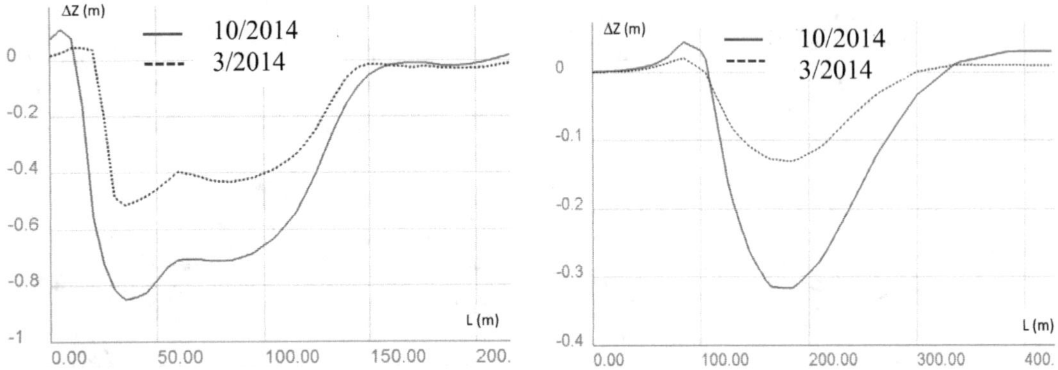

Figure 12. Comparison of bed morphology evolution during the rainy season (continuous line) and dry season (dash line). Left: section 1-1 (north); right: section 3-3 (south). Note: $\Delta Z < 0 \rightarrow$ erosion, $\Delta Z > 0 \rightarrow$ accretion.

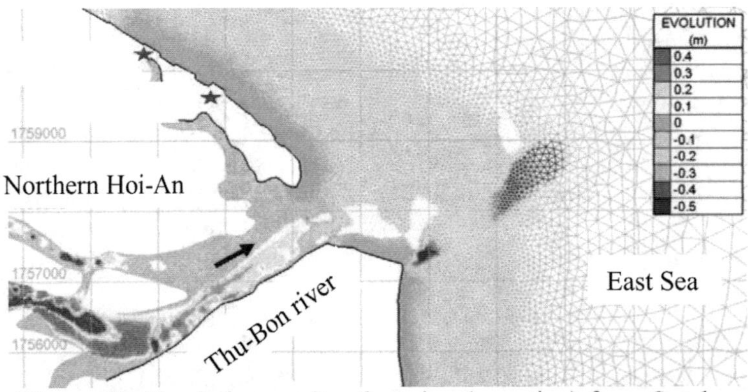

Figure 13. Simulation results of erosion (accretion) from October 1, 2014 to November 15, 2014 without the presence of waves

5. Conclusion

The simulation results show that the main trend of coastal currents caused by tides and waves tends to go southward, leading to coastal erosion in the north coast, especially during the northeast monsoon season with a rate up to one meter in october 2014. This is not the case for estuarine area with a sediment transport of 0.0002 m^2/s and sedimentation of about 0.3m is observed. In addition, the model also shows the crucial role of waves in shoreline erosion, with the degree of erosion in the north coast near Cua-Dai being more severe than the southern coast, through the formation of local eddy flow on the north coast.

Acknowledgements

The simulations have been performed at the High Performance Computing Center (HPCC), Ho Chi Minh City University of Technology (HCMUT).

This research is funded by Vietnam National University Ho Chi Minh City (VNU-HCM) under grant number C2021-20-43. We would like to thank HCMUT, VNU-HCM for the support of time and facilities for this study.

References

[1] Wiberg L and Harris K 1994 Ripple geometry in wave-dominated environments *Journal of*

Geophysical Research **99** pp 775-89. DOI: 10.1029/93JC02726 /American Elsevier p 517

[2] Ruessink G and Kroon A 1994 The behaviour of a multiple bar system in the nearshore zone of Terschelling, the Netherlands: 1965–1993 *Marine Geology* **121** pp 187–97

[3] Dyer R and Huntley A 1999 The origin, classification and modelling of sand banks and ridges *Continental Shelf Research* **19** pp 1285–330. DOI: 10.1016/S0278-4343(99)00028-X

[4] Van de Meene H, VanRijn C 2000 The shoreface-connected ridges along the central Dutch coast—part 1: Field observations. *Continental Shelf Research* **20** pp 2295–323 DOI:10.1016/S0278-4343(00)00048-0

[5] Viet T, Hoang C, Hai D, Giap V and Tanaka H 2015 Analysis on erosion of beaches adjacent to Cua Dai River mouth, central Vietnam. In: Sana, A.; Hewawasam, L., and Waris, M.B. (eds.), *Proceedings of the 5th International Conference on Estuaries and Coasts* (Muscat, Oman) pp 153-58

[6] Tuan Q, Tin C, Quang L and Tuan T 2017 Historical Monitoring of Shoreline Changes in the Cua Dai Estuary, Central Vietnam Using Multi-Temporal Remote Sensing Data. *Geosciences* **7** p 72 DOI:10.3390/geosciences7030072

[7] Thanh T and Son H 2020 Understanding Shoreline and Riverbank Changes Under the Effect of Meteorological Forcings. In: Trung Viet N., Xiping D., Thanh Tung T. (eds) APAC 2019 *Springer* Singapore DOI: 10.1007/978-981-15-0291-0_177

[8] Hoang C, Viet T and Tanaka H 2015 Morphological change on Cua Dai Beach, Vietnam: Part II theoretical analysis *Tohoku Journal of Natural Disaster Science* **51** pp 87-92

[9] Hung T, Vinh T, Nam Y and Lee L 2017 Cause analysis of erosion-induced resort washout on Cua Dai Beach, Vietnam. In: Lee, J.L.; Griffiths, T.; Lotan, A.; Suh, K.-S., and Lee, J. (eds.), The 2nd International Water Safety Symposium. *Journal of Coastal Research*, Special Issue No. **79**, pp 214-8 Coconut Creek (Florida) ISSN 0749-0208 DOI: 10.2112/SI79-044.1

[10] Hervouet M 2007 Hydrodynamics of Free Surface Flows modelling with the finite element method WILEY

[11] Pawlowicz R, Beardsley B and Lentz S 2002 Classical tidal harmonic analysis including error estimates in MATLAB using T_TIDE *Computers and Geosciences* **28** pp 929-37 DOI: 10.1016/S0098-3004(02)00013-4

Application of GIS on building the spatial database for supporting drainage management in Ninh Kieu District

Nguyen Thanh Ngan[1,2*], Dinh Diep Anh Tuan[3], Nguyen Hieu Trung[3]

[1]Faculty of Environment, Ho Chi Minh City University of Natural Resources and Environment (HCMUNRE)

[2]College of Environment and Natural Resources, Can Tho University (CTU)

[3]Research Institute for Climate Change, Can Tho University

Corresponding author's e-mail address: ntngan@hcmunre.edu.vn

Abstract. Ninh Kieu District is located in the center of Can Tho City, established under the Government's Decree No. 05/2004/ND-CP. This is a modern and rich district, with a history of formation and development associated with the Mekong Delta. According to statistics in 2019, Ninh Kieu District has an average population of 280,792 people, with a population density of 9,605 people/km^2 and an area of 29.23 km^2. In the current period, the urbanization process in Ninh Kieu District takes place at a fast and strong pace. This has made it impossible for the urban drainage system to keep up with this change, thereby creating great difficulties for the drainage management here. In order to deal with these enormous challenges, the managers in this area need to incorporate new technologies into drainage management. This paper presents the main result of building a spatial database to support drainage management for Ninh Kieu District. This is considered as a fundamental source of data for application of GIS in urban drainage research and management in Ninh Kieu District.

Keywords: GIS, spatial database, drainage management, Ninh Kieu District.

1. Introduction

Ninh Kieu District is one of the important economic, political and socio-cultural centers of the Mekong Delta in general and Can Tho City in particular, located on the right bank of the Hau River [1]. This district was officially established in 2004 under the Government's Decree No. 05/2004/ND-CP [2]. According to the Statistics Office of Can Tho City, Ninh Kieu District has an average population of 280,792 people, an area of 29.23 km^2 and a population density of 9,605 people/km^2 in 2019 [3]. The Can Tho City Department of Construction pointed out that the receiving source of the Ninh Kieu District's drainage system includes a system of rivers and lakes, canals, sewers and pumping stations. The whole district has a network of rivers and canals with a density of about 2.0 km/km^2 [4]. As of 2016, Ninh Kieu District's drainage system is still using a common network of pipes and culverts to collect and drain rainwater and wastewater, which has not been treated and discharged directly into rivers and canals [4]. Most of the sewer system in this area has been damaged and eroded, greatly hindering the ability to drain water [4]. In the current period, Ninh Kieu District is facing the process of population growth and urbanization taking place at a rapid and complex pace. These processes have created great pressure on drainage management in this area. This pressure has revealed many inadequacies in the district's drainage management, among which are the limitations in terms of policy mechanisms and the lack of coordination among stakeholders. For this reason, management agencies need to apply new technologies to improve the efficiency of drainage management in this area. One of the new technologies that can be applied to Ninh Kieu District to solve this problem is the application of a Geographic Information System (GIS).

This advanced technology has been used to support drainage management in many parts of the world and has yielded practical and positive results [5-10]. Shamsi (1996) integrated a lumped parameter hydrologic model (Penn State Runoff Model) with a planning level GIS in implementing a stormwater management plan on a basin in Pennsylvania [5]. Northcott et al. (2000) used remote sensing and GIS technology to locate and delineate existing Subsurface Drainage Systems in the Midwest USA [6]. Nie et al. (2002) developed a GIS-based hydrological approach to generate urban watersheds, compute their features and delineate flow networks at a locality in Norway [7]. Singh (2018) researched the applications of GIS and remote sensing techniques in managing the problems of salinization and drainage of water resources in irrigated areas, thereby providing suitable solutions to map the degree and severity of drainage-induced land salinization [8]. Martin et al. (2020) developed a tool on the basis of Smart GIS technology that allows the integration of heterogeneous data sources to facilitate the digitization of the integrated urban drainage system [9]. Shedekar and Brown (2020) introduced steps to collect and process related GIS and GPS data, methods for applying that data to drainage planning and design, and guidelines for system management drainage system [10].

In previous periods, the application of GIS to support drainage management has also been implemented by some Vietnamese scientists in many localities in the country [11-14]. Trung et al. (2014) applied GIS technology to support urban water management to adapt to climate change in Can Tho City, the result of the project is that the related data and information is summarized into a GIS database and displayed as a Map Book and WebGIS [11]. Dang and Kumar (2017) used remote sensing techniques combined with hydrological modeling based on GIS technology to identify flood risk in urban areas in District 8 of Ho Chi Minh City [12]. Nguyen and Dao (2018) researched GIS technology to analyze and map flood warning services for flood prevention, rescue and traffic, providing people with forecasts on the change of travel routes through flood areas and apply functional processes of planning management models to support zoning planning management [13]. Dung et al. (2021) developed a calculation model of Sustainable Urban Drainage System (SuDS) integrating GIS and remote sensing technology to assess and forecast the level of flooding in the Nhieu Loc – Thi Nghe basin of Ho Chi Minh City [14].

To be able to apply GIS to drainage management, one of the indispensable factors is the spatial data component. This paper introduces the main results of using GIS to build a detailed spatial database to support drainage management in Ninh Kieu District. The main research area of this topic is the whole district with all 11 wards.

2. Methods and data

2.1. Research methods

In order to build a spatial database to support drainage management, a directional approach using Quantum GIS (QGIS) software is deployed. QGIS is chosen because it is open source software, not only free of purchase but also free of copyright, users can freely copy, publicly study, work without having to ask for permission. Meanwhile, commercial GIS software such as ArcGIS, MapINFO, AutoCAD Map 3D have to cost quite a lot to buy the license when implementing research projects, but the functions are similar to QGIS. Besides, QGIS is also used with many different types of GIS data formats and is easy to deploy for small projects such as drainage management for an urban district. In this research, the solution of building spatial database on single computer is chosen because of the simple implementation process, moderate amount of data, small number of managed objects and low cost when implemented in practice. The process of building this spatial database consists of six main stages: (1) collecting relevant data, (2) processing data and digitizing maps, (3) building spatial database, (4) evaluating the accuracy of digital map data layers, (5) creating thematic maps, (6) drawing conclusions and making recommendations. Of the above five stages, two stages (2) and (3) are considered the most important because these stages directly affect the accuracy and reliability of the spatial database. The selected format for this spatial database is SpatiaLite for the following reasons: spatial data storage is topology relational, allows multiple objects to be used at the same time, has the ability to compress reduce storage capacity, can be used on many software, high data security features. The data layers in this spatial database are compared with published official maps of Can Tho City to evaluate the accuracy of geospatial data. The detailed research process of the project is shown in Figure 1.

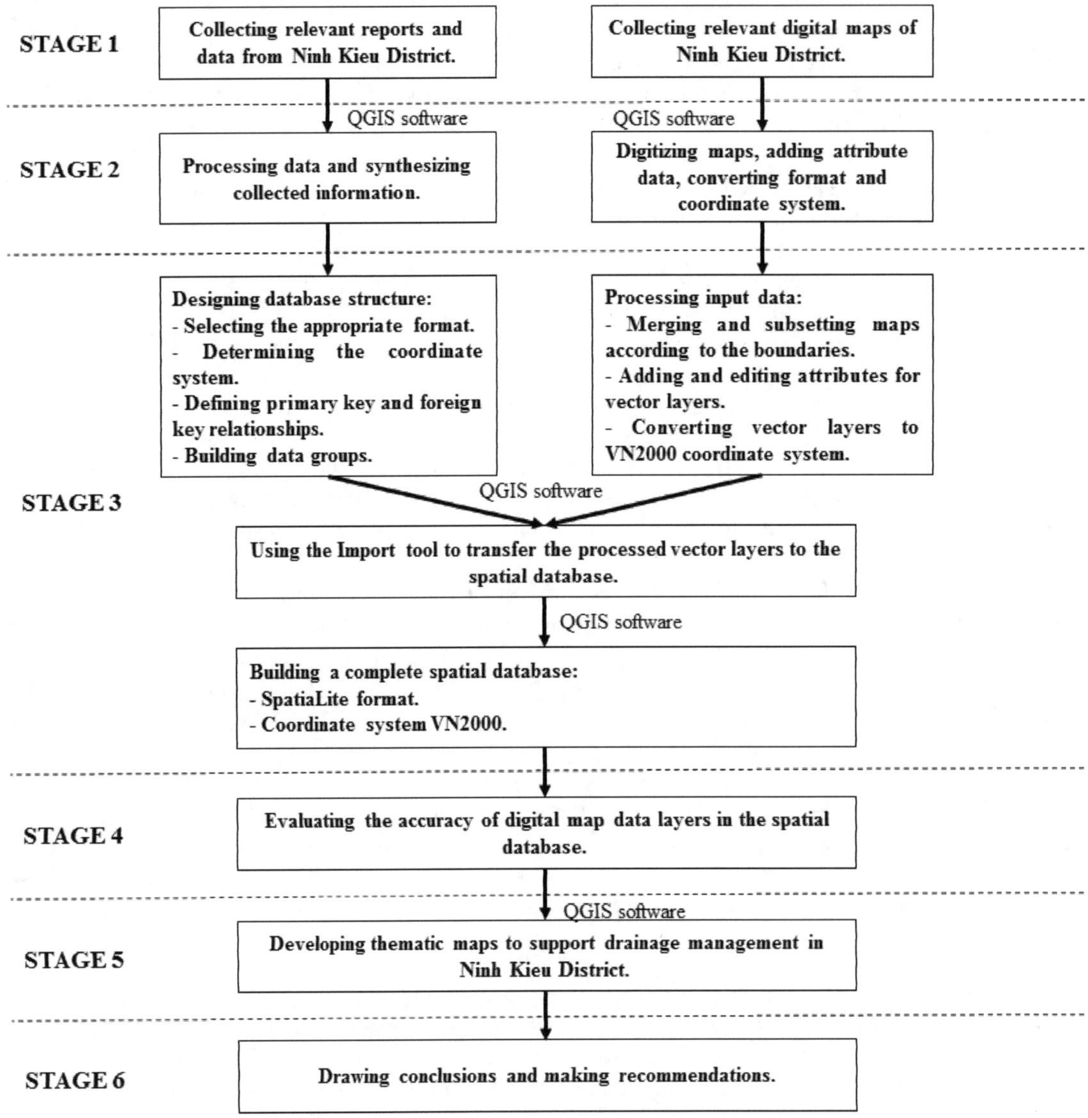

Figure 1. The detailed flowchart of the research.

2.2. Research data

Two main types of data are used to build a spatial database to support drainage management in this research including: (1) reports and statistics on drainage management in Ninh Kieu District, (2) digital map data of Ninh Kieu District. The digital map data of Ninh Kieu District is collected from many sources with three main formats: CAD file (*.dwg), Shapefile (*.shp) and TAB file (*.tab). These digital maps are collected from two sources, the Mekong Delta Geographic Information Systems Project and management agencies of Can Tho City. They are divided into six categories: (1) administrative map, (2) traffic map, (3) topographic map, (4) housing map, (5) land use map, (6) drainage network map. The drainage network maps are built in the form of AutoCAD drawings by the Can Tho City Department of Construction. The administrative boundaries of the study area are shown in Figure 2.

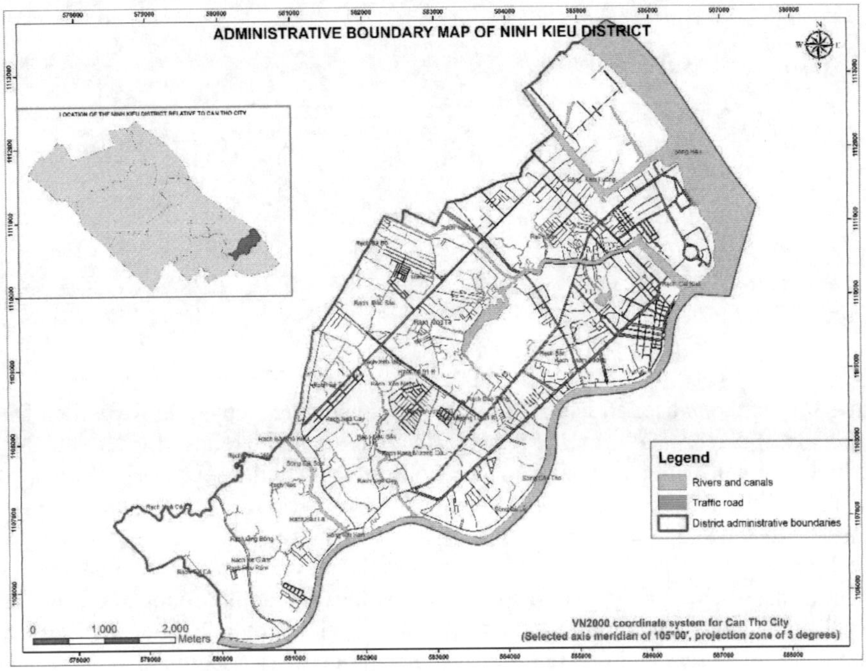

Figure 2. The map of the study area - Ninh Kieu District.

3. Results and discussion

3.1. Results of building spatial database to support drainage management

Two tools Create Layer and DB Manager of QGIS software are used to build a spatial database to support drainage management for Ninh Kieu District. The main data source for building this spatial database is the digital map of the study area with vector data model. The base maps were developed mainly in 2018, while the drainage network maps were built in 2016. These digital maps are converted to the same format and geographic coordinate system before processing and import into the spatial database.

The spatial database to support drainage management in Ninh Kieu District is designed based on the relational model, with the degree of data abstraction is physical level. This spatial database is divided into two main components: terrain data and thematic drainage data. The terrain data component is designed based on a standardized structure published by the Environmental Protection Agency in 2006 [15]. Thematic drainage data component is designed based on the specific professional requirements of the management agencies in Ninh Kieu District. This spatial database is built with the SpatiaLite format, a spatial extension of SQLite that provides vector geodatabase functionality. With the choice of SpatiaLite format, this spatial database can be used flexibly across a variety of GIS software. This spatial database is built on the VN2000 coordinate system for Can Tho City with the selected axis meridian of 105°00' and the projection zone of 3 degrees (k = 0.9999) to comply with Vietnamese regulations. The overview diagram of the spatial database structure of drainage management is shown in Figure 3.

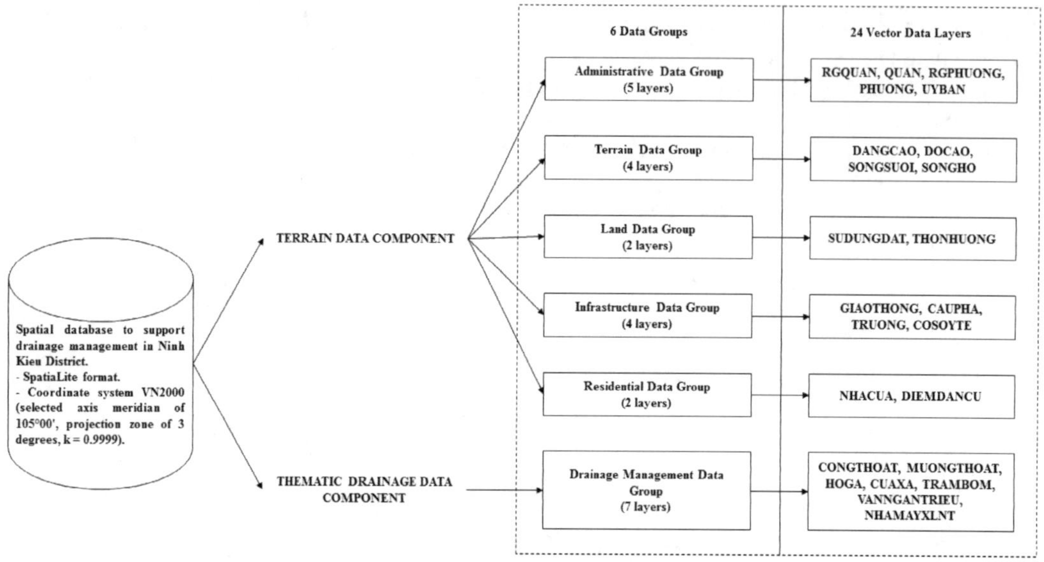

Figure 3. Spatial database structure to support drainage management in Ninh Kieu District.

The entire spatial database of drainage management has six data groups, divided into 24 vector data layers. In which, thematic data group on drainage management includes seven vector data layers: sewers, ditches, manholes, discharge gates, pumping stations, tidal valves, wastewater treatment plants. This spatial database can be used for subjects such as government officials as well as to support scientists in research on drainage systems in Ninh Kieu District. The data layers in this spatial database are also the basis for creating many different types of maps describing the topographic, population and infrastructure characteristics of the study area. Figure 4 shows the map of elevation points, while Figure 5 shows the map of main residential areas and Figure 6 shows the map of houses and buildings of the study area.

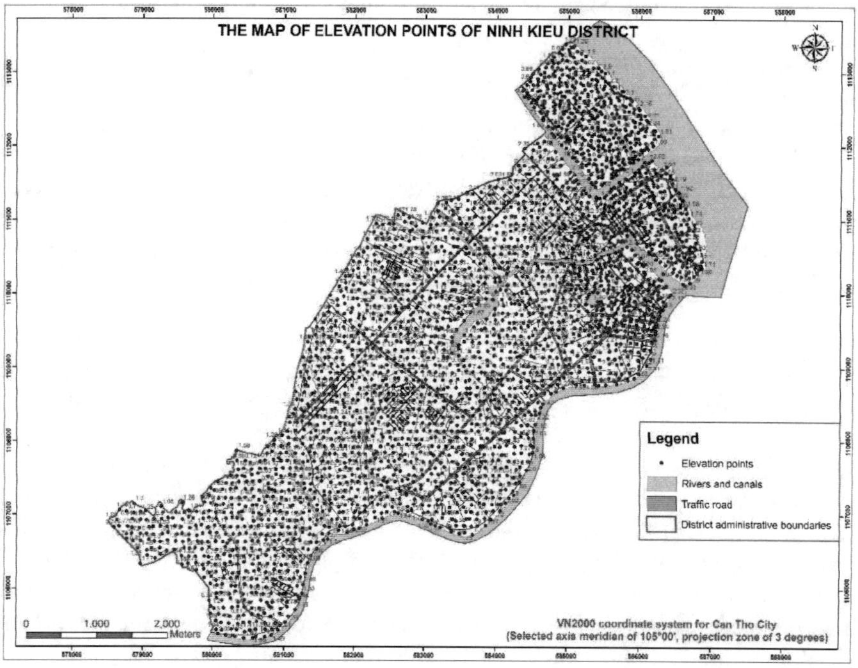

Figure 4. The map of elevation points of Ninh Kieu District.

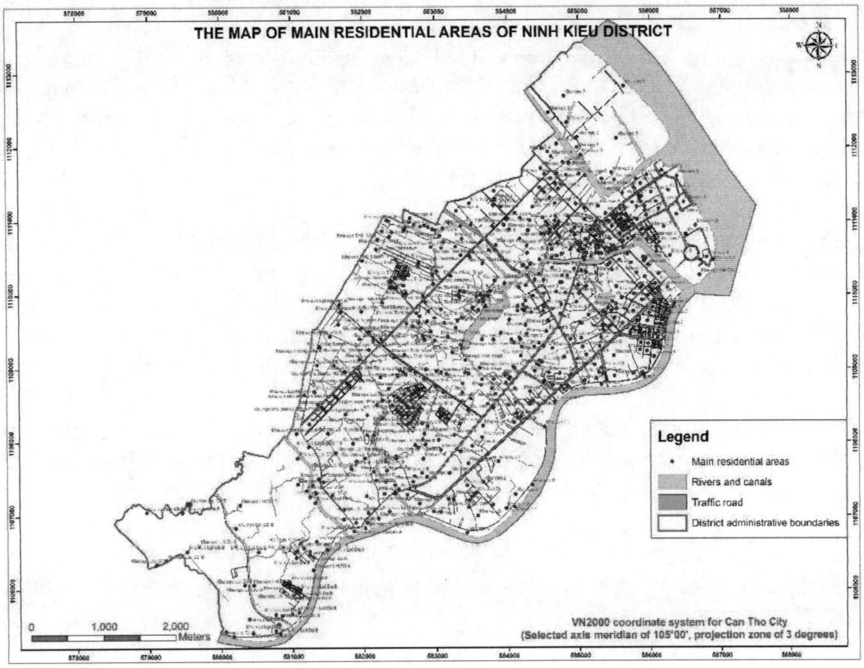

Figure 5. The map of main residential areas of Ninh Kieu District.

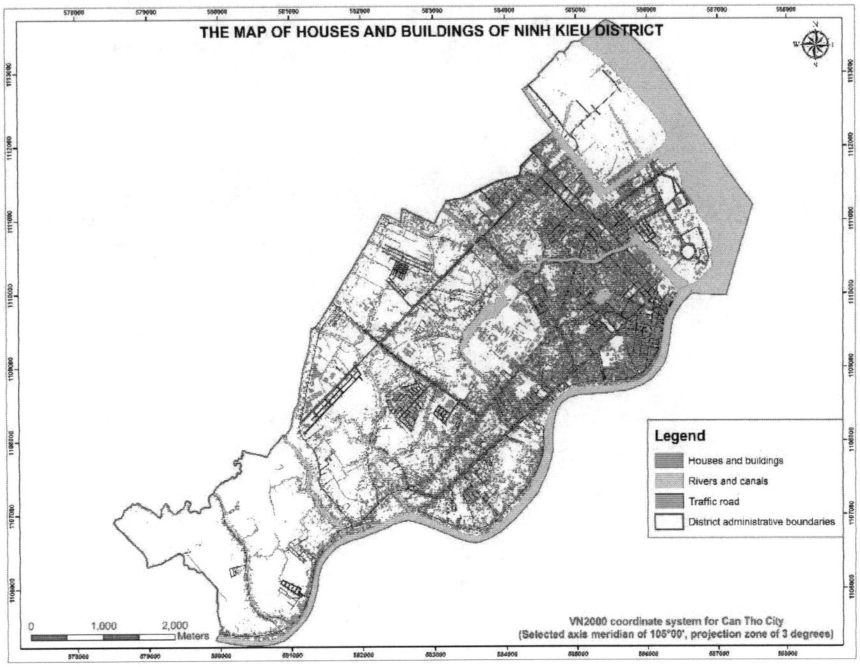

Figure 6. The map of houses and buildings of Ninh Kieu District.

In addition to building the spatial database component, a system of processes to exploit this database is also drafted to serve the professional work of the drainage management agencies in the study area. The main processes for exploiting spatial database to support drainage management include: adding new data, updating attributes, editing data, querying data, analyzing data, creating maps, extracting reports. These processes make it easy for users to access the data in the spatial database.

IOP Conf. Series: Earth and Environmental Science **964** (2022) 012002 doi:10.1088/1755-1315/964/1/012002

3.2. Results of building thematic maps to support drainage management

After the stage of designing and building a complete spatial database to support drainage management, the data layers contained in this spatial database are used to establish thematic maps on drainage management in the study area. The role of these digital maps is to assist the authorities in Ninh Kieu District in managing assets as well as solving problems arising in the process of drainage management. These maps are designed on standard A1 paper, stored in digital image format JPG (*.jpg) with a resolution of 500 DPI.

These thematic maps are built according to two main contents related to drainage management in Ninh Kieu District. The first content is main drainage sewers and ditches within the administrative boundaries of Ninh Kieu District, includes nine types: Sewers D200, Sewers D300, Sewers D400, Sewers D500, Sewers D600, Sewers D800, Sewers D1000, Sewers D1200, Ditches. Figure 7 and Figure 8 show the thematic maps of main drainage sewers and ditches in the study area. The second content is discharge gates and tidal valves within the administrative boundaries of Ninh Kieu District. Figure 9 and Figure 10 show the thematic maps of discharge gates and tidal valves in the study area. As mentioned in the previous section, the drainage management data layers in the spatial database are built based on AutoCAD drawings established in 2016 by the Can Tho City Department of Construction.

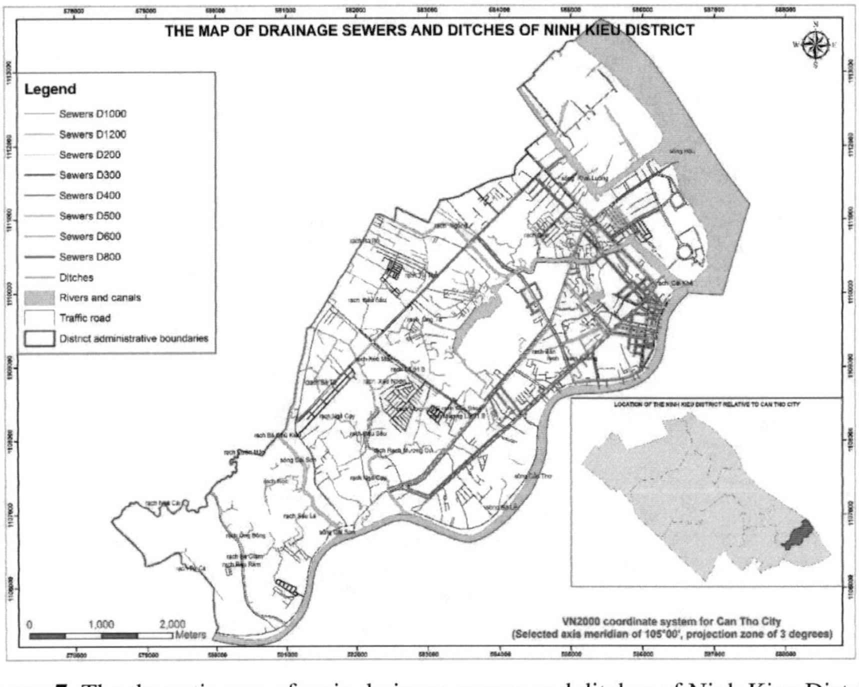

Figure 7. The thematic map of main drainage sewers and ditches of Ninh Kieu District.

IOP Conf. Series: Earth and Environmental Science **964** (2022) 012002 doi:10.1088/1755-1315/964/1/012002

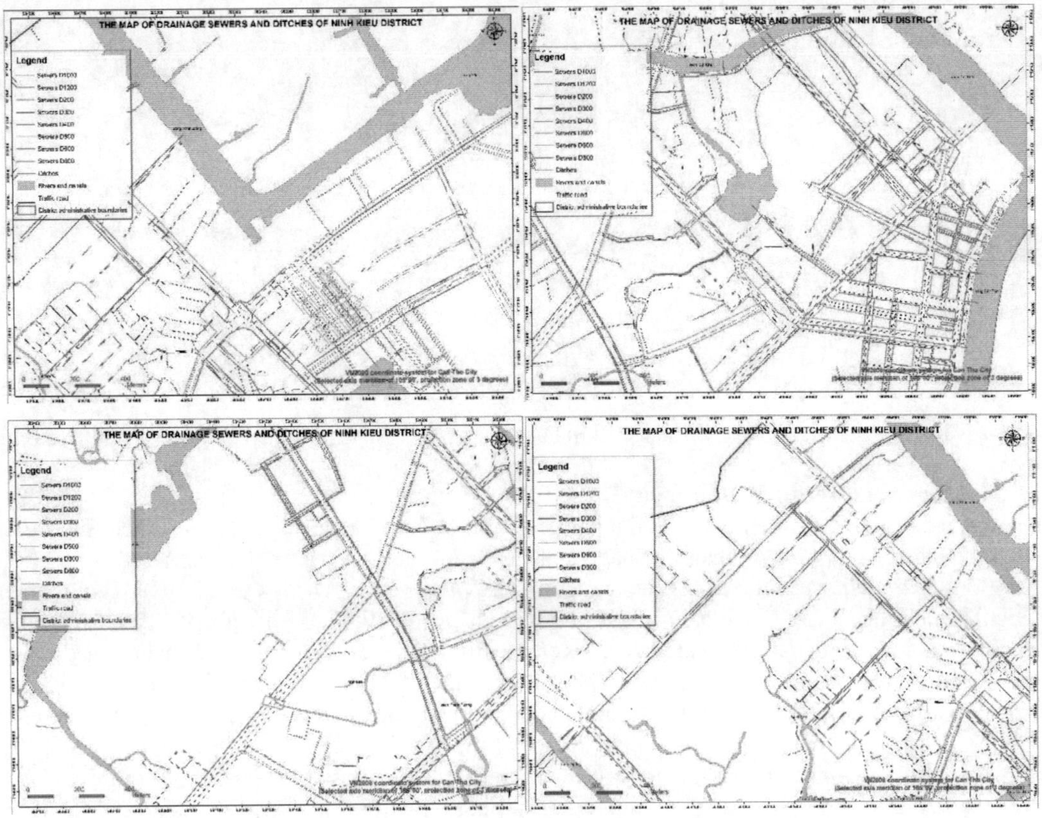

Figure 8. The detailed maps of the locations of main drainage sewers and ditches in the study area.

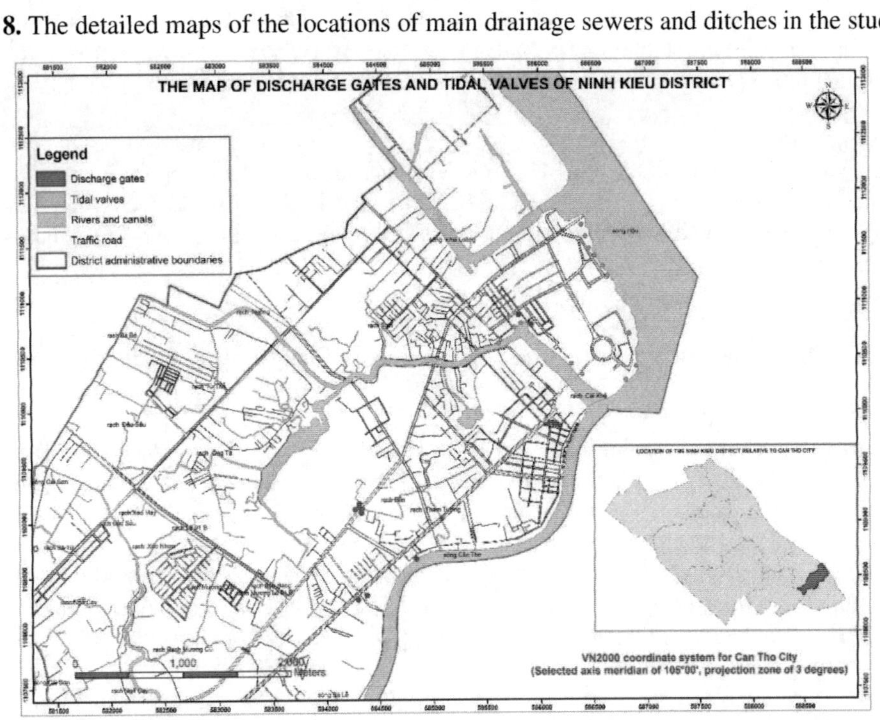

Figure 9. The thematic map of discharge gates and tidal valves of Ninh Kieu District.

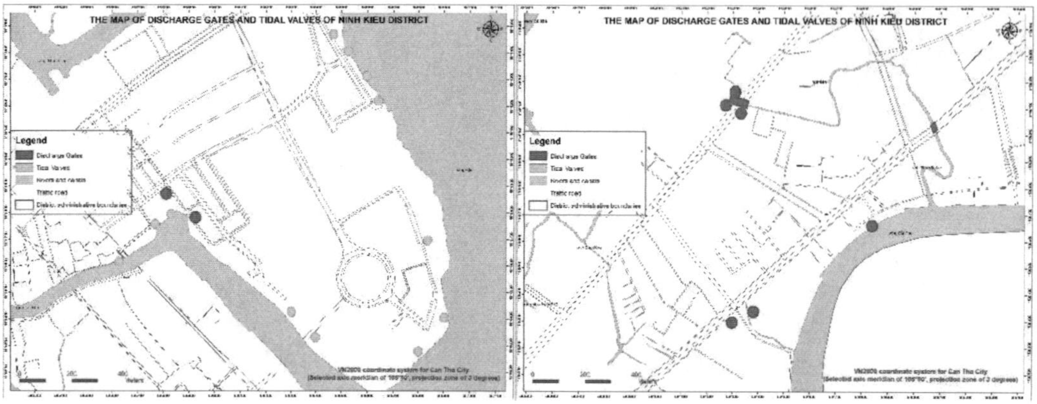

Figure 10. The detailed maps of the locations of discharge gates and tidal valves in the study area.

4. Conclusions and recommendations

The obtained results show that the research has accomplished the initial set of tasks, which is to build a spatial database to support drainage management in Ninh Kieu District with a structure suitable for regulations of the Ministry of Natural Resources and Environment as well as professional requirements of management agencies. This spatial database is built on the VN2000 coordinate system, with six groups of data, divided into 24 vector data layers, including seven thematic data layers on drainage, which can meet requirements of drainage management in the study area. Because it is built in SpatiaLite format, this spatial database can be used both on commercial GIS software as well as open source GIS software. This spatial database is an important scientific basis for the application of GIS at different levels to drainage management in Ninh Kieu District. This spatial database is considered as a technological solution for managers in Ninh Kieu District to solve the existing difficulties and challenges of drainage management. The implementation of this study has enabled the authors to develop a detailed procedure for establishing a spatial database from various types of digital maps as shown in Figure 1. It can be said that this will be a valuable reference for future studies in the field of drainage.

Acknowledgments

The authors would like to thank the ValBGI Project for sharing as well as supporting and introducing to collect data from the authorities of the Mekong Delta, the Mekong Delta Geographic Information Systems Project (MGIS) and the 100 Resilient Cities Program (100RC) for providing data support for the research.

References

[1] Department of Natural Resources and Environment of Can Tho City 2019 *Map: Administrative map of Can Tho City 2012* Access date 25/07/2021 http://stnmt.cantho.gov.vn/map/bdhc2012.htm

[2] Vietnam Government 2004 *Decree on establishing Ninh Kieu, Binh Thuy, Cai Rang, O Mon districts, Phong Dien, Co Do, Vinh Thanh, Thot Not districts and communes, wards, towns of Can Tho City under the Central* Number: 05/2004/ND-CP pp 1-6.

[3] Can Tho City Statistics Office 2020 *Can Tho City Statistical Yearbook 2019* Statistical Publishing House Publishing registration confirmation number: 2075-2020/CXBIPH/05-14/TK pp 21-85.

[4] Can Tho City Department of Construction 2016 *Summary of Can Tho City Drainage Planning to 2030, vision to 2050: Chapter 3 Current Status of Drainage* General explanation Can Tho City pp III-2-III-53.

[5] Shamsi U M 1996 *Storm-water management implementation through modeling and GIS* Journal of Water Resources Planning and Management **122(2)** pp 114-127 doi:10.1061/(ASCE)0733-9496(1996)122:2(114).

IOP Conf. Series: Earth and Environmental Science **964** (2022) 012002 doi:10.1088/1755-1315/964/1/012002

[6] Northcott W J, Verma A K and Cooke R A 2000 *Mapping Subsurface Drainage Systems using Remote Sensing and GIS* Paper presented at 2000 ASAE Annual International Meeting, Technical Papers: Engineering Solutions for a New Century, Milwaukee, WI., United States, 7/9/00 - 7/12/00 pp 2625-2634.

[7] Nie L, Schilling W, Killingtveit Å, Sægrov S and Selseth I 2002 *GIS based urban drainage analyses and their preliminary applications in urban stormwater management* In Ninth International Conference on Urban Drainage (9ICUD), Global Solutions for Urban Drainage pp 1-13 doi:10.1061/40644(2002)117.

[8] Singh A 2018 *Managing the salinization and drainage problems of irrigated areas through remote sensing and GIS techniques* Ecological indicators **89** pp 584-589 doi:10.1016/j.ecolind.2018.02.041.

[9] Martin C, Kamara O, Berzosa I and Badiola J L 2020 *Smart GIS platform that facilitates the digitalization of the integrated urban drainage system* Environmental Modelling and Software **123** 104568 doi:10.1016/j.envsoft.2019.104568.

[10] Shedekar V S and Brown L C 2020 *GIS and GPS applications for planning, design and management of drainage systems* Precision Conservation: Goespatial Techniques for Agricultural and Natural Resources Conservation **108** pp 209-229.

[11] Trung N H, Tuu N T, Doan T C, Van Thinh L, Tuan D D A and Nguyen M 2014. *Application of GIS to support urban water management in adapting to a changing climate: A case study in Can Tho city* Vietnam Report number: UCCRN Case Study 14.

[12] Dang A T N and Kumar L 2017 *Application of remote sensing and GIS-based hydrological modelling for flood risk analysis: a case study of District 8, Ho Chi Minh city, Vietnam* Geomatics, Natural Hazards and Risk **8(2)** pp 1792-1811 doi:10.1080/19475705.2017.1388853.

[13] Nguyen A T and Dao H S 2018 *GIS applications in urban management* In MATEC Web of Conferences EDP Sciences **193** p 01004 doi:10.1051/matecconf/201819301004.

[14] Dung L T, Phu V L, Lan N H M, Tien N T C and Hiep L D 2021 *Sustainable Urban Drainage System Model for The Nhieu Loc – Thi Nghe Basin, Ho Chi Minh City* In IOP Conference Series: Earth and Environmental Science IOP Publishing **652(1)** p 012012 doi:10.1088/1755-1315/652/1/012012.

[15] Environmental Protection Agency 2006 *Developing a database standard set to serve the development of Vietnam's environmental map, Part "Upgrading the database structure of the Environmental Geographic Information System"* Ministry of Natural Resources and Environment, Report on performance of tasks pp 45-204.

IOP Conf. Series: Earth and Environmental Science **964** (2022) 012003 doi:10.1088/1755-1315/964/1/012003

Remote sensing and GIS application in assessing the adaptability of some key crops in Dak Nong province

Nguyen Thi Thanh Huong[1*], Ho Dinh Bao[1], Cao Thi Hoai[1], Phan Thi Hang[1], Ngo The Son[1], Dang Thi Thuy Kieu[1], Nguyen Thi Ngoc Quyen[1], Nguyen Duc Dinh[1], Tran Thi Xuan Phan[1], Pham Tuan Anh[2], Pham Thi Mai Thy[3], Nguyen Thi Hong Diep[4]

[1]*IRMG, Faculty of Agriculture and Forestry, Tay Nguyen University, 567 Le Duan Street, Buon Ma Thuot City, Dak Lak Province, Viet Nam*

[2]*Department of Agriculture and Rural Development of Dak Nong province, Tran Hung Dao Street, Gia Nghia City, Dak Nong province, Viet Nam*

[3]*HCMC Space Technology Application Center (STAC), Vietnam National Space Center (VNSC), Vietnam Academy of Science and Technology (VAST), 01 Mac Dinh Chi Street, Ben Nghe Ward, District 1, Ho Chi Minh City, Vietnam*

[4] *College of Environment and Natural Resources, Can Tho University, Campus II, 3/2 street, Ninh Kieu district, Can Tho city 900000, Vietnam*

*Corresponding author email: huongthanh.frem@ttn.edu.vn

Abstract. Remote sensing (RS) and Geographic information system (GIS) is widely applied in the world and gradually affirms its role in Vietnam in managing agricultural and forest resources. This application is highly effective, providing information timely for managers to make decisions and build development strategies. In this study, RS and GIS were integrated to assess suitability for key crop species in Dak Nong province including coffee, rubber, cashew, and durian based on their suitability to site conditions such as soil (soil type, soil texture, soil thickness), topography (elevation, slope) and climate (temperature, precipitation). Using the restrictive method and overlapping map layers of natural factors, classified into adaptive levels according to FAO (1976). Results show that most land areas in Dak Nong province have different levels of potential suitability for key crop species ranging from non-adaptive to less-adaptive and moderately adaptive. However, most suitable areas for key crops are only at low (accounting for a large proportion) and the average adaptation level. The findings from the study are the scientific information for managers to make decisions regarding the structure of major crops in the province.

1. Introduction

In land resources planning and management, assessing soil adaptation is the key step that provides the basic information for land use planning and crops structure. The Food and Agriculture Organization of the United Nations - FAO (1985) [7] pointed out the need for the appropriate use of each type of soil for different crops in a rational way to continue sustainable cultivation, meet social need as well as conserve fragile ecosystems. One method of soil assessment that was widely used by experts is the land suitability assessment method developed by FAO (1976) [6]. This method assesses the suitability of land for a particular use by comparing potential uses with current uses, through land assessment we can know land resources have been used optimally or not, or how to adjust the land use purpose in the most optimal way. GIS was used by Ekanayake and Dayawansa (2003) [4] to identify potential areas for afforestation, the results showed that most of the land in the study area did not conform to the given criteria. Studies have also shown that multi-layer GIS analysis will make it easier to develop optimal use of land, increase productivity while preserving environmental conditions as well as handling a large amount of accessible information about soils (Patil et al. (2001) [11], Dengiz et al. (2003) [3]).

Like other localities, the reality of agricultural land use in Dak Nong province is strongly influenced by the impact of prices, moreover, with fertile soil favorable for the development of many crops, especially industrial crops as well as fruit trees. Because of that, it leads to rampant farming without planning, or lack of planning that leads to waste of natural resources and human resources. Therefore, it is necessary to analyze the potential adaptability of crops as a basis to minimize the risks as well as the loss of resources, economy of the people and society. Moreover, in the context of climate change, the selection of crops as well as cultivated methods also needs to be proactive to adapt to climate changes that have a significant impact on production efficiency.

Information on adaptive soil units that need to be determined on a spatial and attribute basis is important in order to be able to determine its property and scope. This information needs to be determined by an aggregate way. For example, to grow a certain crop, it is indispensable for information on plant ecology such as soil requirements, temperature, water, altitude, etc.; In contrast, in areas with high slopes, far from surface water sources, frequent droughts, etc., the selection of agro-forestry options is very important. On a larger scale, aggregated information about natural and social conditions will help to make planning more accurate, thereby supporting planning as well as decision making and directional development strategy. This is the basis for the conversion or solutions to improve crop production efficiency. In addition, land use information is also an important component in understanding the interaction between human activities and the environment.

Geographic information system (GIS) is widely applied in the world and gradually asserts its role in Vietnam in term of managing agro-forestry resources. This application brings high efficiency, provides timely and objective information for managers to make decisions and build development strategies. However, through the review of the agricultural database system of the provinces in general, Dak Nong in particular still has many limitations, mainly due to the lack of digital data on the tree distribution map which were grown in different soil conditions. This causes many difficulties in management, strategic planning, as well as calling for investment. Therefore, this research has used GIS and Remote sensing (RS) as the effective tools to create land adaptation units for some key crops, which is important information as a basis for choosing or transforming crops in Dak Nong province.

The objective of this research were analyzing potential adaptation and reality land use of some key crops in Dak Nong province based on integrating multi factors analysis.

2. Study area and methods

2.1. Study area

The study was conducted in Dak Nong Province in the Central Highlands of Vietnam (Figure 1). It located in the geographical coordinates from 11^{0}45' to 12^{0}50' North latitude, 107^{0}13' to 108^{0}10 East longitude. The average elevation is 700m above sea level. The mean temperature is 24 degrees Celsius. The climate conditions produce general characteristics of a subequatorial tropical monsoon climate. The province has characteristics of humid tropical highland climate and is affected by dry-hot southwest monsoons. There are two distinct annual seasons: the rainy season starts in April and ends in November, and the dry season starts in December and ends in March the following year. The average annual rainfall is 2,500mm, of which 90% occurs during the rainy season which supplies enough water for agriculture. However, it also causes erosion, landslide, flood which affect the human life and productive activities in the areas; In dry season average rainfall takes only 10% of the year. Because of that, some areas have severe drought especially in Cu Jut district, some communes in the East of Dak Mil and the North of Krong No district. The main wind direction in the rainy season is southwest, in the dry season is northeast, the average wind speed is 2.4 - 5.4 m/s. There are almost no storms so it does not affect fragile crops like coffee, rubber, and pepper.

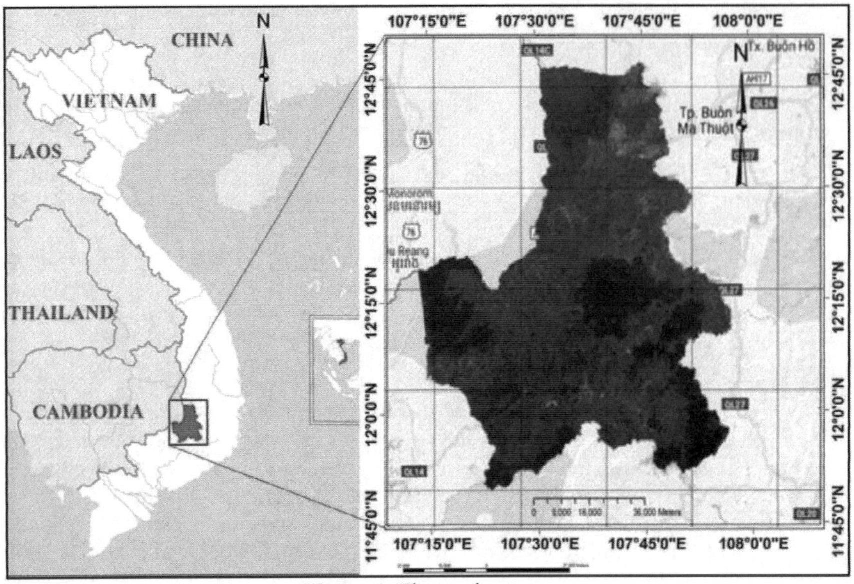

Figure 1. The study area

The study area is 650,927 ha in which, agricultural land is 366,401ha takes a proportion of 56.29% total natural area. Almost the agricultural lands are perennial crops and forestry land; annual crops are rice, corn and beans (Dak Nong statistical yearboook-2019) [2].

2.2. Methods

Data collecting: All data such as Sentinel 1A and 2A images, digital elevation model (DEM), land use, soil, weather, and geographic map collected to build maps assessing potential land of key crops.

- Sentinel images, which resample of 10x10 m resolution acquired in 2020 and DEM achieved from Google Earth Engine (GEE) were used for creating slope, elevation, and crop maps
- Remote sensing and GIS used for creating the land use map in 2020
- Soil type, soil texture, and soil thickness were extracted from land map which derived by Sub National Institute of Agricultural Planning and Projection in Central Vietnam in 2005.
- Precipitation, temperature maps were produced based on weather data of Central Highland Hydro-Meteorological Centre in 2020. IDW tool in ArcGIS was used to interpolate rainfall and temperature maps.

Building key crops map: Sentinel-2A images were processed at level-1C including radiometric and geometric corrections. Orthor-rectification and spatial registration on a global reference system with sub-pixel accuracy (Sentinel-2 user handbook) [5]. Cloud cover was removed automatically in GEE; Sentinel-1A image was collected by GEE and processed at level-1 including Thermal noise removal, Radiometric calibration, speckle filtering using Enhanced Lee filter; It was also analysed image structure using Grey level co-occurrence matrix (GLCM) method to improve the classification. The pre-processed Sentinel-2A and Sentinel-1A images were combined to form a single image to serve as input data for image classification and identification of land cover/ land use types (LULC). 5,158 sample points which were collected using Google Earth (GE), available data, field collected were used to discriminate LULC. In which, 60% of the sample were used for classification, the remaining was used for accuracy assessment. The two algorithms of Random Forest (RF) and Super vector machine (SVM) were used to classify the image. In this case, the validation shown that although the efficiency of the two methods is almost the same, the SVM took a long time (nearly 10 times comparing to RF), so RF was employed to classify the image into different LULC types. Using Object based image analysis (OBIA) method together with RF algorithm to determine LULC for the province. First, satellite images were segmented into objects using multiresolution segmentation method in eCognition software. Then, the image consisting of segmented objects is extracted in TIF file with the average values of the image channels and the normalized difference vegetation index (NDVI). Finally, the image containing segmented objects is classified using the RF based on the training dataset.

Accuracy assessment: The study employed sample points in the testing data based on the criteria of overall accuracy, producer accuracy and user accuracy to assess the classification accuracy. In addition, the Kappa coefficient also indicates the accuracy of the classification results compared to reality. The Kappa index values varies between 0 to 1 where, the closer it is to one, the closer the accuracy level of the map obtained from classification to reality (Abdi, 2005). The remaining 40% of the sample dataset employed to assess the accuracy of the classified image.

Assessing potential adaptation of land use unit (LUU) for key crops based on multi-factors: Overlaying factors map according adaptive level to form LUU. The key crops identified in this area are those that have a large area and/or contribute mainly to the household's income. Accordingly, the key crops were selected in this area including coffee and pepper, rubber, and cashew. Restrictive method was used to assess the adaptability LUU for each key crops. This method used the least suitable factor as restrictive factor (FAO, 1993) [9]. According to FAO (1993), natural adaptation of crops divided into four levels including non-suitable (N), very high suitable (S1), moderate suitable (S2) and least suitable (S3). One LUU that is seen as very high suitable for a certain crop when all factors are very high suitable (S1) for that crop. In contrast, if a LUU has at least one factor that non suitable (N) for a crop, the LUU is not adaptable for that crop. If not in these two cases, the LUU is considered to be less suitable for a crop when there is at least one element of the least adaptive class (S3). Finally, the LUU is considered as moderately adapted when it does not fall into the above three cases.

Determining the adaptability of key crop species based on multiple factors: Crop plants growing depends on many environmental factors such as soil, water, temperature, slope etc. In addition, each crop requires certain condition of environmental factors. Using four levels of crops' natural adaptation such as N, S1, S2, and S3 which have defined in the previous step to determine suitable areas for each crop. Although the crops depend on many elements during their growing, some main ecological requirements of crops such as soil type, soil texture, soil thickness, slope, elevation, temperature and rainfall were used under the current study due to available relevant data in this area. All multi-factors were divided into four levels as described above for the entire province.

Using Overlay Intersect function to build the map of potential adaptation for each crop. Areas of intersection between objects on factor maps will form new objects with all 7 attributes (soil type, soil thickness, soil texture, elevation, slope, maximum temperature, and average rainfall) of the factor maps.

Assessing the current status of the key crops in the province based on adaptation levels using land unit suitable to each crop with single factors, and integrated factors. The next step is to overlay the adaptive maps with the crop maps to determinate the state of the cultivated main crops (illustrated as Figure 2) as a foundation for proposing conversion. Figure 3 presents the diagram of research steps.

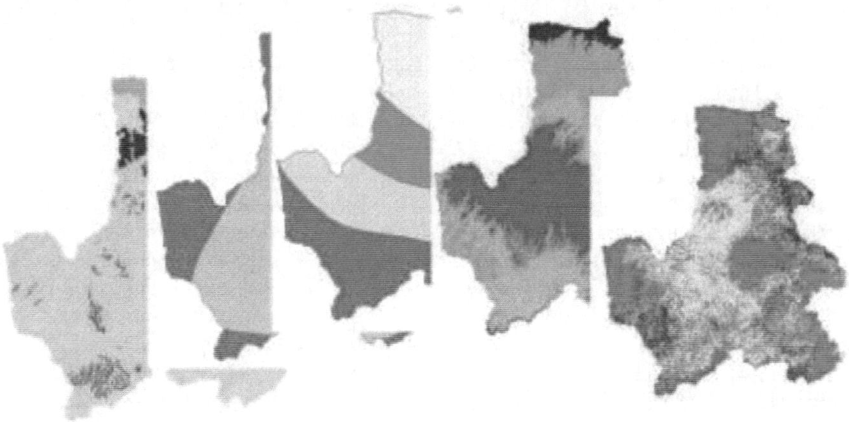

Figure 2: Overlay adaptive factors map with current LULC map

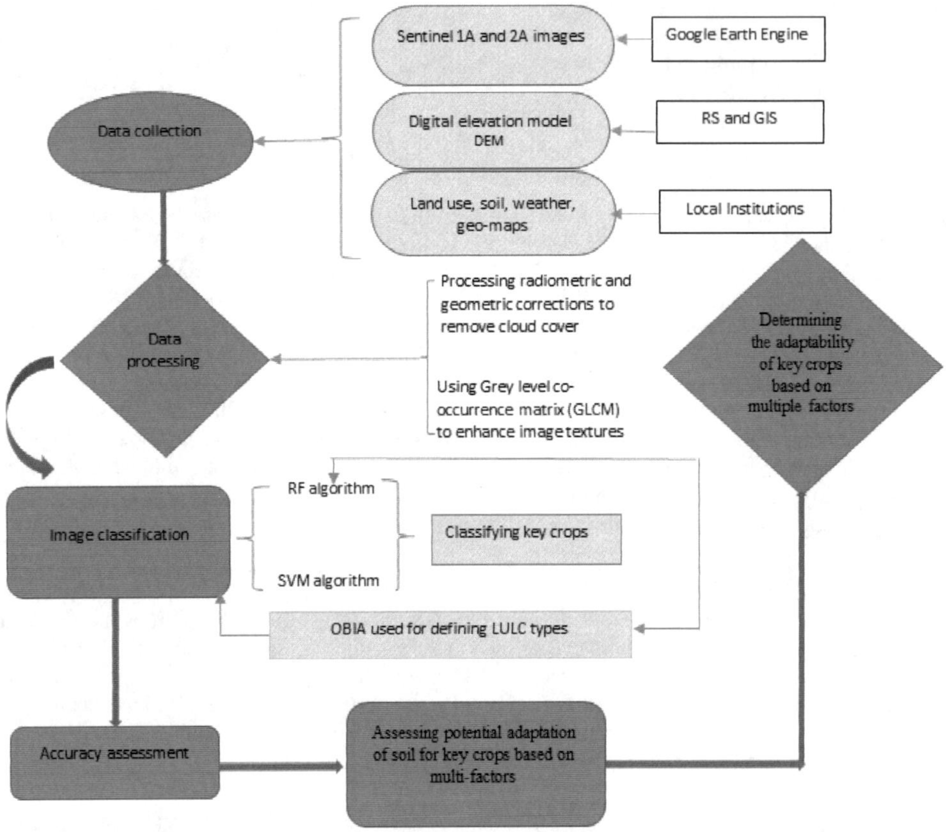

Figure 3: Methodology flow-chart

3. Results and discussion

3.1. Accuracy assessment

Ten of LULC were distinguished in the study using the Sentinel satellite images. They composed of natural forest (1), plantation forest (2), rubber (3), cashew (4), coffee and pepper (5), other crops (6) (e.g. fruit, cassava,...), annual crops (7), residential area (8), water surface (9), and other lands (10). Table 1 shows the results of overall accuracy (OA), Kappa index, Producer Accuracy (PA) and User Accuracy (UA). The OA of the classification was 75.57 % with a kappa of 0.70, indicating a strong agreement between the classification results and validation data. Most LULC obtained more than 60 % of accuracy, except for other crops, which had PA of about 48.38 %. It is found that the water surface was very distinguishable compared to other LULC classes due to its feature, so this class derived the most accurate result with PA = 100 % and UA =89,38 %. The following received from natural forests with PA=92.71% and UA=84.79%. The natural forests are easy to distinguish from other types because of their relatively high homogeneity and less fragmentation. Among the key crops coffee and pepper class presented the best accuracy with the values more than 70% both PA and UA; thought UA of rubber class was more than 80%, the UA obtained below 70%. In practice, the ages of industrial crops such as coffee, pepper, cashew, rubber vary widely. Although the same plant species, but their age varies from new plants (e.g. a few pairs of leaves) to mature trees (harvested, closed canopy). Moreover, the intercropping with 2-3 species of plants in the same garden plot is very common in the area. This makes it difficult to clearly distinguish these crops from other land covers, resulting in not really high in accuracy.

Table 1. Overall Accuracy (%), Kappa coefficient, Producer's Accuracy (%), and User's Accuracy (%) index for the classification image

LULC	1	2	3	4	5	6	7	8	9	10	UA%
1	1857	46	76	118	53	22	18	0	0	0	84,79
2	25	251	14	8	4	0	13	5	0	0	78,44
3	30	6	451	7	23	0	0	0	0	0	87,23
4	59	6	32	626	147	54	0	0	0	21	66,24
5	24	5	59	175	1369	174	61	1	0	32	72,05
6	8	15	9	22	84	284	44	8	0	23	57,14
7	0	0	4	7	12	12	193	34	0	14	69,93
8	0	0	7	0	1	8	24	121	0	0	75,16
9	0	0	0	5	5	0	7	0	143	0	89,38
10	0	8	0	3	38	33	0	15	0	175	64,34
											OA= 75,57%
PA%	92,71	74,48	69,17	64,47	78,86	48,38	53,61	65,76	100,00	66,04	Kapa= 0,70

1: Natural forest, 2: plantation, 3: rubber, 4: cashew 5: coffee and pepper, 6: other crops, 7:annual crops, 8: residential area, 9: water surface, 10: other lands

3.2. Tabular of LULC

In order to harmonize with LULC data existing in the Dak Nong province, the LULC which detected based on satellite image were also adjusted after confirming with localities through seminars and comparing with the 2019 and 2020 statistical yearbooks. The final results are filtered, adjusted and vectored. Among the LULC classes, the forest area covers as the largest portion of the total area (35.09%), key crops such as coffee area cover 25.20%, rubber is of 3.65%, and cashew is of 2.47%. The illustration is shown in Figure 3.

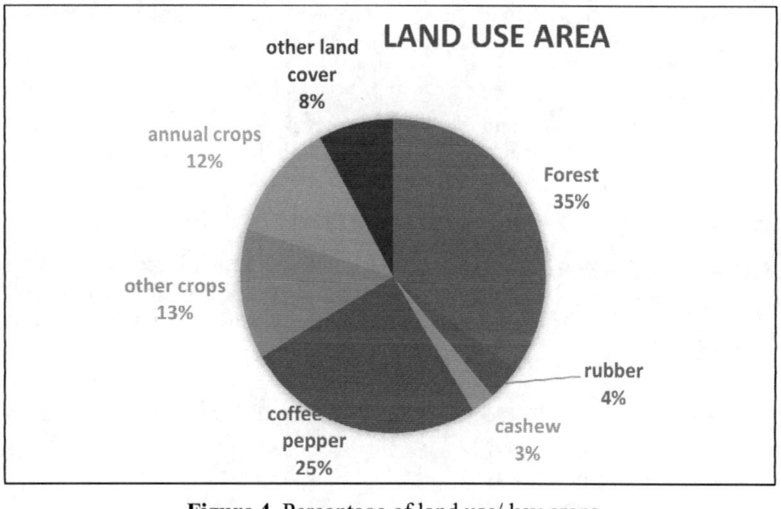

Figure 4. Percentage of land use/ key crops

Forestry land has the largest area, with an area of 228.43 thousand hectares, accounting for 35.09%. Among other land use types (including water surface), coffee and pepper class has the largest area of 164.01 thousand ha, accounting for 25.20%, widely distributed in all districts of the province, of which Dak Song and Dak Mil has the largest area. Followed by other crops and annual crops accounted for 13.46% and 12.24%, respectively. Cashew has been planted about 16.09 thousand hectares, accounting for 2.47%, mainly in Cu Jut, Tuy Duc and Krong No districts. Rubber area is mainly in Cu Jut, Tuy Duc, Dak R'lap and Krong No districts with about 23,74 thousand ha.

3.3. Adaptive ability of key crops

Soil map

The map of soil types was collected from Sub National Institute of Agricultural Planning and Projection in Central Vietnam. There are eleven soil types within the area including newly modified soil, eroded soil, red soil, grey soil, alluvial soil, dark brown soil, compacted clay, black soil, cracked soil, grey soil, brown soil (Table 2).

Table 2. Soil type statistics

No.	Sign	Soil type	Area	(%)
1	CM	Newly modified soil	11,125.00	1.71
2	E	Eroded soil	9,389.19	1.45
3	Fd	Red soil	391,837.00	60.32
4	GL	Gley soil	5,303.46	0.82
5	P	Alluvial soil	2,669.88	0.41
6	PH	Dark brown soil	27,387.00	4.22
7	PL	Compacted clay soil	338.59	0.05
8	R	Black soil	1,292.98	0.20
9	VR	Cracked soil	2,681.74	0.41
10	X	Grey soil	185,809.00	28.61
11	XK	Brown soil	11,730.80	1.81

Soil thickness

Soil thickness was classified into 5 levels including: <30cm, 30-50cm; 50-70cm; 70-100cm; >100cm (FAO, 1993b) [8]. Almost soil thickness in the area was above 100cm with the area about 449,000ha (68.86%). This characteristic is suitable for industrial crops such as coffee, pepper, cashew, rubber, etc. The second level was < 30cm that cover about 711,124.3ha (10.91% of land in total).

Soil texture

Soil texture was classified into four levels including loam, sandy, silty loam and clay. Soil with loam and clay texture take a large proportion of the area which are suitable for industrial crop such as coffee and rubber plantation. Of which, loam soil is of 67.35% and clay soil with 25.61%.

Slope

Slope is varying from zero to 60.32^0 and divided into 6 levels: <3^0, 3-8^0, 8-15^0, 15-20^0, 20-25^0, >25^0. Slope level 3-8^0 and level 8-15^0 occupy the largest area with the proportion of 32.85% and 32.59% respectively.

Annual rainfall

Annual rainfall in Dak Nong province is from 1,400 – 3,000mm. Rainfall hierarchy map was built with 4 levels: 1,482 – 1,700mm, 1,700 – 2,100mm, 2,100 – 2,500mm, >2,500mm. Accordingly, the rainfall level of 2,100 – 2,500mm has the largest area with 32.46%, the other take equivalent proportion from 21 – 24% (Table 3).

Table 3. Area of rainfall levels

No.	Rainfall level (mm)	Area (ha)	(%)
1	1,482 – 1,700	161,416.25	24.80
2	1,700 – 2,100	138,760.25	21.32
3	2,100 – 2,500	211,308.25	32.46
4	>2,500	139,442.25	21.42
	Total	650,927.00	100.00

Temperature

The annual average maximum temperature goes from 25 – 32^{0}C in this area. The temperature was classified into three levels: 25 – 27^{0}C, 27 – 30^{0}C, >30^{0}C for the entire area. In which, temperature level 27 – 30^{0}C has the largest proportion with 61.87% and the second is over 30^{0}C with 24.25%. This high temperature is very suitable for coffee, rubber and cashew.

The overlaying multi-factor maps and current key crop maps was conducted to identify adaptability levels of each. Simultaneously, adaptation analysis was also carried out on the entire area of the province with some popular crops such as citrus, pomelo, durian, avocado, etc...to serve as a basis for conversion of non-adapted areas of the key crops or crop planning more appropriate. The result The LULC map was overlaid on the multi-factor maps to determine the current state. The results indicated that out of 164,000 ha of coffee, about 92.5 thousand ha have been grown in less-adapted to moderate-adapted areas, about 57% respectively, and un-adapted area is about 70.01 thousand ha, equivalent to 43% of the province's coffee area. Comparison of adaptability to other species as mentioned above, about 32.55 thousand ha of these can be converted to other crops such as rubber, cashew, durian, oranges, tangerines and pomelos.

Similarity, some areas of 11.07 thousand ha of un-adapted cashew cultivation can be converted to other crops such as rubber, coffee durian, oranges, tangerines and pomelos; with 12.06 thousand ha of rubber cultivation under the non-adapted category, only about 34.51% of the area can be converted to other crops such as cashew, coffee and fruit trees.

In all three cases of crops, there are many non-adapted areas for which suitable crops have not been identified. For these areas, it needs studies and trials of new plants to make recommendations, or afforestation with native trees, afforestation of raw materials, multi-purpose forestry trees in the form of pure planting. or agroforestry to increase the cover to contribute to climate change mitigation.

For the natural area of the province, adaptability analysis for the key crops was indicated in Table 4. However the planning should be considered to specific areas which exclude other LULCs such forestry land, residential area, etc.

Table 4. Suitable land for coffee and pepper, rubber, and cashew in Dak Nong province

Adapted levels	Coffee, pepper		Rubber		Cashew	
	Area (ha)	(%)	Area (ha)	(%)	Area (ha)	(%)
Non suitable	356,125.00	57.06	424,059.00	65.84	579,170.8	89.93
Least suitable	194,728.00	31.20	207,597.00	32.23	64,833.35	10.07
Moderate suitable	73,287.60	11.74	12,371.20	1.92	8.78	0.0014

The result also shows that suitable lands for some key crops in Dak Nong province are distributing in different places and the most adapted lands are concentrating in the central and the south of the province (Figure 4). It indicates that the average proportion of suitable lands for growing these species is quite low, especially cashew trees comparing to other crops.

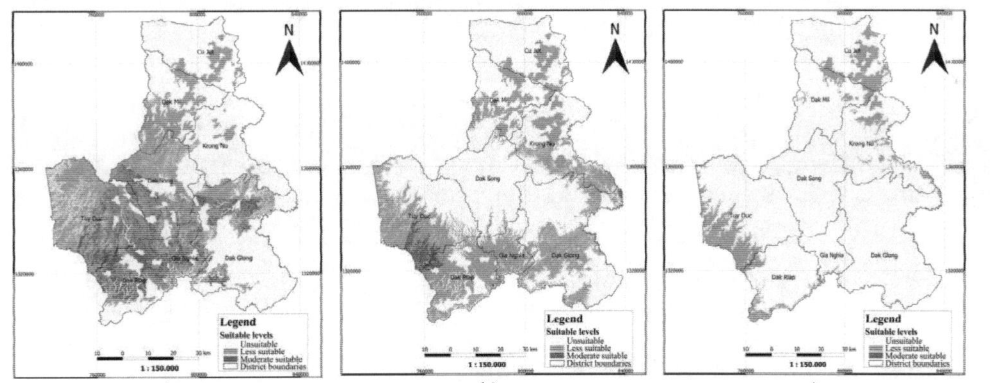

a) b) c)

Figure 5. Potential adaptation soil for key crops in Dak Nong province: a) coffee and pepper, b) rubber, and c) cashew

4. Conclusions

Integration of remote sensing and GIS application is useful to analyse the adaptability of crops. The analysis provide basis for determining the current cultivation and recommending to convert the areas which are under crop requirements. In practice, the cultivation of a certain crop on a less or unsuitable area is often compensated by highly intensive farming, using a lot of chemical fertilizers, which results in residues in the product and contaminates the environment. In addition, the production efficiency may be low because of the high investment. Therefore, determining the area suitable for crops contributes more effectively in farming not only for economic, society, and for environment as well in some ways. Although the OA is quite high in the present study, the PA and/or UA of the main crop is not really high, so additional approaches are needed to improve the classification accuracy. In addition, some other factors such as humidity, altitude, water source, etc. should also be considered in the analysis of LUU adaptation for crops.

IOP Conf. Series: Earth and Environmental Science **964** (2022) 012003 doi:10.1088/1755-1315/964/1/012003

References

[1] Congalton, R.G. and Green, K. 1999. Assessing the accuracy of remotely sensed data: Principles and practices, Lewis Publishers, Boca Raton FL.
[2] Dak Nong Department of Statistic, 2020. Statistic yearbook 2019
[3] Dengiz, O., Bayramin, I. and Yuksel, M., 2003. Geographic information system and remote sensing based land evaluation of Beypazarı area soils by ILSEN model. Turk. J. Agric. 27:145-153.
[4] Ekanayake, G. K. and Dayawansa, N. D. K., 2003. Land Suitability Identification for a Production Forest through GIS Technique. University of Peradeniya. http://www.gisdevelopment.net
[5] ESA, 2015. Sentinel-2 User handbook
[6] FAO, 1976. A Framework for Land Evaluation. International Institute for Land Reclamation and Improvement, The Netherlands.
[7] FAO, 1985. Land Evaluation for Irrigated Agriculture Guidelines. FAO Soil Bulletin No. 55. Food and Agriculture Organization of the United Nations, Rome, Italy, 219 p.
[8] FAO,1993b. An international framework for evaluating sustainable land managemen, Rome, Italy.
[9] FAO, 1993. Guidelines for land use – planning, Rome.
[10] Navulur K. (2007). Multispectral Image Analysis Using the Object-Oriented Paradigm, CRC Press Taylor & Francis Group.
[11] Patil, A., Prathumchai, K., Samaakoon, L. and Honda, K., 2001. Evaluation of land utilization for regional development- A GIS approach. 22nd Asian Conference on Remote Sensing. Singapore.

Investigating Sea Surface Temperature and Coral Bleaching in the Coastal Area of Khanh Hoa Province

Tran Thi Van[1,2*], Nguyen Trinh Duc Hieu[1,2,3], Nguyen Huu Huan[3,4], Nguyen Phuong Lien[5]

[1] Department of Environmental and Resources Information System, Faculty of Environment and Natural Resources, Ho Chi Minh City University of Technology, (HCMUT), 268 Ly Thuong Kiet Street, District 10, Ho Chi Minh City, Vietnam
[2] Vietnam National University Ho Chi Minh City, Linh Trung Ward, Thu Duc District, Ho Chi Minh City, Vietnam
[3] Institute of Oceanography, Vietnam Academy of Science and Technology (VAST)
[4] Graduate University of Science and Technology, Vietnam Academy of Science and Technology (VAST)
[5] Coastal Branch, Vietnam – Russia Tropical Centre

* Corresponding author: tranthivankt@hcmut.edu.vn

Abstract. Khanh Hoa Coastal area is considered the area with the most diverse coral reefs in the west of the East Sea. With the trend of increasing global temperature, the coral reefs here are affected, including the phenomenon of bleaching. This paper uses a Multi-scale Ultra-high Resolution Sea Surface Temperature (MUR SST) data source to evaluate the possible relationship between sea surface temperature and coral bleaching in the period 2010-2019 in Khanh Hoa coastal area is based on two parameters: Hot Spot (HS) and Degree Heating Months (DHM). Research results show that in the past 10 years, corals in Khanh Hoa coastal area may experience heat stress in 6 years, including 2010, 2013 and the years from 2016 to 2019. The phenomenon of heat stress starts to occur in May of each year, and the level of heat stress in 2010 was stronger than in other years. Within 6 years, there have been heat stress, coral bleaching events due to temperature only occurred in 2010; while in 2013, 2016, 2017, 2018 and 2019, corals suffered from heat stress in the watch level, meaning an increase in temperature is not yet capable of causing coral bleaching. The cause of coral bleaching in 2010 was the combined effect of the increase in sea surface temperature and the suppression of upwelling during the southwest monsoon.

Keywords: coral bleaching, degree heating months, hot spot, sea surface temperature

1. Introduction

Coral reef ecosystems are characterized by high diversity, biological productivity and are home to many marine species. Thus they are considered the "rainforest" of the sea [1]. Today, coral reefs are considered an essential marine resource and are closely linked to the lives of many communities living in and around the reef. However, recent statistics show that the world coral reefs have lost about 19%. About 15% of reefs are in a state of serious threat, and they disappear within the next 10-20 years. 20% of reefs are threatened and likely disappear within 20-40 years [2]. Since the 1970s, coral bleaching has become more severe and frequent with the increase in CO_2 content and sea surface temperature [3]. In the future, the frequency and extent of bleaching are predicted to become more severe as the earth temperature gets hotter and hotter, which become a threat to the world coral reefs; because an increase in sea temperature is considered to be the main cause of mass coral bleaching [4]. Tropical coral reefs live just below their maximum temperature limit, and even slight increases above historical values can cause bleaching [5].

According to the US National Oceanic and Atmospheric Administration (NOAA) statistics, three severe coral bleaching events have been recorded worldwide due to global warming [6]. The first global bleaching event, combined with a strong El Niño, raised ocean temperatures by about 2°C and destroyed about 16% of the world corals in 1997-1998 [7, 8]. The second bleaching event was during a mild El Niño in 2010 that raised sea temperatures and caused mass bleaching of coral reefs in many parts of the world [6]. The third bleaching event was a strong El Niño-induced global warming in 2015-2016, which NOAA announced as the most prolonged and most extensive bleaching event on record [6, 9]. In addition, the bleaching event in 1982-1983 killed 95% of corals in the Galapagos Islands [10], but NOAA did not record this event.

Today, sea surface temperature data collected from satellites are used to monitor coral heat stress [11]. NOAA has developed a reef monitoring program to monitor coral bleaching events caused by increases in global sea surface temperature (https://coralreefwatch.noaa.gov/). Through this monitoring program, NOAA has developed a monthly or seasonal coral bleaching warning scheme on a global scale. These schemes are widely used and adapted in all tropical countries to predict and monitor bleaching phenomena. However, these products have a rather large resolution (5 km x 5 km) for a detailed evaluation of bleaching events at islets or reefs [6]. In addition, the NOAA method uses Degree Heating Weeks (DHW) to assess coral bleaching. Meanwhile, the majority of outputs from climate models are average monthly temperatures. Therefore, some recent studies have used Degree Heating Months instead of Degree Heating Weeks [12].

In the world, there have been many studies using DHM/DHW to assess coral bleaching levels from SST data collected from satellite devices. In 2009, Donner forecast heat stress on coral reefs worldwide using DHM over four months based on SST data over the period 1985-2000 [13]. The results of their study shows that coral bleaching occurs when the DHM is 1°C greater than the maximum monthly mean (MMM). When DHM is greater than MMM 2°C, bleaching becomes severe and can be fatal to corals. In 2014, the team of authors Yumiko Yara and colleagues evaluated the impact of increased sea surface temperature on coral bleaching caused by global warming. They used 21st century monthly SST data collected from 23 climate models to determine the DHM index to predict coral bleaching [12]. They predicts that corals will be severely bleached or dead in large areas of tropical and subtropical oceans by the middle of this century. In 2016, the author group Sam Wouthuyzen and colleagues compared the increase in SST and its effect on coral bleaching in 2010 and 2016 in Indonesian waters using the temperature parameter such as HS and DHW from the 4-km-resolution SST data source, collected from the MODIS Aqua satellite during the period 2002-2017 [6]. Their study shows that the normal temperature threshold to which corals can adapt is 29.1°C. Heat stress for corals occurred in 2010 and 2016. Both coral bleaching events in these two years started and ended in the same period (starting March and ending June). However, the bleaching intensity of 2016 was stronger than that of 2010, with an average SST of about 0.4°C higher in May - June.

This paper uses the monthly average SST data source to determine HS and the DHM to examine the possible relationship between sea surface temperature and coral bleaching events in Khanh Hoa coastal area in 2010-2019.

2. Data and method

2.1. Study area

The study area is located in the coastal area of Khanh Hoa province, limited to the range of coordinates from 11.80° – 12.84°N and 109.05° – 110.00°E (Figure 1). The study area is dominated by a tropical monsoon climate in winter (northeast monsoon) and summer monsoon (southwest monsoon). The winter monsoon has the prevailing wind direction from north to northeast. The active period of the winter monsoon coincides with the rainy season in Khanh Hoa province, especially when the northeast monsoon combines with the activity of the equatorial trough. The path, the tropical convergence zone, storms, and tropical depressions cause heavy rain to create floods for Khanh Hoa province. The summer monsoon usually occurs in early summer, with the prevailing wind direction being southwest. This wind brings hot, dry air, high temperature and low humidity, making the weather over the basin during this period the hottest and driest of the year [14].

Khanh Hoa is a coastal province with 385 km of coastline, with a particular position facing the East Sea with favourable natural conditions for high biodiversity [15]. An overview from previous studies has shown that the coastal area of Khanh Hoa has high biodiversity. It is home to many habitats such as coral reefs, mangroves, seagrass beds, islands and beaches with various marine species. Among them, coral reefs are distributed in shallow water, mainly in Nha Trang Bay and Van Phong Bay and outside Cam Ranh Bay, with an area of more than 3,000 hm^2. Therefore, the coastal waters of Khanh Hoa are considered to be the area with the most diverse coral reefs in the western part of the East Sea [16]. However, marine economic development in Khanh Hoa generates several environmental problems such as overexploitation, loss and degradation of ecosystems, pollution, and landscape degradation [15]. Previous research showed that coral reefs in Khanh Hoa coastal area are being affected by bleaching events related to current climate change [17]. Coral bleaching has been recorded in Khanh Hoa and southern Vietnam in 1998, 2010, 2013 and 2014. Besides, coral reefs in Khanh Hoa have been severely affected by the integrated effects of rising sea surface temperature and other anthropogenic activities [16]. The above data shows that the increase in sea surface temperature may affect the coral reef system in Khanh Hoa coastal area.

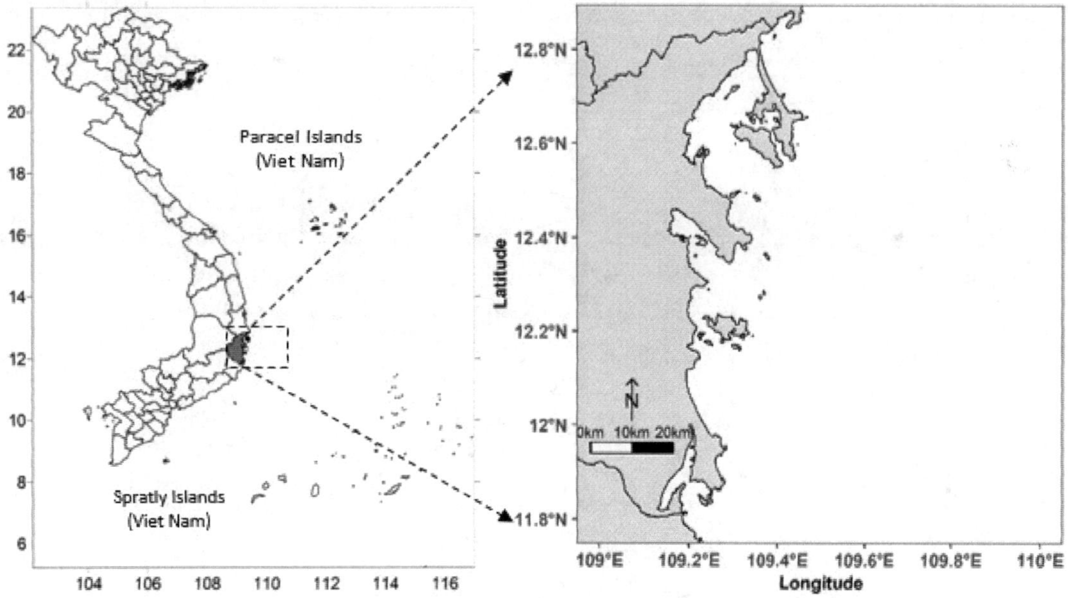

Figure 1. Map of the study area

2.2. Data
This study uses monthly mean data, spatial resolution 0.01 degrees (Latitude) x 0.01 degrees (Longitude) from the dataset MUR SST (Multi-scale Ultra-high Resolution Sea Surface Temperature). The data was mined between January 2010 and December 2019. The MUR SST level 4 dataset is analyzed based on sea surface temperature observations from several satellites of the National Aeronautics and Space Administration (NASA), NOAA, and field SST observations from the NOAA iQuam project. This dataset is sponsored by NASA, provided by the Physical Oceanography Distributed Active Archive

Center (PO.DAAC) of NASA Jet Propulsion Laboratory (JPL). The data is collected free of charge from the website https://coastwatch.pfeg.noaa.gov/erddap/files/.

2.3. Method to assess coral bleaching risk using sea surface temperature
Consider the possible relationship between SST and coral bleaching events during the past 10 years in Khanh Hoa coastal area based on two parameters, HS and DHM. These two indices are used to assess the heat stress leading to coral bleaching. The procedure for determining HS and DHM is described in detail in Figure 2 and equations (1) and (2) [18].

From the SST data series, calculate the 10-year MMM for the period 2010-2019. MMM is considered as the value of sea surface temperature under normal conditions, and corals can adapt over a long period of 10 years [6]. After determining the MMM, calculating the difference between the average monthly SST in each year and the MMM, this value is called HS, and is determined by equation (1) [6, 18]:

$$\text{HS} = \text{Average monthly SST for each year} - \text{MMM} \quad (1)$$

After calculating the HS, the DHM value was determined based on the monthly HS accumulation > 0°C at any pixel over 3 consecutive months according to Equation (2) [4]:

$$\text{DHM}_t = \sum_{i=t}^{t+2} \text{HS}_{ij} \quad (2)$$

In which, t is the start month of the forecast.

Based on the calculated HS and DHM values from equations (1) and (2), classify the heat stress of corals according to Table 1 [4, 18]. Combining the calculation results of HS, DHM and published documents on coral bleaching events in Khanh Hoa, we will consider the possible relationship between coral heat stress and SST at the time of the bleaching event.

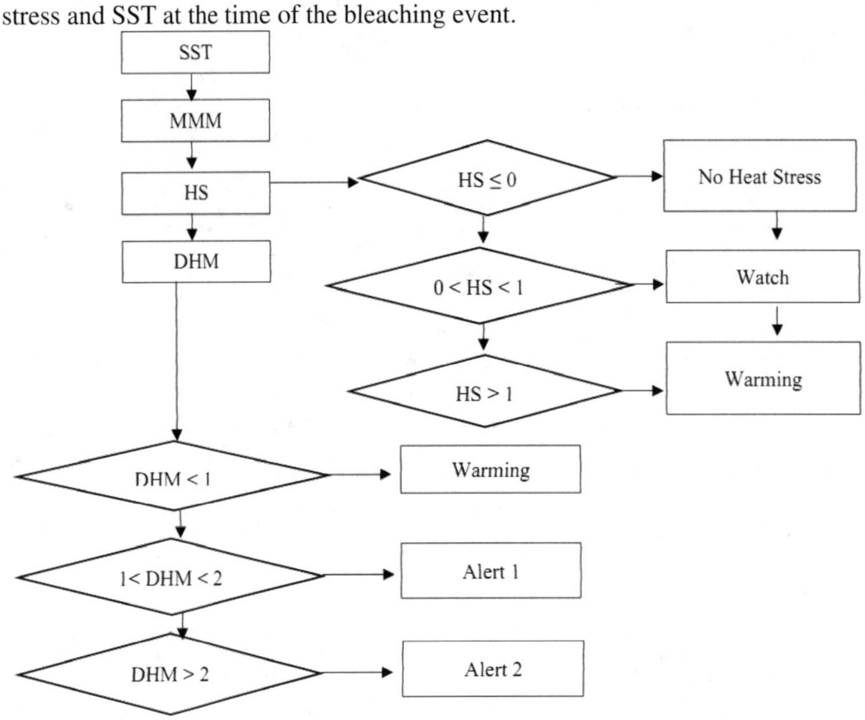

Figure 2. Procedure for determining the heat stress of corals

Table 1. Classification of coral heat stress levels based on HS and DHM

Stress level	Definition	Effect	Color
No	$HS \leq 0$	No heat stress	

Stress level	Definition	Effect	Color
Watch	0 < HS < 1	Low level heat stress, no effect on coral yet, observed bleaching	
Warming	HS ≥ 1 DHM < 1	Heat stress at cumulative level, coral bleaching possible	
Alert 1	HS ≥ 1 1 ≤ DHM < 2	Strong heat stress, capable of causing partial bleaching of corals	
Alert 2	HS ≥ 1 DHM ≥ 2	Heat stress is severe, causing widespread coral bleaching, corals are likely to die	

3. Result

Statistical calculation results show that the SST in the period 2010-2019 in Khanh Hoa coastal area fluctuates in a wide range from 22.5 to 30.8°C, with the average value reaching 27.6 ± 1.6°C (Figure 3). The difference between the maximum and minimum SST values in Khanh Hoa coastal area is about 8.3°C, two times lower than the fluctuation range in the Vietnam South Central Coast (17.4 - 35.0°C) and nearly 3 times lower than in the East Sea (10.1 - 35.0°C) [19]. The value of the 95% highest density interval (HDI) in the SST data series ranged from 24.3 to 30.5°C. SST values that fall within the 95% HDI range (blue area in Figure 3) have higher probability density and confidence than SST values outside the 95% HDI range (red area in Figure 3). Based on the probability density graph, the SST data series has a right skewed distribution, so the mean value of SST (27.6°C) is smaller than the median value (28.0°C). SST values common in the Khanh Hoa coastal area range from 28.6 to 29.6°C with probability density greater than 40% and accounts for more than 28% in the data set (Figure 3).

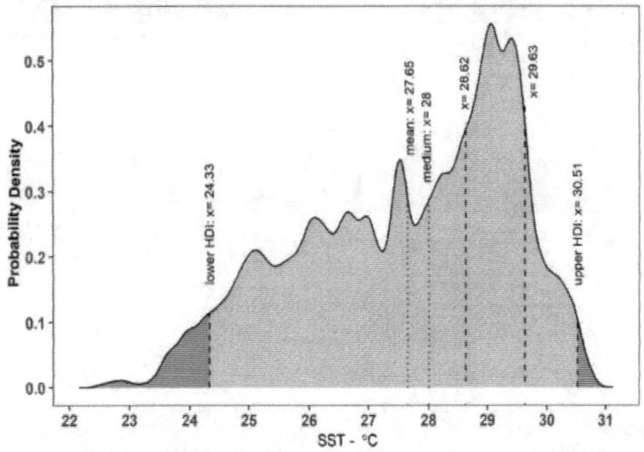

Figure 3. Probability density graph of SST values in Khanh Hoa coastal area

Table 2 and Figure 4 present the monthly average sea surface temperature distribution by year, used to determine the MMM value in Khanh Hoa coastal area in the period 2010-2019. Statistical results show that SST usually peaks in May (or June) each year, then tends to decrease in June, July, and August; continued to increase in September. In Khanh Hoa coastal areas, during the prevailing southwest monsoon (June to August), upwelling usually activates at moderate or weak levels [20, 21]. Therefore, it is possible that due to the influence of upwelling, the SST value does not increase continuously from May to September but tends to decrease in June, or July, or August each year. From the monthly average SST data, the calculation results show that the MMM value reaches 29.6°C, which means that the threshold at which corals can adapt to sea surface temperature is below 29.6°C. Generally, the optimal SST for coral reefs to grow is between 22°C and 28°C [22]. However, the temperature range in which corals can survive is between 18°C and 36°C [23]. Thus, the MMM value in Khanh Hoa coastal area is outside the optimal temperature range for coral growth but still within the allowable range for coral survival. When the SST exceeds the threshold of 29.6°C, coral heat stress occurs with the stress level depending on the difference between the average monthly SST and the MMM. This difference is defined as the HS heat index and is determined as follows: HS = monthly mean SST (at times of heat stress) – 29.6°C.

The charts in Figure 4 illustrate a comparison of the average monthly SST value for the period 2010-2019 with the MMM value. The results showed that the SST value increased and exceeded the MMM threshold at the following 12 times: from May to July 2010, May and June 2013, May and September 2016, May and September 2017, May 2018, May and June 2019. This result indicates that in the past 10 years, corals in Khanh Hoa coastal area may be subjected to heat stress in 2010, 2013 and the years from 2016 to 2019. In particular, the phenomenon of heat stress begins to occur in May of each year. Previous studies have also recorded coral bleaching in Khanh Hoa coastal area in 2010 and 2015 with different degrees of bleaching [16, 24].

Table 2. Determination of MMM from SST

Year	Jan	Feb	Mar	Apr	May	Jun	Jul	Aug	Sep	Oct	Nov	Dec
2010	25.5	26.2	26.3	28.8	30.3	30.4	30.1	29.4	29.6	29.2	26.8	26.4
2011	24.6	25.6	25.3	26.5	28.3	28.2	27.5	29.3	29.3	27.1	26.2	24.1
2012	24.0	25.1	26.4	27.7	29.4	28.5	28.6	28.1	29.1	28.2	27.9	26.9
2013	25.2	26.1	27.6	28.9	29.6	29.8	29.4	28.7	28.9	27.5	26.7	24.8
2014	23.5	24.3	26.3	28.6	29.4	28.0	28.5	28.5	29.5	28.5	27.7	25.9
2015	24.5	24.8	26.4	27.5	29.5	29.1	27.6	29.1	29.4	29.1	28.4	26.9
2016	26.2	25.1	25.3	26.8	29.8	28.2	28.6	29.0	29.7	29.2	27.6	26.2
2017	25.2	24.8	25.7	27.1	29.8	29.3	29.4	29.3	29.7	28.8	27.0	25.7
2018	24.6	24.3	26.2	27.7	30.2	28.6	28.6	28.9	29.5	28.5	27.6	27.3
2019	25.7	26.8	27.7	29.5	30.0	30.5	28.7	28.8	29.0	28.8	27.0	25.2
MMM	24.9	25.3	26.3	27.9	29.6	29.1	28.7	28.9	29.4	28.5	27.3	25.9

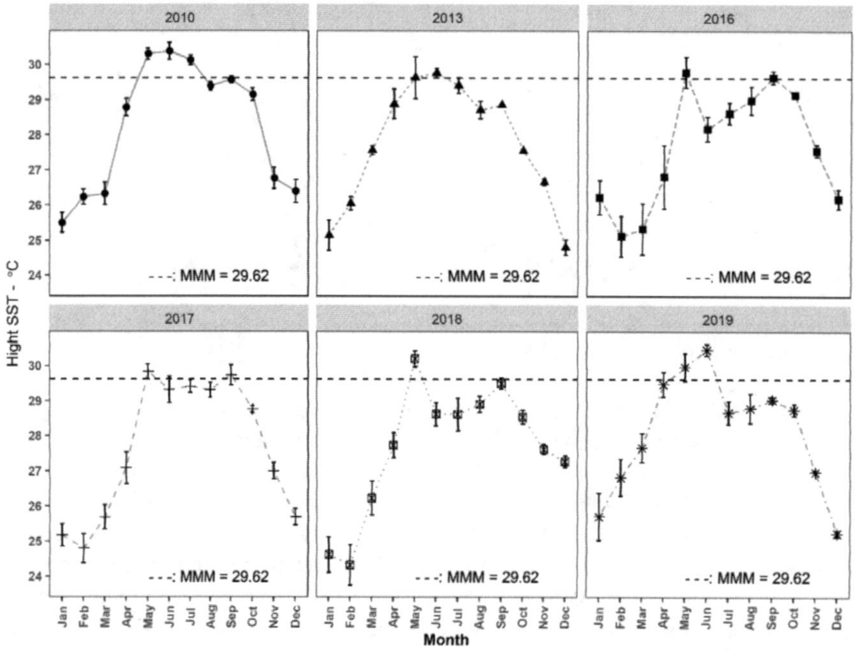

Figure 4. Comparison of average SST per month and MMM

In May 2010, Khanh Hoa coastal area experienced heat stress at three different levels, including alert 2, alert 1 and watch with area ratios of 48%, 51% and 1%, respectively (Table 3). Heat stress in coastal areas (109.4°E back to shore) mainly occurs at alert 1 level – potentially causing partial bleaching of corals. However, in Van Phong Bay, there is a alert 2 level- coral bleaching and possibly dying (Figure 5). Research on the resilience of coral reefs in the Khanh Hoa coastal area shows that most of the coral reefs in the northern area of Khanh Hoa, especially the coral reefs in the nearshore waters in Van Phong Bay, are more sensitive to rising sea surface temperature. Coral reefs in Nha Trang Bay and the southern region are less vulnerable to heat [16]. In June 2010, heat stress levels also occurred with 3 different levels, including alert 1, warming and watch with area ratios of 4%, 24% and 72%, respectively. However, only heat stress occurred at the watch level in the coastal waters (109.4°E back to shore), i.e. there was heat stress but no effect on coral reefs (Figure 5). Meanwhile, in July 2010, heat stress only occurred at the watch level in Khanh Hoa coastal area (Table 3 and Figure 5).

At times from 2013, 2016 to 2019, although there was a heat stress phenomenon in Khanh Hoa coastal areas, most of them were only watch level, i.e. an increase in temperature is not likely to affect coral reefs (Table 3). Specifically, in May 2013, the warning status appeared in the offshore area of Cam Ranh Bay with the rate of 2%; the watch and non-heat stressed states accounted for 47% and 51%, respectively. Table 3 also shows that heat stress in the watch state accounted for 83% in June 2013, 66% in May 2016, 59% in September 2016, and 82% in May 2017, 67% in September 2017, 99% in May 2018, 78% in May 2019 and 84% in June 2019. In addition, in June 2019, heat stress at warming level also appeared in the offshore waters (109.6°E and offshore) in Van Phong and Nha Trang Bays, with a rate of 16%.

Table 3. Percentage (%) of areas with heat-stressed corals in Khanh Hoa coastal areas

Classified	May 2010	Jun 2010	Jul 2010	May 2013	Jun 2013	May 2016	Sep 2016	May 2017	Sep 2017	May 2018	May 2019	Jun 2019
Alert 2	48	0	0	0	0	0	0	0	0	0	0	0
Alert 1	51	4	0	0	0	0	0	0	0	0	0	0
Warming	0	24	0	2	0	0	0	0	0	0	0	16
Watch	1	72	100	47	83	66	59	82	67	99	78	84
No	0	0	0	51	17	34	41	18	33	1	22	0

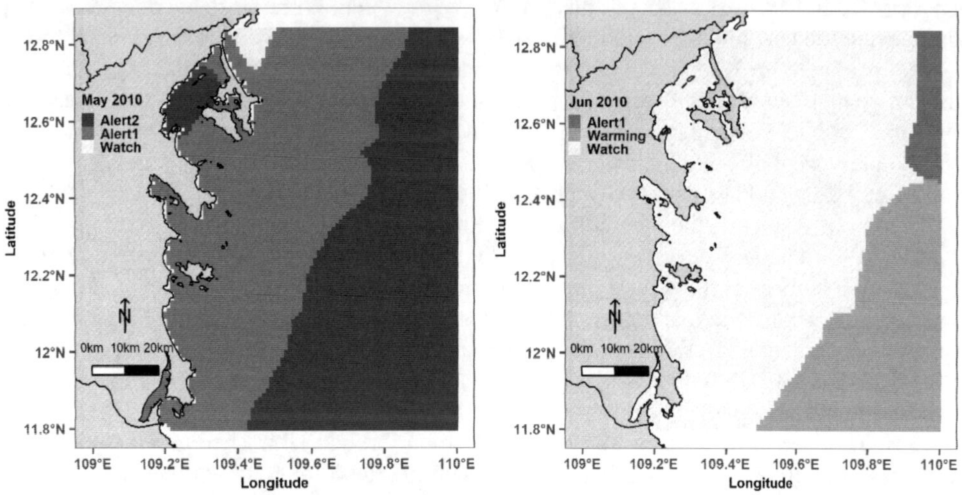

Figure 5. Heat stress on corals in Khanh Hoa coastal area in May and June 2010

4. Discussion

Research results show that, in Khanh Hoa coastal areas, heat stress for corals occurred in 2010. Globally, a mild El Niño in 2010 increased SST and caused mass bleaching of coral reefs in many parts of the world [6]. In an overview report on coral bleaching in 2010 in Southeast Asia, SST exceeded the bleaching threshold of 2°C from May to July 2010. The increase in SST has caused coral bleaching in some areas of the South Central Coast, such as Van Phong, Nha Trang, Cam Ranh and Ninh Hai [25]. In addition, the field survey results also showed that coral bleaching was recorded in Nha Trang Bay in May 2010 [26].

An overview report on the status of coral reefs in the East Asian Sea in the years 2015-2017 shows that, in 2015, coral bleaching phenomenon with a mild degree was detected in shallow water (2-7m deep) in Van Phong Bay, Nha Trang Bay and Cam Ranh Bay. However, there was no record of bleaching in these three Bays in 2016 and 2017 [24]. This study also shows that in 2016 and 2017, only the watch state - temperature increase was not likely to affect the reef and the absence of heat stress. Thus, it can be seen that, although a strong El Niño wave occurred in 2016, which increased SST and caused coral bleaching on a global scale [6], it did not affect coral reefs in Khanh Hoa coastal area. Recent research shows that 2016 is a year where two phases co-occur: the end of El Niño, the preparation for the La Niña period; and 2016 is considered the time to bleach corals like 1998 and 2010. However, in July 2016, the upwelling phenomenon appeared with the centre moving closer to the shore than in normal years. The offshore area of Ninh Thuan - Khanh Hoa was strongly affected by upwelling [27]. When upwelling water is active, cold deep water layers move vertically to the surface, reducing sea surface temperature. This can help reduce heat stress for corals in Khanh Hoa coastal area in 2016. Previous studies also showed that the occurrence of upwelling during the southwest monsoon season helped reduce heat stress and coral bleaching in Khanh Hoa coastal area [28, 29].

The above analysis indicates that the occurrence of upwelling during the southwest monsoon helps to reduce the heat stress for corals, however in 2010, coral bleaching was recorded in Khanh Hoa. Research on the upwelling water system in the South Central Coast shows that in June-July 2010, there was a decrease in long-shore currents in the north-northwest direction from the north of Ninh Thuan to the sea of Binh Dinh. This current system eliminated the anti-cyclonic vortex off the coast of Ninh Thuan - Binh Thuan, causing the upwelling phenomenon to be eliminated in 2010 [27]. Thus, due to the combined effect of the increase in SST (above 30°C) and the suppression of the upwelling phenomenon, coral bleaching in Khanh Hoa coastal area in 2010.

Although the results from this study suggest no heat stress on corals in 2015, coral bleaching has been recorded in Khanh Hoa as presented in the overview report on the status of coral reefs in the East Asian Sea. Reviewing previous studies, it can be seen that, in addition to being affected by the increase in SST, coral reefs in Khanh Hoa coastal area are also degraded by other causes such as increased sediment content, outbreaks of starfish and local tourism activities [29, 30]. Some recent studies have shown that, in Nha Trang Bay, coral degradation is not due to an increase in sea surface temperature. Instead, corals were affected mainly by outbreaks of thorns starfish, with average densities increasing from 1.7 to 4.2 individuals per 100m^2 between 2013-2019 [28, 31].

The analysis results on the distribution over time of SST and the effect of SST on corals show that, in summer, Van Phong Bay (north of Khanh Hoa) has a higher sea surface temperature; and corals are subjected to a higher degree of heat stress than in the southern part of Khanh Hoa. Meanwhile, in the Van Phong Bay area, a thermal power plant with a capacity of 1,320 MW has been started [32]. A recent heat spread calculation study shows that if this thermal power project is operated at total capacity as planned, about 15,000,000 m^3 of water (sea) has a higher temperature than water every day. The sea in the receiving area, about 7°C, is discharged into the northwest of the My Giang - Hon Do - Bo Co sea area [33]. The area of reefs from My Giang extending through Hon Theo to Bo Co is home to many coral reefs. Therefore, the most concerned pollutant in Van Phong Bay in the future will be the heat from the cooling water of thermal power projects, and management agencies should have harmonized solutions for the goals of economic development and environmental protection in the locality.

5. Conclusion

Research results show that in the past 10 years, corals in Khanh Hoa coastal area can be subjected to heat stress in 6 years, including 2010, 2013 and the years from 2016 to 2019. Heat stress events all start to occur in May of each year, and the level of heat stress in 2010 was stronger than in other years. Of the six years of heat stress, the temperature-induced coral bleaching event occurred only in 2010. Meanwhile, in 2013, 2016, 2017, 2018 and 2019, corals suffered from heat stress in the watch state, which means an increase in temperature is not yet capable of causing coral bleaching.

In summer, the Van Phong Bay area (north of Khanh Hoa) has higher sea surface temperatures and higher heat stress corals than the southern part of Khanh Hoa. The cause of coral bleaching in 2010 was the combined effect of the increase in sea surface temperature and the suppression of upwelling during the southwest monsoon.

Acknowledgements

We would like to thank Ho Chi Minh City University of Technology (HCMUT), VNU-HCM for the support of time and facilities for this study. We would also like to thank the project of the Institute of Oceanography under contract number 2702/ĐTCS2020 for providing the data for us to complete this study. This paper is a contribution to celebrate the 100 years Anniversary of the Institute of Oceanography, Vietnam Academy of Science and Technology.

References

[1] Connell JH 1978 Diversity in tropical rain forests and coral reefs *Science* **199**(4335) 1302-1310.
[2] Wilkinson C 2008 *Status of Coral Reefs of the World: 2008* (Global Coral Reef Monitoring Network and Reef and Rainforest Research Centre, Townsville): Australia.
[3] Junzhe Liu 2020 *Possible Measurement to Coral Bleaching from the Perspective of Climate Change - Taking South China Sea as an Example*. Paper presented at the IOP Conf. Series: Earth and Environmental Science. q
[4] Spillman C, Alves O, Hudson D 2013 Predicting thermal stress for coral bleaching in the Great Barrier Reef using a coupled ocean–atmosphere seasonal forecast model. *International Journal of Climatology* **33**(4) 1001-1014.
[5] Goreau T, Hayes RL 2005 Global coral reef bleaching and sea surface temperature trends from satellite-derived hotspot analysis. *World Resource Review* **17**(2) 254-293.
[6] Wouthuyzen S, Abrar M, Lorwens J 2018 *A comparison between the 2010 and 2016 El-Ninō induced coral bleaching in the Indonesian waters.* Paper presented at the IOP Conference Series: Earth and Environmental Science.
[7] Wilkinson C 2004 *Status of coral reefs of the world: 2004.* Australian Institute of Marine Science. Townsville, Australia.
[8] Hoegh-Guldberg O 1999 Climate change, coral bleaching and the future of the world's coral reefs. *Marine and Freshwater Research* **50**(8) 839-866.
[9] Hughes TP, Kerry JT, Álvarez-Noriega M, Álvarez-Romero JG, Anderson KD, Baird AH, Babcock RC, Beger M, Bellwood DR, Berkelmans R 2017. Global warming and recurrent mass bleaching of corals. *Nature* **543**(7645) 373-377.
[10] Glynn PW 1984 Widespread coral mortality and the 1982–83 El Niño warming event. *Environmental Conservation* **11**(2) 133-146.
[11] Putra R, Suhana M, Kurniawn D, Abrar M, Siringoringo R, Sari N, Irawan H, Prayetno E, Apriadi T, Suryanti A 2019. *Detection of reef scale thermal stress with Aqua and Terra MODIS satellite for coral bleaching phenomena.* Paper presented at the AIP Conference Proceedings.
[12] Yara Y, Fujii M, Yamano H, Yamanaka Y 2014 Projected coral bleaching in response to future sea surface temperature rises and the uncertainties among climate models. *Hydrobiologia* **733**(1) 19-29.
[13] Donner SD 2009 Coping with commitment: projected thermal stress on coral reefs under different future scenarios *PLoS One* **4**(6) e5712.
[14] Institute of Irrigation Planning 2016 *Adjusting and supplementing the irrigation development planning of Khanh Hoa province for the period 2015-2025 and orientation 2035* Thematic Report on Hydrometeorology.
[15] Vo Si Tuan 2016 Conservation of marine biodiversity to develop green economy in Khanh Hoa province, Vietnam. *Collection of Marine Research Works* 22.

[16] Vo Si Tuan, Hua Thai Tuyen, Phan Kim Hoang 2019 A study of coral reef resilience and implications of adaptive management and rehabilitation in Khanh Hoa Province, Vietnam. *Acta Oceanologica Sinica* **38**(1) 112-117.

[17] Nguyen Van Long, Vo Si Tuan, Phan Kim Hoang, Hua Thai Tuyen, Nguyen An Khang, Thai Minh Quang, Phan Kim Hong 2014 Coral reefs in Van Phong bay, Khanh Hoa province: status and management perspectives. *Collection of Marine Research Works* 14.

[18] Liu G, Eakin C, Rauenzahn J, Christensen T, Heron S, Li J, Skirving W, Strong A, Burgess T 2012 *NOAA Coral Reef Watch's decision support system for coral reef management.* Paper presented at the Proceedings of the 12th International Coral Reef Symposium, Cairns, Australia.

[19] Vu Van Tac 2016 Temperature characteristics and surface salinity in the South Central Sea of Vietnam. *Journal of Marine Science and Technology* 17(1).

[20] Bui Hong Long 2011. *Handbook of research on natural conditions, ecological environment, economy, society and integrated management of the South Central Coastal zone*. Natural Science and Technology Publishing House.

[21] Han Trong Dat 2019 *Study on the influence of some oceanic factors on the distribution and change of resources of the scad group (Decapterus spp.) in the upwelling area of South Central Vietnam.* (Master Thesis), University of Science - Vietnam National University, Hanoi

[22] Hubbard D 1997 Reefs as dynamic systems. *Life and death of coral reefs* 43-67.

[23] Khalil I 2019 *Satellite observation of trends in sea surface temperature and coral bleaching in the Indo-Pacific region.* University of Southampton.

[24] Kimura Tadashi, Tun Karenne, Chou Loke Ming 2018 *Status of coral reefs in East Asian Seas Region* Ministry of the Environment of Japan and Japan Wildlife Research Center, Tokyo, Japan.

[25] Tun, K., L. Chou, J. Low, T. Yeemin, N. Phongsuwan, N. Setiasih, J. Wilson, A. Amri, K. Adzis, and D. Lane 2011 *The 2010 coral bleaching event in Southeast Asia - A regional overview*. 22nd Pacific Science Congress.

[26] Vo Si Tuan 2013 Negative changes of coral reefs due to the natural catastrophes recorded recently in south Viet Nam. *Collection of Marine Research Works* 19.

[27] Tran Van Chung, Bui Hong Long 2016 Studies on the variability of upwelling system in the South Central Vietnam waters under the impact of ENSO events and its impact on hydrographic conditions of the Ninh Thuan-Binh Thuan waters. *Journal of Marine Science and Technology* 19.

[28] Tkachenko, K.S., T.A. Britayev, N.H. Huan, M.V. Pereladov, and Y.Y. Latypov 2016 Influence of anthropogenic pressure and seasonal upwelling on coral reefs in Nha Trang Bay (Central Vietnam). *Marine Ecology* 37(5), 1131-1146.

[29] Eliza C Heery, Bert W Hoeksema, Nicola K Browne, Reimer JD, Put O Ang, Danwei Huang, Daniel A Friessi, Loke Ming Chou, Lynette HL Loke, Poonam Saksena-Taylor, Nadia Alsagoff, Thamasak Yeemin, Makamas Sutthacheep, Si Tuan Vo, Arthur R. Bos, Girley S. Gumanao, Muhammad Ali Syed Hussein, Zarinah Waheed, David JW Lane, Ofri Johan, Andreas Kunzmann, Jamaluddin Jompa, Suharsono, Daisuke Taira, Andrew G. Bauman, & Peter A. Todd 2018 Urban coral reefs: Degradation and resilience of hard coral assemblages in coastal cities of East and Southeast Asia. *Marine pollution bulletin,* **135**, 654-681.

[30] Nguyen Van Long, Vo Si Tuan 2013 Degradation trend of coral reefs in the coastal waters of Vietnam. *Galaxea, Journal of Coral Reef Studies* **15**(Supplement) 79-83.

[31] Tkachenko, K.S., N.H. Huan, N.H. Thanh, and T.A. Britayev 2020 Extensive coral reef decline in Nha Trang Bay, Vietnam: Acanthaster planci outbreak: the final event in a sequence of chronic disturbances. *Marine and Freshwater Research*.

[32] MBVPEZ - The Management Board of Van Phong Economic Zone 2019 Groundbreaking ceremony of Van Phong 1 BOT thermal power plant, https://vanphong.khanhhoa.gov.vn/article/tin-van/le-dong-tho-nha-may-nhiet-dien-bot-van-phong-1.html.

[33] Nguyen Ky Phung 2015. *Investigate the hydrographic conditions and dynamics in the area of My Giang - Hon Do - Bai Co (Ninh Phuoc commune, Ninh Hoa town) to serve the management and environmental protection of Van Phong economic zone*. Summary report of science and technology results of Khanh Hoa provincial Department of Science and Technology

Assessing the applicability of Fusion Landsat-MODIS data for mapping agricultural land use - A case study in An Giang Province

P K Diem[1*], N K Diem[1], N T Can[2], V Q Minh[1], H T T Huong[1], N T H Diep[1], P C Tao[1]

[1]College of Environment and Natural Resources, Can Tho University, Can Tho, Vietnam

[2]The Joint Graduate School of Energy and Environment, King Mongkut's University of Technology Thonburi, Thailand

*Corresponding author: pkdiem@ctu.edu.vn

Abstract. This study aimed to evaluate the applicability of using time-series data of spatiotemporal fusion Landsat-MODIS imagery for mapping agricultural land use in An Giang province, Vietnam. The Spatial and Temporal Adaptive Reflectance Fusion Model (STARFM) was adopted for fusion techniques to integrate the relatively high spatial resolution of Landsat (30 meters) and frequently revisit time of MODIS (MOD09Q1, 8-days). The Maximum Likelihood Classifier (MLC) was then used to classify the land cover categories based on variations of NDVI (Normalized Difference Vegetation Index) time-series over the observation period. The overall accuracy is about 84.9%, and a kappa coefficient of K=0.7, which revealed the effectiveness of using Fusion Landsat-MODIS NDVI data in land cover classification at the provincial scale. The current of the agricultural land use was finally mapped, including seven categories, namely built-up areas (10.49%), double rice crops (4.8%), triple rice crops (68.24%), perennial tree/orchards (4.08%), annual crops (7%), water surfaces (3.07%), and forest (2.32%). The results indicate that the agricultural land use cover can be detected in detail using Fusion Landsat-MODIS imagery. The classification is dramatically higher compared to the map classified by a conventional method of solely Landsat 8 image analysis (overall accuracy of 67.3% and Kappa coefficient K=0.35). The research outcomes will support the detailed information for managers in evaluating the impact of climate change on the rice cropping system toward sustainable agriculture development.

Keywords: Agricultural land use, An Giang province, fusion, Landsat-MODIS, STARFM.

1. Introduction

An Giang is one of the provinces with the main rice production areas of Vietnam's biggest "rice bowl" in the Vietnam Mekong Delta. The rice production area of An Giang province was about 620 thousand hectares, accounting for 8.3% of the country' rice cultivation area in 2019. The rice production in 2019 reached nearly 4 million tons, accounting for 9.2% of rice production in Vietnam [1]. The transformation of farming structure and land-use changes need to be continuously monitored and updated. However, the implementation requires much time, cost, human resources, and cooperation between different management levels. The government issued Decree No 03/2019/NDD-CP on 04 January 2019 [2] on the exploitation of remote sensing data, which mentioned the critical role of remote sensing in mapping the current agricultural land use.

Nowadays, remote sensing data have been utilized effectively in monitoring the change of agricultural cultivation in many areas around the world [3, 4, 5]. However, most of these studies used satellite images with moderate spatial resolution such as Landsat (30 m), Sentinel 2 (10 m) at a single time, or low-resolution data of multi-temporal such as MODIS MOD09A1, MOD09Q1, and MOD13A1. Yet, the above studies have limitations of detecting detailed agricultural land use such as triple rice crop, double rice crop, and other crops with seasonal growth characteristics. In addition, the percentages of cloud cover on the optical satellite imagery are very high, which is the cause of the low capacity of mapping seasonal crops. On the other hand, the multi-temporal MODIS images have a spatial resolution of no less than 250 m. One pixel of MODIS data is equivalent to 6.25 hectares in the field, including many mixed land cover types. Therefore, it reduces the detail of classification, and it is only suitable for regional and national studies.

The above analysis has shown the limitations of using single moderate-resolution imagery and low resolution of multi-temporal imagery. Integration of multi-temporal and high spatial resolution is proved more efficient in many studies as it gives better solutions for detecting, mapping the distribution of rice crop seasonality [6], monitoring the current land-use changes [7], vegetation cover changes [8, 9, 10]. The spatial and temporal adaptive reflectance fusion model (STARFM) [11] is the most widely used data fusion algorithms for Landsat and MODIS imagery [12, 13], beneficial for detecting gradual changes over large land areas [11, 14]. Therefore, this study was conducted to evaluate the applicability of using time-series data of fusion Landsat-MODIS imagery based on STARFM algorithms to map agricultural land use in An Giang province. Simultaneously, a comparison with mapping land use using single Landsat imagery was also be conducted to confirm the effectiveness of using fusion Landsat-MODIS imagery.

2. Study Area and Methods

2.1. Study area

An Giang is located in the Southwestern of Vietnam within a latitude of 10^{0}10’30” - 10^{0}37’50” N and longitude of 104^{0}47’20” - 105^{0}35’10” E. The Northwest borders with Cambodia, the East and Northeast borders Dong Thap province, the South and Southwest borders Kien Giang province, the Southeast borders Can Tho city (Figure 1).

2.2. Data collection

An Giang province is covered by three separate Landsat scenes. The Landsat 8 data with a low cloud cover rate of less than 10% were collected from the US Geological Survey (USGS) website (https://earthexplorer.usgs.gov/). These scenes were captured on 10/01/2019, 11/02/2019, 27/02/2019, 14/07/2019, and 18/10/2019. This data is used for fusion with MODIS MOD9Q1.

The MODIS MOD09Q1 data (8-day composite product with 250-meter spatial resolution) collected using the GEE platform was captured from January 2019 to December 2019 with 46 images in total. In addition, the other maps were also collected, including the land inventory map of An Giang province in 2019, the administrative map of An Giang province, inventory data, and the provincial statistical yearbook in 2019.

IOP Conf. Series: Earth and Environmental Science **964** (2022) 012005 doi:10.1088/1755-1315/964/1/012005

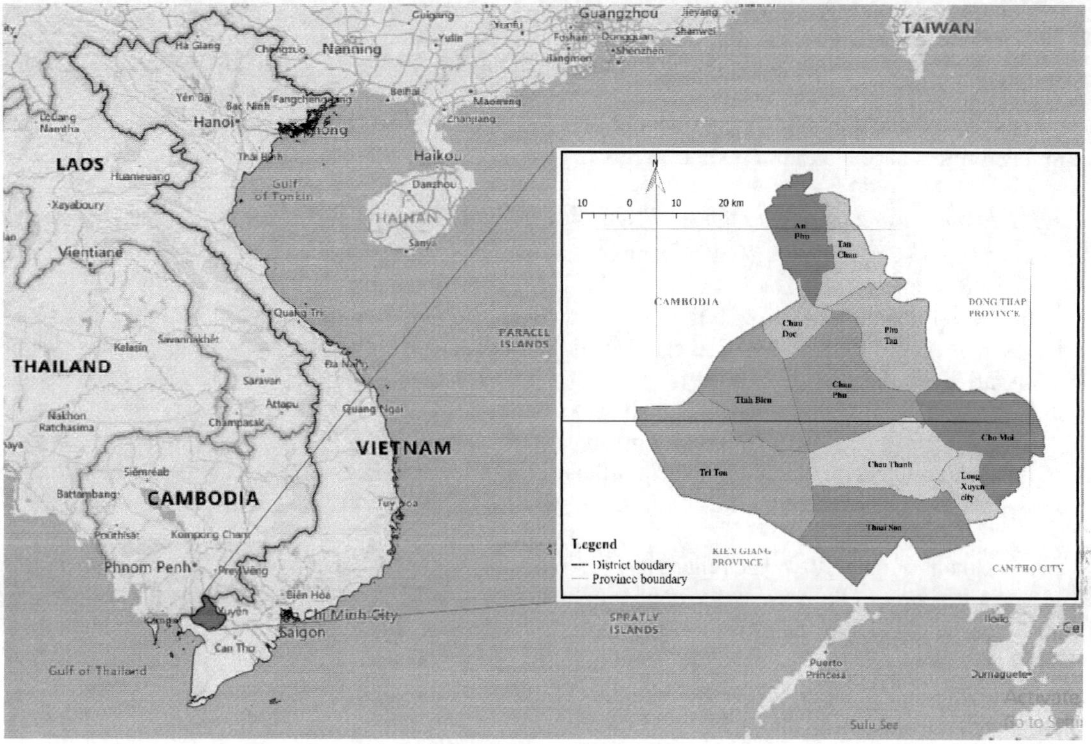

Figure 1. Location map of the study area (An Giang province)

2.3. Methods

2.3.1. Preprocessing and calculating NDVI images

The collected Landsat 8 and MODIS MOD09Q1 imagery were preprocessed using the Google Earth Engine (GEE) platform to optimize performance. These data were adjusted coordinate system, spatial subset, estimating the Normalized Difference Vegetation Index (NDVI) as formula (1) [15]:

$$NDVI = (NIR - RED)/(NIR+RED) \quad (1)$$

where NIR and RED are Near Infrared and Red wavelengths, respectively.

2.3.2. Landsat-MODIS fusion

Fusion of Landsat and MODIS datasets were processed by using the Spatial-temporal adaptive reflectance fusion model (STARFM) [11, 16]. STARFM was used for simulation between multi-temporal MODIS images (250 m spatial and 8-day temporal resolutions) and a Landsat image (30 m resolution) to create new datasets at 30 m spatial and 8-day resolutions directly for NDVI images using equation (2). Images of MODIS and Landsat acquisition on the same day are paired to create the model. MODIS pixels map to their corresponding Landsat pixels and a weighting window is created based on the central Landsat pixel and the surrounding Landsat pixels that affect the MODIS values. The MODIS pixels are weighted based on the values of the pixels they cover, both spatially and in relation to the amount of time from one image to the next.

$$L\left(x_{\omega/2}, y_{\omega/2}, t_0\right) = \sum_{i=1}^{\omega}\sum_{j=1}^{\omega}\sum_{k=1}^{\omega} W_{ijk} \times \left(M\left(x_i, y_j, t_0\right) + L\left(x_i, y_j, t_k\right) - M\left(x_i, y_j, t_k\right)\right) \quad (2)$$

where L and M are Landsat and MODIS surface reflectance, respectively; ω is the searching window size, $(x_{\omega/2}, y_{\omega/2}, t_0)$ is the center pixel of the moving window; (x_i, y_j) is a given pixel location for a Landsat and MODIS image pair; t_k is the acquisition date for the image pair; t_0 is the acquisition date for a simulated date; and W_{ijk} is the weight deciding the influence of each neighboring pixel to the simulated reflectance of central pixel $(x_{\omega/2}, y_{\omega/2})$.

2.3.3. Remote sensing interpretation method and reliability assessment

Landsat-MODIS classification: The Maximum Likelihood Classifier (MLC) was used to classify a time series of 46 fused MODIS-Landsat images. The MLC is the most effective method in classifying land use/land cover, whereas the probability of each pixel is calculated and then assigned to the class name with the highest probability [17]. The classification samples are selected based on the characteristics of NDVI variations. In the rice cultivation areas, the NDVI variation graph is sinusoidal, the maximum value of NDVI corresponding to the growth peaks of rice [18].

Landsat 8 classification: The Maximum Likelihood Classification (MLC) was applied on a single Landsat 8 image dated 11/02/2019. The classification aims to evaluate the effectiveness of using a single Landsat image compared to the Fusion Landsat-MODIS time-series in mapping the agricultural land use.

Assessment of reliability: The reliability of classification was assessed by the Kappa coefficient (K) and overall accuracy (T) based on the confusion matrix [19, 20]. The classified by single Landsat 8 and fusion Landsat-MODIS time-series were then reclassed into the same number of classes and class types. The total number of survey points for each land use type was decided depending on the proportion of land use types. There are 450 ground truth points collected through a field trip survey (250 points) and virtual tours on Google Earth (200 points).

3. Results and Discussion

3.1. Analyzing the variations of NDVI time-series for different land use types

Each land use type is characterized by variations of NDVI time-series over a period of time. In general, NDVI values vary from −0.39 to 0.9 throughout the year. The detail of NDVI variations for each land use type in this study is shown in Figure 2.

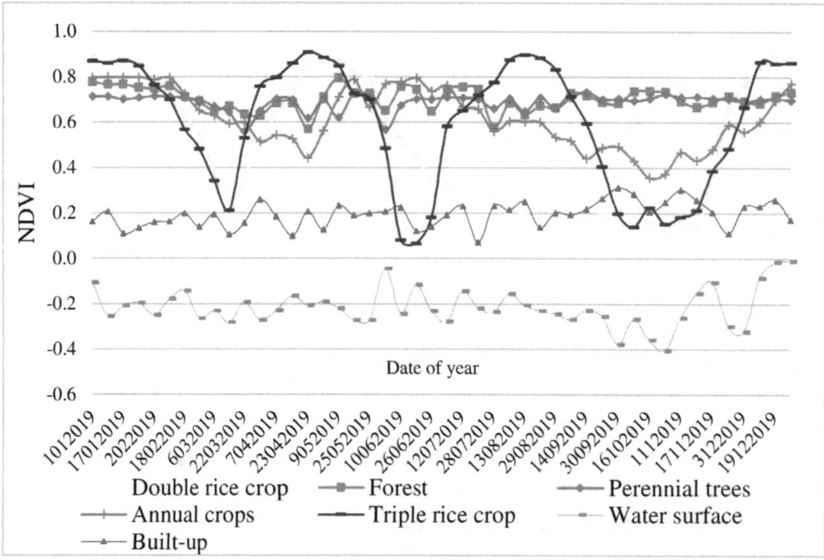

Figure 2. The variations of NDVI time-series extracted from fused Landsat-MODIS imagery for different land use types.

NDVI Landsat-MODIS time-series provides the information for monitoring the changes of vegetation cover over time. In the rice crop systems, the pixel has the curve of NDVI value following a sinusoidal pattern with two times reaching maximum value corresponding to double rice crop, and the curve with sinusoidal pattern with three times reaching the maximum value (maximum value of NDVI about 0.9) presented for triple rice crop. The NDVI time-series of forest is quite stable at high range values (about 0.61 - 0.8) during the year. The forest in An Giang is classified as the mixed forest, and melaleuca forest (based on the field survey). These forest types are general evergreen forests with no dry season defoliation. Therefore, there is no significant change in NDVI value over different seasons in one year. For the perennial trees, NDVI is also stable at a high-value range, while the NDVI time-series of annual crops are seasonally variable depending on characteristics of growing length and specific species such as corn, bean, and other upland vegetables. As the curve of the perennial trees and forest above (pink line and green lines) show a quite similar pattern, the classification of these classes, therefore, was differentiated by combining with the distribution sites in An Giang inventory map. In a different way, the NDVI values of the water surfaces are always lower than zero due to the passing through water of incident energy. Built-up areas, including residential land, headquarters, offices, and roads, are generally identified with NDVI values between 0.13 and 0.26 during the year.

3.2. Assessing of fusion imagery applicability for mapping agricultural land use

There are 450 ground truth points collected in total for assessing the classification accuracy. It was split into small subsets, including 20 perennial trees points, 64 annual crops points, 32 forest points, 33 built-up points, 32 double rice crop points, 247 triple rice crop points, and 22 water surface points. The survey locations for each land use type were randomly selected and which is shown in Figure 3.

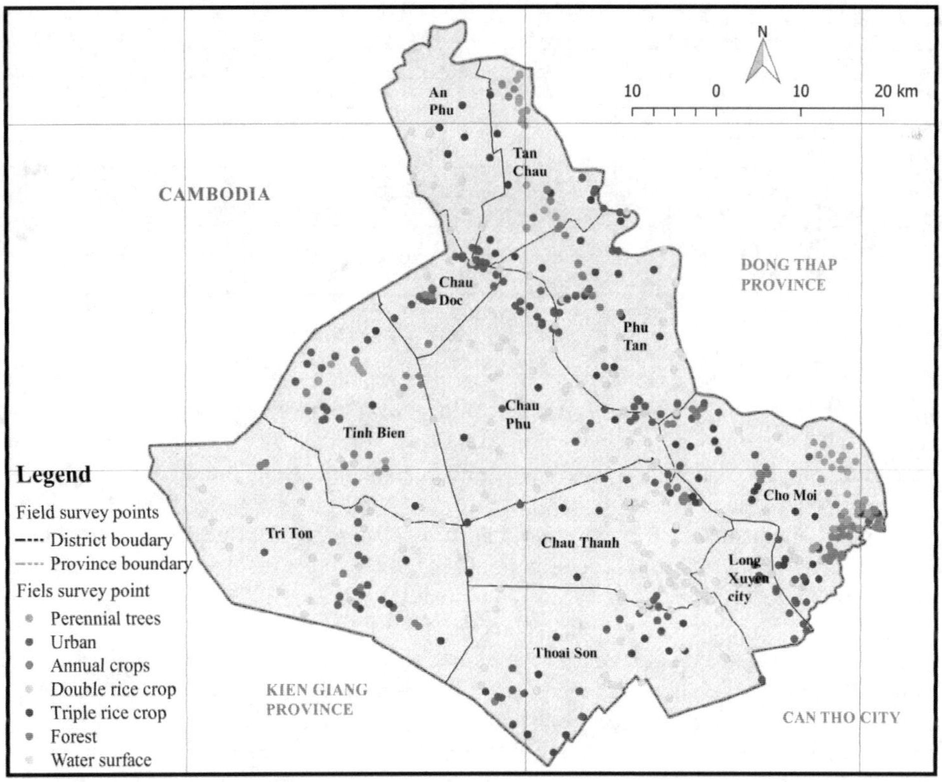

Figure 3. The location of the ground truth points for accuracy assessment.

The results of reliability assessment for the fused Landsat-MODIS multi-time-series and the single Landsat 8 image are shown in the confusion matrix Table 1, respectively.

Table 1: The classification error matrix using fused Landsat-MODIS time series and single Landsat 8 image.

Fused Landsat-MODIS									
		Predicted classes							Accuracy (%)
	Land use types	Triple rice crop	Double rice crop	Perennial trees	Annual crops	Forest	Built-up	Water surface	
Actual classes from survey	Triple rice crop	220	4	2	15	0	4	2	89.1
	Double rice crop	4	23	2	2	1	0	0	71.9
	Perennial trees	2	0	15	3	1	0	0	71.4
	Annual crops	5	0	2	50	6	1	0	78.1
	Forest	0	1	2	0	25	4	0	78.1
	Built-up	1	0	1	1	0	30	0	90.9
	Water surface	2	0	0	0	0	0	20	90.9
	Reliability (%)	94.0	82.1	62.5	70.4	75.8	76.9	90.9	84.9
	Kappa index	0.7							

Landsat 8 data								
		Predicted classes						
		Rice crops	Perennial trees	Annual crops	Forest	Built-up	Water surface	Accuracy (%)
Actual class from survey	Rice crops	188	2	10	2	4	1	90.8
	Perennial trees	23	14	8	0	0	0	31.1
	Annual crops	51	2	41	4	1	0	41.4
	Forest	3	2	1	13	4	0	56.5
	Built-up	20	0	4	1	34	4	54.0
	Water surface	0	0	0	0	0	13	100.0
	Reliability (%)	65.5	63.6	64.1	75.0	79.1	72.2	67.3
	Kappa index	0.35						

These results show that there are 07 types of agricultural land use, including double rice crop, triple rice crop, perennial trees, annual crops, forest, built-up, and water surface. The classified map using the fused Landsat-MODIS time-series achieved an overall accuracy of 84.9% and a kappa coefficient of 0.7. Meanwhile, only 06 land use types have been recognized using the single Landsat 8 image, including rice crops, perennial trees, annual crops, forest, built-up, and water surface, with an overall accuracy of 67.3% and kappa coefficient of 0.3. In order to compare the accuracy scientifically between the fusion Landsat-MODIS time-series and the single Landsat 8, the double rice crop and triple rice crop were grouped into rice crops type to ensure consistency with Landsat 8 classified map. The result indicates that the grouped classified overall accuracy of 86.9% and Kappa coefficient of 0.74. The above results show that the achievement of current agricultural land use maps in An Giang province using Landsat-MODIS multi-time-series is higher reliability than single Landsat imagery. However, the misclassification of the perennial trees and annual crops remains (Table 1) due to small areas of cultivation in reality that could not be detected with the use of 30 m spatial resolution imagery. In addition, the remaining thin cloud pixels in the Landsat 8 input senses (Scenes dated 14/7/2019, 27/02/2019) lead to a sudden drop in NDVI value of perennial trees in the fused Landsat-MODIS time-series compared to the other period in this year. However, this research has shown the feasibility of fusion Landsat-MODIS images in mapping agricultural land use. Further studies on the fusion of higher temporal and spatial resolution satellite images need to be taken into consideration for highly applicable monitoring and management of agriculture at the provincial level.

3.3. The current agricultural land use map

The agricultural land use in An Giang province in 2019 was mapped with high reliability, which included 07 land use types as perennial trees, annual crops, forest, double rice crop, triple rice crop, water surface, and built-up (Figure 4).

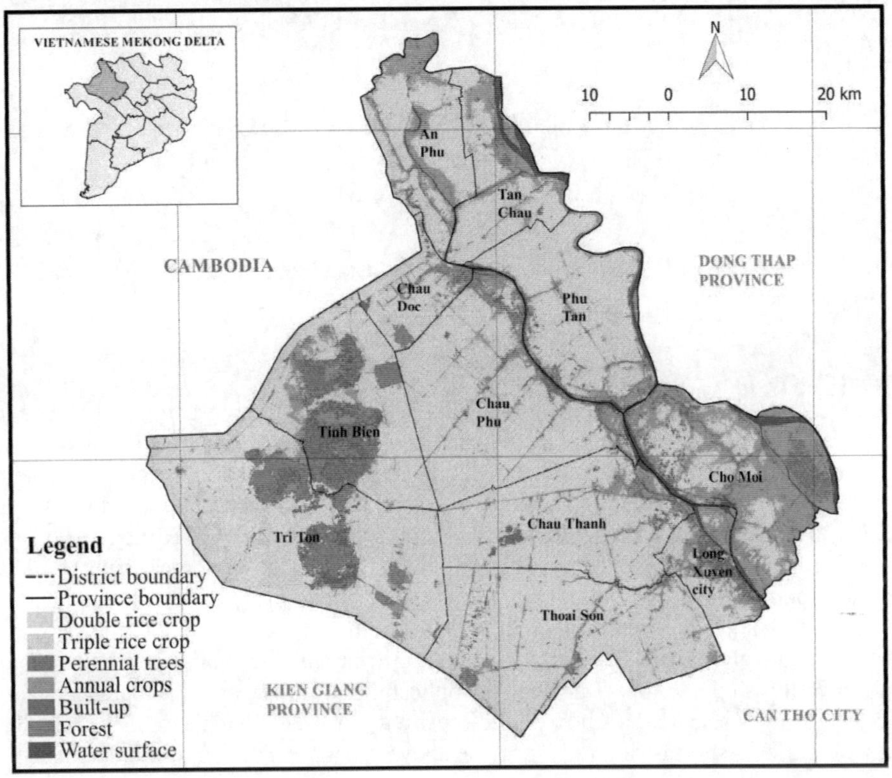

Figure 4. The current agricultural land use map classified by fusion data in An Giang in 2019.

In 2019, the agricultural land area in An Giang was about 299,679.20 ha, accounting for 84.30% of the total area. In which, the rice cultivation area accounts for the largest proportion, mainly triple rice crop with an area of 236,582.71 ha (68.24%), double rice crop with 16,647.08 ha (4.80%). The perennial trees take up only 14,128.09 ha (4.08%). Annual crops such as watermelon, corn, sweet potato distributed in alluvial soil located along the coastal river occupy about 24,263.04 ha (7.00%). A small area is occupied by mixed and melaleuca forests, about 8,058.12 ha (2.32%). The rest area is built-up, and the water surfaces jointly account for 13.56% of the total land area (Figure 5).

The distribution of agricultural land use is summarised as follows: the triple rice crop accounts for the highest proportion of An Giang province. It is distributed in almost all districts. The double rice crop is distributed in the high terrain of Tri Ton district within a small area, mostly around the foothills with a limited water supply for farming systems than the lowland areas. The perennial trees and annual crops are concentrated in the islets in Cho Moi, Tan Phu district, Tan Chau Township, where the presence of alluvial soil is suitable for most cultivation models. Most of the forest is distributed in mountainous areas in Chau Doc, Tinh Bien, Tri Ton, and Thoai Son districts.

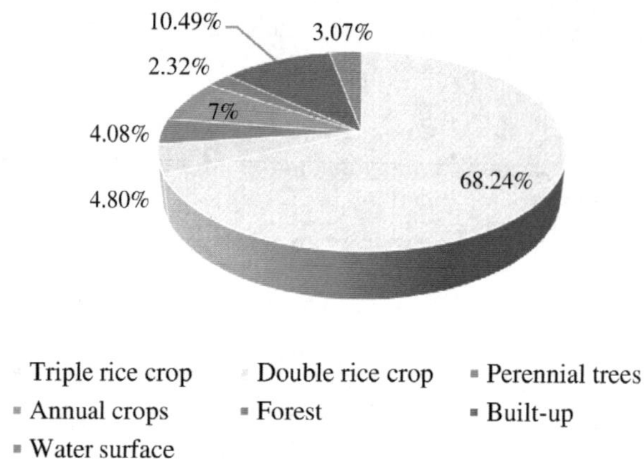

Figure 5. Proportion of land use types classified by fused Landsat-MODIS imagery in An Giang province in 2019.

3.4. Research limitations and outlook

This study improved agricultural mapping at relatively high resolution compared to previous studies toward monitoring modest dynamics related to agricultural transformation. It has become a special matter of concern in the past few years in the Mekong Delta when many spontaneous conversion models have been emerged purely based on the subjective opinions of the farmers. Brackish aquaculture and lotus fields in the ineffective rice cultivation areas of upstream Mekong River are typical examples. These models might disrupt the general plan as well as the local ecological balance. Therefore, we attempt to support the local agricultural managers in this work more precisely. Yet, technology transfer for this fusion technique is currently challengeable. Even though we tried to optimize performance using the start-of-the-art platforms (GEE, R), cross-platform is still the main barrier, especially for data preparation. In this study, we only verified the applicability of fusion data compared to standard data of Landsat 8. In further research, a study on the sensitivity of small changes in land use/land cover in a small region during a short period regarding the above-mentioned agricultural transformations may be a worthy work expected to perform.

Additionally, we conducted Landsat-MODIS fusion to obtain time-series data at 30 meters resolution. Nevertheless, it should be noted that there are free-of-charge data of Sentinel-2 presently data being comparatively similar in terms of characteristics and covering wavelengths with three-time better in spatial resolution (10 meters). Hence, fusion data between Sentinel-2 and MODIS is a reasonable expectation that might be fulfilled, yet we might need adjustments in parameters while processing these two data. It will be meaningful for monitoring and management, especially for local dynamics in land use/land cover changes.

4. Conclusion

The current agricultural land use in An Giang province in 2019 was mapped with an overall accuracy of 84.9% and a kappa coefficient of 0.7 using fused time-series imagery. In particular, the STARFM technique was successfully executed for creating Landsat-MODIS multi time-series (30 m spatial resolution, 8-day temporal resolution). The results revealed that the fused Landsat-MODIS data is more effective in classifying and mapping the agricultural land use in comparison to using a single Landsat image. The research findings provide a ground for further research in assessing and monitoring long-term changes in land use and rice cropping patterns under the condition of climate change.

References

[1] Department of Statistics of An Giang 2020 The provincial statistical yearbook in An Giang in 2019.

[2] The Vietnam government 2019 Decree No. 03/2019/NĐ-CP dated 04 January, 2019 on remote sensing operations.

[3] Huynh Thi Thu Huong, Vo Quang Minh, Ho Van Chien 2013 Monitoring the progress of rice sowing and cropping calendar for early warning rice pests in the Mekong river Delta based on remote sensing images *Proc. GIS Conference* 128-139.

[4] Huynh Van Chuong, Nguyen Ngoc Thanh 2015 Analysis the application of MODIS TERRA/AQUA of remote sensing image to observe and manage Summer-Autumn paddy crop in Thua Thien Hue province in the summer 2014 *The Scientific Journal of Hue University* **4**(103) 33-44

[5] Hoang Phi Phung, Lam Dao Nguyen 2019 Monitoring rice crop in the Mekong Delta ad Red river Delta using Sentinel 1 data. *Proc. of Conference: basis research in Earth and Environmental Science* DOI: 10.15625/vap.2019.000123.

[6] Li, L., Zhao, Y., Fu, Y., Pan, Y., Yu, L., Xin, Q. 2017 High Resolution Mapping of Cropping Cycles by Fusion of Landsat and MODIS Data. *Remote Sens.* **9** 1232 doi:10.3390/rs9121232

[7] Lu, Y., Wu, P., Ma, X., & Li, X. 2019 Detection and prediction of land use/land cover change using spatiotemporal data fusion and the Cellular Automata–Markov model *Environmental Monitoring and Assessment* **191**(2) doi:10.1007/s10661-019-7200-2

[8] Gao, F., Hilker, T., Zhu, X., Anderson, M., Masek, J., Wang, P., & Yang, Y. 2015 Fusing Landsat and MODIS Data for Vegetation Monitoring *IEEE Geoscience and Remote Sensing Magazine* **3**(3) 47–60 doi:10.1109/mgrs.2015.2434351

[9] Xin, Q., Olofsson, P., Zhu, Z., Tan, B., and Woodcock, C. E. 2013 Toward near real-time monitoring of forest disturbance by fusion of MODIS and Landsat data. *Remote Sensing of Environment* **135** 234–247

[10] Zhang, F., Zhu, X., and Liu, D. 2014 Blending MODIS and Landsat images for urban flood mapping *International Journal of Remote Sensing* **35**(9)

[11] Gao, F., Masek, J., Schwaller, M., & Hall, F. 2006 On the blending of the Landsat and MODIS surface reflectance: Predicting daily Landsat surface reflectance. *IEEE Transactions on Geoscience and Remote Sensing* **44**(8) 2207–2218 http://dx.doi.org/ 10.1109/TGRS.2006.872081

[12] Emelyanova, I. V., McVicar, T. R., Van Niel, T. G., Tao Li, L., & Van Dijk, A. I. J. M. 2013 Remote sensing of environment assessing the accuracy of blending Landsat — MODIS surface reflectances in two landscapes with contrasting spatial and temporal dynamics : A framework for algorithm selection *Remote Sensing of Environment* **133** 193–209 http://dx.doi.org/10.1016/j.rse.2013.02.007

[12] Singh, D. 2011 Generation and evaluation of gross primary productivity using Landsat data through blending with MODIS data *International Journal of Applied Earth Observation and Geoinformation* **13**(1) 59–69 http://dx.doi.org/10.1016/j.jag.2010. 06.007.

[13] Hilker, T., Wulder, M. a., Coops, N. C., Linke, J., McDermid, G., Masek, J. G., et al. 2009 A new data fusion model for high spatial- and temporal-resolution mapping of forest disturbance based on Landsat and MODIS *Remote Sensing of Environment* **113**(8) 1613–1627 http://dx.doi.org/10.1016/j.rse.2009.03.007.

[13] Tucker, C. J. 1979 Red and photographic infrared linear combinations for monitoring vegetation *Remote Sensing of Environment* **8**(2) 127–150 https://doi.org/https://doi.org/10.1016/0034-4257(79)90013-0

IOP Conf. Series: Earth and Environmental Science **964** (2022) 012005 doi:10.1088/1755-1315/964/1/012005

[14] Nguyen-Thanh Son, Chi-Farn Chen, Ly-Yu Chang, Cheng-Ru Chen, Shin-Ichi Sobue, Vo-Quang Minh, Shou-Hao Chiang, Lam-Dao Nguyen & Ya-Wen Lin 2016 A logistic-based method for rice monitoring from multitemporal MODIS-Landsat fusion data *European Journal of Remote Sensing* **49**(1) 39-56 DOI: 10.5721/EuJRS20164903

[15] Le Van Trung 2015 *Remote sensing* Ho Chi Minh National University Publisher.

[16] Ali, A. M., Savin, I., Poddubskiy, A., Abouelghar, M., Saleh, N., Abutaleb, K., El-Shirbeny, M., Dokukin, P. 2020 Integrated method for rice cultivation monitoring using Sentinel-2 data and Leaf Area Index *The Egyptian Journal of Remote Sensing and Space Science* doi:10.1016/j.ejrs.2020.06.007.

[17] Cogalton, R.G. 1991 A review of assessing the accuracy of classifications of remotely sensed data. *Remote sensing of Environment* **37**(2) 35-46.

[18] Cogalton, R.G., and Green, K. 2009 Assessing the accuracy of Remotely sensed data: Principles and Practices *The Photogrammetric record* (Second edit, vol.2). Taylor and Francis Group.

Assessment the impacts of climate change on drought in the Ba River basin, Central Vietnam using Landsat remote sensing data

Han T N Tham[1*], Thy T M Pham[1], Thi N K Truong[1], Huong T T Nguyen[2], Nguyen D Lam[1] and Trung V Le[3,4]

[1] Ho Chi Minh City Space Technology Application Center, Vietnam National Space Center, Vietnam Academy of Science and Technology, 01 Mac Dinh Chi Street, Ben Nghe Ward, District 1, Ho Chi Minh City, Vietnam
[2] IRMG, Faculty of Agriculture and Forestry, Tay Nguyen University, 567 Le Duan Street, Buon Ma Thuot City, Dak Lak Province, Vietnam
[3] Faculty of Environment and Resources, Ho Chi Minh City University of Technology (HCMUT), 268 Ly Thuong Kiet Street, District 10, Ho Chi Minh City, Vietnam
[4] Vietnam National University Ho Chi Minh City (VNU-HCM), Linh Trung Ward, Thu Duc District, Ho Chi Minh City, Vietnam

*E-mail: ttnhan@vnsc.org.vn

Abstract. Sustainable management of the river basin is a profound challenge for environmental management in the context of climate change. Drought situations in a basin occur in relation to meteorological, hydrological, agricultural factors and climate change as well. In this study, remote sensing technology was applied to assess the impacts of climate change on drought in the Ba River basin, Central Vietnam. Drought in the basin has been created by land use/land cover changes in recent years, which has resulted in a sharp decrease in forest area in the period 1989 to 2019 (-41.5%) and a significant increase of agricultural land with 38.2%. Following that, the area of drought agriculture rose by 28.8%. The remarkably high drought areas in agricultural land were in El Nino years, 2016 (99.2%) and 2019 (87.3%), which indicated that under climate change impacts, a drought occurred more severely. Moreover, drought also appeared in the forest. The forest area deceased but the drought levels in the forest increased slightly since 2005 and hit a peak drought value in 2016 with 97.0% of forest area. During El Niño years, the precipitation, atmospheric moisture, and water flow in the basin were all lower than in previous years.

1. Introduction

South-East Asia has long endured severe droughts, which occur on average every five years. The prolonged 2015 and 2018 droughts were the worst on record for two decades [1]. According to the United Nations, drought risk is already intensifying across the region. The drivers of drought risk in South-East Asia are inherently created by various climatic drivers, mainly the El Niño-Southern Oscillation (ENSO). Moreover, average temperatures have risen due to climate change which means drought severity will increase if the climate gets warmer.

Drought is a complex natural phenomenon caused by the imbalance of precipitation and evapotranspiration. It usually occurs with a lack of precipitation and results in a decrease in soil moisture content [2]. Drought, in general, may negatively affect agriculture, fisheries, power production and water supply in river basins. The drought in 2019-2020 heavily devastated the South Central and Central Highland regions in Vietnam [3] including the disrupted livelihoods and food security and created forest fire and haze. It is crucial to understand how recurrent droughts exacerbate in a changing climate. Then adaptive policy interventions need to be promoted to reduce and prevent droughts from occurring, prepare and respond to droughts when they happen, and restore and recover after a drought has passed [1]. The latest developments in science and technology will reinforce the successful scale-up of drought management interventions. With advances in remote sensing systems, droughts can be monitored on a regional and continental scale. To date, the research on drought based on the images from space technology is flourishing in every corner of the world as well as in Vietnam [4-8], which shows the advantages of remote sensing technology in environmental studies.

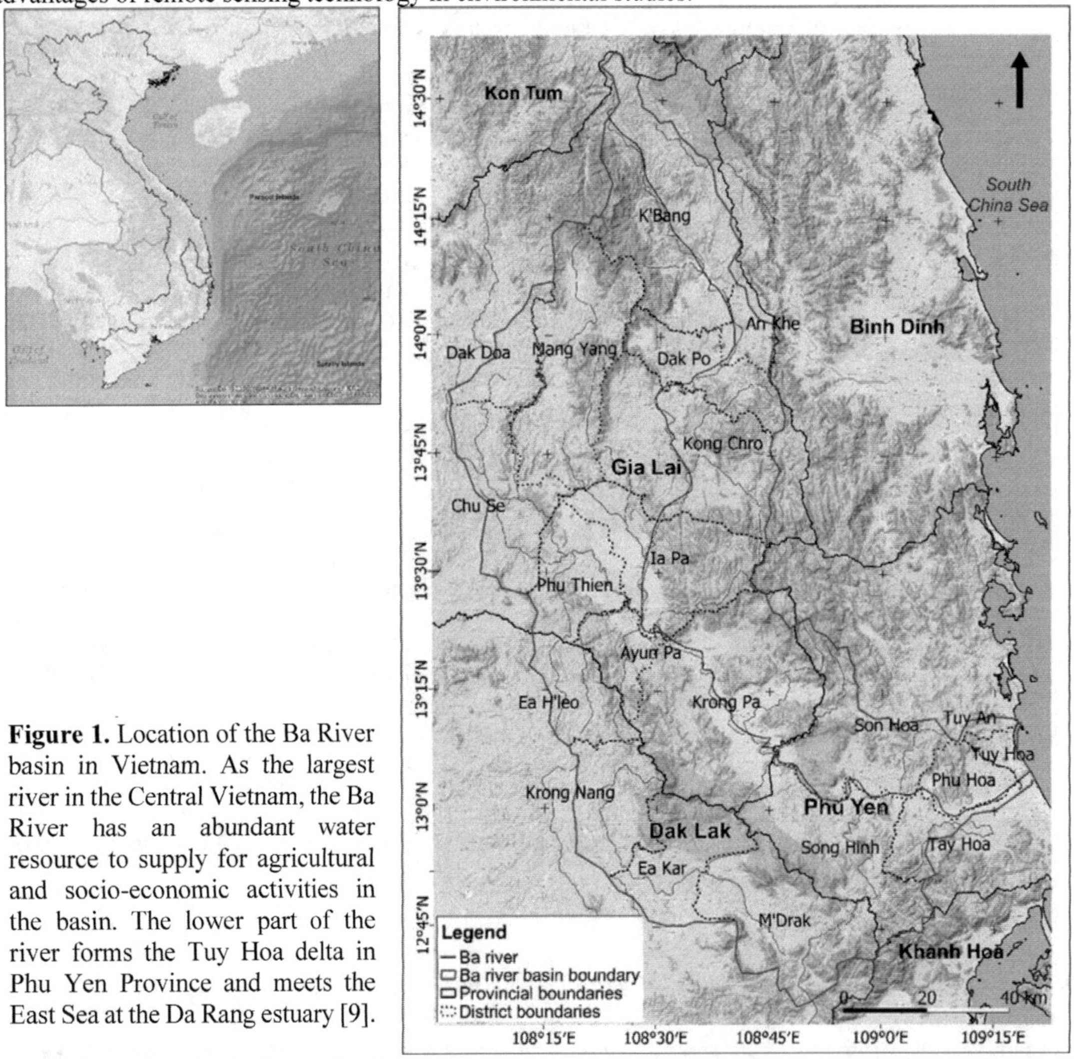

Figure 1. Location of the Ba River basin in Vietnam. As the largest river in the Central Vietnam, the Ba River has an abundant water resource to supply for agricultural and socio-economic activities in the basin. The lower part of the river forms the Tuy Hoa delta in Phu Yen Province and meets the East Sea at the Da Rang estuary [9].

The Ba River basin is on the South-Central Coast of Vietnam with an area of 13,900 square kilometers, consisting of Gia Lai, Dak Lak and Phu Yen Province. It extends from 12°35′N to 14°38′N latitude and from 108°00′E to 109°55′E longitude (Fig.1). The basin has several advantages in the land, climate, and water resources for the development of agriculture and local agricultural products have had a high export value since 2010 [10] such as coffee, rubber, pepper, cashew nuts. Socio-economic growth and climate change, however, have been putting more pressure on the environment in the basin. Nowadays, land use/land cover (LULC) in the basin has been changed to meet human demand. Furthermore, drought destroyed crops and affected water resources in the basin in recent years [11-13]. Especially, in the background of global warming and during the specific timing of El Niño, drought events occurred more seriously.

Typically, the 2015-16 El Niño caused scorching temperatures and prolonged drought in the Central and Central Highlands in Vietnam, which affected agriculture activities [14]. As reported by the Central Highlands Regional Hydrometeorological Center in 2018-2019, the Eastern and Southeastern districts of Gia Lai Province were experiencing an intense drought. Not only did the precipitation in region just reach the amount of approximately below 60% of the annual rainfall, but the Ka Nak Hydroelectric Reservoir, the largest one in the upper basin, was also in a state of being the driest ever [15].

Hence, this work focuses on the examination of climate change effects on drought in the Ba River basin by applying space-based technology. A paper first reviews the changes in land use/land cover (LULC) from 1989 to 2019. Then, soil moisture is assessed in each type of LULC to reveal drought variability across the basin. The paper also lends a hand to understand how land resources here are being affected by climate change, which helps the local people use natural resources economically and sustainably.

2. Materials and Methodology

2.1. Materials

Figure 2 provides a descriptive workflow of the study. The information from remote sensing data is practical in depicting how the LULC and soil moisture have changed in the basin. A part of the data process was managed in Google Earth Engine and an overlay technique was implemented to calculate the drought affected LULC area. After that, from the collected climate data, the effects of climate change on drought were assessed.

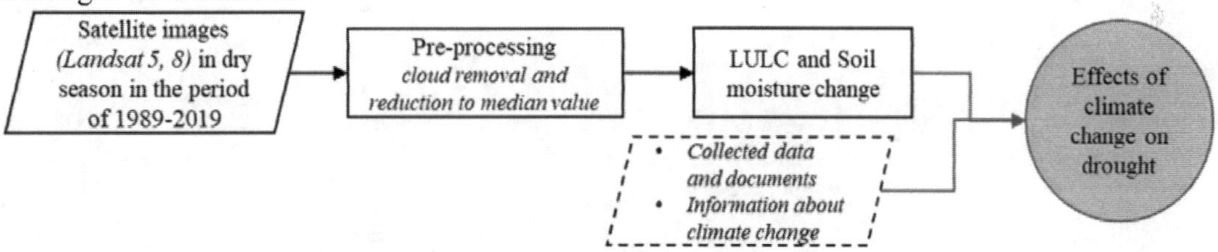

Figure 2. Workflow of the study

The materials were used in this study include satellite images, field data and the relevant documents from authorities in the basin. Landsat-5 and Landsat-8 images were applied to extract information about LULC and reveal the soil moisture through TVDI (Temperature Vegetation Dryness Index) in the dry season (from January to May) in 1989, 2000, 2005, 2010, 2017, and 2019. Meanwhile, field data was collected in 2017 and 2019 to assess the accuracy of LULC classification. In addition, documents and data regarding hydrometeorology, drought or climate change in the Ba River basin were also aggregated to consider generally the research's feasibility compared to the actual situation.

2.2. Land use/land cover classification

In this study, the classification method was used to obtain information about LULC change over time from a source of remote sensing images. A supervised classification algorithm was conducted to classify land surface entities into three categories including agriculture, forest, and other lands.

The supervised classification is a method where taxonomies are established based on sampling regions and use decision rules based on the appropriate algorithms for labeling pixels for specific coverage areas. The supervised classification method with Maximum Likelihood Classifier (MLC) was applied in the study. This method takes into consideration each LULC class in each spectrum channel to have a standard distribution. Then, pixels will be classified into the class that has the highest probability.

Accuracy of classification is evaluated based on error matrix and norms such as overall accuracy and Kappa coefficient [16]. Overall accuracy is computed by the number of correctly classified pixels divided by the number of pixels used for classification. The high accuracy of commonly accepted classification is over 85%. Kappa coefficient (K) is used as a measure of the accuracy of image classification. It shows the basic difference between what is real about the error of the deviation of the matrix and the total number of changes indicated by the row and column. The Kappa coefficient is usually between 0 and 1. The high accuracy of the accepted classification is commonly with $K > 0.8$.

2.3. Soil moisture - Temperature Vegetation Dryness Index

TVDI is a simplified land surface dryness index developed by Sandholt et al. [17], which is based on an empirical parameterization of the relationship between Ts (Land Surface Temperature) and NDVI (Normalized Vegetation Difference Index). NDVI can monitor vegetation status and stress, specifically in water shortage sectors which reflects vegetation drought conditions [18, 19], and Ts rises rapidly with water stress [17, 20]. Hence, soil moisture could be estimated through the relationship between Ts and NDVI and this method only uses information derived from satellite images [17].

Figure 3 depicts the scatter plots of the conceptual Ts/NDVI space that often results in a trapezoidal shape with dry edge and wet edge [21, 22]. The points closer to the dry edge reflect many water-stressed surfaces with lower soil moisture and higher Ts. On the other hand, the wet edge indicates the evapotranspiration capacity and soil moisture become higher due to under no-water-lack conditions.

TVDI is one of the best-documented predictors of drought as vegetation index (NDVI) and other surface parameters (Ts) derived from satellite data have been widely used to monitor drought. Several studies show that integrating NDVI and Ts can offer more complete information on drought with requirements from bare soil to complete vegetation coverage [2, 23, 24].

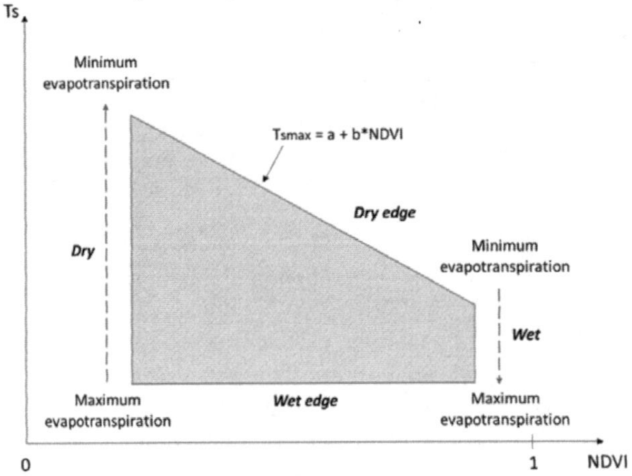

Figure 3. Definition of TVDI

The TVDI for a given NDVI is estimated using Ts, Ts_{min} and Ts_{max} shown in Equation (1) [17]:

$$TVDI = (Ts - Ts_{min}) / (Ts_{max} - Ts_{min}) \quad (1)$$

On the dry edge, Ts_{max} can be represented by straight-line relations with the NDVI as Equation (2). For a given pixel, Ts_{max} is the value of Ts on the dry edge corresponding to the value of NDVI in that pixel.

$$Ts_{max} = a + b*NDVI \quad (2)$$

where the parameters a and b are estimated based on pixels from an area large enough to represent the entire range of surface moisture contents.

The NDVI value is calculated from the near-infrared (NIR) and red bands as follows:

$$NDVI = (NIR - Red) / (NIR + Red) \quad (3)$$

3. Results and Discussion

3.1. The changes in agricultural and forest areas

In 1989 and 2019, LULC in the basin has changed dramatically (Fig.4). The accuracy assessment of classification results of 2017 and 2019 using field data indicated overall accuracies of 89% with a kappa coefficient of 0.77 and 0.71, correspondingly.

Over 30 years, the West and Southeast areas of the Ba basin were the most dramatic LULC changes. Most of the forest area was transformed into agricultural and other lands. Only the forest area in the North of the basin remained since the region is a part of Kon Ka Kinh National Park and Kon Chu Rang Nature Reserve. Besides that, ranges of dangerous mountains in the boundary between Phu Yen and Dak Lak (M' Drak and Song Hinh districts) and in the East areas of Gia Lai province (Kong Chro, Ia Pa and Krong Pa districts) (Fig.1) were the unchanged-forest area.

Figure 5 and Table 1 show agricultural area increased from 176,969 ha to 686,526 ha (+287.9% with the rate +9.3% per year) from 1989 to 2019, while forest area diminished by 553,321 ha (–41.5% with the rate –1.38% per year) and declined fastest between 1989 and 2005. From 2005 to 2019, the agricultural area more than doubled with a 108% increase or rose by 7.2% per year, which proved to be the greatest increase since 2005. On the other hand, the basin in 1989 was covered by 79.2% of forest (1,055,049 ha), but nearly half of the forest area disappeared after 31 years (plunged to 501,728 ha) and the agricultural accounted for over ½ area of the basin (51.5%) in 2019.

Moreover, there was a slight growth in the area of other lands including water bodies, impervious surface, and bare land, from 7.5% (1989) to 10.8% (2019). A part of this is due to urbanization and the construction of hydropower and irrigation reservoirs since 1995 such as Ayun Ha and An Khe-Kanak (Gia Lai Province), Song Ba Ha and Song Hinh (Phu Yen Province), and Krong H'Nang (Dak Lak Province).

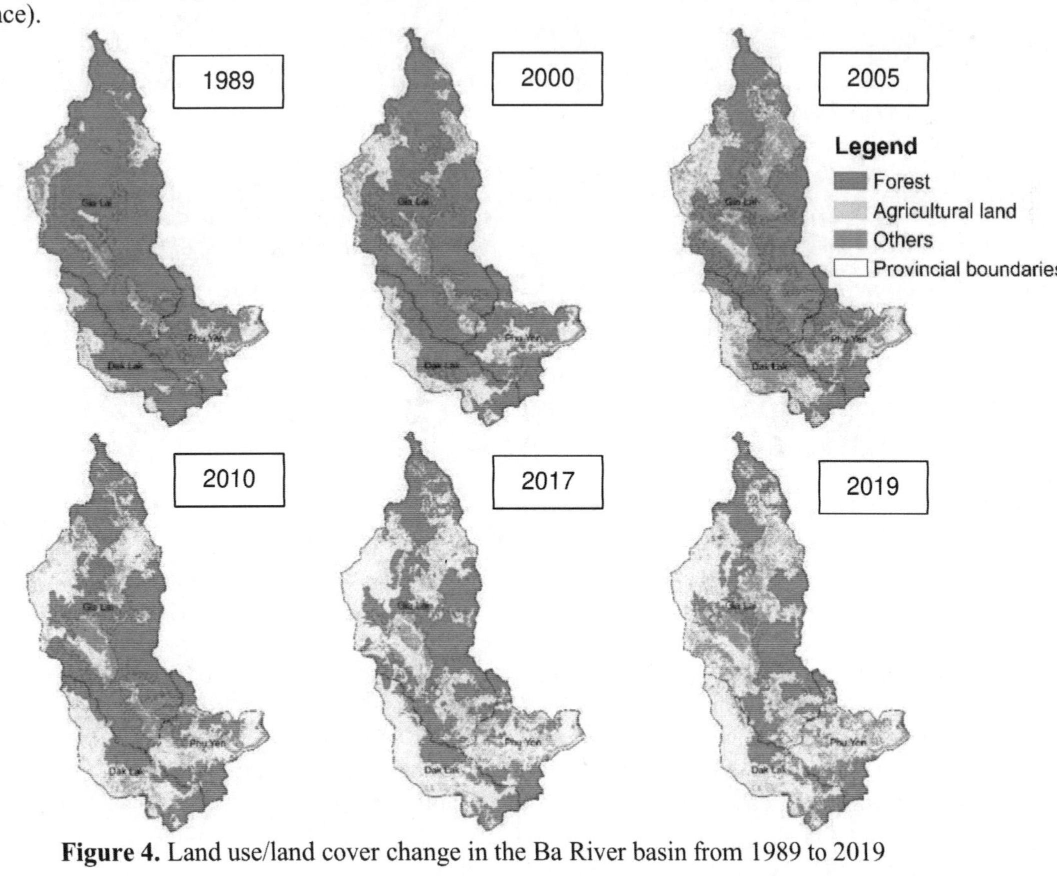

Figure 4. Land use/land cover change in the Ba River basin from 1989 to 2019

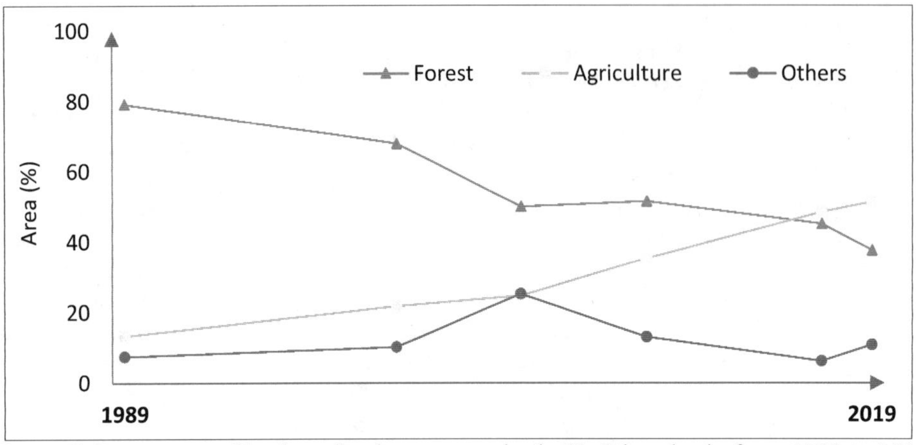

Figure 5. Percentage of land use/land cover area in the Ba River basin from 1989 to 2019

Table 1. Land use/land cover area (ha) in the Ba River basin from 1989 to 2019

Class name	1989	2000	2005	2010	2017	2019
Forest	1,055,049	906,138	667,612	687,526	602,353	501,728
Agriculture	176,969	290,630	329,713	472,084	648,305	686,526
Other lands	100,592	135,842	335,285	173,000	81,952	144,356
Total	1,332,610	1,332,610	1,332,610	1,332,610	1,332,610	1,332,610

3.2. The change of soil moisture

In line with previous studies [26, 27], TVDI ranging in [0, 1] is defined in five levels, which are labeled as follows: very wet (0.0 < TVDI ≤ 0.2), wet (0.2 < TVDI ≤ 0.4), balanced (0.4 < TVDI ≤ 0.6), dry (0.6 < TVDI ≤ 0.8), and very dry (0.8 < TVDI ≤ 1.0).

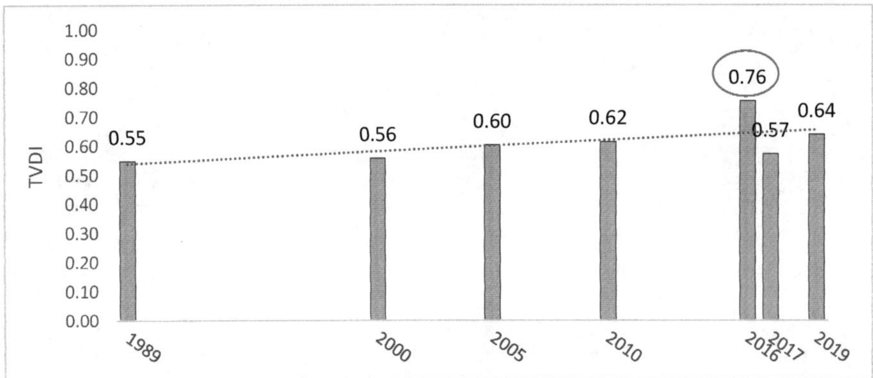

Figure 6. Average TVDI values of the dry season in the Ba River basin extracted from Landsat data acquired in the period of 1989 - 2019

Figure 6 showed the average TVDI values increased slightly from 2005 to 2019. It means soil moisture values of the basin changed from a balanced (0.55 in 1989) to a dry level (0.64 in 2019). Especially, the TVDI in 2016 got the highest value (0.76) which was very dry. That points to the fact that dryness in the basin might continue to escalate in the following years. In order to follow the soil moisture changes in each LULC, soil moisture distributing spatially were also depicted in Figure 7.

IOP Conf. Series: Earth and Environmental Science **964** (2022) 012006 doi:10.1088/1755-1315/964/1/012006

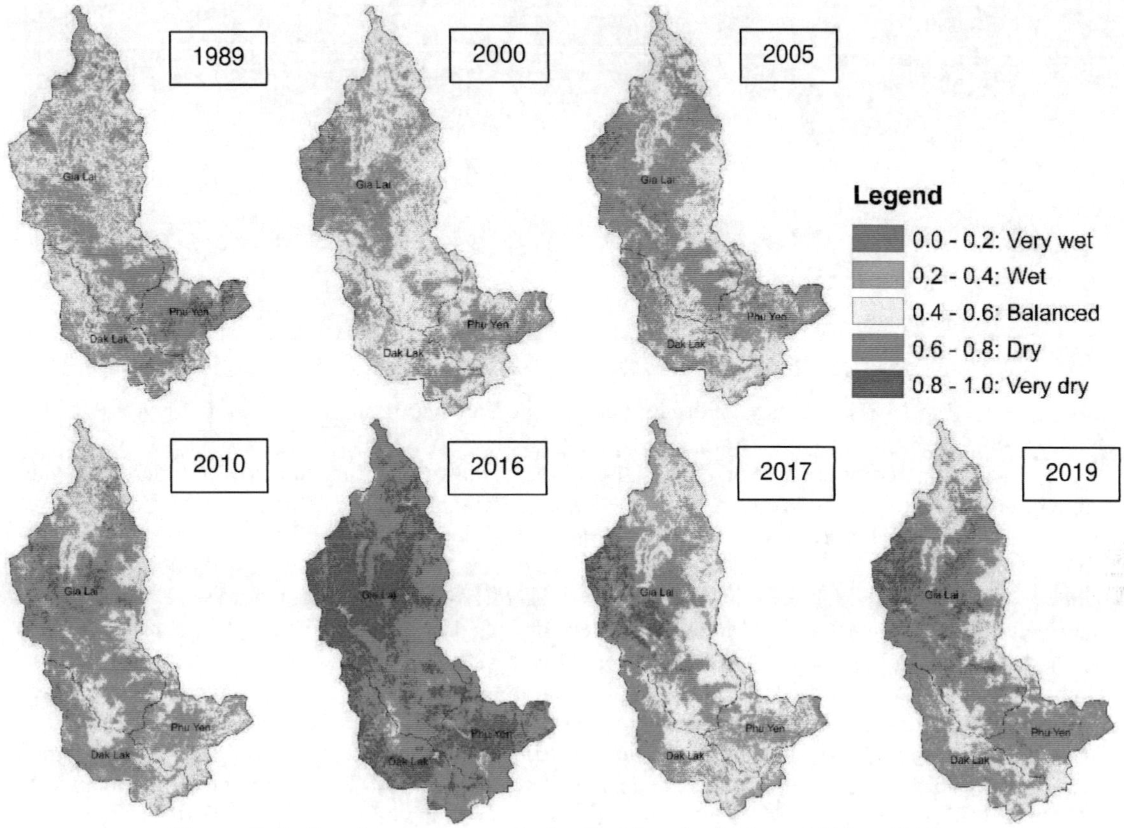

Figure 7. Soil moisture changes in the Ba River basin from 1989 to 2019

3.3. *The impacts of climate change on drought*

El Niño is described as one of the most unspoken climate risks in East Asia and the Pacific [25]. El Niño's impacts harm society, the economy, and the agricultural sector with droughts and water scarcity. From the regional scale to the whole of Vietnam, there is always a high exposure probability to climate shocks and the South Central Coast as well as Central Highlands are ones of the country's most vulnerable regions to ENSO. The research investigated historical data which show there were five in six El Niño years causing severe droughts: 1997-1998, 2004-2005, 2010, 2014-2016, and 2018-2019. In these El Niño periods, the 2014-2016 (the longest El Niño) and 2018-2019 droughts were witnessed to be the most terrible [1].

From 1989 to 2019, the drought area in the basin has increased markedly in two levels with the dry level rising from 40.3% to 59% and the very dry going from 1.6% up to 5.6% (Fig.8). The most noticeable change in drought was in 2016. The basin in 2016 was almost in the dry and very dry level which were 66.8% and 30.2%, respectively. This was known as the highest values in the basin over the past two decades. The second-highest percentage of drought area was in 2019 with 59% in the dry level and 5.6% in the very dry.

The result of calculating drought through TVDI index showed that the drought situation in the basin was occurring more severely, especially in 2016 and 2019, caused by El Niño. That coincides with the UNESCAP report [1] that the prolonged droughts in the periods of 2014-2016 and 2018-2019 were the worst on record for two decades in South-East Asia.

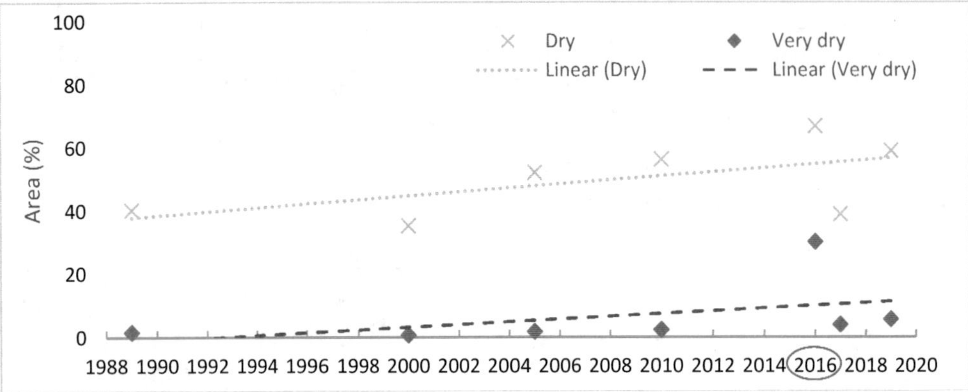

Figure 8. Percentage of drought area in the Ba River basin from 1989 to 2019

The variation of drought area on each LULC in the basin was scrutinized below in order to consider in more detail the change of drought under the impact of climate change.

As described in Figure 9, drought area in agriculture had a rising tendency from 1989 to 2019, increasing by 495,662.56 ha (Table 2). Especially, due to El Nino, the highest drought areas were indicated in 2016 (643,176.84 ha) and in 2019 (599,124.17 ha), comprising 99.2% and 87.3% agricultural land, respectively. Moreover, the very dry level in 2016 reached a peak during a 30-year period, accounting for 44.8% of the total drought area (288,027.92 in 643,176.84 ha). That led to a substantial number of plants drying during 2016 and the year after 2017 such as annual crops, coffee, pepper, rice plants. Furthermore, according to the statistics of the Departments of Agriculture and Rural Development in the Ba River basin, the drought in 2015-2016 ravaged more than 2,100 billion Vietnam Dong in the agriculture sector due to crop death and yield loss of 30-70% compared to previous years. This was also a consequence of a slow incline in agricultural land after 2016 (2017-2019) (Fig.5).

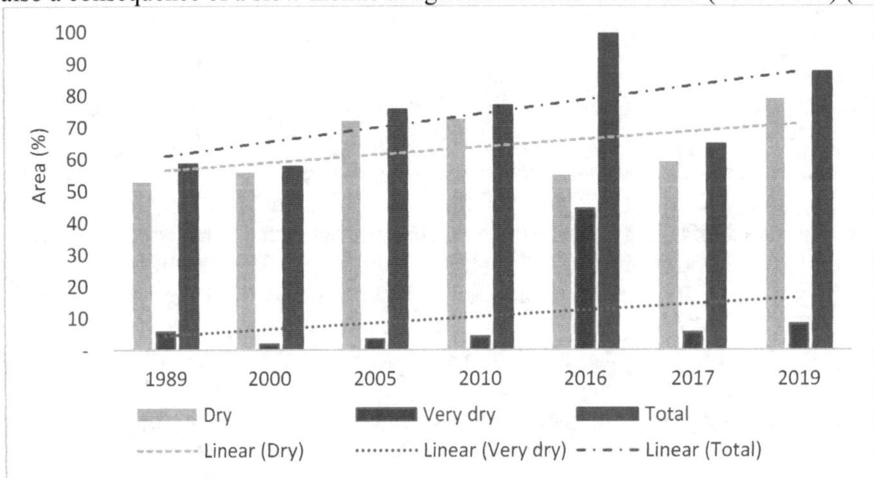

Figure 9. Percentage of drought area in agricultural land in the Ba River basin from 1989 to 2019

Table 2. Drought area in agricultural land (ha) in the Ba River basin from 1989 to 2019

	1989	2000	2005	2010	2016	2017	2019
Dry	93,390.41	162,150.74	237,581.29	342,710.31	355,148.93	383,026.57	543,042.82
Very dry	10,071.20	5,233.58	11,513.44	20,079.12	288,027.92	35,750.86	56,081.35
Total drought	103,461.61	167,384.31	249,094.74	362,789.43	643,176.84	418,777.43	599,124.17
Agriculture area	176,969.00	290,630.00	329,713.00	472,084.00	648,305.00	648,305.00	686,526.00

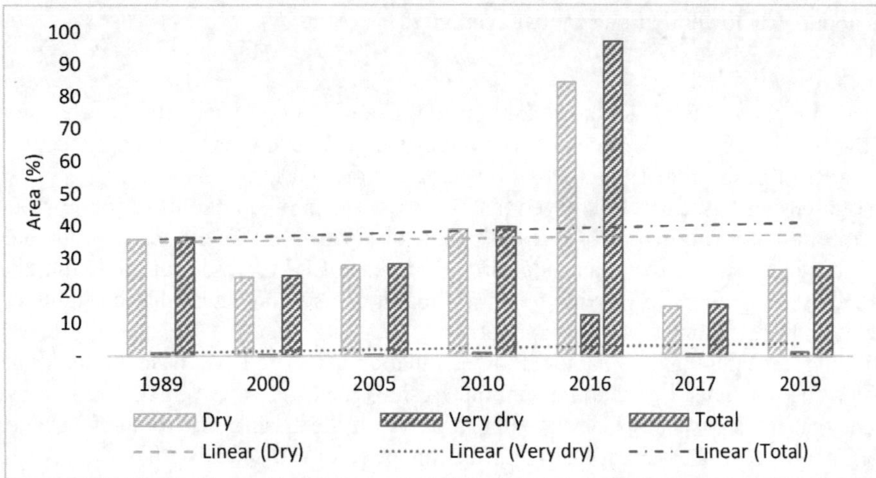

Figure 10. Percentage of drought area in forest land in the Ba River basin from 1989 to 2019

Table 3. Drought area in forest land (ha) in the Ba River basin from 1989 to 2019

	1989	2000	2005	2010	2016	2017	2019
Dry	377,701.41	220,239.50	186,759.27	267,379.81	508,850.01	91,847.11	132,994.70
Very dry	8,877.63	3,344.99	1,793.07	5,603.92	75,280.86	3,376.66	5,780.69
Total drought	386,579.04	223,584.50	188,552.34	272,983.73	584,130.87	95,223.77	138,775.39
Forest area	1,055,049.00	906,138.00	667,612.00	687,526.00	602,015.00	602,353.00	501,728.00

In regard of drought in the forest, although the drought area went down from 386,579.04 ha in 1989 to 138,775.39 ha in 2019 (Table 3) owning to the cutting-down forest to convert into cultivated areas, the drought area of forested land also increased but more slightly than that of agriculture from 1989 to 2019 (Fig.10). During El Niño years, 2005, 2010, 2016 and 2019, the total drought area in these years was all higher than the others, accounting for 28.2%, 39.7%, 97.0% and 27.7% forest area, correspondingly. It is obvious that the very dry area in the forest started increasing from 2005, whereas there was a rather quick decrease in forest area since that year (Fig.5). That signified the basin is already being ravaged by drought and drought could aggravate human's life as well as the ecosystem in the basin.

Under aggressive greenhouse gas emission, extreme El Niño may increase in frequency and amplitude by the end of the 21st century. That means in a warming climate when strong El Niño increases, rainfall extremes are projected to shift eastward along the equator in the Pacific Ocean during El Niño events [28]. From these statements, the results also demonstrated that the drought situation in the Ba River basin is occurring with increasing frequency and severity, especially during El Niño years, which IS drought in the basin is being affected by a climate transition.

In El Niño years (2004-2005, 2010 and 2014-2016), the recorded data from meteorological stations in the basin was witnessed to be lower than in the other years, such as lower precipitation, atmospheric moisture, and water flow. For example, in the 2004-05 El Niño, the Central Highlands and South Central Coast had low rainfall and the rainy season ended 1 to 1.5 months sooner than the average. Then, in 2010 El Niño, the Central region in Vietnam was the most severely affected area and the rainfall was exceptionally low, at only about 70% of the average. For the next El Niño event (2016), the rainfall and river flows in the Central Highlands and South Central Coast were only 50 to 70% of the average, and in some areas, rainfall reduced to only 20% of the average [1]. As the report by the General Department of Hydrometeorology revealed that the natural reserve of water in the Central Highland was not abundant in recent years as the forest had been cleared and the rainfall in the region during the 2019 dry season was about 50% lower than the annual average [29].

4. Conclusions

The paper had presented the impact of drought on LULC and how seriously climate change affects the drought in the Ba River basin. The study analyzed the change of LULC as well as the change in drought area in the Ba River basin from 1989 to 2019 by using remote sensing.

The remote sensing-based results showed that there was a significant modification in LULC over 30 years with the most transformation was from forest to agricultural land and the soil moisture was becoming drier. Furthermore, the changes of drought in each LULC were prominent that the levels of drought had a rise, especially in El Niño years. Hence, it can be concluded that droughts in the basin change due to both the change of climate and LULC.

These findings are encouraging that the drought situation in the Ba River basin under climate change could threaten the residents' livelihood and food. Besides that, the nature of ENSO is a cyclical and natural event that has consistently detrimental impacts on the basin's environment, economies, and agriculture. Therefore, it is essential that the government and local scientists have to take actions and address undeniable problems related to the severity of drought.

Acknowledgement

The study was conducted within the framework of the research project "Study and evaluate the impacts of climate change and socio-economic activities on the Ba/Da Rang River Basin environment by using remote sensing and GIS technology," project number VT-UD.10/18-20, which is part of the National Program on Space Science and Technology (2016–2020), Vietnam Academy of Science and Technology.

We would like to thank Ho Chi Minh City University of Technology (HCMUT), VNU-HCM for the support of time and facilities for this paper.

References

[1] UNESCAP, "*Ready for the Dry Years: Building Resilience to Drought in Southeast Asia*," 2020 [Online]. Available: https://www.unescap.org/op-ed/ready-dry-years-building-resilience-drought-southeast-asia# [Accessed: June 2021]

[2] Shi, Siqi, Fengmei Yao, Jiahua Zhang, and Shanshan Yang. "Evaluation of temperature vegetation dryness index on drought monitoring over Eurasia." *IEEE Access* 8, 30050-30059, 2020.

[3] Viet Nam News, "*PM calls for urgent action on drought, saltwater intrusion*," 2020 [Online]. Available: https://vietnamnews.vn/environment/717097/pm-calls-for-urgent-action-on-drought-saltwater-intrusion.html [Accessed: June 2021]

[4] Huicong Jia, Fang Chen, Jing Zhang, and Enyu Du, "Vulnerability Analysis to Drought Based on Remote Sensing Indexes," *Int. J. Environ. Res. Public Health*, 17, 20, 7660, 2020. doi: https://doi.org/10.3390/ijerph17207660

[5] O. Rojasa, A. Vrielingb, and F. Rembolda, "Assessing drought probability for agricultural areas in Africa with coarse resolutionremote sensing imagery," *Remote Sensing of Environment*, 115, 2011, 343-352, 2011.

[6] Su Z., He Y., Dong X., and Wang L., "Drought Monitoring and Assessment Using Remote Sensing," In *Remote Sensing of Hydrological Extremes*, Lakshmi V. eds.: Springer Remote Sensing/Photogrammetry, Springer, Cham, 151-172, 2017.

[7] Vu Hien Phan, Vi Tung Dinh, and Zhongbo Su, "Trends in Long-Term Drought Changes in the Mekong River Delta of Vietnam," *Remote Sens.*, 12, 18, 2974, 2020. doi: https://doi.org/10.3390/rs12182974

[8] Hoa Thi Tran, James B. Campbell, Tri Dinh Tran, and Ha Thanh Tran, "Monitoring drought vulnerability using multispectral indices observed from sequential remote sensing (Case Study: Tuy Phong, Binh Thuan, Vietnam)," GIScience & Remote Sensing, 54, 2, 167-184, 2017. doi:

https://doi.org/10.1080/15481603.2017.1287838
[9] N. D. Tuan, Integrated Management of Water Resources in the Ba River Basin, Thuyloi University, Vietnam, 2007.
[10] OECD, "*Food and Agriculture Reviews: Agricultural Policies in Viet Nam 2015*," 2015 [Online]. Available: http://dx.doi.org/10.1787/9789264235151-en [Accessed: August 2020]
[11] Tran Thanh Xuan, Tran Thuc, and Hoang Minh Tuyen, *Impact of climate change on water resources in Vietnam*, Vietnam Institute of Meteorology, Hydrology and Environment, Ha Noi, Vietnam, 2011.
[12] Phan Thi Thanh Hang, "The trend of climate and runoff changes in the Ba river basin," *VIETNAM JOURNAL OF EARTH SCIENCES*, 36, 1, 31-40, 2014.
[13] Ministry of Natural Resources and Environment Electronic Newspaper, "Miserable due to drought: The risk of losing everything," 2019 [Online]. Available: https://baotainguyenmoitruong.vn/khon-don-vi-kho-han-nguy-co-mat-trang-249580.html [Accessed: December 2019]
[14] FAO, "El Niño event in Viet Nam: Agriculture, food security and livelihood need assessment in response to drought and salt water intrusion," 2016 [Online]. Available: http://www.fao.org/resilience/resources/resources-detail/en/c/430047/ [Accessed: June 2021]
[15] Minh Tan, "Water shortage in the upper Ba river," *Da Nang City police news*, February 2019 [Online]. Available: http://www.cadn.com.vn/news/99_202438_kho-khat-noi-dau-nguon-song-ba.aspx [Accessed: July 2019]
[16] Hofmann Peter, Josef Strobl, and Ainura Nazarkulova, "Mapping green spaces in Bishkek—how reliable can spatial analysis be?," *Remote Sensing*, 3, 6, 1088-1103, 2011.
[17] Sandholt Inge, Kjeld Rasmussen, and Jens Andersen, "A simple interpretation of the surface temperature/vegetation index space for assessment of surface moisture status," *Remote Sensing of Environment*, 79, 2-3, 213-224, 2002.
[18] S. K. Jain, R. Keshri, A. Goswami, A. Sarkar, and A. Chaudhry, "Identification of drought - vulnerable areas using NOAA AVHRR data," *Int. J. Remote Sens.*, 30, 10, 2653-2668, 2009.
[19] L. Geng, M. Ma, W. Yu, X. Wang, and S. Jia, "Validation of the MODIS NDVI Products in Different Land-Use Types Using In Situ Measurements in the Heihe River Basin," *IEEE Geoscience and Remote Sensing Letters*, 11, 9, 1649-1653, 2014.
[20] Goetz, S. J, "Multi-sensor analysis of NDVI, surface temperature and biophysical variables at a mixed grassland site," *International Journal of Remote Sensing*, 18, 1, 71-94, 1997.
[21] Moran, M. S., T. R. Clarke, Yoshio Inoue, and A. Vidal, "Estimating crop water deficit using the relation between surface-air temperature and spectral vegetation index," *Remote sensing of Environment*, 49, 3, 246-263, 1994.
[22] Xin Jingfeng, Guoliang Tian, Qinhuo Liu, and Liangfu Chen, "Combining vegetation index and remotely sensed temperature for estimation of soil moisture in China," *International Journal of Remote Sensing*, 27, 10, 2071-2075, 2006.
[23] M. E. Holzman, R. Rivas, and M. C. Piccolo, "Estimating soil moisture and the relationship with crop yield using surface temperature and vegetation index," *Int. J. Appl. Earth Obs.*, 28, 181-192, 2014.
[24] N. Son, C. Chen, C. Chen, L. Chang, and V. Minh, "Monitoring agricultural drought in the Lower Mekong Basin using MODIS NDVI and land surface temperature data," *Int. J. Appl. Earth Obs.*, 18, 417-427, 2012.
[25] William R. Sutton, Jitendra P. Srivastava, Mark Rosegrant, James Thurlow, and Leocardio Sebastian. "Striking a Balance: Managing El Nino and La Nina in Vietnam's Agriculture," *The World Bank*, 103 pages, 2019.
[26] Gao, C. and Li, X.. "Precipitation thresholds of drought disaster for maize in areas in front of Bengbu sluice, Huaihe river basin, China," *Water*, 10, 10, p.1395, 2018.
[27] Wang, C., Qi, S., Niu, Z. and Wang, J.. "Evaluating soil moisture status in China using the temperature–vegetation dryness index (TVDI)," *Canadian Journal of Remote Sensing*, 30, 5, 671-679, 2004.

[28] NOAA Research, "How will climate change change El Niño and La Niña?," 2020 [Online]. Available: https://research.noaa.gov/article/ArtMID/587/ArticleID/2685/New-research-volume-explores-future-of-ENSO-under-influence-of-climate-change [Accessed: September 2021]
[29] General Department of Hydrometeorology, "Water and dry season in Central Highlands," 2020 [Online]. Available: http://www.kttvqg.gov.vn/public/index.php/tin-tuc-tai-nguyen-nuoc-va-moi-truong-114/nuoc-va-mua-kho-tay-nguyen-5934.html [Accessed: June 2021]

Application of multi-temporal Sentinel-1 SAR data for yield estimation of rice crops in An Giang, Vietnam

Hoang Phi Phung[1,2], Lam Dao Nguyen[1*], Nguyen Van Anh Vu[1], Nguyen Kim Thanh[1] and Le Van Trung[3,4*]

[1] Ho Chi Minh City Space Technology Application Centre, Vietnam National Space Centre, Vietnam Academy of Science and Technology, 01 Mac Dinh Chi Street, Ben Nghe Ward, District 1, Ho Chi Minh City, Vietnam
[2] Graduate University of Science and Technology, Vietnam Academy of Science and Technology, 18 Hoang Quoc Viet Street, Cau Giay, Hanoi, Vietnam
[3] Faculty of Environment and Resources, University of Technology (HCMUT), 268 Ly Thuong Kiet Street, District 10, Ho Chi Minh City, Vietnam
[4] Vietnam National University Ho Chi Minh City (VNU-HCM), Linh Trung Ward, Thu Duc District, Ho Chi Minh City, Vietnam.

*Corresponding author: ldnguyen@vnsc.org.vn; lvtrung@hcmut.edu.vn

Abstract. Rice is one of the main agricultural crops and plays an important role in food security. Therefore, it is essential to propose a method for monitoring the distribution of rice yield. Radar remote sensing data sources provide a sustainable solution for rice monitoring challenges in the countries located in the tropical monsoon region like Vietnam. The SAR (Synthetic Aperture Radar) remote sensing data from the Sentinel-1 satellite provided by the European Space Agency (ESA) is free of charge, has a large coverage and high spatial-temporal resolution. In this paper, rice growing areas in the An Giang province of Vietnam Mekong Delta were analyzed, which demonstrates the potential applications of multi-temporal data and proposes a method to estimate rice yield for agricultural management. The analysis results showed that in 2018 the Winter-Spring rice crop has the highest yield, and the Autumn-Winter crop has the lowest yield. Accurate and timely estimation of rice yield and production can provide important information in terms of spatial distribution and seasonal yield for government and decision-makers in policy making related to import and export.

1. Introduction

Rice is one of the five important grains and is the staple food for most of the world's population. In 2018, about 161.6 million hectares of the world's area was used for rice cultivation with a global production of about 728 million tons [1]. In which, Vietnam produced about 6% of the world's rice production, with a rice growing area of about 7.76 million hectares and a production of about 44.88 million tons [1]. Rice yield and production monitoring has an important significance in agricultural management [2]. Besides, food security has become important due to the rapid population growth of developing countries in Asia including Vietnam, with the majority of the population work mainly in the agricultural sector. Being the 5th rice producing country in the world (in 2018) [3], the information related to rice production becomes even more urgent. Therefore, it is necessary to have an effective solution to estimate rice yield accurately and timely to support agricultural production management and ensure food security.

Nowadays, most applications of radar remote sensing use backscatter characteristics of rice plants according to growth stages to determine rice growing areas. Applying the backscatter threshold value from the C-band polarizations of RADARSAT-2 imagery in the study of Chen et al. [4] in Taiwan area, Oyoshi et al. [5] in Japan proved the effectiveness in rice mapping. In addition, X-band and L-band radar image data have also been studied [2,6–9] for their ability to provide information to support rice monitoring. In terms of rice monitoring using C-band data with single polarization like ERS-1&2 and RADASAT-1, the classification method is based on temporal changes of backscatter intensity which is extracted from multi-temporal images [10–12]. Using HH and VV dual polarization data of ENVISAT ASAR images (C-band) gives better classification results than using only single polarization [13,14]. In particular, the recent Sentinel-1A and Sentinel-1B satellites, which acquire images with a 12- or 6-day period (received from two satellites in some areas), have become a source of free C-band radar images with high spatial resolution (20 m for Interferometric Wide (IW) swath mode). This is an important data source in agricultural monitoring applications, it is especially beneficial for developing countries in the tropics with limited funds, and the ability of radar remote sensing to acquire images is very little affected by weather conditions.

As for the application of radar imagery, recent studies have demonstrated the ability of radar remote sensing in rice monitoring [15] and yield estimation [16]. For estimating rice yield and production, many studies have been conducted using agro-meteorological modeling methods like ORYZA2000 [17] [16] and experimental regression models between radar backscatter values and in situ rice yield [14]. However, when applying the ORYZA model, a lot of ground data is needed. Moreover, there are some other methods using radar remote sensing data as the main input such as: experimental regression [14,18,19], neural network [20], random-forest [21]. In which, the commonly used method is the experimental regression model. Shao et al. [18] used a self-recommended model for the RADARSAT data, based on three different times in three rice growth stages, and achieved an accuracy of 91%. The rice yield estimation model using ENVISAT ASAR APP data [14] and TerraSAR-X StripMap [22] achieved the results with high accuracy when compared with statistical data. In addition, there were also studies that have applied this model to estimate rice yield and production using X-band COSMO-SKYMED and C-band RARDASAT-2 [23]. Lam Dao [14] applied an experimental regression model to estimate rice yield from C-band ENVISAT ASAR data. The study then used dual polarization TerraSAR-X data to estimate rice yield and calculate rice production [19]. The main objective of this study was to apply an experimental regression model to estimate rice yield based on rice age [24], which was developed for the study area using C-band multi-temporal Sentinel-1 data for rice yield estimation and for analyzing the spatial and seasonal rice yield distribution in 2018 in An Giang province.

2. Study area, materials, and methods

2.1. Study area

The study area is in the region from latitude 10°12' to 10°57'N and longitude 104°46' to 105°35'E. An Giang is one of the leading provinces regarding rice production area in Vietnam Mekong Delta. (Figure 1). Therefore, this province was selected to experiment the application of radar remote sensing in rice growing area monitoring and yield estimation. Table 1 presents the main rice crop calendar in the study area with three crops including Winter-Spring, Summer-Autumn and Autumn-Winter crops. The study area has two distinct seasons: the dry season from December to April, with rainfall accounting for only 10% of the total annual rainfall; the rainy season lasts from May to November, accounting for nearly 90% of the total annual rainfall [25].

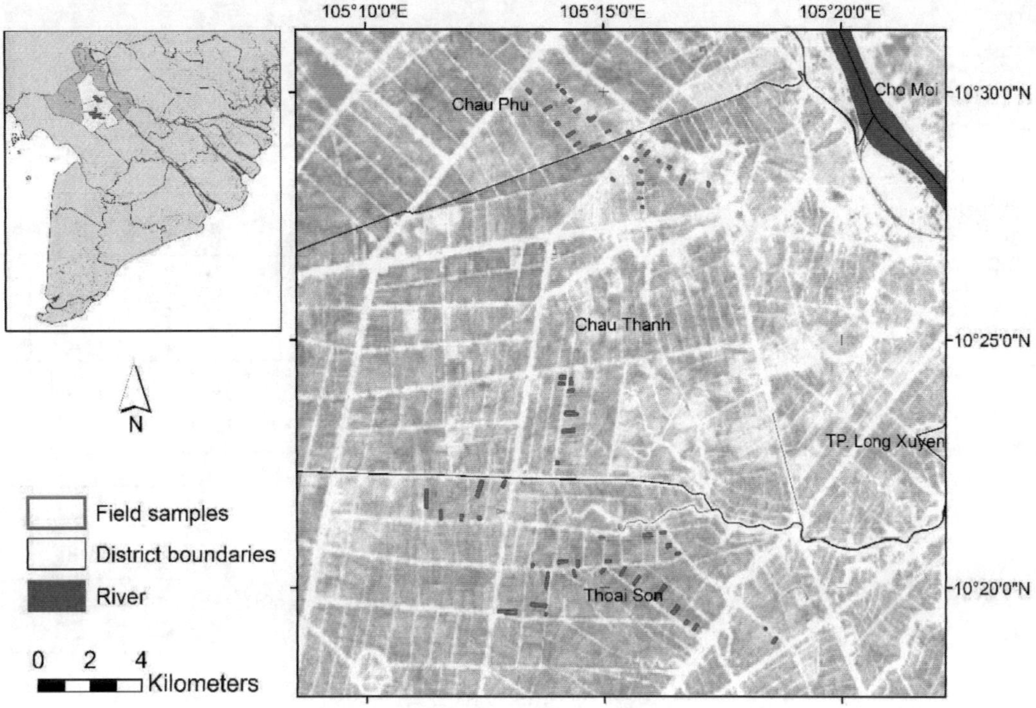

Figure 1. Location of the study area, An Giang province and sample fields.

Table 1. Main rice crop calendar in An Giang, Vietnam (WS: Winter Spring crop; SA: Summer Autumn crop; AW: Autumn Winter crop).

Month	10	11	12	1	2	3	4	5	6	7	8	9	10	11	12
Rice season				WS				SA					AW		

2.2. Materials

Data used are VH-polarized Sentinel-1 images with IW acquisition mode, spatial resolution of 20 m. For the study area, Sentinel-1A&1B satellites have a repeat acquisition time of 6 days. The acquisition period of Sentinel-1 data for rice crops in 2018 was from November 1st, 2017 to November 30th, 2018. These data are used to map rice growing area and estimate rice yield. In addition, the study also collected in situ data in 60 sample fields in the study area. The data collected at the sample fields included sowing date, harvest date, rice yield, etc. These data were divided into two datasets including: training dataset consisting of 41 samples (~70 %) and validation dataset with 19 samples (~30%).

2.3. Methods

The study evaluated the ability to estimate rice yield in the 2018 crops by analyzing VH polarized Sentinel-1 data with in-situ data as shown in Figure 2. Radar images were collected during one rice crop, then go through the following steps: a) Pre-processing [26, 27]; b) Rice mapping: applying threshold method for multi-temporal images acquired during the crop season [14, 28]; c) Extracting radar backscatter values according to rice age at the sample fields [29, 30]; d) Regression analysis and mapping estimated rice yield [24].

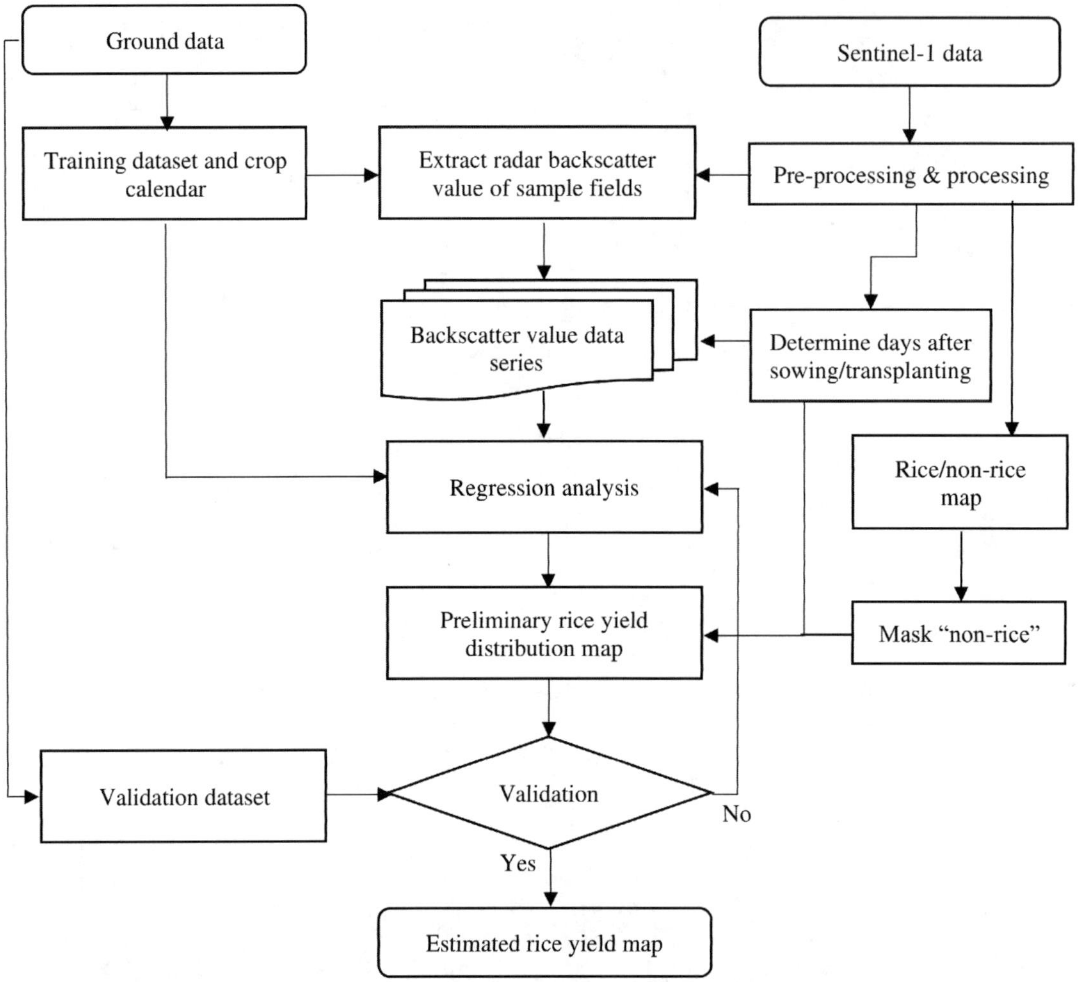

Figure 2. Process of mapping estimated rice yield using Sentinel-1 remote sensing data.

3. Results

The result of mapping estimated rice yield distribution for the 2018 rice crops was established based on an experimental regression model [24]. Sentinel-1 data were collected in each crop season to estimate rice yield. Figure 3 shows maps of rice yield distribution for the Winter-Spring crop, Summer-Autumn crop, and Autumn-Winter crop in 2018. These results show that it is possible to estimate rice yield of all crop seasons using multi-temporal Sentinel-1 data.

The map of estimated rice yield distribution in the 2018 Winter-Spring crop (Figure 3a) was created by an experimental regression model which based on the correlation between Sentinel-1 VH polarization data and in situ rice yield. The estimated yield distribution result shows that the yield was mostly in the range of 6 to 8 tons/ha. Similarly, the map of estimated rice yield distribution in the 2018 Summer-Autumn crop is shown in Figure 3b. Figure 3b shows that the estimated rice yield is fairly evenly distributed, concentrated in the range of 5 to 7 tons/ha. Finally, the map of estimated rice yield distribution for the 2018 Autumn-Winter crop is shown in Figure 3c, which shows that the estimated rice yield is also relatively evenly distributed, mainly in the range of 4 to 6 tons/ha.

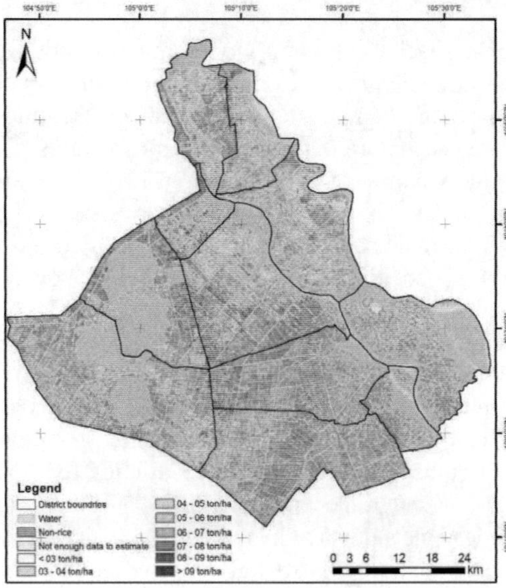

(a) 2018 Winter-Spring crop

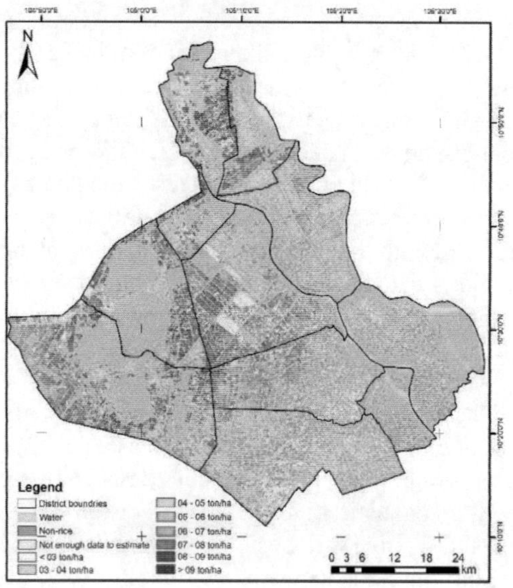

(b) 2018 Summer-Autumn crop

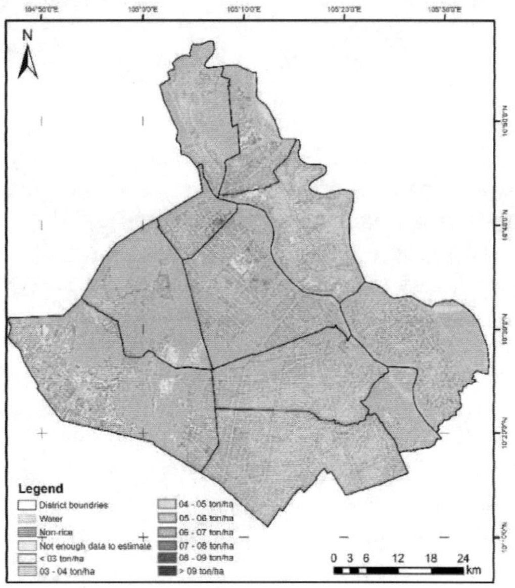

(c) 2018 Autumn-Winter crop

Figure 3. Map of estimated rice yield distribution of Winter-Spring crop (a), Summer- Autumn crop (b), and Autumn Winter crop (c) in 2018 in An Giang province.

4. Discussion

The maps of estimated rice yield distribution by crop season in 2018 were established after validating the classification results of radar remote sensing images based on in situ data collected in the study area and statistics. Validation dataset was collected from 19 samples to validate the model. The results of the analysis showed that the estimated rice yield of the Winter-Spring crop in 2018 and the in-situ yield is similar, 6.5 and 6.6 tons/ha, respectively, and the RMSE (root mean square error) was 0.8 tons/ha [24]. Therefore, this model shows potential for estimating rice yield in the study area using Sentinel-1 data.

An Giang province, like other provinces in the Mekong Delta (Vietnam's rice granary), produces up to 3 rice crops per year and is one of the largest rice producing provinces in the country, just behind Kien Giang province. By 2018, the total cultivated area of An Giang in the whole year reached more than 680 thousand hectares; in which the rice growing area was more than 623 thousand hectares. The average rice yield in 2018 reached 6.3 tons/ha with a production of nearly 4 million tons [31]. The result of the estimated rice yield using Sentinel-1 remote sensing data showed a relative error of 12% when compared with the statistics for the 2018 Winter-Spring crop. This demonstrates the possibility of using Sentinel-1 data to map estimated rice yield and estimate rice production for An Giang province, towards building a process for applying remote sensing in near real-time to map rice yield for the whole Mekong Delta.

The results of the yield distribution map of the three rice crops showed that most of them have a yield of 5 to 8 tons/ha. Based on the assessment of An Giang Provincial People's Committee [32], after 10 years of implementing the project "Development of agricultural and forestry plant varieties, livestock breeds and aquatic strains" under Decision No. 2194/QD-TTg dated December 25th, 2009 of the Prime Minister, the rice varieties of An Giang province have undergone a drastic change. In the 2000s, the rate of using high-quality varieties was still low, then by 2018 about 70% of the area used high-quality varieties such as: OM6976, OM4218, OM5451, Jasmine, etc. [32], which increases the production and value of rice for farmers. Therefore, monitoring rice yield and production becomes a necessity in agricultural management to ensure sustainable rice production and food security.

The results of the rice yield distribution map of the 2018 Autumn-Winter crop (Figure 3c), showed that the rice yield was mostly in the range of 4 to 6 tons/ha, lower than the other two crops. Meanwhile, the 2018 Winter-Spring crop had a high rice yield mainly about 6 to 8 tons/ha (Figure 3a). Previously, the Mekong Delta in general and An Giang province in particular had two rice crops, Winter-Spring and Summer-Autumn, the Winter-Spring crop was the main crop due to its high yield, low cost, and the most commodity rice for export. This crop is sown around mid-October of the lunar calendar, when the weather is very favourable, in the dry season – no rain from sowing to harvest – the number of sunny days is high, the temperature is high, so the rice yields are high [33]. As for the third rice crop (also known as the Autumn-Winter crop), sowing is done around July - August of the lunar calendar. This crop is in the rainy season, so sowing is very difficult, when it rains, all seeds must be sown again, the number of rainy days is high (lower number of sunny days), the rice growing is not as good as the Winter-Spring crop, the yield is lower, the cost is higher due to the need to sow/transplant, care and harvest the damp rice; and it also requires to spend money to pump water out, and the money for dike construction and for the reinforcement every year is not yet included. An Giang province developed a system of dikes to prevent floods [34], which help increasing the area of rice cultivation in the Autumn-Winter crop during the rainy season.

This study demonstrated the potential of multi-temporal Sentinel-1 SAR data in monitoring the spatial distribution of rice yields for each crop season in 2018 at the provincial scale. The results show that the rice yield of each crop in the year is different, it depends on the cultivation conditions of each crop, so the average yield is different. This method of estimating rice yield is based on a special rice farming model for An Giang province in the Mekong Delta. The method can be modified to each study area, the calibration and validation steps need to be performed with the local conditions relevant to those areas, especially for the areas that have a rice farming model using different rice varieties, with different climatic and environmental conditions. This study also shows the potential for monitoring rice yield distribution for each crop season using C-band SAR time series data, which is consistent with previous studies [22,24].

5. Conclusion

The study showed that Sentinel-1 radar remote sensing data can be used to estimate rice yield, which is important information for agricultural management and food security. The rice yield assessment model using Sentinel-1 data in one crop season proved to be an effective method for estimating rice yield at the provincial scale. The results of rice yield estimation compared with in situ yield data are reliable, demonstrating the applicability of multi-temporal SAR Sentinel-1 data to estimate rice yield for agricultural management. The results of the rice yield distribution map show that in 2018, the Winter-Spring crop had the highest rice yield and the crop with the lowest rice yield was the Autumn-Winter crop. These are important information in agronomic management to ensure sustainable rice production and food security. At the same time, the study also points out the necessity to expand the scope of research such as the Mekong Delta to validate the regional rice yield estimation model.

Acknowledgements

The study was carried out within the framework of the research project "Applied research on optical and radar remote sensing data for rice planted area monitoring and rice yield, production estimation in the Mekong Delta and Red River Delta", the project number: VT-UD-08/17-20, which is part of the National Program on Space Science and Technology (2016–2020), Vietnam Academy of Science and Technology.

The authors also acknowledge the support of time and facilities from Ho Chi Minh City University of Technology (HCMUT), VNU-HCM for this study.

References

[1] IRRI. World Rice Statistics Online Query Facility – IRRI. Available online: http://ricestat.irri.org:8080/wrsv3/entrypoint.htm (accessed on Oct 5, 2020).

[2] Koppe, W.; Gnyp, M.L.; Hütt, C.; Yao, Y.; Miao, Y.; Chen, X.; Bareth, G. Rice monitoring with multi-temporal and dual-polarimetric TerraSAR-X data. Int. J. Appl. Earth Obs. Geoinf. 2013, 21, 568–576, doi:10.1016/j.jag.2012.07.016.

[3] FAO Top 20 Countries Production of Rice, paddy 2018. FAO Available online: http://www.fao.org/faostat/en/#rankings/countries_by_commodity (accessed on Oct 5, 2020).

[4] Chen, J. Bin; Chen, C.F.; Son, N.T.; Chang, L.Y.; Chen, C.R. Rice crop mapping with RADARSAT-2 data in the middle of Taiwan. In Proceedings of the ACRS 2015 - 36th Asian Conference on Remote Sensing: Fostering Resilient Growth in Asia, Proceedings; 2015.

[5] Oyoshi, K.; Tomiyama, N.; Okumura, T.; Sobue, S.; Sato, J. Mapping rice-planted areas using time-series synthetic aperture radar data for the Asia-RiCE activity. Paddy Water Environ. 2016, doi:10.1007/s10333-015-0515-x.

[6] Ling, F.; Li, Z.; Chen, E.; Tian, X.; Bai, L.; Wang, F. RICE areas mapping using ALOS PALSAR FBD data considering the Bragg scattering in l-band SAR images of rice fields. In Proceedings of the International Geoscience and Remote Sensing Symposium (IGARSS); 2010.

[7] Feilong, L.; Zeng-yuan, L.; Li-na, B.; Xin, T.; Er-xue, C.; Yong-tian, Y. Rice mapping using ALOS PALSAR dual polarization data. J. Remote Sens. 2011.

[8] Lopez-Sanchez, J.M.; Ballester-Berman, J.D.; Hajnsek, I. First Results of Rice Monitoring Practices in Spain by Means of Time Series of TerraSAR-X Dual-Pol Images. IEEE J. Sel. Top. Appl. Earth Obs. Remote Sens. 2011, 4, 412–422, doi:10.1109/JSTARS.2010.2047634.

[9] Inoue, Y.; Sakaiya, E.; Wang, C. Potential of X-band images from high-resolution satellite SAR sensors to assess growth and yield in paddy rice. Remote Sens. 2014, 6, 5995–6019, doi:10.3390/rs6075995.

[10] Ribbes, F.; Le Toan, T. Rice field mapping and monitoring with RADARSAT data. Int. J. Remote

Sens. 1999, 20, 745–765, doi:10.1080/014311699213172.

[11] Nguyen, L.D.; Toan, T. Le; Floury, N. The use of SAR data for rice crop monitoring a case study of mekong river delta - Vietnam. In Proceedings of the Asian Association on Remote Sensing - 26th Asian Conference on Remote Sensing and 2nd Asian Space Conference, ACRS 2005; 2005.

[12] Le Toan, T.; Ribbes, F.; Li-Fang Wang; Floury, N.; Kung-Hau Ding; Jin Au Kong; Fujita, M.; Kurosu, T. Rice crop mapping and monitoring using ERS-1 data based on experiment and modeling results. IEEE Trans. Geosci. Remote Sens. 1997, 35, 41–56, doi:10.1109/36.551933.

[13] Bouvet, A.; Le Toan, T.; Tan, B.; Li, B.; He, W.; Zhang, P. Assessment of ENVISAT ASAR alternating polarisation data for rice mapping in Jiangsu Province, China. In Proceedings of the Proc. 2005 Dragon Symposium "Mid-Term Results", Santorini, Greece; 2005.

[14] Nguyen, L.D. Rice Crop Monitoring Using New Generation Synthetic Aperture Radar (SAR) Imagery, University of Southern Queensland, Toowoomba, Australia, 2009.

[15] Bouvet, A.; Le Toan, T. Use of ENVISAT/ASAR wide-swath data for timely rice fields mapping in the Mekong River Delta. Remote Sens. Environ. 2011, 115, 1090–1101, doi:10.1016/j.rse.2010.12.014.

[16] Shen, S.; Yang, S.; Li, B.; Tan, B.; Li, Z.; Le Toan, T. A scheme for regional rice yield estimation using ENVISAT ASAR data. Sci. China Ser. D Earth Sci. 2009, 52, 1183–1194, doi:10.1007/s11430-009-0094-z.

[17] Pazhanivelan, S.; Kannan, P.; Christy Nirmala Mary, P.; Subramanian, E.; Jeyaraman, S.; Nelson, A.; Setiyono, T.; Holecz, F.; Barbieri, M.; Yadav, M. Rice crop monitoring and yield estimation through COSMO Skymed and TerraSAR-X: A SAR-based experience in India. In Proceedings of the International Archives of the Photogrammetry, Remote Sensing and Spatial Information Sciences - ISPRS Archives; International Society for Photogrammetry and Remote Sensing, 2015; Vol. 40, pp. 85–92.

[18] Shao, Y.; Fan, X.; Liu, H.; Xiao, J.; Ross, S.; Brisco, B.; Brown, R.; Staples, G. Rice monitoring and production estimation using multitemporal RADARSAT. Remote Sens. Environ. 2001, 76, 310–325, doi:10.1016/S0034-4257(00)00212-1.

[19] Lam-Dao, N.; Hoang-Phi, P.; Huth, J.; Cao-Van, P. Estimation of the rice yield in the Mekong Delta using dual polarisation TerraSAR-X data. VNU J. Sci. Earth Sci. 2012, 28, 20–28.

[20] Chen, C.; Mcnairn, H. A neural network integrated approach for rice crop monitoring. Int. J. Remote Sens. 2006, 27, 1367–1393, doi:10.1080/01431160500421507.

[21] Clauss, K.; Ottinger, M.; Leinenkugel, P.; Kuenzer, C. Estimating rice production in the Mekong Delta, Vietnam, utilizing time series of Sentinel-1 SAR data. Int. J. Appl. Earth Obs. Geoinf. 2018, 73, 574–585, doi:10.1016/j.jag.2018.07.022.

[22] Lam-Dao, N.; Hoang-Phi, P.; Huth, J.; Cao-Van, P. Estimation of the rice yield in the Mekong Delta using SAR dual polarisation data. VNU J. Sci. Earth Sci. 2012, 28, 20–28.

[23] Lam-Dao, N.; Le-Toan, T.; Hoang-Phi, P.; Pham-Thi-Mai, T. Rice crop monitoring in the Mekong Delta, Vietnam using time series remotely sensed data. In Proceedings of the International Symposium on Remote Sensing; Tainan, Taiwan, 2015.

[24] Hoang-Phi, P.; Nguyen-Kim, T.; Nguyen-Van-Anh, V.; Lam-Dao, N.; Le-Van, T.; Pham-Duy, T. Rice yield estimation in An Giang province, the Vietnamese Mekong Delta using Sentinel-1 radar remote sensing data. IOP Conf. Ser. Earth Environ. Sci. 2021, 652, 012001, doi:10.1088/1755-1315/652/1/012001.

[25] Luu Van Ninh; Nguyen Minh Giam. Climate characteristics of An Giang province (in Vietnamese). Vietnam Journal of Hydro – Meteorology. 2017, 12, 18–26.

[26] Quegan, S.; Le Toan, T.; Yu, J.J.; Ribbes, F.; Floury, N. Multitemporal ERS SAR analysis applied to forest mapping. IEEE Trans. Geosci. Remote Sens. 2000, 38, 741–753, doi:10.1109/36.842003.

[27] Lopes, A.; Touzi, R.; Nezry, E. Adaptive speckle filters and scene heterogeneity. IEEE Trans. Geosci. Remote Sens. 1990, 28, 992–1000, doi:10.1109/36.62623.

[28] Phung, H.P.; Nguyen, L.D. Rice Crop Monitoring in the Mekong Delta, Vietnam Using Multi-temporal Sentinel-1 Data with C-Band. In ICSCEA 2019, Lecture Notes in Civil Engineering; Reddy, J.N., Wang, C.M., Luong, V.H., Le, A.T., Eds.; Springer Nature Singapore Pte Ltd, 2020; pp. 979–986 ISBN 9789811551437.

[29] Phung, H.-P.; Nguyen, L.-D.; Thong, N.-H.; Thuy, L.-T.; Apan, A.A. Monitoring rice growth status in the Mekong Delta, Vietnam using multitemporal Sentinel-1 data. J. Appl. Remote Sens. 2020, 14, 1, doi:10.1117/1.JRS.14.014518.

[30] Hoang-Phi, P.; Lam-Dao, N.; Pham-Van, C.; Chau-Nguyen-Xuan, Q.; Nguyen-Van-Anh, V.; Gummadi, S.; Le-Van, T. Sentinel-1 SAR Time Series-Based Assessment of the Impact of Severe Salinity Intrusion Events on Spatiotemporal Changes in Distribution of Rice Planting Areas in Coastal Provinces of the Mekong Delta, Vietnam. Remote Sens. 2020, 12, 3196, doi:10.3390/rs12193196.

[31] GSO General Statistics Office of Vietnam Available online: https://www.gso.gov.vn/default_en.aspx?tabid=778 (accessed on Sep 4, 2021).

[32] Portal_AG An Giang leads the country in the socialization of rice varieties Available online: http://www.angiang.gov.vn/wps/portal/Home/home/xem-chi-tiet/an-giang-di-dau-ca-nuoc-trong-cong-tac-xa-hoi-hoa-giong-lua (accessed on Sep 4, 2021).

[33] Nguyen Ngoc De. Rice lecture (in Vietnamese); Viet Nam National University Ho Chi Minh City, Vietnam, 2008.

[34] Duc Tran, D.; van Halsema, G.; Hellegers, P.J.G.J.; Phi Hoang, L.; Quang Tran, T.; Kummu, M.; Ludwig, F. Assessing impacts of dike construction on the flood dynamics of the Mekong Delta. Hydrol. Earth Syst. Sci. 2018, 22, 1875–1896, doi:10.5194/hess-22-1875-2018.

An assessment of groundwater quality for drinking and agricultural purposes in Ca Mau peninsula, Vietnamese Mekong Delta

Quang Khai Ha[1,2*], Mai Van Le Thi[3], Phu Le Vo[1,2], Hong Quan Nguyen[2,4,5], Abhijit Mukherjee[6]

1. Faculty of Environment and Natural Resources, Ho Chi Minh City University of Technology (HCMUT), 268 Ly Thuong Kiet St., Dist. 10, Ho Chi Minh City, Vietnam

2. Vietnam National University Ho Chi Minh City, Linh Trung Ward, Thu Duc District, Ho Chi Minh City, Vietnam

3. Centre for Water Resources Monitoring and Forecast, National Centre for Water Resources Planning and Investigation, 93/95 Vu Xuan Thieu St., Sai Dong Ward, Long Bien Dist., Ha Noi City, Vietnam

4. Institute for Circular Economy Development, Linh Trung Ward, Thu Duc city, Ho Chi Minh City

5. Center of Water Management and Climate Change (WACC) Institute for Environment and Resources, Linh Trung Ward, Thu Duc city, Ho Chi Minh City, Vietnam

6. Department of Geology and Geophysics, Indian Institute of Technology Kharagpur, Kharagpur 721 302, India

* Corresponding author email: *quangkhai02@hcmut.edu.vn*

Abstract. Groundwater serves as the major source for drinking water and the only fresh water source in the dry season in Ca Mau province, the Vietnamese Mekong Delta. The study is performed to assess groundwater quality in the area. Groundwater Quality Index (GWQI), magnesium hazard (MH), sodium percentage (Na%), sodium adsorption ratio (SAR), permeability index (PI) combined with geostatistical tools were conducted by using groundwater quality data from 400 samples collected from private wells. GWQI values suggest that 7%, 63%, 28%, 2%, 1% of total samples are classified into excellent, good, poor, very poor and unsuitable for drinking water. High Cl and TDS are the main cause of high GWQI values suggesting that saltwater intrusion deteriorates groundwater quality. For irrigation purposes, 4%, 17%, 36%, 43% of total samples classify into Good, Permissible, Doubtful, Unsuitable respectively by Na% indicator; 55.0%, 30.9%, 11.9%, 2.2% of total samples is Excellent, Good, Doubtful and Unsuitable according to SAR, 54% samples are unsuitable by MH index, and all samples are good by PI. This study provides useful information on groundwater quality for various purposes in Ca Mau province, that could help water management and water users to identify the necessary approach for groundwater protection and water-treatment methods before utilising the groundwater source for specific purposes in the study area.

1. Introduction

Groundwater is widely used over the world as the most important resources for drinking and irrigation purposes in coastal area due to seawater intrusion and pollution of surface water. Lack of water supply system and easily accessible are also the reason for many people prefer using groundwaters. However, groundwater quality is deteriorating due to natural forces such as geochemical, weathering processes, tectonic, and human activities such as overexploitation, urbanisation, fertiliser application [1]. In coastal aquifers, groundwater quality is additionally impacted by seawater level rise and seawater intrusion [2]. Consumption of contaminated drinking water can lead to human health problems [3]. Using poor groundwater quality for irrigation can also cause adverse impacts on soil, plant, and production yield [4, 5].

Assessment of groundwater quality suitability for various purposes of water uses are important for managers, planners, policy makers and user to address appropriate solutions to groundwater management and protection either technological treatment or alternative water resources for avoiding the adverse effects on human health and the environments [6]. Poor groundwater quality may be caused by a single or multiple (bio)chemical elements that concentrations exceed the recommendations for using such as drinking or irrigation water quality. It is difficult to directly stack the pollution effects brought by different elements due to the different dimensions of elements. Moreover, information about groundwater quality also needs to be reported in a simple language and understandable manner for different stakeholders in practices. Therefore, integrated index methods such as groundwater quality index (GWQI) is widely used for evaluation of groundwater quality for drinking purpose [6]. The GWQI method standardises different indicators and then converts them into a single value. This makes it easy to express the comprehensive influence of various indices on groundwater and facilitate the comparison between samples collected from different regions. GWQI has been applied as an effective approach to communicate water quality information to decision makers and is widely used in many areas such as India [7], Iran [8], Palestine [9], USA [10]. The Geographic Information System (GIS) can support water quality assessment to understand clear picture of spatial variability of groundwater quality. Thus, integration of the GIS and the GWQI can help the decision-maker to take appropriate steps for monitoring and conservation of the groundwater quality for sustainable management of the aquifer system [7].

Groundwater is crucial resources for water supply in Vietnam. Especially, groundwater provides 62% of total water supply for domestic use in Mekong Delta [11]. In Ca Mau Province, groundwater is the only source for drinking water [12]. Over abstraction of groundwater induces negative effects such as land subsidence [13], saltwater intrusion [14]. Even though many groundwater geochemical issues are noticed in other provinces of the Vietnamese Mekong Delta such as groundwater arsenic contamination in An Giang, Dong Thap [14], seawater intrusion and freshening in Soc Trang Province [15] or acidic groundwater [16]; groundwater quality in Ca Mau still remains unknown. Therefore, this study aims to investigate the suitability of groundwater quality for drinking and irrigation practices for the Ca Mau province. The information will contribute to help decision makers, planner and users in preparing appropriate water treatment techniques and protection measures for groundwater resource in the area.

2. Study area

Ca Mau is a coastal province in the Vietnamese Mekong Delta and locate in the south most part of Vietnam. The study area is a peninsula, bound by Gulf of Thai Lan in the west and East Sea in the east. The province area is about 5,300 km^2 and population of 1.4 million people. The area elevation varies from 0 to 2m. Therefore, the area is susceptible to climate change and sea level rise. Climate has two distinct seasons: the rainy season starts from May to October and the dry season lasts from November to April in the following year. According to monitoring data in Ca Mau province from 2015 to 2017, annual rainfall is about 2140 mm and 91% of the total annual rainfall observed in rainy season. Surface water in the area controlled by dense canal and tidal regime. The land use in the area is mostly for agricultural and aquacultural practices.

Due to climate change, dam construction in the upstream part of Mekong Delta, expansion of shrimp farms, water resources are facing with many challenges including seawater intrusion, contamination, and severe droughts. Due to poor quality, surface water is generally not used for drinking water and industrial works. No surface water is available in cannel systems. Water supply in the area is mostly dependent on groundwater resource. Groundwater even uses for irrigation and dilution of saltwater in shrimp farm illegally, in particularly in the dry weather.

The area has 7 confined aquifers, namely Holocene (qh), Upper Pleistocene (qp^3), Upper-middle Pleistocene (qp^{23}), lower Pleistocene (qp^1), middle Pliocene (n_2^2), lower Pliocene (n_2^1) and Upper Miocene (n_1^3). According to hydrogeological investigation of DWRPIS [12], average of hydraulic conductivity values of qp^3 aquifer k= of 27.8 m/day and k=20.4 m/day for $qp^{2,3}$, K= 18 m/day for qp1, K = 22.4 m/day for n_2^2, K = 17.5 m/day for n_2^1. Total 175,000 wells installed in aquifer systems and groundwater abstraction rate is about 430,000 m^3/day. The highest abstraction volume is observed in aquifer qp23 (~224,000 m^3/day). The aquifer qp_1, n_2^2, n_2^1, groundwater abstraction rates are 92,000, 107,000 and 3,000 m^3/day, respectively. Groundwater abstraction rate in Holocene and Upper Pleistocene are negligible due to saline groundwater. Due to very deep below surface, not many groundwater abstraction wells were installed in n_2^1 and n_1^3 aquifers.

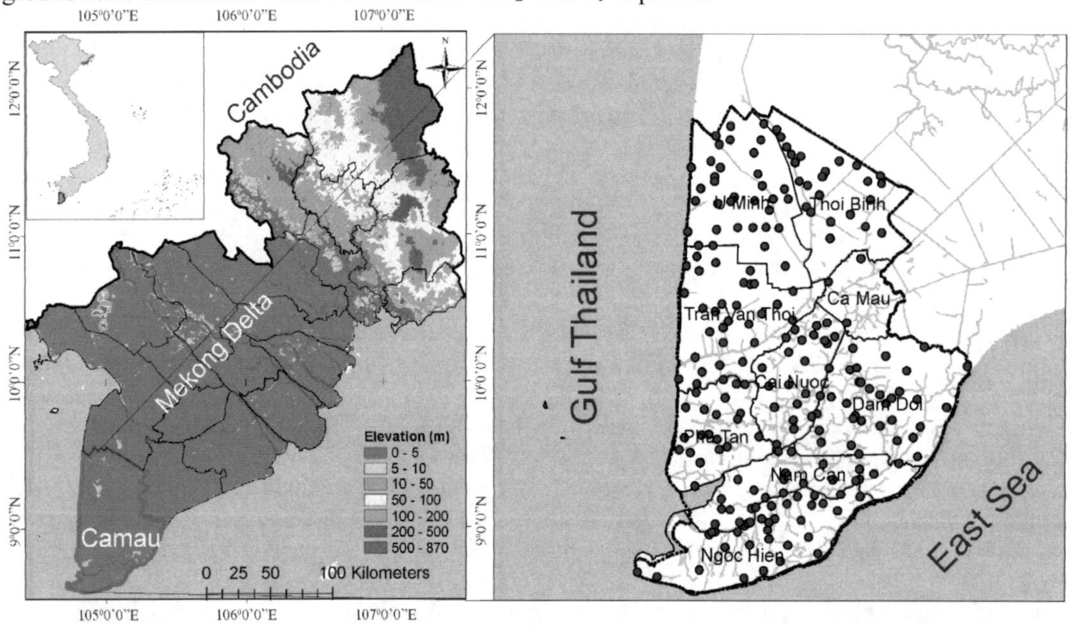

Figure 1. Location of Ca Mau plotted on topography map and groundwater sampling area

3. Method

3.1. Dataset

The study utilized the dataset of about 246 groundwater samples collected from 224 abstraction wells during 2013-2016 in Mekong Delta. The groundwater samples were collected by Division for Water Resources Planning and Investigation for the South of Vietnam (DWRPIS, [12]). Among 224 investigated wells, groundwater samples were doubly collected from 22 selected wells for Quality Assurance (QA) of sampling and chemical analysis. The groundwater samples were taken by submersible pump that install directly in the wells. The private wells are pumped every day for water consumption of local people. Water in the wells were pumped out about 10 min, before sampling. The samples were analyzed after sampling by Division for Water Resources Planning and Investigation for the south of Vietnam following water National Water Chemical Analysis Standards for Na^+, K^+, Ca^{2+}, Mg^{2+}, NH_4^+, SO_4^{2-}, Cl^-, HCO_3^-, NO_3^-, and NO_2^-.

3.2. Drinking water assessment

This study applied GWQI method for drinking water evaluation. Calculation of GWQI was based on TDS, pH, K^+, Ca^{2+}, Na^+, Mg^{2+}, Cl^-, HCO_3^-, SO_4^{2-} and NO_3^-. The weight of each ion for calculating the GWQI was assigned according to their relative importance on drinking water quality and human health. The weight normally assigned by local expert's opinions; therefore, it is difficult for comparison of the GWQI in global scale. To overcome this issue, this study will summaries weight values from previous studies to generate the general values for the area (Table 1).

Table 1. Weight of each ion for GWQI calculation in previous studies

Chemical parameter	India				Iran	Bangladesh	Ghana	Pakistan	Turkey	This study
	[17]	[18]	[19]	[7]	[20]	[21]	[22]	[23]	[24]	
pH	4		5	4	3	4	4	4	4	4
TDS	4	5	5	5	3	4	5	4	5	5
TH	4			2						4
Ca^{2+}	2	3	2	3	3	2	3	2	3	3
Mg^{2+}	2	3	2	3	3	2	3	2	3	3
Na^+	4	4	4		4	3		3	4	4
K^+	1	2	1		2	2		2	2	2
HCO_3^-	1	1	3	1	1	1	1	3	1	2
Cl^-	5	5	4	5	5	3	5	4	5	4
SO_4^{2-}	5	5	5	4	5	4	5	5	5	5
NO_3^-	5	5	5	5	5	5	5	5	5	5

The relative weight for each ion was calculated as per Eq. (2)

$$W_i = \frac{w_i}{\sum_{i=1}^{n} w_i} \qquad (1)$$

Where, Wi, wi and n represent the relative weight, the weight for each ion and the number of ions to be evaluated, respectively.

The quality rating scale (qi) for each ion was calculated as equation:

$$q_i = \frac{C_i}{S_i}\ x\ 100 \qquad (2)$$

Where, Ci and Si represent the actual concentration of each ion and their concentration in drinking water limit (DWL) recommended by WHO [3], respectively.

Table 2. General groundwater chemistry and standard values for drinking water

Parameter	Mean	SD	Min	Max	DWL [3]	No of sample exceeded DWL
pH	8.1	0.4	6.8	9.2	6.5-8.5	23
TDS (mg/L)	991.6	529.2	202.0	3210.7	1000.0	85
Na^+(mg/L)	278.5	168.3	51.2	833.3	200.0	134
TH (mg/L $CaCO_3$)	193.7	170.7	10.0	1164.8	300.0	35
K^+ (mg/L)	12.8	7.3	1.3	58.6	no limit listed	
Ca^{2+} (mg/L)	29.0	26.1	2.0	245.5	75.0	10

Parameter	Mean	SD	Min	Max	DWL [3]	No of sample exceeded DWL
Mg^{2+} (mg/L)	29.4	28.0	1.2	182.4	50.0	33
NH_4^+ (mg/L)	0.47	0.70	0.00	3.68	3.0	5
Fe (mg/L)	0.10	0.11	0.00	0.53	0.3	15
HCO_3^- (mg/L)	516.2	158.9	91.5	870.1	120.0	223
Cl-(mg/L)	227.9	260.3	4.3	1262.0	250.0	74
SO_4^{2-} (mg/L)	51.7	62.2	0.0	499.4	250.0	4
NO_3^- (mg/L)	6.6	5.3	0.0	40.4	50.0	0

The sub index SI value was calculated for each ion as equation below:

$$SI_i = W_i \, x \, q_i \qquad (3)$$

The GWQI value is calculated by

$$GWQI = \sum_{i=1}^{n} SI_i \qquad (4)$$

The standards of classifying drinking water quality in terms of GWQI are shown on Table 3

Table 3. Classification of groundwater quality

Water index	Index values	Water condition	No samples	Percentage
Drinking				
Groundwater quality index (GWQI)	<50	Excellent	19	7%
	50-100	Good	176	63%
	100-200	Poor	77	28%
	200-300	very Poor	5	2%
	>300	Unsuitable	2	1%
Irrigation				
Sodium Adsorption Ratio (SAR)	<10	Excellent	153	55%
	10-18	Good	86	31%
	18-26	Doubtful	33	12%
	>26	Unsuitable	6	2%
Sodium percentage (Na%)	<20	Excellent	0	0%
	20-40	Good	12	4%
	40-60	Permissible	47	17%
	60-80	Doubtful	99	36%
	>80	Unsuitable	120	43%
Permeability index (PI)	>75	Suitable	239	86%
	25-75	Good	40	14%
	<25	Unsuitable	0	0%
Magnesium Hazard (MH)	<50%	Suitable	128	46%
	>50%	Unsuitable	150	54%

3.3. Irrigation water assessment

The chemical composition of irrigation water directly impacts on plants yield in terms of toxicity/deficiency or indirect by changing nutrient availability. Previous studies proposed various indicator for assessment of irrigation water.

Sodium percent (Na%)

$$Na\% = \frac{Na^+}{Ca^{2+}+Mg^{2+}+Na^++K^+}x100 \text{ (concentration of parameters meq/l)}$$

Sodium adsorption ratio (SAR)

$$SAR = \frac{Na^+}{\sqrt{\frac{Ca^{2+}+Mg^{2+}}{2}}}x100$$

Permeability index

$$PI = \frac{Na^+ + \sqrt{HCO_3^-}}{Ca^{2+}+Mg^{2+}+Na^+}x100$$

According to Doneen (1964), PI can be categorized in three classes: class I (>75%, suitable), class II (25–75%, good) and class III (<25%, unsuitable). Water under class I and class II are recommended for irrigation.

Magnesium hazard

$$MH = \frac{Mg^{2+}}{Ca^{2+}+Mg^{2+}}x100$$

According to this, high magnesium hazard value (50 %) has an adverse effect on the crop yield as the soil becomes more alkaline.

4. Results and discussion

4.1. General groundwater characteristic

General characteristics of groundwater chemistry in Ca Mau Province is shown on Table 2. Groundwater pH ranges from 6.75 to 9.18, and 8.06 in average indicating that chemical characteristic of the groundwater samples is neutral to alkaline. The drinking water pH values are recommended within the range of 6.5–8.5 [3]. TDS ranges from 202 to 8089 mg/L, about 36% of total groundwater samples show TDS greater than 1000 mg/L (Drinking Water Limit; [3]).

The ionic dominance trend for cations was $Na^+ > Mg^{2+} > Ca^{2+} > K^+$, while $HCO_3^- > Cl^- > SO_4^{2-} > NO_3^- > NO_2^-$ for anions. The mean concentrations of sodium and bicarbonate were the highest among all the cations and anions, respectively. The sodium and bicarbonate ions were widely dispersed in groundwater, which might be generated by weathering and cation exchange occurred by freshening process the coastal aquifers. Na^+ has concentration levels of 51 to 1,280 mg/L (average of 280 ± 175). The maximum Na^+ concentration in drinking water is 200 mg/L according to guidelines of WHO [3]. About 61% of groundwater samples have Na^+ concentration exceeding the drinking water guidelines. Sodium is important for human body. However, drinking of high sodium concentration in water may be cause adverse impacts on human health problems such as hypertension, heart disease or kidney for some people.

The concentration of Ca^{2+}, and Mg^{2+} were 29.5 and 32.4 mg/L in average, respectively. Generally, there is no limitation for Ca and Mg concentration in drinking water. High contents of Ca^{2+} and Mg^{2+} may be good for human healthy. However, high concentration of Ca^{2+} and Mg^{2+} induce high in total hardness. Hard water is generally not harmful to human health but can create serious problems in industrial settings. Calcium and magnesium carbonates precipitation on the surfaces of pipes and the surfaces of heat exchangers can reduce the effective function of machine and clogging of pipeline. Water hardness can even cause breaks in equipment that handles water such as boilers, cooling towers. The hardness of water is also suggested as the cause of the non-formation of soap suds and making difficulties in cleaning and laundering. In the study area, groundwater hardness varied from 10 to 1164.8

mg/L (as $CaCO_3$ concentration) and 193.7 mg/L in average. About 16% of groundwater samples even have hardness greater than recommended for drinking water (300 mg/L; [3]).

K^+ is a necessary element for the human body, no recommendation for K concentration in drinking water. Fe and NH_4^+ concentrations are up to 0.5 and 11 mg/L, respectively. Generally, the Fe and NH_4^+ concentration in drinking water will not induce health risks, but they can cause laundry problems and smelly due to iron precipitation and high NH4. The concentration of Fe and NH_4^+ in drinking water are suggested for 0.3 and 3 mg/L, respectively.

Among all anions in groundwater samples, the mean and standard deviation concentrations of HCO_3^-, Cl^-, SO_4^{2-}, NO_2^-, NO_3^-, were found at 515±150, 240±380, 54.8±80.7, 1.77±3.42, and 5.86±5.22 mg/L, respectively. About 31 and 2% of groundwater samples have Cl^- and SO_4^{2-} concentrations are greater than those values in DWL for 250 mg/L for Cl and SO_4 [3]. Even though NO_3 concentrations is up to 40.4 mg/L, smaller than the maximum concentration of 50 mg/L recommended for drinking water by WHO (2011), the NO_3 concentration are much higher than observed NO3 in groundwater from others province in the Mekong Delta. According to data from National Groundwater Monitoring Network for southern Vietnam [13], NO_3 concentration was generally low or not detected in Long An, Tra Vinh, An Giang, Ben Tre, Vinh Long and Dong Thap. The high NO_3 concentrations are normally observed high in Ho Chi Minh City, Tay Ninh, Binh Duong and Dong Nai due to high permeable of surface sediment [14]. Unconfined aquifers are generally susceptible to NO_3 contamination [25]. However, the aquifer in Ca Mau is mostly confined, aquifer is protected by thick clay layers [12]. Therefore, observation of high NO_3 concentration is very abnormal. The NO_3 may be introduced from upper part through well casing or other pathways to deep aquifers [16]. However, finding the origin of NO_3 in groundwater of this area is beyond scope of this study, it needs to have further investigation.

Generally, the contamination of TDS, TH, Na and Cl was the most concern in the groundwater of Ca Mau Province. The high concentration of TDS, Cl is observed in coastal areas such as Nam Can, Phung Hiep, Tran Van Thoi, Cai Nuoc, and U Minh may relate to saltwater intrusion (Figure 2). The groundwater samples have Na concentration exceeding that value in DWL observed in most part of province except Thoi Binh and Ca Mau City. In contrast, the extremely hardwater observed mainly in Tran Van Thoi and U Minh districts.

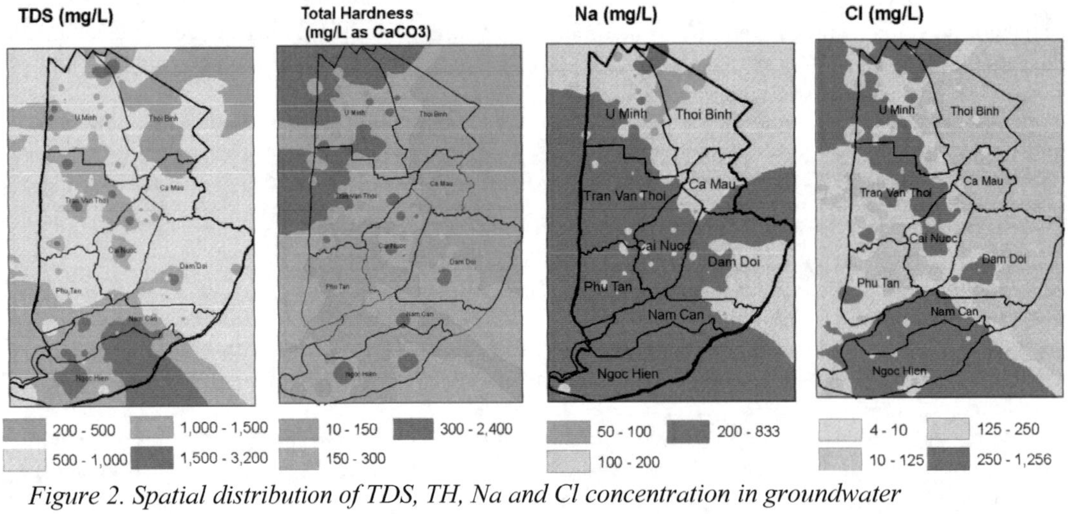

Figure 2. Spatial distribution of TDS, TH, Na and Cl concentration in groundwater

4.2. Drinking water assessment

As mentioned above groundwater quality index is widely applied for evaluation of groundwater quality for drinking purpose. Calculated GWQI values in this study range from 25.2 to 585 (average of 87.9). The groundwater quality belongs 4 groups including good, poor, very poor and unsuitable quality for drinking purpose. About 31% of groundwater samples show poor to unsuitable groundwater quality and mostly observed in south and western part of the study area (Figure 3). The high GWQI is associated to high TDS, Cl and Na concentrations suggesting salinisation is the main cause of deterioration of groundwater quality.

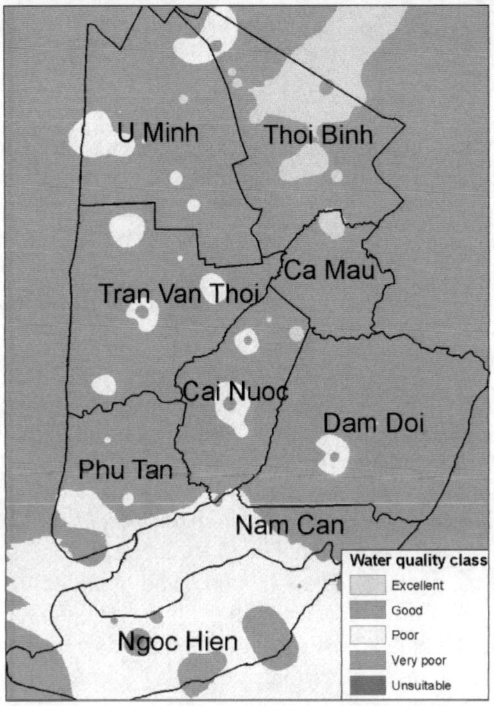

Figure 3. Groundwater quality map of Ca Mau Peninsula

4.3. Irrigation water assessment

Higher salinity will reduce the osmotic activity of plants and prevent water from reaching the branches and leaves of plants, resulting in inferior production [4,5]. Electricity conductivity (EC) is a typical measurement of salinity. In general, EC should be less than 2250 uS/cm in irrigation water. Higher concentration of sodium in irrigation water will affect the soil permeability and make the texture of soil hard to plough. SAR and %Na are good measures of alkali/sodium hazard to crops. Na% should be lower than 60 % in irrigation water, and SAR lower than 18 is suitable for irrigation water but may be acceptable if it is lower than 26. The results listed in Table 1 show that the SAR values of all samples vary from 1.2 to 29.6. Only 6 samples show SAR greater than 26, they are in high saline groundwater zone. According to PI index, all groundwater samples can be used for irrigation. In contrast, about 78% of groundwater samples have Na% greater than 60, indicating unsuitability of groundwater for irrigation. According to MH index only 46% of groundwater can be used for irrigation.

5. Conclusion

Groundwater in Ca Mau province is facing with high TDS, TH, Na and Cl concentrations. 61 % of samples show exceeded Na concentration for drinking water. Overall, about 31% of groundwater samples in the area were classified into poor and unsuitable for drinking purpose. High TH level in groundwater may also indue problems for industrial and domestic uses. For irrigation, even though groundwater quality is suitable for irrigation according to SAR and PI index, the 79% and 54% of groundwater samples are unsuitable for irrigation by Na% and MH index, respectively. This finding suggest that groundwater needs to be treated before using. Additionally, over exploitation of groundwater in this area is known as the underlying cause for land subsidence and contamination. Therefore, there is the need for seeking alternative water sources for water supply, especially in coastal areas.

Acknowledgement
This research is funded by the Vietnam Ministry of Science and Technology (MOST) under grant ID: ĐTĐL.CN-45/18. We also acknowledge the support of time and facilities from Ho Chi Minh City University of Technology (HCMUT), VNU-HCM for this study.

References

[1] Appelo, C.A.J. and Postma, D. 2004 Geochemistry, groundwater and pollution. CRC press.
[2] Ferguson, G. and Gleeson, T., 2012. Vulnerability of coastal aquifers to groundwater use and climate change *Nature Climate Change* **2**(5) 342-345.
[3] WHO 2011 Guidelines for Drinking-Water Quality, 4th ed World Health Organization, Geneva, Switzerland
[4] Yasuor, H., Yermiyahu, U. and Ben-Gal, A. 2020 Consequences of irrigation and fertigation of vegetable crops with variable quality water: Israel as a case study. *Agricultural Water Management* **242** 106362.
[5] Che, Z., Wang, J. and Li, J. 2021 Effects of water quality, irrigation amount and nitrogen applied on soil salinity and cotton production under mulched drip irrigation in arid Northwest China *Agricultural Water Management* **247** 106738.
[6] Uddin, M.G., Nash, S. and Olbert, A.I., 2021. A review of water quality index models and their use for assessing surface water quality. *Ecological Indicators* **122** 107218.
[7] Verma, P., Singh, P.K., Sinha, R.R. and Tiwari, A.K. 2020 Assessment of groundwater quality status by using water quality index (WQI) and geographic information system (GIS) approaches: a case study of the Bokaro district, India *Applied Water Science* **10**(1) 1-16.
[8] Amiri, V., Kamrani, S., Ahmad, A., Bhattacharya, P. and Mansoori, J. 2021 Groundwater quality evaluation using Shannon information theory and human health risk assessment in Yazd province, central plateau of Iran *Environmental Science and Pollution Research* **28**(1) 1108-1130.
[9] El Baba, M., Kayastha, P., Huysmans, M. and De Smedt, F. 2020 Evaluation of the groundwater quality using the water quality index and geostatistical analysis in the Dier al-Balah Governorate, Gaza Strip, Palestine *Water* **12**(1) 262.
[10] Bangia, K., August, L., Slocombe, A. and Faust, J. 2020 Assessment of contaminants in California drinking water by region and system size *AWWA Water Science* **2**(5) e1194.
[11] Bui, T.V., Dang, T.L., and Le, T.M.V. 2015 Groundwater issues and hydrogeological survey of the Mekong River basin in Vietnam. In: Ha, K., Nguyen, T.M.N., Lee, E., and Jayakumar, R. (Eds.), Current Status and Issues of Groundwater in the Mekong River Basin. KIGAM-CCOP-UNESCO Bangkok Office, pp. 93–121.
[12] Pham, K.T., Nguyen, T.H., Vu, B.M., Nguyen, V. N., Dang, V.T., Cao, X.V. 2019 Groundwater assessment and investigation in Vietnamese Mekong Delta Key economic region. Division for Water Resources Planning and Investigation for the South of Vietnam.
[13] Minderhoud, P.S.J., Erkens, G., Pham, V.H., Bui, V.T., Erban, L., Kooi, H. and Stouthamer, E., 2017. Impacts of 25 years of groundwater extraction on subsidence in the Mekong delta, Vietnam. *Environmental Rresearch Letters* **12**(6) 064006.
[14] Ha, Q.K., Kim, K., Phan, N.L., Phung, T.H., Lee, J., Nguyen, V.K. and Phan, C.N. 2019 A hydrogeological and geochemical review of groundwater issues in southern Vietnam *Geosciences Journal* **23**(6) 1005-1023.
[15] Ha, Q.K., Le Vo, P., Phan, C.N., Pham, V.H. and Nguyen, V.K. 2021 Identification of freshwater-saltwater interface in coastal areas using combination of geophysical and geochemical methods: A case study in Mekong Delta, Vietnam *IOP Conference Series: Earth and Environmental Science* **652** 012006.
[16] Ha, Q.K., Choi, S., Phan, N.L., Kim, K., Phan, C.N., Nguyen, V.K. and Ko, K.S. 2019 Occurrence of metal-rich acidic groundwaters around the Mekong Delta (Vietnam): A phenomenon linked to well installation. *Science of the Total Environment* **654** 1100-1109.

[17] Adimalla, N. 2019 Controlling factors and mechanism of groundwater quality variation in semiarid region of South India: an approach of water quality index (WQI) and health risk assessment (HRA). *Environmental Geochemistry and Health* 1-28.
[18] Vasanthavigar, M., K. Srinivasamoorthy, K., Vijayaragavan, K. Rajiv Ganthi, R., Chidambaram, S., Anandhan, P., Manivannan, R. and Vasudevan, S. 2010 Application of water quality index for groundwater quality assessment: Thirumanimuttar sub-basin, Tamilnadu, India *Environmental Monitoring and Assessment* **171**(1) 595-609.
[19] Rao, N.S., Sunitha, B., Rambabu, R., Rao, P.N., Rao, P.S., Spandana, B.D., Sravanthi, M. and Marghade, D. 2018 Quality and degree of pollution of groundwater, using PIG from a rural part of Telangana State, India. *Applied Water Science* **8**(8) 1-13.
[20] Abbasnia, A., Yousefi, N., Mahvi, A.H., Nabizadeh, R., Radfard, M., Yousefi, M. and Alimohammadi, M. 2019 Evaluation of groundwater quality using water quality index and its suitability for assessing water for drinking and irrigation purposes: Case study of Sistan and Baluchistan province (Iran). *Human and Ecological Risk Assessment: An International Journal* **25**(4) 988-1005.
[21] Islam, A.R.M.T., Siddiqua, M.T., Zahid, A., Tasnim, S.S. and Rahman, M.M. 2020 Drinking appraisal of coastal groundwater in Bangladesh: An approach of multi-hazards towards water security and health safety *Chemosphere* **255**126933.
[22] Boateng, T.K., Opoku, F., Acquaah, S.O. and Akoto, O. 2016 Groundwater quality assessment using statistical approach and water quality index in Ejisu-Juaben Municipality, Ghana *Environmental Earth Sciences* **75**(6) 1-14.
[23] Khanoranga and Khalid, S. 2019 An assessment of groundwater quality for irrigation and drinking purposes around brick kilns in three districts of Balochistan province, Pakistan, through water quality index and multivariate statistical approaches *Journal of Geochemical Exploration*, **197** 14-26.
[24] Varol, S. and Davraz, A., 2015. Evaluation of the groundwater quality with WQI (Water Quality Index) and multivariate analysis: a case study of the Tefenni plain (Burdur/Turkey) *Environmental Earth Sciences* **73**(4) 1725-1744.
[25] Chae, G.T., Kim, K., Yun, S.T., Kim, K.H., Kim, S.O., Choi, B.Y., Kim, H.S., and Rhee, C.W. 2004 Hydrogeochemistry of alluvial groundwaters in an agricultural area: an implication for groundwater contamination susceptibility *Chemosphere* **55** 369–378.

IOP Conf. Series: Earth and Environmental Science **964** (2022) 012009 doi:10.1088/1755-1315/964/1/012009

Climate trends and climate change scenarios in Ho Chi Minh City

Anh Ngoc Le[1], Thi Nguyen Vo[1], Van Hong Nguyen[1] and Dang Mau Nguyen[1]

[1] Viet Nam Institute of Meteorology, Hydrology and Climate Change

E-mail address: leanhngoc.sihymete@gmail.com

Abstract. This paper reviews the trends of climate and climate change scenarios in Ho Chi Minh City (HCMC). The linear regression method is used to determine the trend and variation of past climate (1980-2019) at Tan Son Hoa station. The annual average temperature tends to increase about 0.024°C/year (r^2=0.54) and the rainfall tends to increase about 6.03 mm/year (r^2=0.67). For temperature scenario, by 2030 the annual average temperature in the whole city will increase from 0.80- 0.81°C (RCP4.5) and 0.92-0.98°C (RCP8.5). By 2050, it will increase 1.23-1.33°C (RCP4.5) and 1.55-1.68°C (RCP8.5). By 2100, it will increase 1.75-1.88°C (RCP4.5) and 3.20-3.55°C (RCP8.5) compared to the base period. Regarding rainfall scenario, in 2030, the city-wide average rainfall will increase by 12-21% (RCP4.5) and by 12-17% (RCP8.5). By 2050, the average rainfall is likely to increase by 13-15% (RCP4.5) and 15-17% (RCP8.5). By 2100, the average rainfall is likely to increase by 18-22% (RCP4.5) and 20-21% (RCP8.5) compared to the base period.

1. Introduction

HCMC has a tropical monsoon climate. A year is divided into two distinct seasons. The northeast monsoon months (dry season) from December to April and the southwest monsoon months (rainy season) from May through November.

The annual average temperature in HCMC is 27°C–28°C, and the hottest and coldest months of the year differ by just 3°C. In recent years, the annual average temperature has increased. The higher temperature increase in HCMC since the 1990s has coincided with the accelerated urbanization in the area. The urban heat island effect, which is likely to have contributed significantly to observed warming and may intensify further as urbanization continues.

Rainfall during the wet season accounts for 90–95% of the total annual. There is already high variation in rainfall, bringing both localized flooding and droughts. HCMC endures droughts, usually in March or April. In recent decades, there were extreme dry seasons in 1993, 1998, and 2002. Dry season drought in the future is likely more frequent.

This paper reviews the trends of climate and climate change scenarios in HCMC to provide current and future climate information for climate change mitigation and production planning, socio-economic development and disaster prevention in HCMC.

2. Materials and methods

2.1. Methods

2.1.1. Climate Trend Analysis Method.
The linear regression method is used to determine the trend and variation of past temperature and precipitation (1980-2019) at Tan Son Hoa Station in HCMC.

Linear regression attempts to model the relationship between two variables. One variable is considered to be an explanatory variable (times), and the other is considered to be a dependent variable (temperature and precipitation). A linear regression line has an equation of the form $Y = a + bX$, where X is the explanatory variable and Y is the dependent variable. The slope of the line is b, and a is the intercept (the value of Y when $X = 0$). A valuable numerical measure of association between two variables is the correlation coefficient (R^2), which is a value between -1 and 1 indicating the strength of the association of the observed data for the two variables.

2.1.2. Dowscalling Dynamic Method.
The climate change scenario in HCMC is calculated on four regional climate models (PRECIS, CCAM, RegCM, clWRF). Thus, there are 12 calculation plans based on results from global climate model.

Based on the evaluation results for 4 climate models, in HCMC, the temperature scenario is calculated according to 8 plans from 3 models (4 plans of CCAM model, 3 plans of PRECIS model and a plan of clWRF model), the precipitation scenario is calculated according to 3 plans of the PRECIS model.

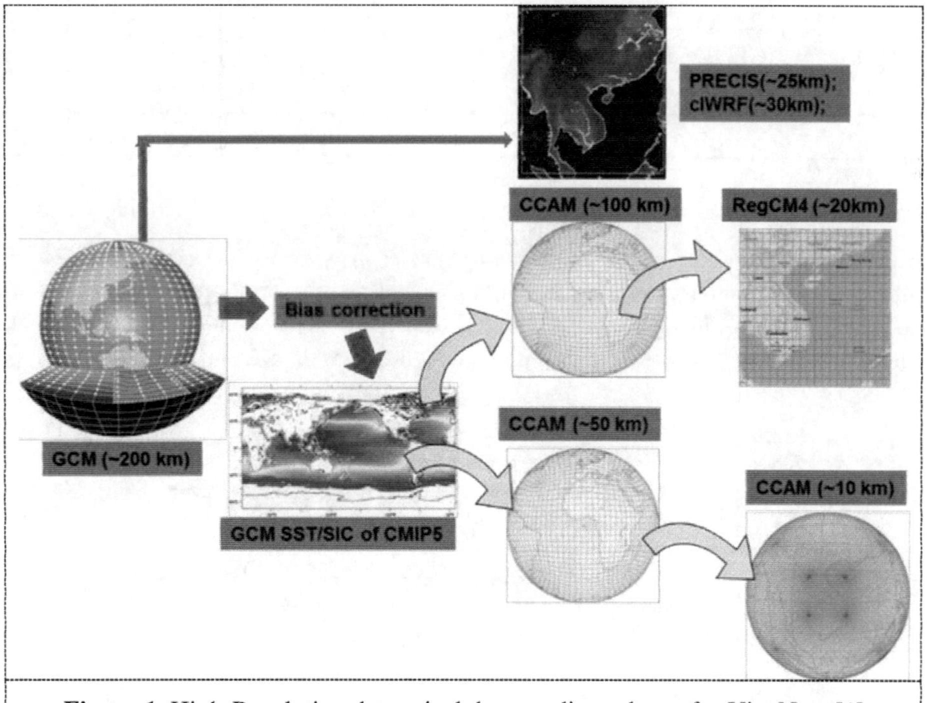

Figure 1. High-Resolution dynamical downscaling scheme for Viet Nam [1]

2.2. Materials
The trend and variation of past temperature and precipitation is based on observed data at Tan Son Hoa Station, with a reliable and long enough data series to calculated in the period of 1980-2019.

Global and regional climate models are the main tools used to forecast future climates, especially climate extremes. The following models is used to calculate high-resolution cc scenarios in HCMC: (i) The PRECIS model of Center's Hadley Meteorology – United Kingdom, (ii) CCAM model of Commonwealth Scientific and Industrial Research Organization (CSIRO), (iii) RegCM model of International Centre for Theoretical Physics (ICTP), and (iv) clWRF model of United States (Table 1).

Table 1. Regional climate models used for updating climate change scenarios [1]

No.	Regional climate models	Boundary conditions (global model)	Resolution	Data period		
				The base period	RCP4.5	RCP8.5
1	CCAM	ACCESS1-0	10km	1970-2005	2006-2099	2006-2099
2		CCSM4				
3		CNRM-CM5				
4		GFDL-CM3				
5		MPI-ESM-LR				
6		NorESM1-M				
7	RegCM	ACCESS1-0	20km	1980-2000	2046-2065 2080-2099	2046-2065 2080-2099
8		NorESM1-M				
9	PRECIS	HadGEM2-ES	25km	1960-2005	2006-2099	2006-2099
10		GFDL-CM3				
11		CNRM-CM5				
12	CLWRF	NorESM1-M	30km	1980-2005	2006-2099	2006-2099

3. Results and discussion

3.1. Trend of temperature and rainfall for the period of 1980-2019

The annual average temperature was 27.9°C. In the period from 1980 to 2019, the temperature at Tan Son Hoa station tended to increase by 0.024°C per year (r^2=0.5416) (Figure 2). In which, the lowest average temperature was 27.0°C (in 1987) while the highest was 28.5°C (in 1999).

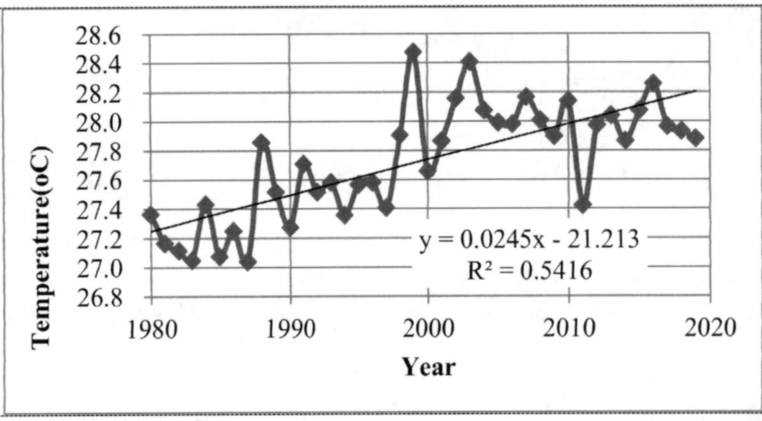

Figure 2. Trend of annual average temperature (°C) at Tan Son Hoa Station (1980-2019).

IOP Conf. Series: Earth and Environmental Science **964** (2022) 012009 doi:10.1088/1755-1315/964/1/012009

January has the lowest average temperature of the year (27.3°C). The temperature then gradually increases, reaching 30.4°C in April. This is the highest temperature of the year. Then the temperature gradually decreases from April to July (28.1°C) and then increased, peaking in September (28.8°C) and then continuing to decrease until December (Figure 3).

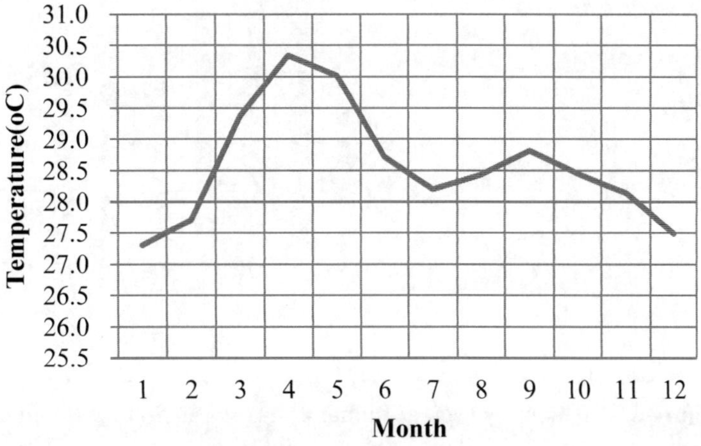

Figure 3. The monthly average temperature (°C) at Tan Son Hoa Station for the period of 1980-2019

In the period from 1980 to 2019, the annual rainfall at Tan Son Hoa Station tends to increase by 6.03 mm per year (r^2=0.669). The average annual rainfall was 1951 mm. The year with the lowest rainfall (1321mm) was 2002 and the year with the highest rainfall (2662 mm) was 2000 (Figure 4).

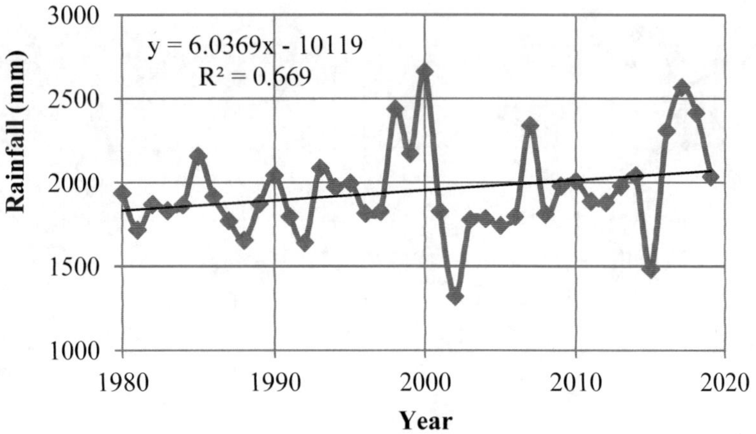

Figure 4. Trend of annual average rainfall (mm) at Tan Son Hoa station for the period of 1980-2019.

The rainy season from May to November gained an average total rainfall of 1798mm, accounting for 92% of the total annual rainfall. The dry season from December to April next year had a total rainfall of 153mm, accounting for only 8% of the total annual rainfall. The month with the highest average rainfall of the year is September (about 310mm) while the month with the lowest average rainfall of the year was February (approximately 10mm) (Figure 5).

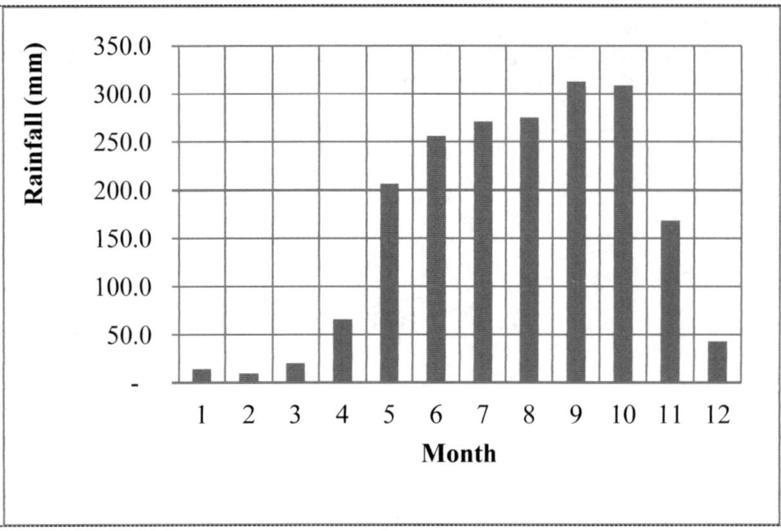

Figure 5. The monthly average rainfall (mm) at Tan Son Hoa station for the period of 1980-2019

3.2. Climate change scenario in Ho Chi Minh City

The average temperature in the base period (1986-2005) is around 26.7 - 27.6°C, the central area has a temperature of about 27.2-27.6°C, the suburban area is about 26.7-27.1°C. Thus, the central area has a temperature 0.5°C higher than the suburban area (Figure 6). The average rainfall is about 1500-1900mm, the highest in the northwest region (Cu Chi district) and decreases in the southeast region (Can Gio district) (Figure 7).

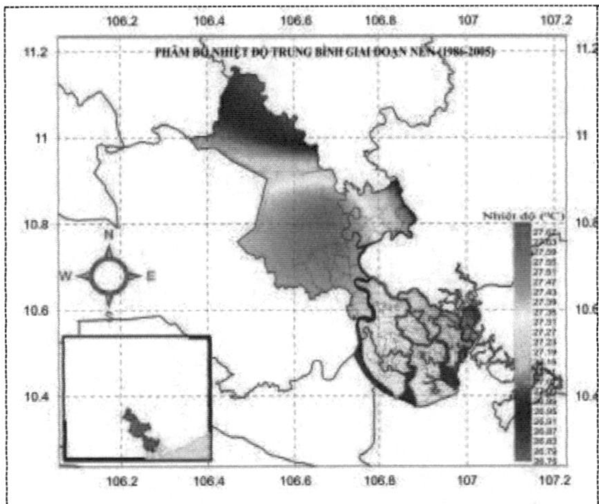

Figure 6. Spatial distribution of the annual average temperature in the base period of 1986-2005

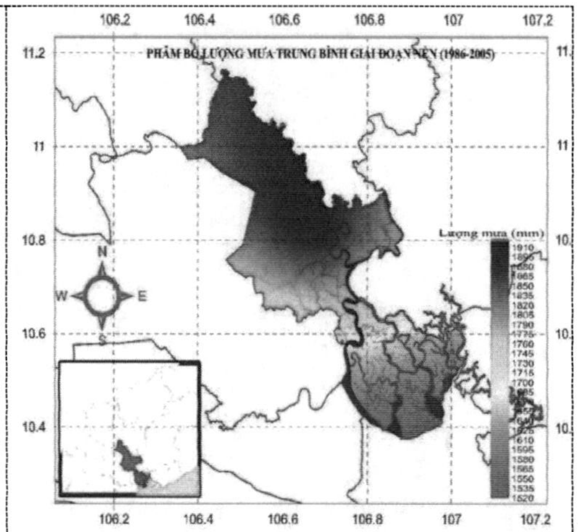

Figure 7. Spatial distribution of the annual average rainfall in the base period of 1986-2005

According to RCP 4.5 scenario, by 2030, the average temperature in HCMC will increase 0.80-0.81°C compared to the base period, the increase in the northeast region (Thu Duc City, District 12) is higher than the rest, by 2050, it will increase by 1.23-1.33°C, and by 2100, it will increase by 1.75-1.88°C, the increase in Can Gio district area is lower than the rest (Figure 8, Figure 9 and Figure 10).

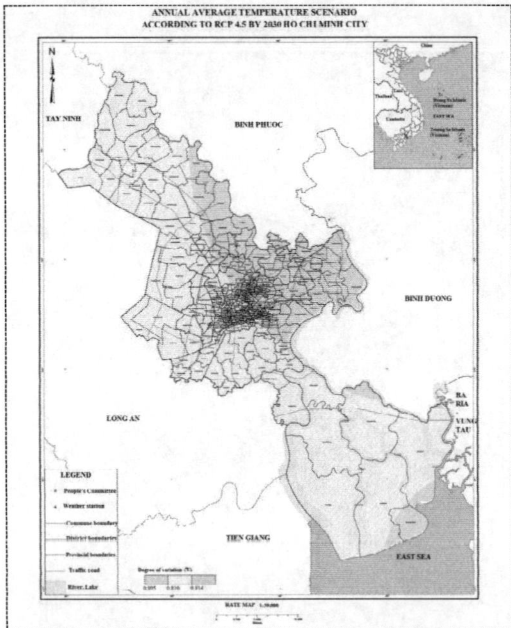

Figure 8. Annual average temperature scenario according to RCP 4.5 by 2030.

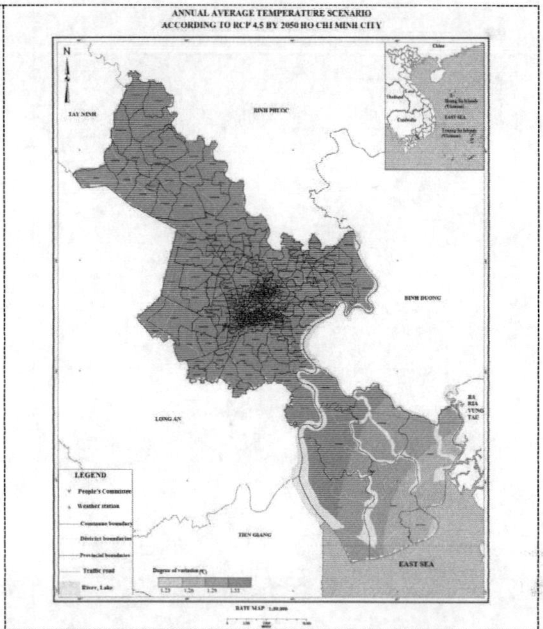

Figure 9. Annual average temperature scenario according to RCP 4.5 by 250.

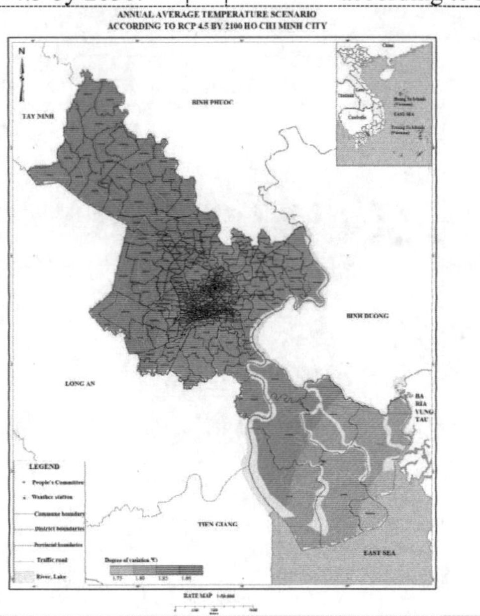

Figure 10. Annual average temperature scenario according to RCP 4.5 by 2100.

According to the RCP 8.5 scenario, by 2030, the temperature increase 0.92-0.98°C, by 2050, the temperature increase 1.55-1.68°C and by 2100, the temperature increase 3.20-3.55°C compared to the baseline period, the increase in Can Gio district area is lower than the rest (Figure 11, Figure 12 and Figure 13).

IOP Conf. Series: Earth and Environmental Science **964** (2022) 012009 doi:10.1088/1755-1315/964/1/012009

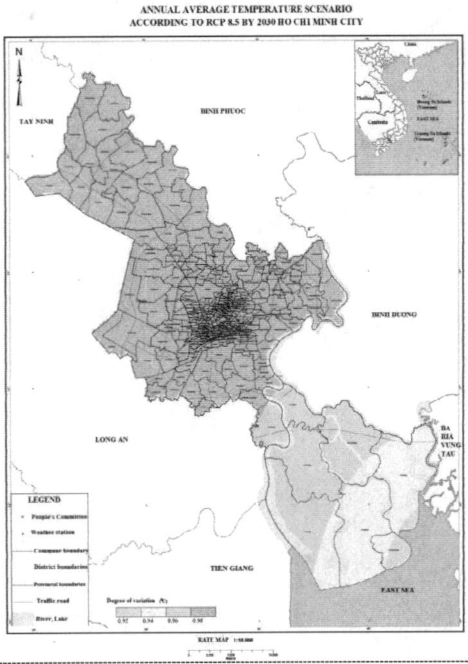

Figure 11. Annual average temperature scenario according to RCP 8.5 by 2030.

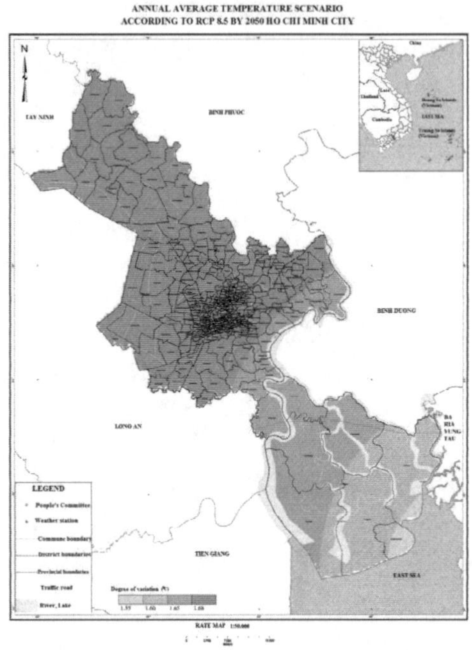

Figure 12. Annual average temperature scenario according to RCP 8.5 by 2050

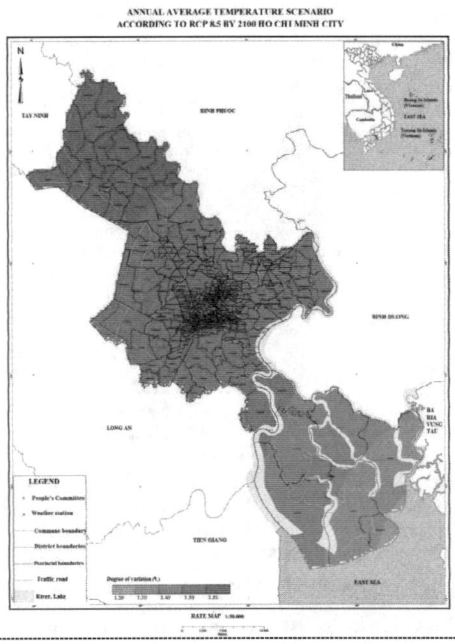

Figure 13. Annual average temperature scenario according to RCP 8.5 by 2100

According to the RCP 4.5 scenario, by 2030, the city-wide rainfall increase by 12-21%, by 2050, the rainfall increase by 13-15% and by 2100, the rainfall increase by 18-22% compared to the base period, the highest change in the region. Southeast (Can Gio district area).

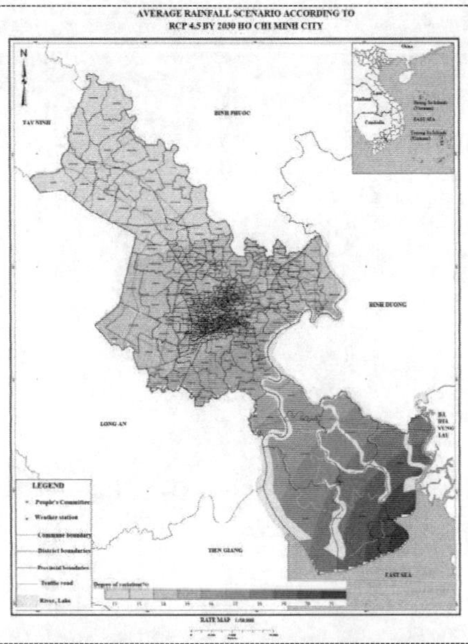

Figure 14. Average rainfall scenario according to RCP 4.5 by 2030.

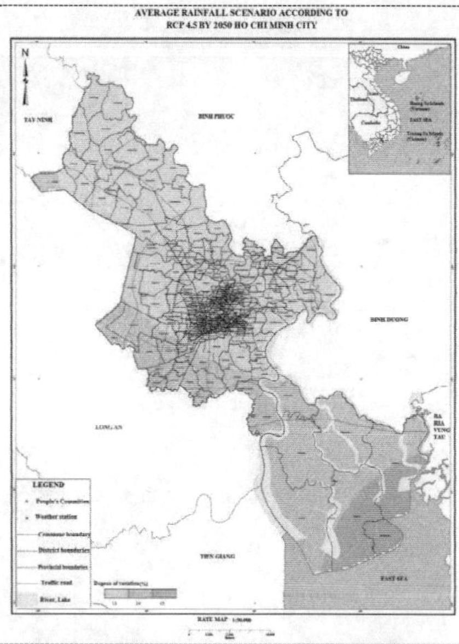

Figure 15. Average rainfall scenario according to RCP 4.5 by 2050.

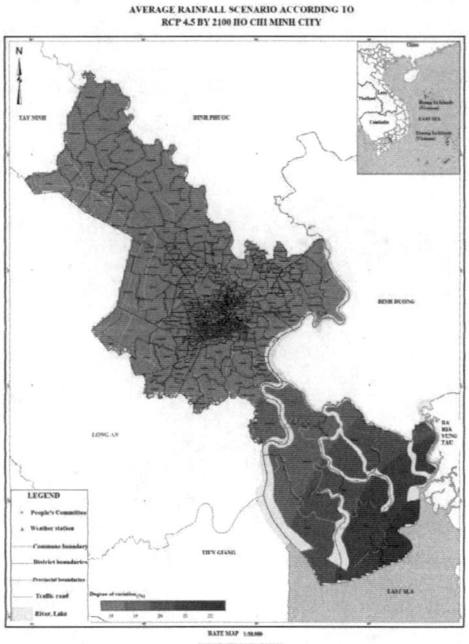

Figure 16. Average rainfall scenario according to RCP 4.5 by 2100.

According to the RCP 8.5 scenario, by 2030, the city-wide rainfall will increase in the range of 12-17%, by 2050, the rainfall will increase from 15% to 17% and by 2100, the rainfall will increase from 20% to 21% compared to the baseline period in which the highest change will appear in the Southeast region (Can Gio district area).

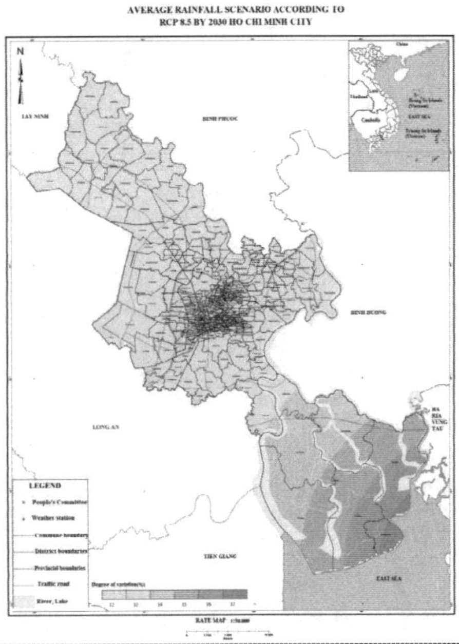

Figure 17. Average rainfall scenario according to RCP 8.5 by 2030.

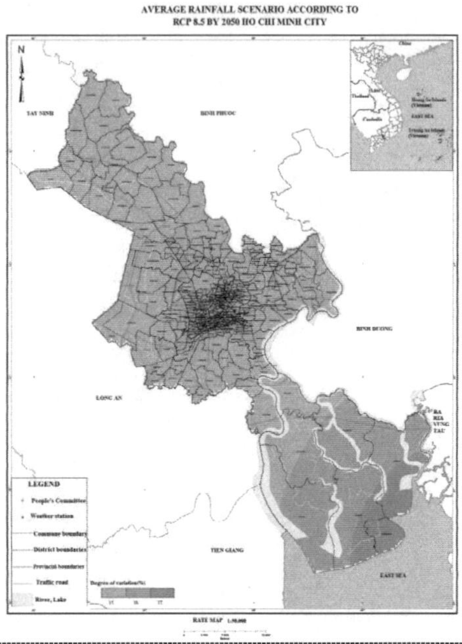

Figure 18. Average rainfall scenario according to RCP 8.5 by 2050.

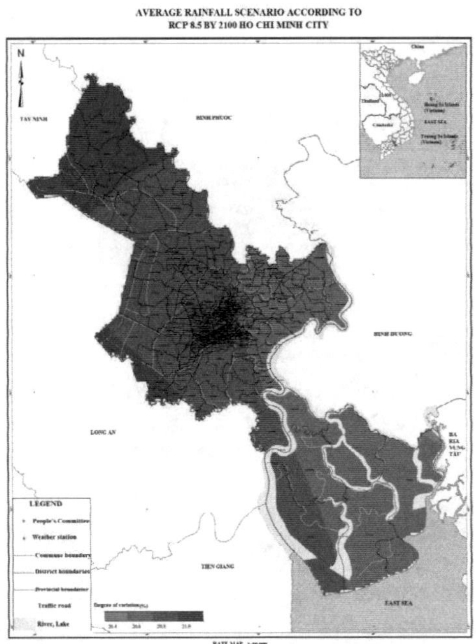

Figure 19. Average rainfall scenario according to RCP 8.5 by 2100.

4. Conclusions

The climate of HCMC has changed in recent years. For the period of 1980-2019, the average annual temperature was 27.9°C and its trend increased by 0.024°C/year. The average temperature of April was the highest, however the temperature of January was the lowest. The annual rainfall was 1951 mm, and its trend increased by 6.03mm/year. The rainy season between from May to November accounted for 92% while the dry season from December to April next year gained only 8% of the total annual rainfall.

According to the RCP 4.5 scenario, by 2030, the temperature in the whole city will increase from 0.80-0.81°C, by 2050, the temperature will increase from 1.23-1.33°C and by 2100, the temperature will increase from 1.75 °C to 1.88°C compared to the base period. The increase in the northern region will be higher than the rest. According to the RCP 8.5 scenario, by 2030, the temperature in HCMC will increase from 0.92 °C to 0.98°C, by 2050, the temperature will increase from 1.55°C to 1.68°C and by 2100, the temperature will increase in the range of 3.20-3.55°C compared to the base period.

According to the RCP 4.5 scenario, the city-wide average rainfall will increase by 12-21% in 2030, by 13-15% in 2050 and by 18-22% in 2100 compared to the base period. According to the RCP 8.5 scenario, the average rainfall will increase in the range of 12-17% in 2030, increase from 15-17% in 2050 and increase by 20-21% in 2100 compared to the base period. For both scenarios, the highest change will take place in the Southeast region (Can Gio District).

References

[1] Ministry of Natural Resources and Environment 2016 *Climate change and sea level rise scenarios for Vietnam,* Vietnam Publishing House of Natural Resources, Environment and Cartography.

[2] Sub-Institute of Hydrometeorology and Climate Change 2019 *Assessing characteristics and developments of meteorological and hydrological factors in south of Vietnam*.

[3] Tuan Ngoc Le 2017 *Research and update climate change scenarios of Ho Chi Minh City according to the new methodology and scenarios of the Intergovernmental Panel on Climate Change (IPCC) and the Ministry of Natural Resources and Environment*, Ho Chi Minh City Department of Science and Technology

[4] National Centre for Hydro-Meteorology Forecasting 2019 *Updated Report on ENSO and Hydrometeorological status.*

[5] Van Den Hurk B J J M, Siegmund P and Klein Tank A 2014 *KNMI 14: Climate Change scenarios for the 21st Century*, A Netherlands perspective: Scientific Report WR2014-01, KNMI, De Bilt, The Netherlands.

[6] Asian Development Bank 2010 *Ho Chi Minh City Adaptation to Climate Change: Summary Report*, ISBN 978-92-9092-108-0, Publication Stock No. RPT101580, The Philippines.

Health risk assessment of arsenic in drinking groundwater: A case study in a central high land area of Vietnam

Mai Nhu Hoang[1,2], Phu Le Vo[1,2*], Trong Vinh Bui[2,3], Pham Hung[1,4], Quang Khai Ha[1,2]

[1] Faculty of Environment and Natural Resources, Ho Chi Minh City University of Technology (HCMUT), 268 Ly Thuong Kiet St., Dist. 10, Ho Chi Minh City, Vietnam

[2] Vietnam National University Ho Chi Minh City, Linh Trung Ward, Thu Duc District, Ho Chi Minh City, Vietnam

[3.] Faculty of Petroleum and Geological Engineering, Ho Chi Minh City University of Technology (HCMUT), 268 Ly Thuong Kiet St., Dist. 10, Ho Chi Minh City, Vietnam

[4.] Department of Natural Resources and Environment, Lam Dong province, Dalat City, Vietnam

Corresponding author: volephu@hcmut.edu.vn

Abstract. Arsenic contaminated groundwaters is a global environmental issue which cause serious problems for human health risks. 188 groundwater samples were collected in private wells of Lam Dong Province, a central highland area, Vietnam to investigate the health risks to the local people by using arsenic contaminated groundwater for drinking purpose. The result showed that the arsenic concentration is average of 14 μg/L and maximum of 500 μg/L. About 12% out of the total groundwater samples have arsenic concentration exceeded that value of 10 μg/L recommended for drinking water by World Health Organization (WHO, 2019). The health risk assessment showed that hazard quotient (HQ) value for adults was up to 60.6 with an average of 1.7 and about 14% of total samples show the HQ values greater than 1. The HQ value for children is average of 4.7 (maximum of 166.7) and about 23% of total groundwater samples show HQ > 1 for children. Cancer risk (CR) values were up to $27x10^{-4}$ (average of $8x10^{-4}$) for adults and $75x10^{-4}$ (average of $21x10^{-4}$) for children. About 26% and 29% of out of the total samples show CR value for adult and children greater than the CR (1×10^{-4}) proposed by the USEPA. The result also indicated that the consumption of arsenic contaminated groundwater may seriously damage the human health. Therefore, groundwater in the area needs to be treated for arsenic removal before drinking to minimize the adverse effect on local communities' health.

Keywords: Arsenic, groundwater quality, health risk assessment, Lam Dong province, central high land area

1. Introduction

Arsenic (As) is one of ten chemicals that considered as the major public health concern by the World Health Organization (WHO). As poisoning in sort period of time can cause muscular weakness and muscle cramps, nausea, vomiting, and diarrhea [1]. Long term exposed to arsenic can lead to diseases like cancers, skin lesions, diabetes, hypertension, etc. Arsenic is a potent genotoxic agent for animals and humans that can damage DNA, induces chromosomal aberrations, sister chromatid exchange and micronuclei formations [1]. As is a natural chemical of the Earth's curst and mobilise to water resources by biogeochemical processes [2,3]. Using arsenic rich groundwater for drinking and domestic use is widely known as the most common reason for long-term exposure.

Worldwide, million people are exposed to arsenic due to using As-contaminated groundwater for daily needs [4]. The current recommended limit of arsenic in drinking-water is 10 µg/L [5]. As rich groundwater (As >10 ug/L) is widely observed in many parts of the world such as Australia [6], Bangladesh [4], China [7], India [8], USA [9], South Korea [10] and Vietnam [11]. Flanagan et al. [4] estimated that about 20 million and 45 million people in Bangladesh were exposed to As concentration greater than 50 µg/L and 10 µg/L respectively and approximately 24,000 deaths annually in the country due to the consumption of As rich groundwater.

Arsenic rich groundwater is also commonly observed in Mekong Delta [11], Red River Delta [12] of Vietnam. Geological condition such as rich organic matter in sediment, low Eh in groundwater is the main cause for As mobilisation in these areas [13]. Consequently, As concentration is up to 1,300 ug/L in groundwater of the Mekong Delta and 3,000 ug/L in groundwater from the Red River Delta [14]. Because groundwater is very important sources for drinking and irrigation purposes, million people in Mekong Delta [15] and Red River Delta could be exposed to As. Berg et al. (2007) observed a significant concentration of As in hair and nails collect from people drinking As rich groundwater in Mekong Delta [15].

High Arsenic concentration up to 800 ug/L was also observed in Lam Dong Province, a central high land area of Vietnam [16,17,18]. Groundwater is also very important for water supply in the area. Several studies have conducted to investigate As concentration in some districts such as Cat Tien [16,17], Da Teh [18] of Lam Dong Province. They found that As rich groundwater is observed even in sedimentary rock aquifers and associate with negative Eh environment. In this study, we will estimate the health risk of resident to As contaminated groundwater. This information will be valuable to water policy makers and planner to find the measures for avoiding health problems of residents.

2. Study area

The study area comprises four Southern districts of Lam Dong province in the central highland region of Vietnam (Figure 1), namely Cat Tien, Da The, Da-Houai, and Bao Loc. The total area is about 9,700 km^2 and population of 1.6 million people. The elevation varies from 500 to 2,300 m above the mean seawater level. This is a mountainous province with a tropical climate characteristic, in which the rainy season starts from April to November and the dry season lasts from December to March in the following year. Annual rainfall is about 1,600 to 2,700 mm. Temperature varies from 16 to 23°C. Land use is mostly forest (up to 70% of total land) and some scatter of urban, resident, and agricultural areas. Lam Dong has a dense network of rivers and streams with an average density of 0.6 km/km^2. Total area of rivers and streams is about 9232 ha. Lam Dong also has a reservoir system with a large water reserve serving people's demands for domestic, irrigation, agriculture, and tourism.

Geological characteristic is dominated by basalt and other rocks such as sedimentary rocks and intrusive rocks. The unconsolidated sediment of Neogene and Quaternary was found in small parts in Cat Tien, Da Teh and Bao Loc. Hydrogeology in Lam Dong province is divided into seven types of aquifers, including Holocene sediments (Qh), Neogene sediments (N_1^3-N_2^1đl), Pleistocene basalt rocks (βQ_1), Pliocene – Pleistocene basalt rocks ($\beta N_2 - Q_1$), Pliocene sedimentary rocks (βN_2), Creta sedimentary rocks (K_2) and Jura sedimentary rocks (J_{1-2}). Groundwater is abstracted from dug well installed in shallow depth (i.e. several meter in depth) and drilled wells installed to deep aquifers (up to 70 m in depth).

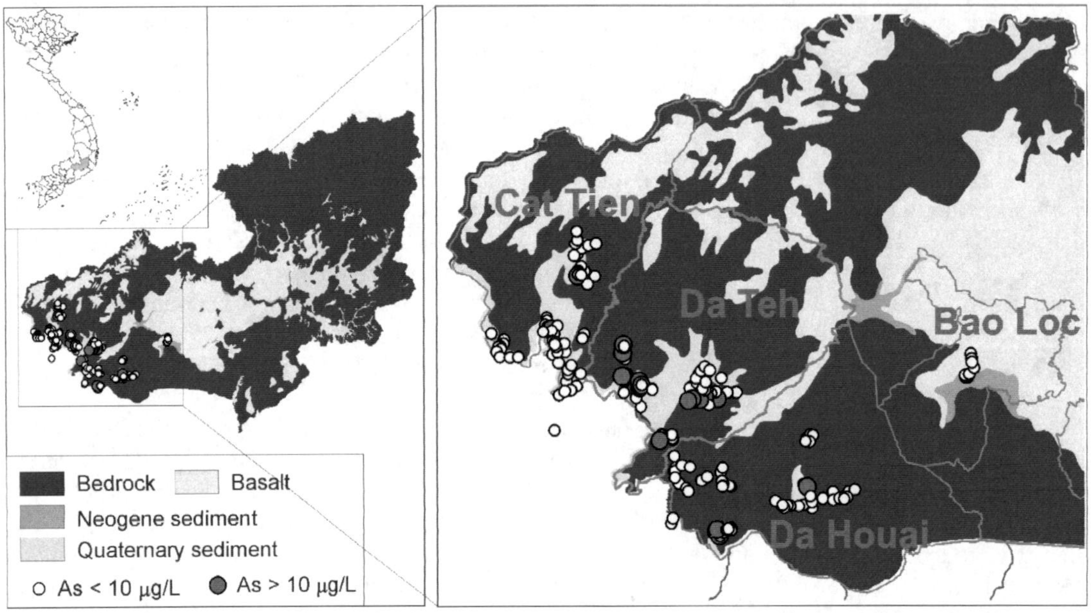

Figure 1. Geology map of Lam Dong province and As concentration in groundwaters

3. Methods

3.1. Dataset

This study utilised dataset that includes information of groundwater samples (As concentration, pH) and wells (depth, types, location) and water treatment tools for drinking purpose of 188 groundwater samples collected from private wells. The groundwater samples and related information were collected by Department of Natural Resources and Environment of Lam Dong Province in 2007 from the Southern area, including Cat Tien, Da Huoai, Da Teh, and Bao Loc districts of Lam Dong province. The location of groundwater samples is shown in Figure 1.

3.2. Health risk assessment

The risk assessment for human health of As in groundwater through drinking and domestic purposes for two groups of residents: children and adults.

Dose assessment and exposure assessment:

The average daily dose (ADD) is calculated as equation (1).

$$ADD=(C\times IR\times EF\times ED)/(AT\times BW) \quad (1)$$

Where: ADD is average daily dose from ingestion (mg/kg.day); C is arsenic content in groundwater samples (mg/L); IR is water intake rate, 1 L/day for children and 2 L/day for adults (Nguyen et al., 2020); EF is exposure frequency (365 days/year); ED is exposure duration, 10 years for children and 70 years for adults. AT is the averaging time for carcinogenic exposure, 3,650 days for children and 25,550 days for adults. BW is the body weight that equals 10 kg for children and 55 kg for adults.

The hazard quotient (HQ) is estimated as following equation 2:

$$HQ=ADD/RfD \quad (2)$$

In which HQ is hazard quotient (unitless); RfD is the reference dose (0.0003 mg/kg.day) [19]. HQ is calculated for non-carcinogenic exposure. If HQ is greater than 1, a non-carcinogenic risk is considered to be possible [19]. On the contrary, no health effects are expected.

The carcinogenic risk (CR) is calculated as equation 3:

$$CR = CSF \times ADD \qquad (3)$$

In which CSF is the cancer slope factor for arsenic (1.5 mg/kg.day) [14]. The highest safe standard for carcinogenic risk is 10^{-4} [19].

4. Results and discussion

4.1. Occurrence of Arsenic rich groundwaters

Generally, arsenic concentration in groundwater samples is up to more than 500 ug/L. About 12% of the total groundwater samples (n=188), As concentration is greater than that values in drinking water guidelines of WHO, 2017 (10 ug/L for As). The high Arsenic concentration is generally observed in Da-Teh district, and some sites in Cat Tien and Da-Huoai districts (Figure 2). About 94% of total investigated wells (n=188) groundwater was pumped for drinking and domestic uses directly without any treatment. Only 7% of the wells' owners simply filter pumped water by sand treatment before using.

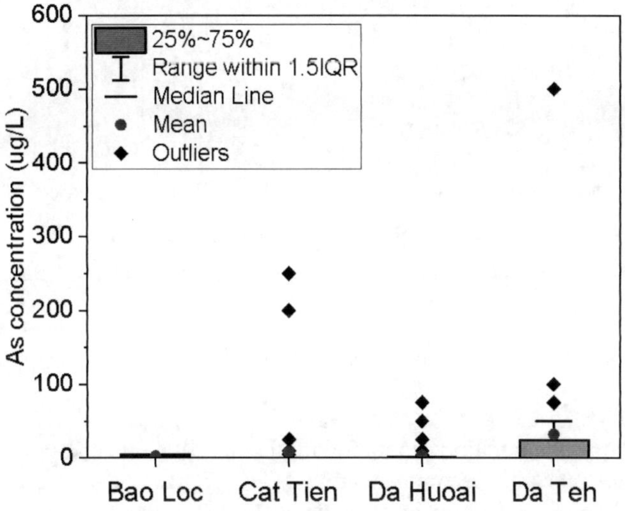

Figure 2. As concentration in groundwater samples from Bao Loc, Cat Tien, Da Huoai, and Da-Teh districts

The groundwater pH varied from 5 to 8 (Figure 3a). High Arsenic concentration generally observed when pH of groundwater is from 6 to 8. The As concentration in groundwater collected from drilled wells (average of 54 ug/L) are much greater than in dug wells (average of 5 ug/L). Generally, As concentration increases as a function of well depth (Figure 3b). The highest concentration of As is observed in the well depth of 40 and 60 m below land surface. This observation is consistent to the investigation of Nguyen and Nguyen [17] and Nguyen and Truong [18] in Cat Tien and Da Teh districts, respectively. These studies show that the highest As concentration in groundwater samples collected from Cat Tien (400 ug/L) and Da Teh districts (820 ug/L) were observed at the well depth of 30 and 180 m below land surface, respectively. Studies in Da-Teh and Cat Tien show that high As rich groundwater occurred in both unconsolidated sedimentary and sedimentary rocks and associated to negative Eh values, high pH (6-8), iron and NH_4 concentrations. The As concentrations in core samples collected in Da Teh [18] and Cat Tien districts [16] were up to 11.54 mg/kg and 11.58 mg/kg, respectively for unconsolidated sediment and up to 28 mg/kg and 154 mg/kg, respectively for sedimentary rocks.

Kim et al. [10] also observed low As concentration in shallow unconsolidated sediment aquifers and higher As concentration in deeper depth wells installed in bedrock aquifers from agricultural area of

South Korea. Kim et al. [10] also suggested that shallow groundwater is influenced by agricultural activities that induced oxidising environment; therefore, groundwater in shallow groundwater have higher Eh levels than deep aquifers. In oxidising environment, As can be easily absorbed on surface of iron (hydro)oxides minerals that could lower As concentration in shallow groundwater. In deep wells, groundwater is may under reducing environment, this is a favourable condition for As mobilisation. As concentration is high when groundwater Eh is negative Eh and high pH (6-8) were commonly observed in As rich groundwater from many area worldwide such as Mekong Delta [11], Red River Delta [12], Bangladesh [4]. Reducing environment and high pH induce desorption of As from iron (hydr)oxides surface [2, 3].

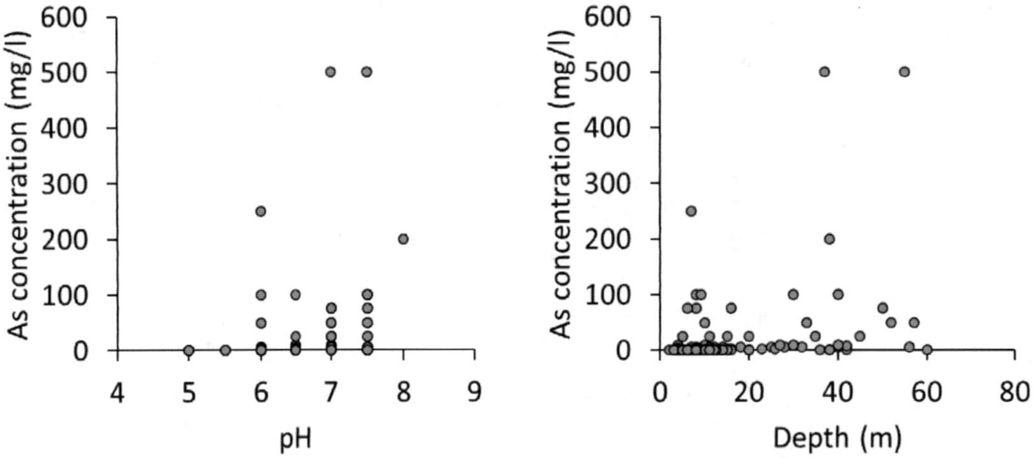

Figure 3. As in groundwater as a function of pH (a) and depth (b)

4.2. Human health risk assessment

The extremely high As concentration (up to 500 mg/L) in groundwater from the study area is 50 folds greater than drinking water limit recommended by WHO [5] (for As 10 ug/L). Most of local residents using the groundwater directly without any treatment. This suggested that the As rich groundwater may threat health of people in the area. The average daily dose (ADD), hazard quotient (HQ), and carcinogenic risk (CR) can be used for health risk assessment. The results indicate that the highest and the lowest values of average CR levels were observed in Da Teh (5.1×10^{-4} for adults and 14×10^{-4} for children) and in Bao Loc district (1.4×10^{-4} for adults and 3.9×10^{-4} for children), respectively. The average CR was already greater than the value of CR 1×10^{-4} recommended by the US EPA [19]. About 70%, 8.3%, 18.3%, and 42.2% CR for Adults and 70%, 8.3%, 26.7% and 45.8% CR for Children in Bao Loc, Cat Tien, Da Huoai and Da Teh districts were higher than CR of 1×10^{-4}, respectively.

Table 1. HQ and CR for Adult and Children from Bao Loc, Cat Tien, Da Huoai and Da Teh districts.

	Bao Loc (n=10)			Cat Tien (n=60)			Da Huoai (n=60)			Da Teh (n=59)		
	Mean	**SD**	**(%)**	**Mean**	**SD**	**(%)**	**Mean**	**SD**	**(%)**	**Mean**	**SD**	**(%)**
HQ_adults	0.32	0.27	0	0.99	4.97	6.7	0.46	1.49	11.7	3.92	11.3	27.1
HQ_Child	0.87	0.74	40%	2.72	13.7	8.3	1.25	4.09	16.7	10.79	31.1	42.4
CR_adults ($x10^{-4}$)	1.4	1.2	70%	4.4	22.4	8.3	2.1	6.7	18.3	17.7	50.9	42.4
CR_Children ($x10^{-4}$)	3.9	3.3	70%	12.3	61.5	8.3	5.6	18.4	26.7	48.5	139.9	45.8

Notes: SD is standard deviation; (%) indicates the percentage of total sample exceeding the US EPA standard for CR and HQ [19].

5. Conclusion

Arsenic contamination in groundwater has been considered as an alarming issue in the Southern area of Lam Dong province with the highest contaminated level of 500 μg/L, 50 times higher than the WHO guidelines for arsenic content in drinking water. 11.7% of total (n=188) samples exceeded the drinking water limit for As (10 ug/L). Among four surveyed areas, Da Teh district were much more polluted than other districts. The high level of arsenic would threaten to human health of residents consuming groundwater for drinking and living activities. The health risk assessment showed that 25.5% and 29.3% of the carcinogenic risk values for adults and children exceeded the highest acceptable value of (10^{-4}) proposed by US EPA. In addition, the average CR value for children was higher than these for adults. These results are warning about the arsenic polluted level in groundwater and the communities' health problems in the study area. The dataset also indicated that 93% of surveyed wells, the groundwater was pumped directly without any treatment and 7% of wells, pumped groundwater was simply treated by sand filtering. This way of using groundwater is unsafe for human health. Therefore, treatment technologies for arsenic removal should be applied for preventing the arsenic exposure and reducing the risk of diseases and cancer in local residents. Simultaneously, it is necessary to monitor the groundwater quality in the area.

Acknowledgement
We acknowledge the support of time and facilities from Ho Chi Minh City University of Technology (HCMUT), VNU-HCM for this study.

References

[1] Mitra A., Chatterjee S., and Gupta, D.K. 2020. Environmental Arsenic Exposure and Human Health Risk. In: Fares A., Singh S. (eds) Arsenic Water Resources Contamination. Advances in Water Security. Springer, Cham https://doi.org/10.1007/978-3-030-21258-2_5

[2] Smedley, P.L. and Kinniburgh, D.G., 2002. A review of the source, behaviour and distribution of arsenic in natural waters. *Applied geochemistry* **17**(5) 517-568.

[3] Amini, M., Abbaspour, K.C., Berg, M., Winkel, L., Hug, S.J., Hoehn, E., Yang, H. and Johnson, C.A. 2008 Statistical modeling of global geogenic arsenic contamination in groundwater. *Environmental Science & Technology* **42**(10) 3669-3675.

[4] Flanagan, S.V., Johnston, R.B. and Zheng, Y. 2012 Arsenic in tube well water in Bangladesh: health and economic impacts and implications for arsenic mitigation. *Bulletin of the World Health Organization* **90** 839-846.

[5] WHO 2011. Guidelines for Drinking-Water Quality, 4th ed. World Health Organization, Geneva, Switzerland.

[6] Appleyard, S.J., Angeloni, J. and Watkins, R. 2006 Arsenic-rich groundwater in an urban area experiencing drought and increasing population density, Perth, Australia. *Applied geochemistry* **21**(1) 83-97.

[7] Wang, H.Y., Göttlicher, J., Byrne, J.M., Guo, H.M., Benning, L.G. and Norra, S. 2021 Vertical redox zones of Fe–S–As coupled mineralogy in the sediments of Hetao Basin–Constraints for groundwater As contamination. *Journal of Hazardous Materials* **408** 124924.
[8] Lone, S.A., Jeelani, G., Mukherjee, A. and Coomar, P. 2021 Arsenic fate in upper Indus river basin (UIRB) aquifers: Controls of hydrochemical processes, provenances and water-aquifer matrix interaction. *Science of the Total Environment* 148734.
[9] Malakar, A., Singh, R., Westrop, J., Weber, K.A., Elofson, C.N., Kumar, M. and Snow, D.D. 2021 Occurrence of arsenite in surface and groundwater associated with a perennial stream located in Western Nebraska, USA. *Journal of Hazardous Materials* **416**, 126170.
[10] Kim, K., Kim, S.H., Jeong, G.Y. and Kim, R.H. 2012 Relations of As concentrations among groundwater, soil, and bedrock in Chungnam, Korea: Implications for As mobilization in groundwater according to the As-hosting mineral change. *Journal of hazardous materials* **199** 25-35.
[11] Ha, Q.K., Kim, K., Phan, N.L., Phung, T.H., Lee, J., Nguyen, V.K. and Phan, C.N. 2019 A hydrogeological and geochemical review of groundwater issues in southern Vietnam. *Geosciences Journal* **23**(6) 1005-1023.
[12] Glodowska, M., Stopelli, E., Straub, D., Duyen Vu Thi, Pham TK Trang, Pham H. Viet, Berg, M., Kappler, A. and Kleindienst, S. 2021 Arsenic behavior in groundwater in Hanoi (Vietnam) influenced by a complex biogeochemical network of iron, methane, and sulfur cycling. *Journal of Hazardous Materials* **407**(2021) 124398.
[13] Buschmann, J., Berg, M., Stengel, C., Winkel, L., Sampson, M.L., Trang, P.T.K., and Viet, P.H. 2008 Contamination of drinking water resources in the Mekong delta flood plains: arsenic and other trace metals pose serious health risk to population. *Environment International* **34** 756–764.
[14] Nguyen, V.T., Tran, T.D., Nguyen, T.B., Dang, B.T. and Bui, X.T. 2020 Arsenic-contaminated groundwater and its potential health risk: a case study in Long An and Tien Giang provinces of the Mekong Delta, Vietnam. *Environmental Science and Pollution Research* 1-14.
[15] Berg, M., Stengel, C., Trang, P.T.K., Viet, P.H., Sampson, M.L., Leng, M., Samreth, S. and Fredericks, D., 2007. Magnitude of arsenic pollution in the Mekong and Red River Deltas—Cambodia and Vietnam. *Science of the Total Environment* **372**(2-3) 413-425.
[16] Nguyen, D.T. 2016 Origin of groundwater As contamination in Cat Tien District, Lam Dong Province. *Science & Technology Development Journal* **1** 101-112.
[17] Nguyen, D.T. and Nguyen, D.T. 2016 Assessment of As concentration in surface and groundwater in Cat Tien District, Lam Dong province. *Science & Technology Development Journal* **4** 99-107.
[18] Nguyen, D.T. and Truong, D.P. 2015 Contamination of Arsenic in groundwater of Da Teh district, Lam Dong Province. *Scinence Journal of Ho Chi Minh City University of Education* **12** 92-101.
[19] US EPA 2005 Guidelines for carcinogen risk assessment. US Environmental Protection Agency, Washington, DC.

Assessing the distribution and variation characteristics of marine primary productivity in the coastal marine area of Vietnam South Centre

Nguyen Trinh Duc Hieu[1,2,3], **Nguyen Huu Huan**[1,4], **Tran Thi Van**[2,3*], **Nguyen Phuong Lien**[5]

[1] *Institute of Oceanography, Vietnam Academy of Science and Technology (VAST)*
[2] *Faculty of Environment and Natural Resources, Ho Chi Minh City University of Technology, (HCMUT), 268 Ly Thuong Kiet Street, District 10, Ho Chi Minh City, Vietnam*
[3] *Vietnam National University Ho Chi Minh City, Linh Trung Ward, Thu Duc District, Ho Chi Minh City, Vietnam*
[4] *Graduate University of Science and Technology, Vietnam Academy of Science and Technology (VAST)*
[5] *Coastal Branch, Vietnam – Russia Tropical Centre*

*tranthivankt@hcmut.edu.vn

Abstract. Primary production (PP) of phytoplankton plays an essential role in food web dynamics, biogeochemical cycles and marine fisheries. It is used as one of the basic information for evaluating marine ecosystems. In this paper, monthly composite PP data on a 4 km x 4 km grid for the period 2003-2020 was used to evaluate the distributional characteristics of PP in the coastal marine area of Vietnam South Centre. The statistical results show that the climatological average of PP in 18 years reached 449.2 $mgC/m^2/day$, ranged from 272.1 to 14,205.4 $mgC/m^2/day$. The PP has seasonal and spatial variations. In time, the lowest value of PP was in spring, and the highest was in winter; in space, PP tended to decrease from shore to offshore, PP was higher in coastal areas than in the open sea areas. During the northeast monsoon season, PP increased by more than 1000 $mgC/m^2/day$ in the coastal area. Meanwhile, in the southwest monsoon season, due to the ecological influence of the upwelling phenomenon, PP increased with a value greater than 1500 $mgC/m^2/day$, distributed along the coastline of Ninh Thuan - Binh Thuan. Primary productivity positively correlated with chlorophyll content but negatively correlated with sea surface temperature with correlation coefficients of 0.9 and -0.6, respectively. There was a weak correlation between PP and ONI with correlation coefficients of -0.23. The temporal-spatial variation of PP was affected by the ENSO (El Niño-Southern Oscillation) phenomenon, the positive phase of ENSO (El Niño conditions) corresponded to lower PP, and the negative phase of ENSO (La Niña conditions) corresponded to higher PP. The research results from this paper can be used as a reference in marine ecosystem management.

1. Introduction

Ocean primary production (PP) plays an essential role in the global carbon cycle, and the earth environmental system constitutes roughly half of the biosphere net PP [1]. The value of PP is used as an indicator of environmental quality and ecological conditions in seawater [2, 3]. Besides that, primary productivity is the base of the food web in marine ecosystems, and it influences the nature of marine food webs and the abundance of marine organisms. It is the ultimate source of energy for all organisms in an ecosystem [4, 5]. However, long-term variability in PP has not been explored [6]. Traditional ship-based in situ measurements are limited in their ability to capture PP large scale spatial and temporal dynamics and are time-consuming and expensive [7, 8]. While remote sensing data can use to observe the dynamics of the ocean surface, providing fundamental means for estimating oceanic PP on large spatiotemporal scales [9]. One of the essential applications of ocean colour data has been in the PP computation at large scales, using remotely sensed fields of phytoplankton biomass [4]. Today, PP in global and regional scales can be estimated by combine remote sensing techniques and mathematic modelling. Many satellite-based PP models have been used to calculate PP in the ocean in recent years, and they are classified into 3 categories: chlorophyll-based, carbon-based and phytoplankton absorption-based models. The Vertically Generalized Production Model (VGPM) is a chlorophyll-based model whose photosynthesis rate is expressed as a function of water depth and photosynthetically active radiation, and it is used to used to estimate regional and global ocean PP [10].

Several studies have used remote sensing data to assess the distribution and variation of primary productivity in the East Sea (ES). Tan and Shi, 2009 [11] examined the spatio-temporal variability of ocean PP in the ES for 1998-2006 using the VGPM model. Liao et al., 2012 [12] investigated the interannual variations of PP associated with two different types of Niño in the ES using more than ten years of satellite-derived PP data. Zhou et al., 2017 [13] estimated the primary and the average monthly data of marine primary productivity in 8 days, a detailed analysis of the differences in the 2006-2015 PP in the traditional fishing areas in the ES on the interannual and seasonal variation and spatial distribution. Xu et al., 2017 [14] used Moderate Resolution Imaging Spectroradiometer (MODIS) data to evaluate spatio-temporal variation of PP for the open ES. These studies have shown that peaks in average PP in the ES appeared in winter and summer, with the winter peak more significant than the summer peak [11]. ENSO and physicochemical environmental conditions influence PP spatial and temporal variability in the ES [11, 12]. The spatial distribution of PP in ES also indicated that the PP values are high in shallow offshore areas and then decreasing with the increasing distance away from the coast [14]. The VGPM model was suitable for PP estimation in the central-western East Sea [13].

Using remote sensing data to calculate the distribution of primary productivity in the coastal marine area of Vietnam South Centre was first published by Nguyen Tac An and Tong Phuoc Hoang Son in 2004 [3]. They used ocean colour images to calculate the distribution of primary productivity in the upwelling region in the Southcentral region of the Vietnam sea. The studied results showed that the distribution trend of primary productivity we have from remote sensing technique is similar to other study results and direct measurements in this sea region. In 2012 and 2014, Phan Minh Thu et al. used MODIS Aqua data to evaluate marine primary productivity in the Vietnam sea. The result showed that the upwelling phenomenon impacted PP distribution suspended sediment from Mekong River nutrients supporting human activities [15, 16]. These studies also indicated that remote sensing data could potentially estimate primary productivity in Vietnamese marine regions.

The coastal marine area of Vietnam South Centre is an area of particular importance to Vietnam marine economy. In addition, this area is affected by upwelling during the southwest monsoon season [17]. During upwelling occur, water that rises to the surface is typically colder and rich in nutrients. These nutrients are used by phytoplankton to increase the primary productivity of the oceanic region. So reasonable fishing grounds typically are found where upwelling is common. Therefore, using remote sensing data sources to monitor the distribution and PP fluctuations in this area is a practical matter to be implemented.

In this paper, empirical spatio-temporal mean and cross-correlation function analysis was conducted to accomplish the purposes: (1) Identifying the distribution and variation characteristics of PP; (2) Determining the correlation between PP and chlorophyll-a (chl-a), sea surface temperature (SST),

ENSO phenomenon in the coastal marine area of Vietnam South Centre for a period of 18 years (2003-2020). The results of this article can be used to provide a scientific basis for socio-economic development planning and propose solutions to protect and develop the marine ecosystem sustainably

2. Study area

The study area is the coastal marine zone of Vietnam South Centre, between latitude 10.30° to 16.25°N and longitude 107.30° to 112.00°E. The studied area is shown in Figure 1. The study area is influenced by winter monsoon (northeast monsoon) and summer monsoon (southwest). The interaction of the atmospheric-oceanic processes with the bottom topography and the coast has formed a substantial upwelling area from Khanh Hoa to Binh Thuan waters during the southwest monsoon season, forming fishing grounds with high benefits in fish and molluscs [18]. During upwelling-active phases, cold water from deep towards the ocean surface, causing the SST to drop significantly [19].

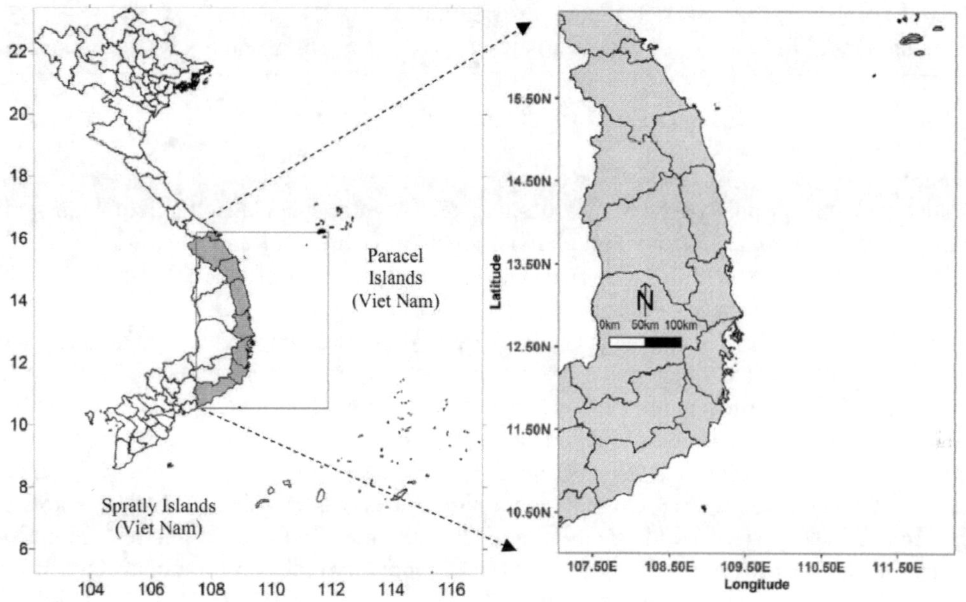

Figure 1. Map showing the study area

3. Data and Methods

The block diagram of the research method in this paper is shown in Figure 2. Accordingly, after determining the research objectives, three types of data are collected as follows: (1) PP data; (2) SST and chl-a data; and (3) Multivariate ENSO Index (MEI) and Oceanic Niño Index (ONI). The collected data source was used to carry out the following two research contents: (1) Analysis of spatial-temporal distribution characteristics; and (2) analyze the possible relationship between PP and chl-a, SST and ENSO phenomenon. To achieve the above two contents, this study uses the following two methods: (1) calculate empirical spatial and temporal mean of PP; and (2) correlation analysis.

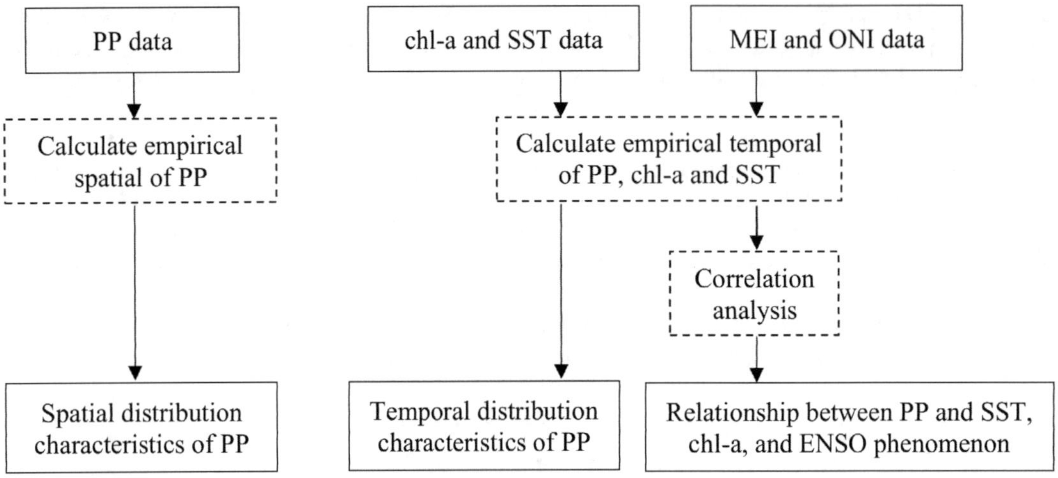

Figure 2. The block diagram of the research method

3.1. Data

In this study, monthly gridded data during 2003-2020 on PP with a spatial resolution of 4 km were from the standard Vertically Generalized Production Model by Behrenfeld and Falkowski (1997) using MODIS Aqua data, provided by National Oceanic and Atmospheric Administration – NOAA (https://coastwatch.pfeg.noaa.gov/erddap).

To discuss the relationship between the oceanographic environment and PP, we used monthly averaged chl-a and SST with 4 km x 4 km spatial resolution from Level-3 Standard Mapped Image products observed by the MODIS Aqua satellite. Chl-a concentration and SST data were provided by the National Aeronautics and Space Administration Ocean Color Web site (https://oceancolor.gsfc.nasa.gov/).

To assess the impact of ENSO on the PP over the coastal marine area of Vietnam South Centre, monthly time series of two climate indices ONI (Oceanic Niño Index) and MEI (Multivariate ENSO Index) during 2003-2020, was derived from National Oceanic and Atmospheric Administration (http://www.esrl.noaa.gov/psd/enso/). The ONI is NOAA primary indicator for monitoring El Niño and La Niña, which are opposite phases of the climate pattern called the El Niño-Southern Oscillation, or "ENSO" for short. NOAA considers El Niño conditions present when the Oceanic Niño Index is +0.5 or higher, indicating the east-central tropical Pacific is significantly warmer than usual. La Niña conditions exist when the Oceanic Niño Index is -0.5 or lower, indicating the region is cooler than usual. The MEI is the bi-monthly time series of the leading combined empirical orthogonal function of five different variables, namely, sea level pressure, sea surface temperature, the surface wind of combined zonal and meridional components, and outgoing longwave radiation, over the tropical Pacific basin (30°S-30°N and 100°E-70°W). Summary of data in this paper is shown in Table 1.

The global PP, chl-a and SST data with netCDF format were preprocessed, including reading, image cropping, calculating average in R programming language, which were cropped by the scope of the studying area.

Table 1. Summary of data in the study

Name	Product type	Time period of data	Data source
PP	4 km resolution monthly grid data	From 1/2003 to 12/2020	https://coastwatch.pfeg.noaa.gov/erddap
chl-a and SST			https://oceancolor.gsfc.nasa.gov/
MEI and ONI	Monthly time series data		http://www.esrl.noaa.gov/psd/enso/

3.2. Methods

3.2.1. Calculate empirical spatial and temporal mean

Assume for the moment that we have observations {PP (s_i; t_j) for spatial locations {s_i: i = 1, 2,......, m} and times {t_j: j = 1, 2,...T}. The empirical spatial mean for location $\mu_{z,s}$ (s_i) is then found by averaging over time by equation 1 [20]:

$$\mu_{z,s}\,(s_i) \;= \frac{1}{T}\sum_{j=1}^{T} PP\ (s_i;\ t_j) \tag{1}$$

where T is the time used to calculate the empirical spatial mean.
The empirical temporal mean for time $\mu_{z,s}$ (t_j), is given by equation 2 [20]:

$$\mu_{z,s}\,(t_j) \;= \frac{1}{m}\sum_{i=1}^{m} PP\ (s_i;\ t_j) \tag{2}$$

where m is the number of pixels of each scene.
Based on the data results of PP, the spatial monthly mean of PP were exported to raster and then mapping and visualization in R software.

3.2.2. Correlation analysis

Pearson correlation analysis was used to measures the statistical relationship between PP and chl-a, SST ONI, MEI, and further to evaluate the impact of ENSO on the distribution and variability of PP. Pearson correlation coefficient (r) in the range from −1 (anti-correlation) to +1 (perfect correlation), between these two monthly time series x and y, with N elements as equation 3 [9]:

$$r \;= \frac{\sum_{i=1}^{N}(x_i-\bar{x})\,(y_i-\bar{y})}{\sqrt{\sum_{i=1}^{N}(x_i-\bar{x})^2}\,\sqrt{\sum_{j=1}^{n}(y_i-\bar{y})^2}} = \frac{Cov(x,y)}{\sigma_x\sigma_y} \tag{3}$$

with Cov being the covariance function, $\bar{x}$ and $\bar{y}$ the average and σ_x and σ_y the standard deviations for x and y, respectively.

Besides that, cross-correlation function (CCF) analysis was also used to describe and model the relationship between two time series. In the relationship between two time series (y_t and x_t), the series y_t may be related to past lags of the x-series. The cross-correlation function helps identify lags of the x-variable that might be useful predictors of y_t [21]. The sample cross-correlation function is defined as the set of sample correlations between x_{t+h} and y_t for h = 0, ±1, ±2, ±3,... A negative value for h correlates the x-variable before t and the y-variable at time t. For instance, consider h = -2. The CCF value would give the correlation between x_{t-2} and y_t . When one or more x_{t+h} , with h negative, are predictors of y_t, x leads y. When one or more x_{t+h}, with h positive, are predictors of y_t, x lags y. In this paper, x represents ONI/MEI, and y represents the PP value. This method used ONI/MEI to forecast fluctuations (increase/decrease) of the PP value.
All the graphs, calculations, and statistical analyses were performed using R software.

4. Result and Discussion

4.1. Spatial-temporal distribution and variation characteristics of PP

The statistical results show that the climatological average of PP in 18 years reached 449.2 mgC/m^2/day, ranged from 272.1 to 14,205.4 mgC/m^2/day. The monthly mean PP varied significantly between months in this study area reached the highest value in January with 578.5 mgC/m^2/day, and the lowest PP in May with the value of 320.3 mgC/m^2/day (Figure 3A). This result is similar to the monthly variation of multiyear averaged PP in the ES reported by Tan and Shi, 2009 [11].

In Figure 3A the PP reached 2 maximum peaks: one strong peak in winter (January) and one weak in summer (August). The winter peak was more significant than the summer, similar to previous studies in the ES [11, 22]. Previous research also showed that in the coastal marine area of Vietnam South Centre, the primary productivity in summer and winter is higher due to the eastern shelf and the Mekong River [13]. This paper said that PP seasonal distribution also indicates that PP reached the highest value in winter and lowest in spring, similar to Xu et al., 2017 [14] and Tan and Shi, 2009 [11].
Figure 3B showed the annual mean values of PP in 2003-2020 with the maximum value in 2011 and the minimum value in 2010 with corresponding values of 500 mgC/m^2/day and 370.9 mgC/m^2/day, respectively. Yearly averaged PP had the lowest values during the summer of 2010. This anomalous event coincided with the 2009-2010 El Niño event. Meanwhile, the maximum PP value in the winter of 2011 coincided with the strong La Niña phenomenon in 2010-2011. Research by Zhou et al., 2017 [13] also showed that in the western part of the ES between 2006 and 2015, PP reached the maximum value in 2011 and the minimum value in 2010.

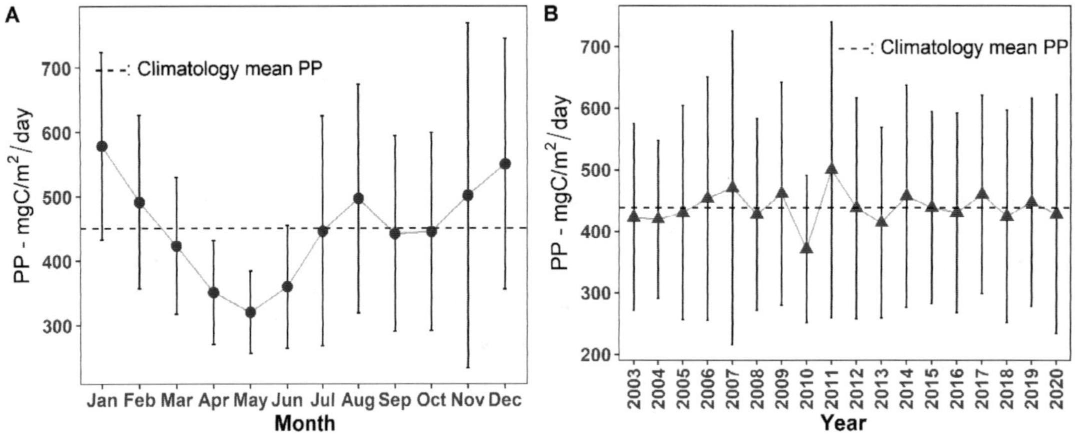

Figure 3. Monthly (A) and yearly (B) average variations of mean PP (mgC/m^2/day) from 2003-2020

The spatial distribution results of the monthly mean value of PP over the past 18 years are listed in Figure 4. It tended to decrease from shore to offshore. PP was higher in coastal areas than in the open sea areas. The lowest levels of PP were identified during spring (March to May), and the region with high PP levels was limited to a very narrow band near the coast in the north of the coastal marine area of Vietnam South Centre (15^0 – 16^{0}N). Spring is the transition period between northeast monsoon to southwest monsoon, low levels of PP may be related to mixed layer depth. According to the discussion of Tan and Shi, 2009 [11] indicated that the mixed layer depth of ES is the shallowest in spring (< 20 m) and deepest in winter (35 - 40 m).
The average PP level in the south of the study area increases significantly in summer, although the average PP level from the coast of Da Nang to the coast of Khanh Hoa was still low. In the southwest monsoon season (June to August), due to the ecological influence of the upwelling phenomenon, PP increased with a value greater than 1,500 mgC/m^2/day, distributed along the coastline of Ninh Thuan - Binh Thuan (Figure 4). High PP appears as a large jet, extending from the coastal waters of Ninh Thuan-Binh Thuan northeastward during the annual southwest monsoon. A previous study showed that this jet of high PP in this study area was associated with coastal upwelling driven by the southwest monsoon and discharges from the Mekong River [11]. The phenomenon of coastal upwelling brought nutrients from the subsurface to the surface, and Mekong River discharge delivers a high flux of nutrients from the continent and into the sea. These nutrients are used by phytoplankton to increase the primary productivity of the coastal marine area of Vietnam South Centre.

The PP levels increased in autumn along the coastline in the study area during autumn (September to November). PP values decreased with the increasing distance away from the coast (Figure 4).
During the northeast monsoon period (December to February of the following year), PP increased with a concentration larger than 1,000 mgC/m^2/day along a narrow strip of shallow water adjacent to the coastline for the whole study area (109.50°E back to shore) (Figure 4). During winter, the northeast monsoon reached its strongest level, increasing the mixed layer depth and bringing nutrient-rich water to the surface, allowing PP to rise to highly elevated levels, the highest in winter [11].

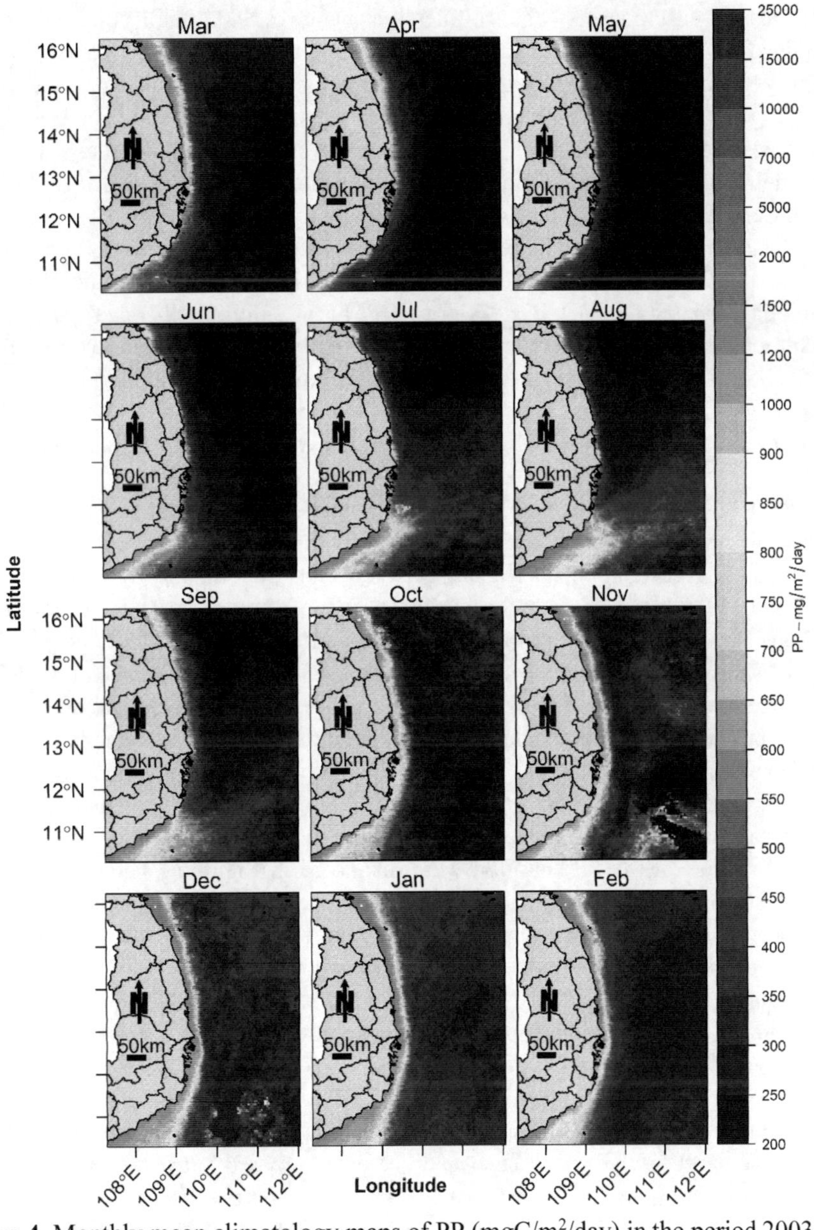

Figure 4. Monthly mean climatology maps of PP (mgC/m^2/day) in the period 2003-2020

4.2. Correlation between PP and SST, chl-a, ENSO

The spatial-temporal variation of PP is influenced by environmental conditions such as chl-a and SST. There exists a negative correlation between PP and SST, with the correlation coefficient reached -0.6 (Pearson test, $P < 0.05$), previous research has also shown that PP and SST showed an excellent inverse correlation with a correlation coefficient equal to -0.8 [11]. This correlation revealed that anomaly high/low SST was associated with anomaly low/high PP levels. Availability of SST is a crucial factor limiting the distribution of PP. However, Tan and Shi, 2009 [11] showed that in the ES, the change of SST is not the cause that directly affects the change of PP. It has shown that, although there exists an inverse relationship between SST and PP, low SST is not the limiting cause of PP. As analyzed above, the southwest and northeast monsoon winds influence a change of SST and PP concentration. The strong monsoon winds affect the movement changes the value of the SST, and affect the mixing of water masses vertically related to the concentration of PP. Some factors such as monsoon winds, upwellings, and river discharges must significantly control phytoplankton biomass and primary productivity by changing the transport and distribution of nutrient-rich water. The above analysis shows that the availability of nutrients can be one of the reasons for limiting PP value in the coastal marine area of Vietnam South Centre.

Meanwhile, PP and chl-a have a significant positive correlation with a correlation coefficient equal to 0.9 (Pearson test, $P < 0.05$). Chlorophyll-a concentration has been invoked to indicate both PP (a flux) and phytoplankton biomass (a scalar). Besides, the VGPM is a "chlorophyll-based" model that estimates net PP from chlorophyll-a content. Thus, an increase/decrease in chl-a level leads to an increase/decrease in PP value. The results of this correlation analysis have shown that in the period from 2003 to 2020, low PP appeared with low chl-a, and high PP appeared with high chl-a.

The results of the correlation analysis also showed that there was no linear correlation between PP and MEI (Pearson test, $r = -0.13$, $P > 0.05$). However, there was a weak correlation between PP and ONI (Pearson test, $P < 0.05$). The correlation of PP and ONI showed a significant negative correlation ($r = -0.23$, $P < 0.05$), which suggested that ENSO was probably the dominant factor that drove PP variation in this study area. The correlation coefficient between ONI and PP is low, indicating that ENSO influence against the PP which seen from the correlation value is not too large. Based on the relationship between ONI and PP, the ONI was used to forecast the fluctuation of the PP value. Cross-correlation function analysis showed that the most dominant cross-correlations occur between $h = -6$ and about $h = 3$ (Figure 5). There are nearly equal maximum values at $h = 0$, -1, and -2, with tapering occurring in both directions from that peak. The correlations in this region ($h = 0, -1, -2$) are negative, indicating that an increase (decrease) in the ONI value leads to a decline (increase) in the PP value with a phase difference of about 0 to 2 months (ONI ahead) (Figure 5). This result indicated that PP values fluctuate (increase or decrease) after the occurrence of the ENSO phenomenon about 0 to 2 months.

The monthly time series of the PP and ONI were displayed in Figure 6. The correlation coefficient between PP and ONI has shown that PP is negatively correlated with ONI. That is, the positive phase of ENSO (El Niño conditions) corresponded to lower PP, and the negative phase of ENSO (La Niña conditions) corresponded to higher PP, the correlation between PP and ENSO in this study is consistent with previous studies in East Sea [23]. This result can be explained because the ENSO phenomenon influences the summer and winter monsoon. The intensity of the East Sea summer monsoon tends to be weaker (stronger) in years associated with a warm (cold) ENSO event [24]. Besides that a weak East Asian winter monsoon also generally links to an El Niño event. At the same time, a strong one corresponds to a La Niña event in the tropical eastern Pacific [25]. The weakened winter monsoon leads to a decrease in the depth of the mixed layer and decreases upward nutrient flux, and the weakened summer monsoon leads to weak upwelling and reduced upward nutrient flux.

The time-series graph of the ONI in Figure 6 indicated that in the period from 2003 to 2020, the ONI value reached 2 minimum peaks corresponding to 2 La Niña times in 2007-2008 and 2010-2011. Observing time series data of PP in Figure 6 showed that PP reached its highest value in November 2007 with a value of about 1,409 $mgC/m^2/day$ due to the influence of strong La Niña in 2007. Besides, the La

Niña event of 2007-2008 also affected the abnormal increase in PP during the period from October 2007 to February 2008. Meanwhile, the La Niña phenomenon in 2010-2011 caused an increase in PP value from October 2010 to March 2011 with a maximum peak of 669 mgC/m^2/day in January 2011.
Besides the 2 minimum peaks, Figure 6 also showed that ONI reached 2 maximum peaks during El Niño in 2009-2010 and 2015-2016. The 2009-2010 El Niño event caused not only a decrease in PP value in the winter of 2009 but also a decrease in PP in the summer of 2010 with minimum values of 417 mgC/m^2/day in February 2010 and 295 mgC/m^2/day in June 2010, respectively. El Niño phenomenon in 2015-2016 also caused an abnormal decrease in PP value from October 2015 to January 2016, with the minimum PP value reaching 333 mgC/m^2/day in November 2015.

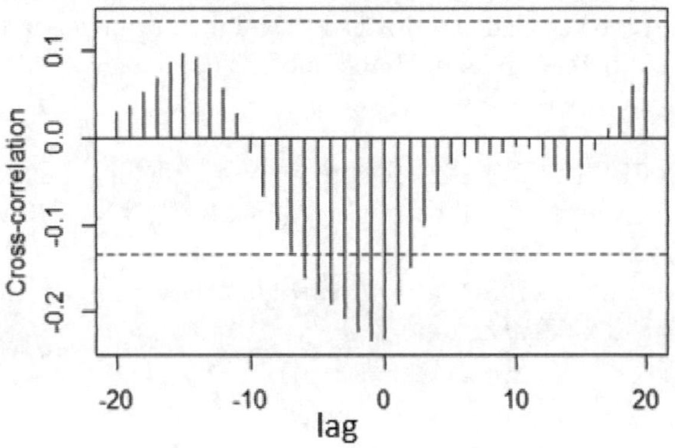

Figure 5. Cross-correlation function of ONI and PP (dotted lines are 95% confidence intervals)

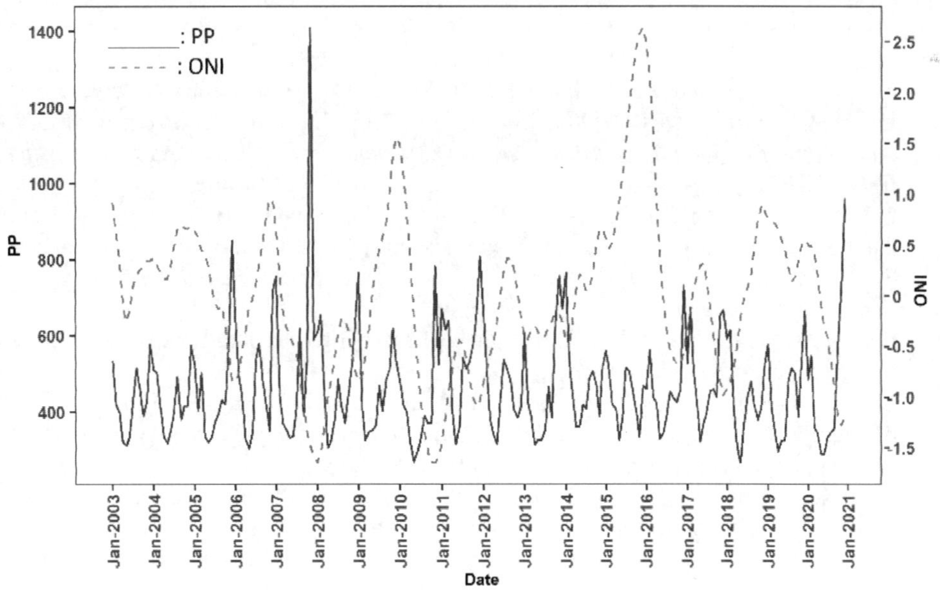

Figure 6. Time series of mean PP (mgC/m^2/day) (blue line) and ONI (red line) in the period of 2003-2020

5. Conclusion

The monthly mean PP varied significantly between months (highest in January, lowest in May), seasons (maximum peak in winter and minimum peak in spring) and years (highest in 2011 and lowest in 2010) in the coastal marine area of Vietnam South Centre during the period of 2003-2020. The spatial distribution of PP tended to decrease from shore to offshore, and PP was higher in coastal areas than in the open sea areas. The increase in PP value in winter and summer is due to the influence of northeast monsoon and southwest monsoon, respectively.
PP was positively correlated with chl-a and negatively correlated with SST with Pearson correlation coefficients of 0.9 and -0.6, respectively. PP also had a weak statistically significant correlation with ONI with a correlation coefficient of -0.23, so that ENSO influence against the PP which seen from the correlation value is not too large. The ENSO phenomenon affected the spatio-temporal variation of PP, the positive phase of ENSO corresponded to lower PP, and the negative phase of ENSO corresponded to higher PP, with a phase difference of about 0 to 2 months (ONI ahead).

Acknowledgements
The authors thank project 01/ĐTT-STB-2021 for the support of facilities for this study. We would like to thank Ho Chi Minh City University of Technology (HCMUT), VNU-HCM for the support of time and facilities for this study.
This paper is a contribution to celebrate the 100 years Anniversary of the Institute of Oceanography, Vietnam Academy of Science and Technology

References

[1] Field, C.B., M.J. Behrenfeld, J.T. Randerson, and P. Falkowski, 1998 Primary production of the biosphere: integrating terrestrial and oceanic components. *Science*, 281(5374), 237-240.
[2] Li, H., H. Peng, M. Sun, J. Liang, and H. Yao, 2009 The study on dynamic changes of NPP in coastal ocean of China. *Remote Sensing of the Ocean, Sea Ice, and Large Water Regions 2009*. International Society for Optics and Photonics.
[3] Nguyen Tac An and Tong Phuoc Hoang Son, 2004 Application of Remote Sensing for Interpretation of Primary Productivity in Bien Dong (Eastern Sea). *Proceedings of the Geo-informatics for Spatial-Infrastructure Development in Earth and Applied Sciences (GIS-IDEAS 2004)*.
[4] Lakshmi, E., D. Pratap, P. Nagamani, K. Rao, T.P. Latha, and S. Choudhury, 2014 Time Series Analysis Of Primary Productivity Along The East Coast Of India Using Oceansat-2 Ocean Colour Monitor (OCM). *The International Archives of Photogrammetry, Remote Sensing and Spatial Information Sciences*, 40(8): p. 1049.
[5] Son, S., M. Wang, and L.W. Harding Jr, 2014 Satellite-measured net primary production in the Chesapeake Bay. *Remote sensing of environment*, 144: p. 109-119.
[6] Liu, G. and F. Chai, 2009 Seasonal and interannual variability of primary and export production in the South China Sea: a three-dimensional physical–biogeochemical model study. *ICES Journal of Marine Science*, 66(2): p. 420-431.
[7] Tin, H.C., M.W. Lomas, and J. Ishizaka, 2016 Satellite-derived estimates of primary production during the Sargasso Sea winter/spring bloom: Integration of in-situ time-series data and ocean color remote sensing observations. *Regional Studies in Marine Science,* 3: p. 131-143.
[8] Carr, M.-E., M.A. Friedrichs, M. Schmeltz, M.N. Aita, D. Antoine, K.R. Arrigo, I. Asanuma, O. Aumont, R. Barber, and M. Behrenfeld, 2006 A comparison of global estimates of marine primary production from ocean color. *Deep-Sea Research Part II: Topical Studies in Oceanography*, 53(5-7): p. 741-770.
[9] Li, W., S.P. Tiwari, H.M. El-Askary, M.A. Qurban, V. Amiridis, K. ManiKandan, M.J. Garay, O.V. Kalashnikova, T.C. Piechota, and D.C. Struppa, 2020 Synergistic Use of Remote Sensing and Modeling for Estimating Net Primary Productivity in the Red Sea With VGPM, Eppley-

VGPM, and CbPM Models Intercomparison. *IEEE Transactions on Geoscience and Remote Sensing*.

[10] Behrenfeld, M.J. and P.G. Falkowski, 1997 Photosynthetic rates are derived from satellite-based chlorophyll concentration. *Limnology and Oceanography*, 42(1): p. 1-20.

[11] Tan, S.C. and G.Y. Shi, 2009 Spatiotemporal variability of satellite-derived primary production in the South China Sea, 1998–2006. *Journal of Geophysical Research: Biogeosciences*, 114(G3).

[12] Liao, X., J. Ma, and H. Zhan, 2012 Effect of different types of El Niño on primary productivity in the South China Sea. *Aquatic Ecosystem Health & Management*, 15(2): p. 135-143.

[13] Zhou, W., H. Xu, A. Li, and S. Ji, 2017 Zone statistics of the oceanic primary productivity for the traditional fishing areas of the open South China sea based on MODIS products. *Applied Ecology And Environmental Research*, 15(3): p. 1013-1024.

[14] Xu, H., W. Zhou, A. Li, and S. Ji, 2016 Similarities and differences of oceanic primary productivity product estimated by three models based on MODIS for the open South China Sea. *International Conference on Geo-Informatics in Resource Management and Sustainable Ecosystem*. Springer.

[15] Thu, P.M., N.H. Huan, L.T. Dung, L.T. Dung, V.H. Thi, T.T.M. Hue, and H.T. Du, 2014 Primary productivity of phytoplankton in Van Phong Bay, Khanh Hoa province, Vietnam. *Collection of Marine Research Works*, 20.

[16] Minh Thu, P., N. Manh Tien, N.H. Thai Khang, and L. Va Khin, 2012 Photosynthesis of phytoplankton in the southern marine regions of Vietnam from MODIS data. *International Symposium on Geoinformatics for Spatial Infrastructure Development in Earth and Allied Sciences 2012.*

[17] Bui Hong Long, 2009 The phenomenon of water upwelling in the sea of Vietnam. *Natural Science and Technology Publishing House.*

[18] Trinh Thi Minh Trang, Nguyen Thi Nguyet Ha, Tran Duc Thanh, 2015 The positional resources of Khanh Hoa coast: Potentials and Prospects. *Journal of Marine Science and Technology*, 15(1): p. 13-24.

[19] Bui Hong Long, 2011 Handbook of search on natural conditions, ecological environment, economy, society and integrated management of the South Central Coastal zone. *Natural Science and Technology Publishing House.*

[20] Cressie, N. and C.K. Wikle, 2015 Statistics for spatio-temporal data. *John Wiley & Sons.*

[21] Shumway, R.H. and D.S. Stoffer, 2017 Time series analysis and its applications: with R examples. *Springer.*

[22] Liu, K.-K., S.-Y. Chao, P.-T. Shaw, G.-C. Gong, C.-C. Chen, and T. Tang, 2002 Monsoon-forced chlorophyll distribution and primary production in the South China Sea: observations and a numerical study. *Deep-Sea Research Part I: Oceanographic Research Papers*, 49(8): p. 1387-1412.

[23] Ma, W., F. Chai, P. Xiu, H. Xue, and J. Tian, 2013 Modelling the long-term variability of phytoplankton functional groups and primary productivity in the South China Sea. *Journal of oceanography*, 69(5): p. 527-544.

[24] Zhou, W. and J.C. Chan, 2007 ENSO and the South China Sea summer monsoon onset. *International Journal of Climatology: A Journal of the Royal Meteorological Society*, 27(2): p. 157-167.

[25] Zhou, W., W. Chen, and D.X. Wang, 2012 The implications of El Nino-southern oscillation signal for south China monsoon climate. *Aquatic Ecosystem Health & Management,* 15(1): p. 14-19.

Acute and chronic toxicity of mixtures of bisphenol A and trace metals (Cd and Pb) to micro-crustacean, *Daphnia magna*

V T Nguyen[1,2], A T Huynh[1,2], T S Dao[1,2,*]

[1] Hochiminh City University of Technology (HCMUT), Ho Chi Minh City, Vietnam
[2] Vietnam National University, Ho Chi Minh City, Vietnam
* Corresponding author: dao.son@hcmut.edu.vn

Abstract. The occurrence of plastic additives and their ecological impacts have attracted much attention in recent years globally. Among plastic additives, the trace metals (e.g., Cd, Pb) are widely used as color pigments and stabilizers, whereas bisphenol A (BPA) is added to enhance the desired physical characteristics of plastic products. However, these additives can easily leach out of plastic materials and enter the aquatic environment causing risks to aquatic ecosystems. Although the toxicity of a single additive on various aquatic organisms has been studied, the responses of zooplankton exposed to the mixed plastic additives have not been fully understood. Therefore, this study aims to evaluate the effects of the binary mixtures (BPA+Cd, BPA+Pb) and trinary mixture (BPA+Pb+Cd) at the metal concentrations of 5 μg/L and BPA level of 50 μg/ L on the life history traits and food feeding rate of the freshwater micro-crustacean, *Daphnia magna*. The results showed exposures to these mixtures for 24h could significantly enhance the food feeding rate of *D. magna* from 2.5 – 5.8 times higher than the control. The survival rate was decreased from 50 – 90% in the organisms exposed to these mixtures after 18 incubated days. We found a synergistic effect of BPA+Pb but an antagonistic effect of BPA+Pb+Cd on the survivorship of *D. magna*. Similarly, the organisms in the exposures delayed their maturity age and reduced their reproduction. The potent impact order of the mixtures on *D. magna* was BPA+Cd > BPA+Pb+Cd > BPA+Pb. Our results evidenced the adverse effects of plastic additive mixtures on aquatic organisms. Therefore, the use and disposal of plastic materials and plastic additives should be paid more attention to protect the environment, ecosystem, and human health. Moreover, our findings proved that the toxicity of multi-contaminants on organisms could be unpredictable even the toxicity of a single contaminant is known.

Keywords. plastic additives, cladoceran, antagonistic effects, synergistic effects

1. Introduction

In the industry of plastic manufacture various chemicals so called plastic additives (e.g., bisphenol A, trace metals, phthalate, among others) are used to form different beneficial characteristics of plastic products such as flexibility, thermal and corrosive resistance, hardening or colors [1]. Among the plastic additives, bisphenol A (BPA) has been widely used in various plastic products such as polyvinyl chloride (PVC), polypropylene, and polyethylene [1,2]. The global consumption of BPA has rapidly increased in recent years reaching approximately 8 million tons in 2016 [3]. Obviously, most plastic additives including BPA are not chemically but physically bound to the polymers, therefore, they can be easily released out of plastic materials entering the aquatic environment then available for organisms [1]. The release of BPA and other additives (e.g., trace metals, phthalate) from plastic products has been reported

IOP Conf. Series: Earth and Environmental Science **964** (2022) 012012 doi:10.1088/1755-1315/964/1/012012

in numerous studies [4, 5, 6]. Also BPA presence in water bodies can be due to the discharges from industrial manufacturing [2]. As a result, BPA is among the plasticizers frequently detected in aquatic environment, in which the highest concentrations found in surface water and in wastewater were > 90 μg/ L and 370 μg/ L, respectively [3]. In Vietnam, the results from the study of Tam et al. [7] indicated that the concentration of BPA in many canals could reach 30 ng/ L, but Vietnam has been listed as among the four countries in the world discharging mismanaged plastic waste into the ocean [8]. Therefore, it could be predicted that the concentration of BPA in water bodies in Vietnam may continuously increase in the next few years. As an endocrine disrupting compound, BPA could not only cause the disruption of intracellular systems but also interfere with the functions of hormones in various organisms that could be considered as a potential risk for aquatic ecosystems and human health [1, 2].

On the other hand, Cd and Pb are among the trace metals frequently detected in aquatic environment at high levels [9, 10]. Although the occurrence of Pb and Cd in the environment is partly due to natural processes, the discharges from anthropogenic activities are mainly responsible for the contamination of these metals in water bodies [11, 12]. Cd is commonly used in nickel-cadmium battery production and special alloys to avoid corrosion due to its non-corrosive properties while Pb plays an important role in battery manufacturing, paints, and gasoline [11, 12]. Moreover, these metals are also used as stabilizers in plastic production, especially PVC [11, 12]. Due to the wide application in many industrial processes, the concentrations of these metals in water bodies have been frequently detected at high levels [9, 10]. In surface water, the concentrations of Cd and Pb could be up to 190 and 430 μg/L, respectively [10], but those could reach higher 4.5 mg/L and 17 mg/L, respectively, in industrial areas [9]. On the other hand, the previous study of Dang et al. [13] recorded the highest concentration of Cd and Pb in the Red River, which is among the main rivers in Vietnam could reach approximately 50 and 480 μg/L, respectively, while the safety concentration of these metals according to the technical regulation for surface water quality of Vietnam are 5 and 20 μg/L, respectively. Seriously, both Cd and Pb are non-essential elements to living things, but they can be highly toxic to aquatic organisms even at a low concentration [14]. Pb has been known as a neurotoxic chemical and exposure to this metal could cause the increase of oxidative stress leading to numerous effects on the behavior, nervous and respiratory systems, reproduction, and enzymatic functions of living things [15, 16]. On the other hand, Cd has been reported to cause detrimental influences on enzymatic systems and induce oxidative stress in organisms [14].

In aquatic ecosystems, zooplankton (e.g., *Daphnia magna*) play a vital role as they have a central position in the food web connecting major primary producers (phytoplankton) with other organisms at higher trophic levels (fish, shrimps) [17]. Besides, they not only have high sensitivity to toxic substances but are also easily maintained in the laboratory conditions. Therefore, *D. magna* was strongly suggested as a model organism in toxicological investigations [18, 19, 20]. There have been numerous studies showed the singly detrimental effects of BPA and the metals Pb and Cd, on *D. magna*. For instance, BPA could cause lethal effects on *D. magna* with the median 48h-EC_{50} value of 8.4 mg/L while those of Cd and Pb have been reported to be 21 and 695 μg/L, respectively [21, 22]. At sub-lethal concentrations, BPA could induce genotoxicity, the alteration in lipid peroxidation levels and enzymatic activities in *D. magna* [23, 24, 25]. Nagato et al. [26] showed a significant change in the metabolisms of *D. magna* exposed to BPA while the adverse influences of BPA on the life history traits of *D. magna*, including survival rate, reproduction, maturation, and growth, have also been reported in numerous studies [3, 23, 24]. Similarly, exposures to Cd and Pb could decline the survivorship and inhibit the reproductive performance of *D. magna* [27, 28, 29, 30, 31]. Moreover, Cd has been reported to cause a decrease in the assimilation rate and biochemical effects in *D. magna* [32] while exposure to Pb could also result in an alteration in the content of enzymes and hemoglobin leading to the impairment in growth and reproduction of the organisms [27].

Although the single effects of BPA, Cd and Pb on *D. magna* have been well documented, the combined toxicity of these chemicals on this organism is understudied. It is important to note that toxicity to organisms in aquatic environments could not result from a single chemical but due to the mixtures of various pollutants. Moreover, previous studies have indicated that the combined effects of the pollutant mixtures on organisms could differ from that of a single pollutant [33, 34, 35]. For instance, Meng et al. indicated the synergistic toxicity of the mixture of five metals (Hg, Cd, Cu, Pb and Cr) on *D. magna* [36]. Similarly, Cooper et al. found a higher toxicity of the binary and ternary mixtures of Pb, Cu, and Zn on reproduction of *Ceriodaphnia dubia* compared to the single effects of each metal [35]. Additionally, the single effects of Cd, Cr and Cu on the marine copepod *Tisbe holothuriae* could be less toxic than the joint effects of the mixtures of these metals [33]. The different responses of the freshwater micro-crustacean *C. dubia* in the exposures to the mixtures of metals (Cu and Zn) and organophosphate compared to those exposed to these pollutants individually could be another typical example [34]. To the best of our knowledge, there is gap on the mixture toxicity of BPA and non-essential metals on freshwater zooplankton upon chronic exposures. Therefore, this study aimed to assess the joint toxicity of BPA and the two trace metals (Cd and Pb) on the life history traits of *D. magna* over a period of 18 days. Besides, the food feeding rate of *D. magna* in an acute exposure to the mixtures of these chemicals were also investigated in our study.

2. Materials and methods

2.1. The test organisms and chemicals for the experiment

The micro-crustacean, *Daphnia magna* was used for the acute and chronic experiments. The organisms were obtained from MicroBio Test (Belgium) and have been maintained in the ISO medium [37] for many generations under laboratory conditions of 24 ± 1 ºC, light intensity < 1000 Lux, and 14h light followed by 10h dark regime APHA [20]. The organisms were fed with the mixture of the green alga *Scenedesmus protuberans* and YTC (yeast, cerrophyl, and trout chow digestion) [38].

The plastic additive Bisphenol A (purification of 99%) was purchased from Aldrich Sigma. The BPA was dissolved in acetone (Merck) at the concentration of 1000 mg/L which was used as the stock solution before spiked into the ISO medium to have a test concentration for experiments. Meanwhile, both the metal solutions $Pb(NO_3)_2$ and $Cd(NO_3)_2$ at the concentration of 1000 mg/L (ICP/MS standard analysis; Merck) were used as stock solutions for the experiments. All the stock solutions were stored at a temperature of 4 ºC before the experiment.

2.2. Experimental setup

2.2.1. Acute exposure on the food feeding rate. The acute experiment was performed to assess the effects of the mixtures of BPA and metals on the food filtering activity of *D. magna*. The food filtering test was designed according to Jesus et al. with minor modifications [39]. Briefly, the test was initiated with *D. magna* of four days old. The organisms were exposed to (1) a mixture of BPA (50 μg/L) and Pb (5 μg/L), abbreviated as BPA+Pb; (2) a mixture of BPA (50 μg/L) and Cd (5 μg/L), abbreviated as BPA+Cd; and (3) a mixture of BPA (50 μg/L), Pb (5 μg/L) and Cd (5 μg/L), abbreviated as BPA+Pb+Cd. Besides, a positive control was also prepared in parallel with the exposures by culturing the organisms in the ISO medium without BPA and the metals. In each treatment (positive control or exposures), 10 organisms (n = 10) were used and each organism was individually raised in a 50 mL glass beaker containing 40 mL ISO medium. In each beaker, one organism was fed with a similar initial amount of *S. protuberans* (52,050±2,158 cells/mL). In addition, to confirm if the *S. protuberans* could grow in ISO medium (containing no nitrogen and phosphorus) or not, we conducted a negative control with ten beakers (n = 10) in which each contained 40 mL ISO medium and a similar initial algal abundance with other exposures but no *D. magna*. The experiment was conducted in the laboratory conditions (temperature, light intensity, and light: dark regime) as mentioned above. After 24h the algal abundance in each beaker was counted [40] with a Sedgewick Rafter counting chamber (Pyser SGI, England) under a microscope (Optika B150, Italy). Then the abundance of *S. protuberans* in each beaker at the start and after 24h was used to assess the food feeding rate (F) of the organisms in control and exposures.

2.2.2. Chronic exposure on life history traits. The chronic experiments to evaluate the toxicity of the mixture of BPA and metals were performed according to APHA with minor modifications [20]. Prior to the experiments, nearly 50 mother *D. magna* were randomly selected and then incubated in a 2 L glass beaker containing 1.5 L of ISO medium. Neonates (< 24h old) from the second brood of mother *D. magna* in the beaker were randomly collected and used for the chronic test [37]. Four treatments including one control and three exposures were prepared including BPA+Pb, BPA+Cd, and BPA+Pb+Cd (as mentioned in food feeding rate experiment above). The BPA and metal concentrations were chosen basing on the concentration of BPA ever detected in the environment and the safe concentrations of Pb and Cd regulated in the Vietnam Technical Regulation for surface water safety (QCVN 08-MT:2015/BTNMT). For each treatment, 2 neonates were incubated together in a 50 mL glass beaker containing 40 mL ISO medium. There were 15 replicates in each treatment (n=15). Besides, the control was conducted in parallel with the exposures by raising organisms in the ISO medium free pollutants (BPA, Cd, Pb). The *D. magna* in each beaker were fed with *S. protuberans* and YTC [38] during an experimental period of 18 days under the laboratory conditions as mentioned above. The medium and food in each beaker was totally renewed three times per week while the life history traits of *D. magna* including survival, total neonates, and maturation age, were carefully recorded daily.

2.3. Data treatment. The abundance of *S. protuberans* in the negative control at the start and the end of the acute test was 52,050±2,158 cells/mL and 49,360±2,797 cells/mL, respectively. This meant the abundance of *S. protuberans* in the negative control did not increase after 24h in ISO medium as we expected. Therefore, the food feeding rate (F, cells/hour) of one *D. magna* was calculated as the equation of $F = V(C_0 - C_t)/t$ [39], where:

V, volume of the test solution (40 mL in this study);
C_0, abundance of *S. protuberans* (cells/mL) at the start of experiment;
C_t, abundance of *S. protuberans* (cells/mL) when the test terminated;
t, exposure time (24 hours in this study).

Sigma Plot Version 12.0 was used for data analyses. The Kruskal-Wallis tests were applied to calculate the statistically significant difference in the food feeding rate and the maturity age of *D. magna* between the control and exposures. A gap of more than 20% in the survival proportion of *D. magna* in the exposures was considered as a significant difference [20].

3. Results and discussion

3.1. Physical and chemical characteristics of the culturing medium

The pH and dissolved oxygen (DO) concentrations of the culturing medium in each treatment were determined at the beginning and before the medium renewal. The pH values of the medium in the control and exposures did not significantly alter during the time of the test and ranged from 7.25 to 7.44 (Metrohm 744) while the DO ranged from 6.60 to 7.28 mg/L (Hanna HI9146). Besides, the hardness and alkalinity of the test medium were analyzed by titration method according to APHA [17] and their values were from 288 - 341 mg $CaCO_3$/L and 32 - 50 mg $CaCO_3$/L, respectively (Table 1). The hardness and alkalinity values revealed that the test media was a medium hardness water [38] whereas the pH and DO values of the media in all the exposures including the control met the standard requirement of APHA [17] for culturing the *D. magna*.

Table 1. The physical and chemical characteristics of the test medium

Exposures	pH	Temperature (°C)	DO (mg/L)	Hardness (mg $CaCO_3$/L)	Alkalinity (mg $CaCO_3$/L)
Control	7.25 – 7.43	23 – 24	7.28 – 7.31	288 – 271	44 – 50
BPA+Pb	7.43 – 7.63	23 – 24	7.06 – 7.14	354 – 296	32 – 42
BPA+Cd	7.41 – 7.44	23 – 24	6.38 – 6.98	337 – 300	32 – 42
BPA+Pb+Cd	7.43 – 7.44	23 – 24	6.60 – 6.63	341 – 300	33 – 41

3.2. The effects on the food feeding rate of Daphnia magna

The results of food feeding rate by *D. magna* were showed in the Table 2. During the experimental period, one *D. magna* in the control fed the *S. protuberans* at a rate of 4,870±2,408 cells/h. However, the feeding rate in the BPA+Pb, BPA+Cd, and BPA+Pb+Cd respectively was 11,967±1,965, 28,043±4,048, and 13,667±7,182 cells/h which was around 2.5, 5.0, and 2.8 fold higher than the control (Table 2). Therefore, the mixtures of BPA and trace metals (Pb, Cd) made the organisms significantly consume more food (*S. protuberans*). Besides, the food feeding rate of *D. magna* in the treatment BPA+Cd was significantly higher than that in the other exposures ($p < 0.001$; Table 2).

Table 2. The food feeding rate (F) of *Daphnia magna* during 24h of experiment. Different letters (a, b, c) indicated the significant difference of the food feeding rate among the treatments ($p < 0.001$) according to the Kruskal-Wallis test.

	Control	BPA+Pb	BPA+Cd	PBA+Pb+Cd
F (mean ± SD; cells/h)	4,870±2,408$^{(a)}$	11,967±1,965$^{(b)}$	28,043±4,048$^{(c)}$	13,667±7,182$^{(b)}$
Proportion to the control	100%	250%	580%	280%

Negative effects of singly BPA or trace metals on food consumption of *Daphnia* were reported. The food assimilation rate of *Daphnia* spp. exposed to Cd (from 10 to 100 µg/L) was reduced which could be the consequences of feeding mechanism decrease and the reduction of the activity of digestive enzymes and food absorption [42]. Liu et al. observed the feeding inhibition in *D. magna* exposed to BPA at the concentration of 178 µg/L [22]. The *D. magna* exposed to BPA (178 µg/L) resulted in the expression of acetylcholine esterase activity consequently activity and behavior impairment in the organism. So far, there has been no report on the influence of Pb on food feeding of *Daphnia*, but Pb, at a certain level, would inhibit the feeding activity of the organism because this metal is a neurotoxic chemical impairing the respiratory system and enzyme activities in zooplankton [16, 43]. In contrast, we found the increase of food feeding rate in *D. magna* exposure to BPA+Cd, BPA+Pb+Cd and BPA+Pb (Table 2). The controversial observations between previous studies and ours could be due two potential reasons. Firstly, we used much lower concentrations of metals (5 µg/L) and BPA (50 µg/L) to test with *D. magna* while the concentrations of chemicals for experiments in previous studies were much higher (10 – 100 µg/L of Cd; and 178 µg/L of BPA). The lower concentrations might not induce the inhibitory effect on food feeding activity of *D. magna* in our study. Secondly, the mixtures of BPA and metals (Pb, Cd) may not have an additive or a synergistic effect, but they may result in an antagonistic effect on feeding rate of *D. magna*.

Lampert [44] proved that *Daphnia* could grow well at the food concentration from 0.5 mg C/L/day which was approximately 35,000 cells/ mL of *Scenedesmus* [45]. The latter authors noted that one *D. magna* grew well upon the providing an amount of 700,000 *Scenedesmus* cells each day as food [45]. In present study, with the feeding rate of 4,870 cells/h (Table 2), one *D. magna* in the control consumed around 117,000 cells/day which was within the suggested range of food for the organism in a previous investigation [45]. Although the mixtures of metals (Pb, Cd) and BPA could impact on the *D. magna* (please see the section 3.2 below) hence the chemicals caused the energy cost in the organisms then the organisms would increase their energy demand to deal with toxic chemical challenge [46], the mixtures might not inhibit the feeding activity in a short exposure time (24h). Besides, the toxic chemicals can not only affect behavior but also the enzyme activities and protein content in *D. magna* [47]. However, in case of protein degradation in the organisms, they could make a compensation by increasing protein synthesis [46]. Taken together, the exposed *D. magna* would require more material and energy to compensate the cost, to deal with the damage and to maintain their normal activities. This would lead to the food feeding rate increase in the exposed *D. magna* as observed in this study. Our study reflects the feeding response of *D. magna* as an adaptive characteristic upon exposures to trace metals and BPA. On the other hand, the influence or the enhancement on feeding rate of *D. magna* by BPA+Cd was strongest among the exposures. This could be link to the more potent impact of Cd than Pb on *D. magna* [30, 31] consequently the organisms in BPA+Cd required more energy and material for compensation and maintenance their normal activities, and physiological and biochemical processes in a short time. Hence the *D. magna* in BPA+Cd exposure fed more food than *D. magna* in other exposures. The similar food feeding rate of *D. magna* in the BPA+Pb and BPA+Pb+Cd (Table 2) reflected that Cd slightly impacted the feeding rate under the presence of BPA+Pb. Because the combined effect of BPA and Cd was very strongly on *D. magna* feeding rate as mentioned above, the Cd seemed to have minor influence on the *D. magna* feeding upon the presence of Pb. This could be related to the competition between Pb and Cd on the same active site in *D. magna* as explained with biotic ligand model by Di Toro et al. [48]. The similar free metal ions (Pb^{2+}, Cd^{2+}) could have the same active binding sites or sites of action in the organism. The competition in binding onto the site lead to the toxic reduction of the mixture metals [48]. Therefore, both Pb and Cd are non-essential and toxic metals to *D. magna* but their mixture did not increase the feeding rate of the organism (compared to the single metal), hence an antagonistic effect, which needs further investigations to clarify.

3.3. The effects on the survivorship of Daphnia magna

After 18 days of the experiment, the survival rate of the control was 100% that strongly met the requirements for a chronic test according to APHA [20]. However, the survival rate of *D. magna* exposed to the mixtures of these pollutants was significantly lower than that in the control. More details, by the end of the test, the survivorship of *D. magna* in the BPA+Pb was approximately 37% while the proportion in the BPA+Cd was 10%. Meanwhile, only 50% of total *D. magna* in the exposure to BPA+Pb+Cd lived at the end of the experiment (Fig 1). We found that in the exposures of BPA+Pb and BPA+Cd the survival rate of *D. magna* was inhibited more than 60% and 90%, respectively, while the mixture of the three pollutants caused a 50% decline in the survival of the organisms.

The decline in the survivorship of *D. magna* in the exposures in this study was the evidence demonstrating the adverse effects of these pollutants on the organisms. Being an endocrine-disrupting chemical, BPA is more likely to cause biological effects such as intracellular disruption as well as enzyme activity alteration, rather than lethal effect on the organisms [3, 22, 23, 24]. For instance, Park and Choi showed that although the survival rate did not decrease, the genotoxic effects were found in *D. magna* exposed to BPA at concentrations from 300 to 30,000 µg/L [23]. Similarly, the mortality of *D. magna* was not different between the control and exposure to BPA at 10,000 µg/ L for 21 days in the study of Jemec et al. [24]. On the other hand, 178 µg/L of BPA could cause the activity decrease of digestive enzymes (trypsin, amylase), antioxidant enzymes (superoxide dismutase, glutathione peroxidase) and the nervous enzyme acetylcholinesterase [22]. Therefore, the BPA at high concentrations (some hundreds to thousands μg/L) could directly or indirectly alter the normal metabolisms and behavior of *D. magna*.

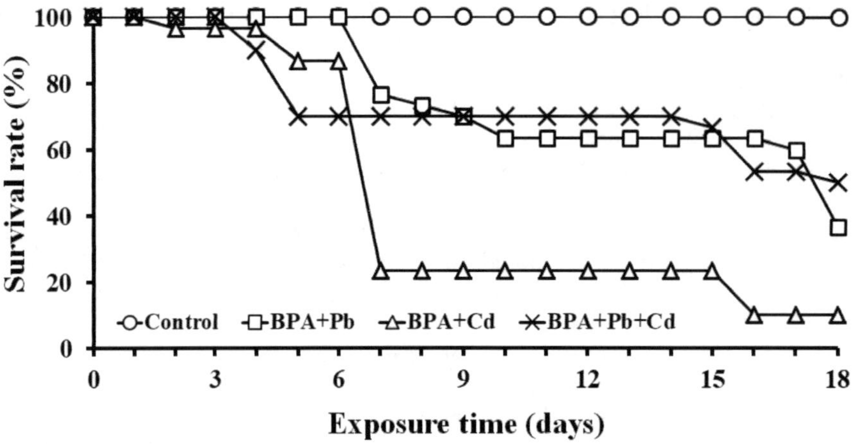

Figure 1. The survival rate of *Daphnia magna* exposed to the mixtures of BPA and trace metals (Pb, Cd) during 18 days of the experiment.

The detrimental effects of Pb on the survivorship of *D. magna* have been showed in the previous study of Luciana et al. [29], in which Pb at the concentration of 90 μg/L could inhibit up to 95% the survival rate of *D. magna* after 15 days of the exposure. Similarly, the survival of *D. magna* was deleteriously impaired in the exposure to 150 μg Pb/L for 21 days [31]. However, survival proportion of C. magna was not decreased in exposure to 5 μg Pb/L [31]. Pb has been known as a neurotoxic chemical, thereby exposures to this metal could lead to various negative effects, such as affecting the respiratory system, inhibiting enzyme activities, and consequently the lethal effects on organisms [16, 43]. Regarding the toxicity of Cd, numerous studies have indicated that Cd could adversely affect the survival of *D. magna* even at low concentrations. Bodar et al. [28] found that *D. magna* exposed to Cd at 10 μg/L for 25 days could lead to a serious decline in the survivorship, up to 60%, whereas the report of Perez and Hoang [30] noted that *D. magna* had high mortality, approximately 80%, in the exposure to 1.5 μg/L of Cd.

It is also important to note that the exposed *D. magna* in our study had a higher food feeding rate than usual (Table 2) so that the organisms could ingest more food then have more material and energy during a short time of incubation (24h). However, upon a long period of time continuously exposing to the pollutants, *D. magna* could be severely stressed in biochemical and physiological alterations as reported elsewhere [22]. Then the energy cost due to damage became larger than the energy in the food consumption by the *D. magna*, consequently mortality from the 4th day of incubation (Fig. 1). Although the joint toxicity of BPA, and Pb and Cd has not been reported, the single adverse effects of BPA, Pb and Cd on the survival of organisms could help to explain the decline in the survival rate of exposed *D. magna* in the current study. As mentioned above, the solely BPA (50 μg/L) or Pb (5 μg/L) did not impair *D. magna* survival [23, 31], but we found that the BPA+Pb (mixture) could inhibit up to 63% of the survival rate of *D. magna* in our experiment that was severer than the individual effect of these two pollutants. It seemed that the mixture of BPA and Pb caused a synergistic effect on the survival of *D. magna* in our study. Further studies are suggested to clarify the synergistic effect of BPA and Pb.

In this study, we exposed *D. magna* to BPA+Cd which inhibited up to 90% in the survivorship of *D. magna*. Compared to the previous study, the used concentration of Cd in this study was higher than the concentration of Cd which reduced 80% survival rate of *D. magna* in the study of Perez and Hoang [30]. Therefore, the decline in the survivorship of *D. magna* in the current study could be mainly due to the toxicity of Cd because BPA at even higher concentration has been demonstrated no effect on the survival of *D. magna* in previous studies [23, 24]. On the other hand, our results showed that the survival rate of *D. magna* exposed to the BPA+Pb+Cd could decline 50% which was less serious than the effect caused by the mixture BPA+Cd, up to 90%. Surprisingly, the occurrence of Pb could reduce the toxicity of the mixture BPA+Cd. This result could be in line with the previous study of Perez and Hoang when the authors found that the occurrence of Zn reduced the toxicity of Cd on the survival of *D. magna* in the exposure lasting 21 days [30]. Similarly, Mahar and Watzin observed the mixture of Zn, Cu, and diazinon causing less toxic than the mixture of the two on the survival of another micro-crustacean species, *C. dubia* [34]. Moreover, the mixture of the three metals including Cd, Cr, and Cu had less toxicity than the mixture of the two metals that have been reported on the copepod *Tisbe holothuriae* [33]. These results could help to explain our observation of the less reduction in the survival rate of *D. magna* exposed to the mixture BPA+Pb+Cd compared to the mixture of BPA+Cd. It was likely that the mixture BPA+Pb+Cd showed an antagonistic effect on the survival of *D. magna*.

3.4. The effects on the maturity age of Daphnia magna

The results on the maturation showed that *D. magna* in the control and in the BPA+Pb matured at a similar age, 4.7±0.4 and 5.4±1.4 days, respectively ($p>0.05$, Kruskal-Wallis test; Fig. 2). Regarding remaining exposures, we found a significant difference in the maturity age of *D. magna* raised in the BPA+Cd and BPA+Pb+Cd compared to the control ($p<0.01$, Kruskal-Wallis test). The maturity ages of the organisms in BPA+Cd and BPA+Pb+Cd were 5.5±0.5 and 5.9±1.0 days, respectively (Fig. 2). Thus the mixtures of BPA+Cd and BPA+Pb+Cd caused a delay in the maturation of *D. magna*.

To the best of our knowledge, the joint effects of BPA+Pb, BPA+Cd, and BPA+Pb+Cd, on the maturation of *D. magna* has not been reported so far. Liu et al. found that BPA (178 µg/L) prolonged the day to first brood of *D. magna* [22]. However, Quach et al. reported that Pb (5 µg/L) did not cause the postponement in the maturation of *D. magna* [31]. Therefore, the BPA concentration (50 µg/L) in the mixture of BPA+Pb was not high enough to induce a negative influence on the maturation of the exposed *D. magna* in our study.

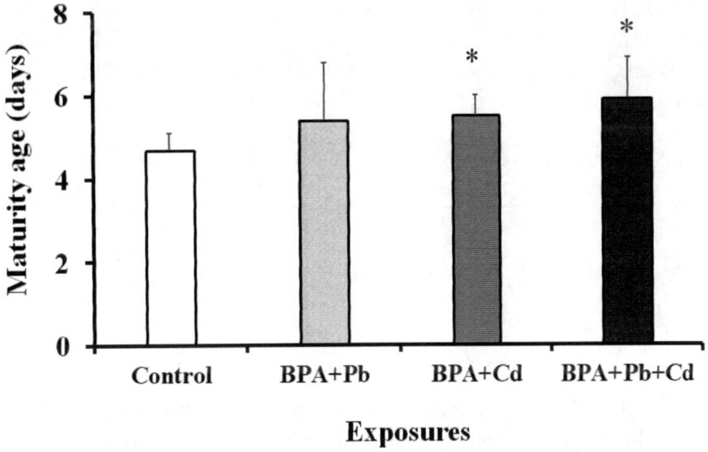

Figure 2. Time to maturity of *Daphnia magna* exposed to the mixtures of BPA and trace metals (Pb, Cd) during 18 days of the experiment. The asterisk indicated the significant difference between control and exposures ($p<0.01$, Kruskal-Wallis test).

Although the impacts of BPA on the maturation of *D. magna* have been not fully understood, this chemical has been known to cause numerous effects related to reproductive activities of the organisms, such as altering expression levels of many genes linking to reproductive processes leading to the inhibition on the reproduction of *D. magna* [25]. Similarly, previous studies have indicated that exposure to Cd could lead to the postponement of the maturity of organisms. For instance, Bodar et al. noted that the Cd at the concentration of 5 µg/L could slightly delay the maturity age of *D. magna* [28]. Moreover, Meng et al. also found that Cd at the same concentration (5 µg/ L) in combination with other metals (Pb, Hg, Cr, and Cu) induced a delay in the maturation of *D. magna* in a chronic test [36]. Besides, energy cost in the exposed *D. magna* as mentioned above (section 3.2) would diminish the egg development inside the organism. Summing up, the impacts of a single pollutant (BPA, Pb, Cd) supported our observation on the postponement in the maturity age of *D. magna* exposed to BPA+Cd and BPA+Pb+Cd. Further studies on the biochemical responses of *D. magna* upon exposures to these mixtures to clarify the joint toxicity on the maturation of the organisms.

3.5. The effects on the reproduction of Daphnia magna

The accumulative neonates of *D. magna* in the control were 1,174 while that in the exposures to the BPA+Pb and BPA+Cd were 1,378 and 301 neonates, respectively. Therefore, the total offspring relative to the control of the mixture BPA+Pb and BPA+Cd were 117% and 26%, respectively (Table 3). On the other hand, the total offspring released from the BPA+Pb+Cd was 923, gaining approximately 79% compared to the control (Table 3). The results on the reproductive performance indicated the influence on reproduction of BPA+Pb+Cd was less than the BPA+Cd but more than the BPA+Pb.

Apparently, there was a strong decrease of population development of offspring *D. magna* exposed to BPA+Cd and BPA+Pb+Cd (Table 3). However, when combined the number of alive mother *D. magna* (Fig. 1), its maturity ages (Fig. 2) and the total offspring (Table 3) we found no inhibition on the reproductive capacity of mother *D. magna* in the exposures. Because mother *D. magna* matured at around 5 – 6 days old (Fig. 2), the *D. magna* would start to reproduce at the 8th – 9th day of the experiment (data not show). Between the 8th day to the end of experimental period, the proportion of mother *D. magna* in BPA+Pb and BPA+Pb+Cd to control was around 50 – 70% (Fig. 1) and the proportion of their offspring (from the BPA+Pb, BPA+Pb+Cd) to the control was from 79 to 117% (Table 3). Hence the fecundity or brood size of exposed *D. magna* was larger than that of the control. Similarly, the 77 – 90% lower number of mother *D. magna* in BPA+Cd (compared to the control; Fig. 1) produced around 74% lower number of offspring (Table 3). Therefore, we concluded that all the binary and trinary mixtures in present study enhanced the reproductive capacity of *D. magna*, and the enhancement order was BPA+Pb > BPA+Pb+Cd > BPA+Cd.

Table 3. The accumulative offspring of *D. magna* in the exposures and their proportion to the control

	Control	BPA+Pb	BPA+Cd	BPA+Pb+Cd
Accumulative offspring	1,174	1,378	301	923
Proportion to the control (%)	100	117	26	79

Previous studies showed that BPA did not adversely inhibit the reproduction of the *D. magna*. Jeong et al. found that BPA at the concentration up to 3,000 µg/L did not significantly reduce the number of offspring *D. magna* [25]. Similarly, the reproduction of *D. magna* exposed to 2,000 µg BPA/L was not influenced over the period of 21 days [3]. Jemec et al. [21] and Brenna et al. [49] observed no detrimental effect of BPA at the concentrations of 1,000 – 1,730 µg/L on the brood size and total neonate of *D. magna*. However, the BPA singly at the concentration of 50 µg/L resulted in an increase of total offspring of *D. magna* compared to control (our unpublished data, from a multigenerational experiment).

Therefore, high concentrations of BPA did not enhance the reproduction of the organisms, but it became possible at a lower concentration. This is similar to the influence of another organic compound and also endocrine disrupting compound, phthalate. The phthalate had negative effects on survival (48h-LC_{50} of 580 μg/L) but the chemical (390 μg/L) strongly stimulated the reproduction in *D. magna*, up to 1.5 times compared to normal [50, 51].

On the other hand, the results of Luciana et al. indicated that the fecundity of *D. magna* was not reduced in the exposure to Pb at the concentration up to 30 μg/L which was much higher than the used concentration of Pb in the current experiment (5 μg/L) [29]. In addition, Quach et al. found that the reproduction of *D. magna* was not inhibited, even slightly stimulated in the exposure to Pb at the same concentration as used in this study (5 μg/L) [31]. In the current study, the used concentrations of BPA and Pb were within the range or even lower than the no-effect concentrations reported in previous studies. Bodar et al. and Perez and Hoang found that the number of offspring *D. magna* in the exposure to Cd (1.5 – 10 μg/L) for 21 days significantly reduce [28, 30].

Therefore, beside the dead organism, the alive *D. magna* exposed to mixtures of BPA and the metals could suffer from the joint effects on reproduction. As mentioned above, both BPA and Pb separately enhanced the reproduction of *D. magna* so their mixture would strongly stimulate the accumulative offspring of *D. magna*. The metal Cd prevented the reproduction of *D. magna* hence the organisms exposed to BPA+Pb+Cd would produce less offspring than those exposed to BPA+Pb did. Then the mother *D. magna* treated with BPA+Cd would produce less offspring as noted in the Table 3. Moreover, the results from the food feeding rate experiment could be an evidence for the energy cost and more material demand for compensation in the exposed *D. magna*. It is reasonable that the more spending energy against the impacts of BPA and metals, and the reproductive capacity maintenance and/or enhancement in the exposed *D. magna* would lead to a higher food demand consequently higher food feeding rates in the exposed organisms as in the Table 2.

The total offspring reduction in the exposures would not be a good outcome because this evidenced for the severe impacts of the mixtures of BPA and the metals. The environmental relevant concentrations of BPA, Pb and Cd not only impaired the survival of mother *D. magna* but also inhibited the population size of offspring *D. magna*. Therefore, in the nature, the influences of mixed pollutants over a longer time (e.g. more than three weeks) and across multi-generations of the *D. magna* would lead to an extinction of the micro-crustacean population, consequently imbalanced aquatic ecosystems. Investigations on the impacts of these pollutants on zooplankton in nature are suggested.

4. Conclusions

Our results revealed that exposures to the mixtures of BPA and the two metals (Cd and Pb) could result in numerous effects on the feeding activity and life history traits of the freshwater micro-crustacean, *D. magna*. In terms of acute exposure, we found that all the mixtures of these chemicals (BPA+Pb, BPA+Cd, and BPA+Pb+Cd) strongly stimulated the feeding activity of the organisms. On the other hand, the survival rate of *D. magna* in the chronic exposure to these mixtures was significantly declined. Besides, although the mixture of BPA and Pb did not adversely affect the maturation and reproductive performance of *D. magna*, exposures to the BPA+Cd and BPA+Pb+Cd could lead to an impairment in the reproduction and a delay in the maturity age of the organisms. Regarding joint toxicity, we found the synergistic effects of BPA+Pb on the survival and reproduction of *D. magna*, but there was an antagonistic effect of Pb+Cd on the survival of the organisms (stronger impacts of BPA+Cd than BPA+Pb+Cd). Our findings indicated that the combinations of the two metals Cd and Pb at the permissible levels (according to the technical regulation for surface water quality of Vietnam; QCVN) and the plastic additive BPA (which has not been included in QCVN) could cause detrimental effects on the organisms. Therefore, further investigations should be conducted to adjust the safety guidelines related to these pollutants for environmental and ecological protection. Our results not only enrich the knowledge on the combined toxicity of BPA and the two non-essential metals (Pb and Cd) but also could be valuable for monitoring and risk assessments of these pollutants in aquatic environment.

Acknowledgement
We acknowledge the support of time and facilities from Ho Chi Minh City University of Technology (HCMUT), VNU-HCM for this study.

References

[1] Hermabessiere L, Dehaut A, Paul-Pont I, Lacroix C, Jezequel R, Soudant P, and Duflos G 2017 Occurrence and effects of plastic additives on marine environments and organisms: a review *Chemosphere* **182** 781-793
[2] Wu NC and Seebacher F 2020 Effect of the plastic pollutant bisphenol A on the biology of aquatic organisms: a meta-analysis *Glob. Chang. Biol.* **26(7)** 3821-3833.
[3] Li D, Chen H, Bi R, Xie H, Zhou Y, Luo Y and Xie L 2017 Individual and binary mixture effects of bisphenol A and lignin-derived bisphenol in *Daphnia magna* under chronic exposure *Chemosphere* **191** 779-786
[4] Lithner D, Nordensvan I and Dave G 2012 Comparative acute toxicity of leachates from plastic products made of polypropylene, polyethylene, PVC, acrylonitrile–butadiene– styrene, and epoxy to *Daphnia magna Environ. Sci. Pollut. Res.* **19** 1763-1772
[5] Bradley EL, Read WA and Castle L 2007 Investigation into the migration potential of coating materials from cookware products *Food Addit. Contam.* **24(3)** 326-335
[6] Ugboka UG, Ihedioha JN, Ekere NR and Okechukwu FO 2020 Human health risk assessment of bisphenol A released from polycarbonate drinking water bottles and carbonated drinks exposed to sunlight in Nigeria *Int. J. Environ. Anal. Chem.* 1-11
[7] Tam LTM, Dan NP, Tuc DQ, Huu HN and Chi DHL 2016 Presence of e-EDCs in surface water and effluents of pollution sources in Sai Gon and Dong Nai River *Sustainable Environment Research* **26(1)** 20-27
[8] Jambeck JR, Geyer R, Wilcox C, Siegler TR, Perryman M, Andrady A, Narayan R and Law KL 2015 Plastic waste inputs from land into the ocean *Science* **347(6223)** 768-771
[9] Rai PK 2009 Heavy metals in water, sediments and wetland plants in an aquatic ecosystem of tropical region, India *Environ. Monit. Assess.* **158** 433-457
[10] Ning L, Liyuan Y, Jirui D and Xugui P 2011 Heavy metal pollution in surface water of Linglong gold mining area, China *Procedia Environ. Sci.* **10** 914-917
[11] Sharma P and Dubey RS 2005 Lead toxicity in plants *Braz. J. Plant Physiol.* **17(1)** 35-52
[12] Hossain MS, Latifa GA, Prianqa and Nayeem AA 2019 Review of cadmium pollution in Bangladesh *J. Health Pollut.* **9(23):190913** 1-10
[13] Dang TH and Alexandra C 2014 Preliminary assessment of dissolved and particulate heavy metals transported by the red river at the Sontay station (in Vietnamese) *Vietnam journal of earth science* **36(3)** 281-288
[14] Jaishankar M, Tseten T, Anbalagan N, Mathew BB and Beeregowda KN 2014 Toxicity, mechanism and health effects of some heavy metals *Interdiscip Toxicol.* **7(2)** 60-72.
[15] Flora G, Gupta D and Tiwari A 2012 Toxicity of lead: a review with recent updates *Interdiscip. Toxicol.* **5(2)** 47-58
[16] Araujo GS, Pavlaki MD, Soares AMVM, Abessa DMS and Loureiro S 2019 Bioaccumulation and morphological traits in a multi-generation test with two *Daphnia* species to lead *Chemosphere* **219** 636-644.
[17] Sterner R 2009 Role of zooplankton in aquatic ecosystems *Encyclopedia of Inland Water Academic Press* 678-688
[18] Adema DMM 1978 *Daphnia magna* as a test animal in acute and chronic toxicity tests *Hydrobiologia* **59(2)** 125-134
[19] Lampert W 2006 *Daphnia*: model herbivore, predator and prey *Pol. J. Ecol.* **54(4)** 607-620

[20] American Public Health Association (APHA) 2012 *Standard Methods for the Examination of Water and Wastewater* (22nd edition) American Public Health Association, American Water Works Association, Water Environment Federation

[21] Kim H, Yim B, Bae C and Lee YM 2017 Acute toxicity and antioxidant responses in the water flea *Daphnia magna* to xenobiotics (cadmium, lead, mercury, bisphenol A, and 4-nonylphenol) *J. Toxicol. Environ. Health Sci.* **9 (1)** 41–49

[22] Liu Y, Zhengyu Y, Zhang L, Deng Z, Yuan J, Zhang S, Chen J and Guo R 2019 Food up-take and reproduction performance of *Daphnia magna* under the exposure of bisphenols *Ecotoxicol. Environ. Saf.* **170** 47-54

[23] Park SY and Choi J 2009 Genotoxic effects of nonylphenol and bisphenol A exposure in aquatic biomonitoring species: freshwater crustacean, *Daphnia magna*, and aquatic midge, *Chironomus riparius Bull. Environ. Contam. Toxicol.* **83** 463-468

[24] Jemec A, Tisler T, Erjavec B, Pintar A 2012 Antioxidant responses and whole-organism changes in *Daphnia magna* acutely and chronically exposed to endocrine disruptor bisphenol A *Ecotoxicol. Environ. Saf.* **86** 213-218

[25] Jeong SW, Lee SM, Yum SS, Iguchi T, Seo YR 2013 Genomic expression responses toward bisphenol-A toxicity in *Daphnia magna* in terms of reproductive activity *Mol. Cell. Toxicol.* **9** 149-158

[26] Nagato EG, Simpson AJ and Simpson MJ 2016 Metabolomics reveals energetic impairments in *Daphnia magna* exposed to diazinon, malathion and bisphenol-A *Aquat. Toxicol.* **170** 175-186

[27] Berglind R, Dave G and Sjobeck ML 1985 The effects of lead on aminolevulinic acid dehydratase activity, growth, hemoglobin content, and reproduction in *Daphnia magna Ecotoxicol. Environ. Saf.* **9** 216-229.

[28] Bodar CWM, Leeuwen CJV, Voogt PA and Zandee DI 1988 Effects of cadmium on the reproduction strategy of *Daphnia magna Aquat. Toxicol.* **12** 301-310

[29] Luciana R, Reno U, Gervasio S, Troiani H and Gagneten AM 2014 Effect of metals on *Daphnia magna* and cladocerans representatives of the Argentinean Fluvial Littoral *J. Environ. Biol.* **35** 689-697

[30] Perez E and Hoang TC 2017 Chronic toxicity of binary-metal mixtures of cadmium and zinc to *Daphnia magna Environ. Toxicol. Chem.* **36(10)** 2739-2749

[31] Quach LT, Vo TMC, Nguyen VT and Dao TS 2019 Chronic effects of lead and arsenic on life history traits of *Daphnia magna VJSTE.* **61(4)** 82-87

[32] Bodar CWM, Sluis IVD, Voogt PA and Zandee DI 1988 Effects of cadmium on consumption, assimilation and biochemical parameters of *Daphnia magna*: possible implications for reproduction *Comp. Biochem. Physiol.* **90C(2)** 341-346

[33] Moraitou-Apostolopoulou M and Verriopoulos G 1982 Individual and combined toxicity of three heavy metals, Cu, Cd and Cr for the marine copepod *Tisbe holothuriae Hydrobiologia* **87** 83-87

[34] Mahar AM and Watzin MC 2005 Effects of metal and organophosphate mixtures on *Ceriodaphnia dubia* survival and reproduction *Environ. Toxicol. Chem.* **24(7)** 1579-1586

[35] Cooper NL, Bidwell JR and Kumar A 2009 Toxicity of copper, lead, and zinc mixtures to *Ceriodaphnia dubia Ecotoxicol. Environ. Saf.* **72** 1523-1528

[36] Meng Q, Li X, Feng Q and Cao Z 2008 The acute and chronic toxicity of five heavy metals on the *Daphnia Magna 2nd Int. Conf. Bioinform. Biomed. Eng.* 4555-4558

[37] Dao TS, Do-Hong LC and Wiegand C 2010 Chronic effects of cyanobacterial toxins on *Daphnia magna* and their offspring *Toxicon* **55** 1244–1254

[38] US. Environmental Protection Agency (US. EPA) 2002 Methods for measuring the acute toxicity of effluents and receiving waters to freshwater and marine organisms (5th ed.) Office of water, Washington, DC

[39] Jesus FT, Martins C and Nogueira AJ 2014 Changes in life-history parameters of *Daphnia longispina* (Cladocera, Crustacea) as a function of water chemistry *J. Limnol.* **73(2)** 340-349

[40] Sournia A 1978 *Phytoplankton Manual* UNESCO, UK, 77

[41] Villavicencio G, Urrestarazu P, Carvajal C, De Schamphelaere KAC, Janssen C, Torres JC and Rodriguez PH 2005 Biotic ligand model prediction of copper toxicity to daphnids in a range of natural waters in Chile *Environ. Toxicol. Chem.* **24(5)** 1287-1299

[42] Gulati RD, Bodar CWM, Schuurmans ALG, Faber JAJ and Zandee DI 1988 Effects of cadmium exposure on feeding of freshwater planktonic crustaceans *Comp. Biochem. Physiol. C Comp. Pharmacol.* **90(2)** 335-340

[43] Hernández-Flores S and Rico-Martínez R 2006 Study of the effects of Pb and Hg toxicity using a chronic toxicity reproductive 5-day test with the freshwater rotifer *Lecane quadridentata Environ. Toxicol.* **21** 533-540

[44] Lampert W 1977 Studies on the carbon balance of *Daphnia pulex* De Geer as related to environmental conditions. II. The dependence of carbon assimilation on animal size, temperature, food concentration and diet species *Arch. Hydrobiol.* **48** 310-335

[45] Gustafsson S and Hansson LA 2004 Development of tolerance against toxic cyanobacteria in *Daphnia Aquat. Ecol.* **38** 37-44

[46] Giraudo M, Douville M and Houde M 2015 Chronic toxicity evaluation of the flame retardant tris(2-butoxyethyl) phosphate (TBOEP) using *Daphnia magna* transcriptomic response *Chemosphere* **132** 159-165

[47] Dao TS, Ortiz-Rodrigues R, Do-Hong LC and Wiegand C 2013 Non-microcystin and non-cylindrospermopsin producing cyanobacteria affect the biochemical responses and behavior of *Daphnia magna Int. Rev. Hydrobiol.* **98** 235-244

[48] Di Toro DM, Allen HE, Bergman HL, Meyer JS, Paquin PR and Santore RC 2001 Biotic ligand model of the acute toxicity of metals. 1. Technical basis *Environ. Toxicol. Chem.* **20** 2383-2396

[49] Brennan SJ, Brougham CA, Roche JJ and Fogarty AM 2006 Multi-generational effects of four selected environmental oestrogens on *Daphnia magna Chemosphere* **64** 49-55

[50] Wang Y, Wang T, Ban Y, Shen C, Shen Q, Chai X, Zhao W and Wei J 2018 Di-(2-ethylhexyl) phthalate exposure modulates antioxidant enzyme activity and gene expression in juvenile and adult *Daphnia magna Arch. Environ. Contam. Toxicol.* **75(1)** 145-156

[51] Seyoum A and Pradhan A 2019 Effect of phthalates on development, reproduction, fat metabolism and lifespan in *Daphnia magna Sci. Total Environ.* **654** 969-977

The potential emission of personal care products derived plastic microbeads: a case study of Ho Chi Minh City, Vietnam

T T T Hoang[1,*], B A Nguyen[2], N N Q Pham[2], N B Nguyen[2], T K T Tran[2], T C L Tu[1], T D Huynh[1], T N T Nguyen[1], T P Q Pham[1], T V Nguyen[1], T L Nguyen[1], Q H Le[1]

[1] *Faculty of Geology and Mineral Resources, Ho Chi Minh City University of Natural Resources, Ho Chi Minh City, Vietnam*

[2] *Faculty of Environment and Labor Safety, Ton Duc Thang University, Ho Chi Minh City, Vietnam*

**Email: httthuy@hcmunre.edu.vn[1]*

ABSTRACT: Plastic microbeads are commonly used in many personal care products and can cause adverse impacts to the environment and ecosystem. The toxicological problem with these pollutants are due to their non-biodegradable materials, which washed down the drain; end up accumulating in the aquatic system causing increased frequency and quantity of items ingested by biota. Several polymers (e.g. Polyethylene) especially those found in plastic microbeads have been reported to be in tandem with other toxic contaminants serving as a vector for their transports in the environment. Thus, the legislative ban for plastic microbeads is used in some developed countries, but many countries including Vietnam do not take any legal action. This present study aimed at potential microbead's existence in the cosmetic market of Ho Chi Minh City (HCMC). The list and ingredients of microbeads containing personal care products (toothpaste, facial cleanser/scrubs and body wash/scrubs) have been checked. The microbeads containing PCPs are common for all explored categories, especially in toothpaste. Data from the online questionnaire survey have shown that 98% of respondents have frequently used at least one microbead containing product. Four polymers (Polyethylene, Acrylates Copolymer Styrene/Acrylates Copolymer and Acrylates/C10-30 Alkyl Acrylate Crosspolymer) have been observed in the product package of several facial cleansers and body scrubs. Thus, the potential negative impacts of this contaminant should not be ignored.

Keywords: *microplastics, microbeads, personal care product, pollution.*

1. Introduction

The pollution caused by microplastics is of growing environmental concern [1]. Microplastics are defined as plastic items sized between 100 nm and 5 mm. Plastic microbeads, which are ingredients of personal care products (PCPs), have been considered as a significant primary source of microplastics in the environment. Microbeads were patented in 1972 [2] as an alternative to the inorganic (e.g. aluminum oxide) and/or natural materials (millet or pumice peels) in a wide range of PCPs such as (but not limited to) soap, shampoo, toothpaste, shaving cream, facial masks and children's bubble bath [3,4]. Plastic ingredients are present in PCPs at different percentages, ranging from a fraction of a percent to more than 90% in some cases. Regarding the frequent usage, the Government of Canada listed 9 common and 11 possible plastic ingredients of PCPs [5]. Major functions of the microbeads were replaced for natural material as abrasive. Its other functions serve as bulking agent, viscosity regulators, film formers, aesthetic agent, and sorbent for delivery of active ingredients. It is noted that the mentioned above functions go beyond the well-known scrubbing effect of microbeads [6].

In the recent literature, PCPs that are categories as facial cleansers and scrubs are the most studied; others, such as toothpaste, and shower gel/body scrubs have received less attention. Actually, there is more to 'microbeads' than meets the eye – while some are large enough to be easily visible to the naked eye, other microbeads on the market for PCP's formulations are as small as 1 μm. Others are even smaller than that (nano-particulates). The microbeads have been extracted from PCPs with filtration after being dissolved with ultra-pure/distilled water. The granular distribution of microbeads could be determined by a combination of different techniques namely microscopic, spectroscopic, utilizing the naked eye and instruments will help to recognize smaller microplastics [7]. Recently single particle inductively coupled plasma mass spectrometry (SP-ICP-MS) has proved to be a feasible methodology for the detection and size characterization of plastic microparticles. Laborda et al. [8] have used SP-ICP-MS to detect microbeads smaller after filtration through 10 μm pore size polycarbonate IsoporeTM membranes. Visual examination by optical microscopy is commonly used for detecting, sorting and counting microplastics over 50 μm [8]. In addition to that, another technique has been developed. For example, Fourier transform infrared (FT-IR) or Raman techniques in combination with optical microscopy (FT-IR and Raman microscopies) will allow for the detection of the chemical and morphological (size and shape) characterization and the number quantification of individual particles down to 10 and 1 μm [9]. The number of extracted microplastics particles from any media is dependent on filter size used in the extraction step so the concentrations of microbeads were relatively low as most their frequent equivalent diameter were 2 to 3 μm [8].

In the PCPs, microbeads have different colors such as white, transparent, opaque, blue, red, orange, etc. Despite the compound "bead" which refers to spherical particles, and the common belief that microbeads are the colored spherical microparticles that can be seen in toothpastes [10,11,12], microbeads are mostly of irregular shape to provide better abrasion [13,14]. In fact, the spherical colored beads are used mainly for decoration and not abrasion [15]. Almost all PCPs have more than one type of bead shape [11] and the addition of irregularly shaped microplastics provides larger specific surface areas and greater friction [16]. In addition, the irregular but typical bead shapes can be attributed to the production process: they are produced by ultrafine grinding in a mill [17].

Microbeads quantity showed wide variation in all of the PCP's categories as standard deviation higher than mean values (Fig 1). Obviously, the higher average concentrations were found in the case of facial and body scrubs (129±571 items/mg) and (333±483 items/mg), respectively. Facial cleanser and shower gel contain significantly lower rates of microbeads (0.22±0.41 items/mg and 0.25±0.43 items/mg, respectively). The data of microbeads in toothpastes was scarce. The value of 1.18 g/mg has been reported with Bråte et al. [18]. The extremely high mean values of facial scrub (3.10×10^{8} items/mL, [8]) could be explained by the different analytical methods.

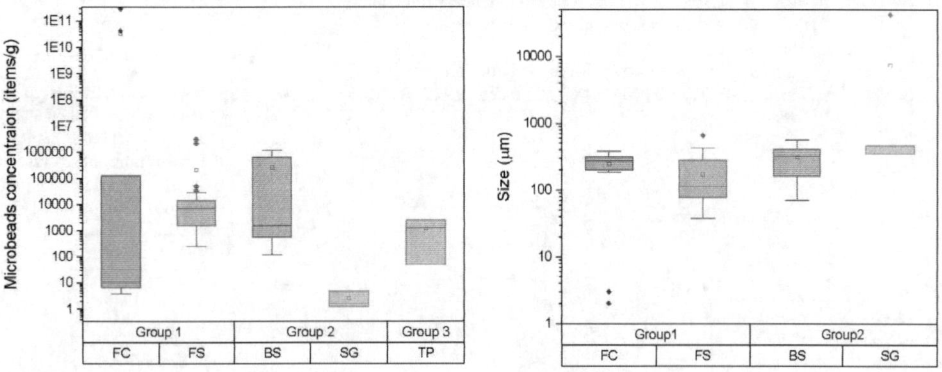

Fig 1. Box-plot of concentration and size of plastic microbeads in PCPs (the detailed data are shown in Appendix B). Three groups have been divided based on the part of human body: (1): FC: Facial cleanser and FS: Facial scrub, (2): Shower gel and body scrub and (3) Toothpaste.

The usage of microbeads deriving PCPs has increased since early 2000s. Fendall and Sewell [10] estimated that every household was using at least one scrub containing microbeads, on a daily or at least on a weekly basis. Plastic microbeads are non-biodegradable materials, and can get washed down the drain and accumulate in receiving water bodies. The first publication that mentioned a threat to the environment due to microbeads released from PCPs was published by Zitko and Hanlon [19]. However at that time, this issue was not recognized globally, because of their seemingly low input into the marine environment at the time (0.2 t/yr for New Zealand) [20]. From households, microbeads are likely to be transported to wastewater treatment plants, where some will be captured in oxidation ponds or sewage sludge [10]. However, due to their small size, they will go straight through the treatment system [21] without getting trapped. Approximately 1.4% [22] to 5% [23] of microplastics are released into the environment after treatment. In developing countries, microbeads can be discharged directly into the receiving area without going through the wastewater treatment system. This significantly increases the amount of microbeads released into the environment. Recently, a much larger input has been noted in many regions and a growing body of study has focused on this emerging pollutant. It is estimated that more than 4,130 tons of microbeads were added to cosmetics in the EU market in 2012, including the Nordic countries [24] while Duis and Coors [25] estimated that 8 g of microplastics from cosmetics per capita per year.

Upon entering the aquatic system, plastic microbeads can be ingested by several organisms, often through the trophic chain [11,16]. The compiled data showed that more than 95% of PCP-derived plastic microbeads have diameter fewer than 350 μm so these could be easily digested by aquatic biota [16]. In addition, like other microplastic, microbeads can act as vectors for toxic pollutants (dioxin, PCBs, PAHs, heavy metals, etc.) and may pose risks to aquatic environments due to their very long presence in marine ecosystems and propensity to be ingested by biota [13,26].

In order to limit and reduce the amount of microbeads discharged into the aquatic ecosystem every year, a number of developed countries have enacted legislation banning the use/production of PCPs containing microbeads, which proves that microbeads will drop quickly in the future. Some countries have enacted laws to ban plastic microbeads in PCPs, such as the UK, USA, Thailand (Fig 2). However, there are still many other developing countries including Vietnam that are still expanding the market for PCPs and have not taken any strict legal action.

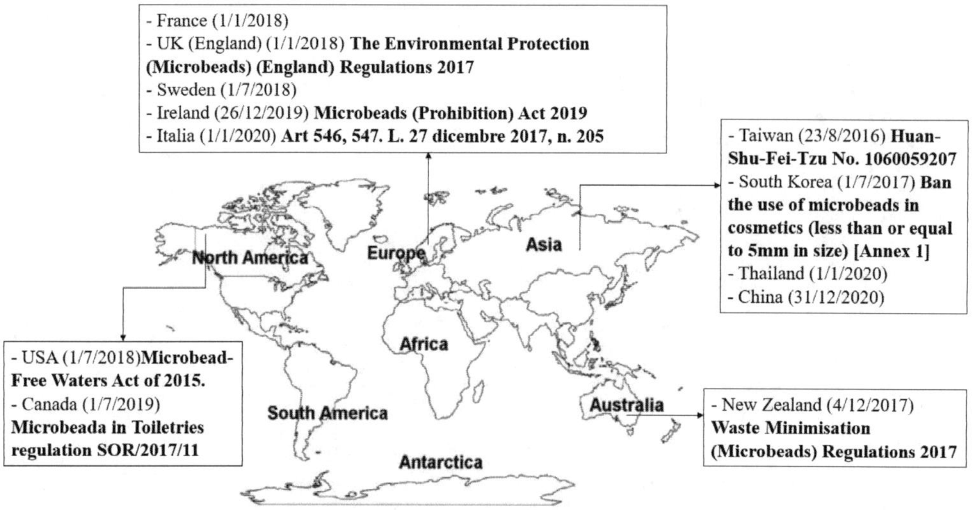

Fig 2. Regulations related with microbeads global.

This paper provides the updated information with regards to the currents situation of PCPs deriving microbeads and the potential risks to residents in Ho Chi Minh city (HCMC). In addition to conclusive remarks, future directions on how to enhance community awareness as a priority measure to reduce microplastic contamination are also suggested.

2. Study area and methods

2.1. Study area

HCMC is an attractive city and located in the south-eastern region of Vietnam. According to the 2019 census, the City has a population of over 8.9 million within the city's border and over 21 million within its metropolitan area. HCMC is the economic center of Vietnam and accounts for a large proportion of the economy of Vietnam. The tropical climate together with the rising living standards, have caused cosmetics to be incorporated into communities. Together with the capital city of Hanoi, 90-95% of the market for international personal care products is centered in these cities [27]. The average expense for skincare products in HCMC is higher than any region in Vietnam and estimated to be about 14 U$ per month. Up to 73% use skincare products once per week or more often [28]. Regularly-used skincare products are facial cleansers. The personal care sector of HCMC has strong growth of 7 percent a year on average. Especially, the rate of municipal wastewater treatment was only 21% of 1.5 million m^3/day that 79% of domestic wastewater discharged directly into the surface water receiving in HCMC [29]. Therefore, HCMC has been selected for the present study.

2.2. Method

2.2.1. Market survey

HCMC is an attractive and expanding market for PCPs, which have been increasingly consumed by the last few decades. To meet the growing demand, a lot of PCPs are imported in addition to local shops. Many multinational companies in HCMC, (namely Proctor & Gamble, Unilever, Johnson & Johnson, etc.) are up against strong competition from local firms, such as Dalan, Myhao, etc (Appendix C). With the miracle effectiveness through extensive advertising by producers together with widespread distribution in the market, microbeads containing PCPs became ordinary in HCMC. A brief overview of the potential microbead's existence in the cosmetic market of HCMC has been explored. Three supermarkets (Lotte Mart, Coopmart, and Bachhoaxanh) in District 7 were twice visited, which are present in almost all HCMC. Due to the Covid 19 pandemic, the survey has been further conducted through the websites of these supermarkets and two other cosmetic shops in HCMC (Guardian and Hasaki). Firstly, beads containing products from four different categories (toothpaste, facial cleanser/scrubs, and body wash/scrubs) were explored. Then, the ingredients will be checked to determine if this product contains plastics or not. However, it is not easy to find out the ingredients of the PCPs since it was not available on the shop's website except the websites from Bachhoaxanh supermarket, where all the ingredients of PCPs are shown.

2.2.2. Community awareness of microbeads in PCPs

Besides the legislative action, community awareness is important because it can increase their knowledge base to influence purchasing decision. Therefore, a quantitative analysis was carried out in the form of an online questionnaire. The survey period extended over five weeks from 12th May to 19th June 2021. The survey was performed with an online survey (Google Forms). The survey asked participants for their opinion about (1) beads/microbeads containing PCPs usage and (2) consumer product decisions. The consumer product decisions are based on the following hypothesis:
- H1. The customer chooses microbeads containing PCPs because of their effectiveness in comparison with other products;
- H2. The microbeads containing PCPs is the new trend of the cosmetic industry;
- H3. The natural ingredients in PCPs are important to make a decision of buying PCPs;
- H4. The usage of microbeads from PCPs could cause environmental pollution.

Five- level Likert scale, ranging from (0) Strongly disagree; (1) Disagree; (2) Fairy Agree; (3) Agree to (4) Strongly agree, has been used (Appendix B). A total of 200 respondents, between 18 to 40 years old, mainly students and officers, have participated in this survey (Table 1). The data from five respondents, who do not use any beads containing PCPs, are not further processed.

Table 1. The data of respondents.

Variable	Number	Percentage (%)
Age (years)		
18–40	156	78
>40	44	22
Gender		
Male	118	59
Female	82	41

2.2.3. Data processing

Statistical analysis has been performed with SPSS Version 11.0 statistical software package. The statistical reliability was first tested to assess the overall consistency of the psychometric questions. The Cronbach's alpha coefficient was used to estimate the internal consistency of the Likert scale.

3. Results and discussions

3.1. The present situation of the beads containing PCPs in HCMC and the potential risks

It has been observed that it is not difficult to find microbeads containing PCPs in the supermarkets and/or shops of HCMC. The microbeads containing PCPs are common for all explored PCPs categories, especially in toothpaste (Table 2, Appendix C). The present study showed that 17 brands from 29 brands of toothpaste, 29 from 123 brands of facial cleansers/scrubs, and 30 from 426 brands body wash/ shower gels, respectively, contain microbeads. The percentages of microbeads containing PCPs in each category could be arranged as follow: Toothpaste (58%) >Facial cleanser/scrubs (18%) >Body wash/Scrubs (6%).

However, for HCMC, not all PCPs list plastic in their ingredient list. In the case of toothpaste, none of the polymers are listed. In the case of facial cleansers and scrubs, 23% of microbeads containing products have listed polymers as Polyethylene, Acrylate Copolymer, and Acrylates/C10-30 Alkyl Acrylate Crosspolymer. For body wash and scrubs, the polymers including Polyethylene, Acrylate Copolymer, Styrene/Acrylates Copolymer, and Acrylates/C10-30 Alkyl Acrylate Crosspolymer have listed in 53% of microbeads containing products (Table 2).

Table 2. The microbeads containing PCPs in HCMC's market.

Categories	Shop/super-market	Number of microbeads containing products/ Total	Type and number of products of polymer
Toothpaste	Coopmart, Lotte Mart, Bachoaxanh	17/29	0
Facial cleanser/scrubs	Coopmart, Lotte Mart, Bachhoaxanh, Hakasi, Guardian	29/123	Polyethylene (3) Acrylates Copolymer (2) Acrylates/C10-30 Alkyl Acrylate Crosspolymer (7)
Body wash/scrubs	Coopmart, Lotte Mart, Bachhoaxanh, Hasaki	30/426	Polyethylene (2) Acrylates Copolymer (4); Styrene/Acrylates Copolymer (7) Acrylates/C10-30 Alkyl Acrylate Crosspolymer (4)

This is not consistent with previous studies, which stated that plastic ingredients are present in all PCPs at different percentages, ranging from a fraction of a percent to more than 90% in some cases. According to physical characteristics, the plastic microbeads could be classified as thermoplastics e.g. Polyethylene, Polypropylene, Polystyrene, Polyamide, Polytetrafluoroethylene (Teflon), and thermoset plastics, e.g. Polyurethanes and certain Polyesters. All thermoset plastics are solid-phase materials. Thermoplastics are also solid materials but they can be melted into liquids when they are heated to temperatures exceeding their melting point (or glass transition temperature). Since the melting points far exceed the temperatures in the marine environment, these plastics are considered solid too. Based on the polymer chain, the microbeads have defined homopolymers and copolymers. Homo-polymers are polymer chains of a single monomer type whereas copolymers are made by polymerizing different monomers in the same chain, either in random order, alternating monomers, or as 'block' copolymers (i.e. monomers clustered into blocks in the polymer chain of the copolymer molecule) [6].

More precisely, polyethylene has been found in 10.3% (3 products made in India and Thailand from 29 microbeads containing products) of facial cleanser/scrub, and 6.7% (2 products made in Malaysia from 30 microbeads containing products) of body wash/scrub (Appendix B). In comparison with previous studies, it can be seen that although microbeads containing PCPs are common in the HCMC market but Polyethylene was not always present in their ingredients. This is different from previous studies. For example, Godoy et al. [30] explored eleven supermarkets in the province of Granada (Spain) and found a higher percentage of products containing Polyethylene particles in ingredients as 50% (2 products; bath gel), 63% (12 products; body treatment), 27.5% (11 products: facial treatment) and 80% (4 products; foot treatment), respectively. Furthermore, Napper et al. [12] investigated six main brands of facial scrubs, based on the prevalence in major supermarkets close to Plymouth UK. The results have shown that all of the products listed in their ingredients contained Polyethylene [12]. According to

Fendall and Sewell [10], all four water-based facial cleansers were purchased at a supermarket in Auckland, New Zealand, were produced by major cosmetic manufactures and containing Polyethylene in their ingredients. Similarly, Lei et al. [31] reported that based on the ingredient information on the packaging, Polyethylene was found in 7.1% (9 products from total 126 products) of facial cleansers, and 2.2% (3 products from total 136 products) of shower gels. Ustabashi et al. [32] has also mentioned that twenty percent of the toothpastes available in Turkey were found to contain Polyethylene at concentrations varying between 0.4 and 1%.

Although only a few names of plastics have been listed in the PCP's sold in HCMC the potential impacts of microplastic could not be neglected. Previous studies have reported that polymers could be present in additional products regardless of whether they are included in the ingredient list. Habib et al. [14] have also noted that two products did not show an ingredients list, which breaches UAE law. However, it was found later that microplastics were isolated from both products.

As mentioned above, microbeads can contribute to the total microplastic load in the aquatic system. However, rarely are these micrometer-sized primary microplastics distinguishable from secondary microplastics when detected in an environmental matrix. The microbeads are not unique; it is not easy to distinguish by shape or other indicators. Other than plastic preproduction pellets and small plastic objects <5 mm that are still recognizable, the exact origins of the tiny microbeads in the aquatic system are untraceable [6].

As indicated in previous studies, microbeads could be littered into the environment via final effluent or biosolids (sewage sludge) from wastewater treatment plants (WWTPs) [33]. Previous studies showed that 95–99.9% of the microbeads may settle out into the sludge, leaving the remaining beads in the effluent [1]. For HCMC, the total volume of domestic wastewater (the major source of microbeads) is estimated to be 1.75 million cubic meters, and 1.3 million cubic meters are collected and treated in WWTPs every day [34]. The low density of microbeads in the effluent may not sound significant but assuming all WWTPs in HCMC operate at maximum capacity and that minimum density of microbeads are found (0.1 per L of effluent, [1]) so at least 130,000 microbeads per day are emitted into the aquatic system of HCMC.

Thus, because of the large volume in production and current end-of-life strategy, microbeads have the potential to be a source of many microplastic particles to aquatic habitats. Like other microplastics, they are ubiquitous and bioavailable for injection by marine organisms [6,15,35]. Hence, it is noted that the release of microbeads of a size suitable for ingestion by marine organisms without degradation and therefore became available. Microplastics can mix with plankton and sediments, causing suspended or detritivorous organisms to ingest them by accident, since they are not always able to distinguish them from food [36]. The danger increases when these small organisms are eaten by larger organisms such as birds, marine mammals, turtles, fishes, etc. which causes microplastics to be transferred along the food chain until they reach human beings [36,37]. Hence, the high sorptive capacity of toxic contaminants has also increased microbeads' toxicity. In HCMC, besides Polyethylene, other abundant polymer types have been indicated in PCPs are Acrylates Copolymer, Styrene/Acrylates Copolymer, and Acrylates/C10-30 Alkyl Acrylate Crosspolymer. However, the toxicity of these polymers was not clear. Thus, the following research needs to explore toxicity when these polymers are released into the environment.

However, the environmental regulations and microbeads management practices in Vietnam are not well developed. Therefore, microbeads released from PCPs could impose an additional burden on the natural environment.

3.2. Community awareness of microbeads in PCPs and future actions

The usage of microbeads containing PCPs in HCMC has been obtained from a questionnaire survey. Since the beads/microbeads containing PCPs are easily found in shops and supermarkets, almost (98%) of the respondents have used at least one product. A list of beads/microbeads containing PCPs including facial cleansers (12 brands), toothpastes (3 brands), body scrubs (4 brands), and shower gel (1 brand), were identified by respondents. In fact, it was discovered the usage of microbeads containing PCPs was

32% (shower gel), 12.5% (facial cleanser), 12% (body scrub), and 92% (toothpaste) of respondents, respectively. There are no differences between gender for the beads containing PCPs' usage, both sexes have increased their demands for quality personal care and cosmetics products which may directly introduce a large number of microplastics (Appendix A).

In statistical analysis, Cronbach's alpha is commonly used to estimate the reliability of the Likert scale. The Cronbach's alpha coefficients were 0.829, which is above 0.60, representing good reliability [38]. Thus, the Likert scales can be considered reliable. The data showed that the majority of consumers placed great importance on product effectiveness ("microbeads containing PCPs are better to use than others") and were highly influenced by new trend of PCPs "microbeads containing PCPs represent a new trend in the cosmetic industry" (69% and 68% of total respondents, respectively). It is expected since most of the respondents are students, so they always keep up-to-date with a new trend of PCPs and therefore will follow to use beads containing PCPs. In addition, the producers always promote the effectiveness of microbeads containing PCPs as totally remove of dead skin cells, deeply embedded impurities but less irritation and damage to the skin (facial scrubs) and remove plaque and stains due to their abrasive action (toothpaste). This is similar to a survey carried out with UC Berkeley students, which showed that majority of students placed great importance on product effectiveness [39]. Furthermore, the respondents specified that they paid attention to natural materials of beads. However, a significant part of respondents (35% of total respondents) were not concerned with the potential negative impacts of beads/microbeads. It is contrary to what was with Malaysia's customers. Praveena et al. 2018 indicated that 62.6% of the respondents acknowledged the very and extremely importance of taking into consideration the environmental impacts resulting from microplastics in personal care products [40].

Therefore, the present study indicated that even students, who belong to a young and dynamic generation of HCMC, do not care about the potential environmental pollution caused by microbeads containing PCPs. In order to reduce the emission of PCP's microbeads, community awareness should be taken into account soon in HCMC. A greater sense of awareness among the general public is to inform the consumers about the negative impacts of microbeads. The campaigns coupled with hashtags such as "#banthebead" could begin in the University to generate awareness among students [39]. Furthermore, legislative action should also go into practice. For example, the consumer should be informed on whether a cosmetic product contains polymeric compounds from its packaging, as it is obligatory from the EU's legislation [41]. Nevertheless, when a cosmetic product is intended for cleansing or abrasion and simultaneously contains a polymeric compound, like Polyethylene, which is indicated high on the INCI list (meaning large quantity), it can be concluded with great certainty that the compound is a microbead [41].

4. Conclusions

The term 'microbeads' is used to describe plastic particles present as ingredients in personal care products and have recently created an environmental threat to ecosystems. In several developed countries (e.g. USA, Canada, France) a legislative ban on plastic microbeads-based PCPs has been proven effective to protect environment. However, other countries including Vietnam, besides the extended market for PCPs, do not take any legal action.

This summary of some primary results related to the potential risks caused by microbeads-based PCPs in HCMC. Approximately 50% of respondents have frequently used facial cleansers and body scrubs, which contain plastic microbeads shown in the product package. Instead of Polyethylene, other plastics such as Acrylates Copolymer, and Styrene/Acrylates Copolymer are recent common polymers in PCP. Hence, it should be noted that although the plastic composition is not shown by the manufacturers, this does not mean that these PCPs are free of microplastics. On the other hand, it is also evident that consumers in HCMC have insufficient knowledge about the negative impacts of plastic microbeads. This fact suggests that source reduction through community awareness is the most efficient method to reducing plastic contamination in this city.

Acknowledgment
Research project granted by Ho Chi Minh City University for Natural Resources and Environment under project "Investigation of microplastic's emission from households in Ho Chi Minh City". The authors thank Ms. Faith Su for English editing.

References

[1] Rochman CM, Kross SM, Armstrong JB, Bogan MT, Darling ES, Green SJ, Smyth AR and Veríssimo D 2015 Scientific Evidence Supports a Ban on Microbeads *Environ Sci Technol.* **49(18)** 10759-61 https://doi: 10.1021/acs.est.5b03909

[2] Beach W 1972 Skin cleaner US3645904A. https://patents.google.com/patent/US3645904.

[3] Sherrington C, Darrah C, Hann S, Cole G, Corbin M 2016 Study to support the development of measures to combat a range of marine litter sources. In *Report for European Commission DG Environment* (Bristol) https://doi.org/10.1001/ jamapediatrics.2013.502

[4] Juliano C and Magrini GA 2017 Cosmetic ingredients as emerging pollutants of environmental and health concern. A mini-review *Cosmetics* https://doi.org/ 10.3390/cosmetics4020011

[5] Government of Canada 2018 Microbeads in toiletries: method 445.0: chapter 1. https://www.canada.ca/en/environment-climate-change/services/canadian-environmental-protection-act-registry/publications/microbeads-toiletries-method-445-0/chapter-1.html. Accessed 5 May 2021

[6] UNEP 2015 Plastics in cosmetics, are we polluting the environment through our personal care? In: *Job Number: DEP/1918/NA* United Nations Environment Programme

[7] Song YK, Hong SH, Jang M, Han GM, Rani M, Lee J, Shim WJ 2015 A comparison of microscopic and spectroscopic identification methods for analysis of microplastics in environmental samples *Mar Pollut Bull* **93** 202–209 https://doi.org/10.1016/j.marpolbul.2015.01.015

[8] Laborda F, Trujillo C and Lobinski R 2021 Analysis of microplastics in consumer products by single particle-inductively coupled plasma mass spectrometry using the carbon-13 isotope. *Talanta* **221** 121486 https://doi.org/10.1016/j.talanta.2020.121486

[9] Xu J-L, Thomas KV, Luo Z, Gowen AA 2019 FTIR and Raman imaging for microplastics analysis: State of the art, challenges and prospects *TrAC Trends in Analytical Chemistry* **119** 15629 https://doi.org/10.1016/j.trac.2019.115629

[10] Fendall LS and Sewell MA 2009 Contributing to marine pollution by washing your face: Microplastics in facial cleansers *Mar. Pollut. Bull.* **58(8)** 1225–28 https://doi.org/10.1016/j.marpolbul.2009.04.025

[11] Cheung P K and Fok L 2017 Characterisation of plastic microbeads in facial scrubs and their estimated emissions in Mainland China *Water Res.* **122** 53–61 https://doi.org/10.1016/j.watres.2017.05.053

[12] Napper IE, Bakir A, Rowland SJ and Thompson RC 2015 Characterisation, quantity and sorptive properties of microplastics extracted from cosmetics. *Mar. Pollut. Bull.* **99(1-2)** 178–185 https://doi.org/10.1016/j.marpolbul.2015.07.029

[13] Guerranti C, Martellini T, Perra G, Scopetani G, Gincinelli A 2019 Microplastics in cosmetics: Environmental issues and needs for global bans Environ. Toxicol. Pharmacol. **68** 75-79 https://doi.org/10.1016/j.etap.2019.03.007

[14] Habib RZ, Salim Abdoon MM, Al Meqbaali RM, Ghebremedhin F, Elkashlan M, Kittaneh WF, Cherupurakal N, Mourad A-H I, Thiemann T, Kindi RA 2020 Analysis of microbeads in cosmetic products in the United Arab Emirate *Environ. Pollut.* **258** 113831. https://doi.org/10.1016/j.envpol.2019.113831

[15] Leslie HA 2014 Review of Microplastics in Cosmetics, 476(July) IVM Institute for Environmental Studies 1–33

[16] Sun Q, Ren S-Y and Ni H-G 2020 Incidence of microplastics in personal care products: An appreciable part of plastic pollution *Sci Total Environ.* **742** 140218. https://doi.org/10.1016/j.scitotenv.2020.140218

[17] Kalčíková G, Alič B, Skalar T, Bundschuh M and Gotvajn AŽ 2017 Wastewater treatment plant effluents as source of cosmetic polyethylene microbeads to freshwater *Chemosphere* **188** 25–31 https://doi.org/10.1016/j.chemosphere.2017.08.131

[18] Bråte ILN, Blázquez M, Brooks SJ and Thomas KV 2018 Weathering impacts the uptake of polyethylene microparticles from toothpaste in Mediterranean mussels (*M. galloprovincialis*). *Sci Total Environ.* **626** 1310–1318 https://doi.org/10.1016/j.scitotenv.2018.01.141

[19] Zitko V, Hanlon M 1991 Another source of pollution by plastics: skin cleaners with plastic scrubbers *Mar. Pollut. Bull.* **22(1)** 41-42 https://doi.org/10.1016/0025-326X(91)90444- W

[20] Gregory MR 1996 Plastic scrubbers' in hand cleansers: a further (and minor) source for marine pollution identified *Mar. Pollut. Bull.* **32(12)** 867–871 https://doi.org/ 10.1016/S0025-326X(96)00047-1

[21] Hidayaturrahman H and Lee TG 2019 A study on characteristics of microplastic in wastewater of South Korea: identification, quantification, and fate of microplastics during treatment process *Mar. Pollut. Bull.* **146** 696–702 https://doi.org/10.1016/ j.marpolbul.2019.06.071

[22] Murphy F, Ewins C, Carbonnier F and Quinn B 2016 Wastewater Treatment Works (WwTW) as a source of microplastics in the aquatic environment *Environ. Sci. Technol.* **50(11)** 5800-8 https://doi.org/10.1021/acs.est.5b05416

[23] European Chemicals Agency 2019 Annex XV Restriction Report. Proposal for a Restriction. Substance Name(s): Intentionally Added Microplastics Ver 1.2. (Helsinki, Finland) https://echa.europa.eu/documents/10162/05bd96e3-b969-0a7c -c6d0-441182893720

[24] Gouin T, Avalos J, Brunning I, Brzuska K, de Graaf J, Kaumanns J, Koning T, Meyberg M, Rettinger K, Schlatter H, Thomas J, van Welie R, Wolf T 2015 Use of microplastic beads in cosmetic products in Europe and their estimated emissions to the North Sea environment *Int J Appl Sci.* **141** 1-33 https://doi.org/10.1016/j.marpolbul.2016.05.046

[25] Duis K and Coors A 2016 Microplastics in the aquatic and terrestrial environment: sources (with a specific focus on personal care products), fate and effects *Environ. Sci. Eur.* **28(1)** 2 https://doi.org/10.1186/s12302-015-0069-y

[26] Torres FG, Dioses-Salinas DC, Pizarro-Ortega CI, De-la-Torre GE 2021 Sorption of chemical contaminants on degradable and non-degradable microplastics: Recent progress and research trends *Sci Total Environ.* **757** 143875 https://doi.org/10.1016/j.scitotenv.2020.143875

[27] https://qandme.net/en/report/research-vietnam-cosmetic-market-2019.html Accessed 30 June 2021

[28] https://viracresearch.com/consumption-of-skincare-industry-for-the-first-six-months-of-2020.html Accessed 30 June 2021

[29] Department of Natural Resources and Environment of Ho Chi Minh City 2018 *Annual report of environment quality monitoring of HCMC* (in Vietnamese)

[30] Godoy V, Martín-Lara MA, Calero M and Blázquez G 2019 Physical-chemical characterization of microplastics present in some exfoliating products from Spain *Mar. Pollut. Bull.* **139** 91–99. https://doi.org/10.1016/j.marpolbul.2018.12.026

[31] Lei K, Qiao F, Liu Q, Wei Z, Qi H, Cui S, Yue X, Deng Y, An L 2017 Microplastics releasing from personal care and cosmetic products in China. *Mar. Pollut. Bull.* **123(1-2)** 122–126 https://doi.org/10.1016/j.marpolbul.2017.09.016

[32] Ustabasi G.S and Baysal A 2019 Occurrence and risk assessment of microplastics from various toothpastes. *Environ. Monit. Assess.* **191(7)** 438 https://doi.org/10.1007/s10661-019-7574-1

[33] Carr SA, Liu J and Tesoro AG 2016 Transport and fate of microplastic particles in wastewater treatment plants. *Water Res.* **91** 174–182 https://doi.org/10.1016/j.watres.2016.01.002

[34] Au NH 2020 Project report *Estimation of wastewater load from Hochiminh City* (in Vietnamese)

[35] https://www.beatthemicrobead.org/get-to-know-microplastics-in-your-cosmetics-2/ Accessed 15 May 2021

[36] Lusher AL, Hollman PCH and Mendoza-Hill JJ 2017 Microplastics in fisheries and aquaculture: status of knowledge on their occurrence and implications for aquatic organisms and food safety. In: *FAO Fisheries and Aquaculture Technical Paper* 615 (Rome, Italy)

[37] GESAMP 2015. Sources, fate and effects of microplastics in the marine environment: a global assessment. In: Kershaw PJ (Ed) *Joint Group of Experts on the Scientific Aspects of Marine Environmental Protection* Reports and Studies (96 pp).

[38] Hair JF, Anderson RE, Tatham RL and Black WC 1998 *Multivariate Data Analysis*, 5th ed. (New Jersey: Prentice Hall Inc.)

[39] Chang M 2015 Reducing microplastics from facial exfoliating cleansers in wastewater through treatment versus consumer product decisions *Mar. Pollut. Bull.* **101(1)** 330–333 https://doi.org/10.1016/j.marpolbul.2015.10.074

[40] Praveena SM, Shaifuddin SNM and Akizuki S 2018 Exploration of microplastics from personal care and cosmetic products and its estimated emissions to marine environment: An evidence from Malaysia. *Mar. Pollut. Bull.* **136** 135–140. https://doi.org/10.1016/j.marpolbul.2018.09.012

[41] Anagnosti L, Varvaresou A, Pavlou P, Protopapa E and Carayanni V 2021 Worldwide actions against plastic pollution from microbeads and microplastics in cosmetics focusing on European policies. Has the issue been handled effectively? *Mar. Pollut. Bull.* **162** 111883 https://doi: 10.1016/j.marpolbul.2020.111883

IOP Conf. Series: Earth and Environmental Science **964** (2022) 012013 doi:10.1088/1755-1315/964/1/012013

Appendix A. The results of statistical analysis.

A.1. T-Test for gender difference

Group Statistics

	gender	N	Mean	Std. Deviation	Std. Error Mean
PCPs	Female	118	2.7966	1.59357	0.14670
	Male	82	3.4146	2.05451	0.22688

Independent Samples Test

		Levene's Test for Equality of Variances		t-test for Equality of Means						
		F	Sig.	t	df	Sig. (2-tailed)	Mean Difference	Std. Error Difference	95% Confidence Interval of the Difference	
									Lower	Upper
PCPs	Equal variances assumed	8.580	0.004	-2.393	198	0.018	-0.61802	0.25828	-1.12736	-0.10869
	Equal variances not assumed			-2.287	145.304	0.024	-0.61802	0.27018	-1.15201	-0.08404

A.2. Reliability of Likert scale.

Case Processing Summary

		N	%
Cases	Valid	195	100
	Excluded[a]	0	0
	Total	195	100

a. Listwise deletion based on all variables in the procedure.

Reliability Statistics

Cronbach's Alpha	N of Items
0.829	4

Item-Total Statistics

	Scale Mean if Item Deleted	Scale Variance if Item Deleted	Corrected Item-Total Correlation	Cronbach's Alpha if Item Deleted
H1	6.4872	8.932	0.722	0.753
H2	6.4410	9.134	0.673	0.776
H3	6.4410	9.145	0.710	0.760
H4	6.2923	10.012	0.527	0.841

A.3. Frequencies.

Statistics

		H1	H2	H3	H4
N	Valid	195	195	195	195
	Missing	0	0	0	0
Mean		2.07	2.11	2.11	2.26
Median		2.00	2.00	2.00	2.00
Std. Deviation		1.21	1.23	1.18	1.23

Frequency Table

H1: The microbeads containing PCPs are better

		Frequency	Percent	Valid Percent	Cumulative Percent
Valid	0.00	24	12.3	12.3	12.3
	1.00	38	19.5	19.5	31.8
	2.00	61	31.3	31.3	63.1
	3.00	45	23.1	23.1	86.2
	4.00	27	13.8	13.8	100.0
	Total	195	100.0	100.0	

H2. The microbeads containing PCPs is the new trend of cosmetic industry

		Frequency	Percent	Valid Percent	Cumulative Percent
Valid	0.00	22	11.3	11.3	11.3
	1.00	41	21.0	21.0	32.3
	2.00	55	28.2	28.2	60.5
	3.00	47	24.1	24.1	84.6
	4.00	30	15.4	15.4	100.0
	Total	195	100.0	100.0	

H3. The natural ingredients in PCPs is important to make decision of buying PCPs

		Frequency	Percent	Valid Percent	Cumulative Percent
Valid	0.00	15	7.7	7.7	7.7
	1.00	51	26.2	26.2	33.8
	2.00	57	29.2	29.2	63.1
	3.00	41	21.0	21.0	84.1
	4.00	31	15.9	15.9	100.0
	Total	195	100.0	100.0	

H4. The usage of microbeads from PCPs could cause environmental pollution

		Frequency	Percent	Valid Percent	Cumulative Percent
Valid	0.00	21	10.8	10.8	10.8
	1.00	28	14.4	14.4	25.1
	2.00	63	32.3	32.3	57.4
	3.00	45	23.1	23.1	80.5
	4.00	38	19.5	19.5	100.0
	Total	195	100.0	100.0	

IOP Conf. Series: Earth and Environmental Science **964** (2022) 012013 doi:10.1088/1755-1315/964/1/012013

Appendix B. Overview of plastic microbead's characteristics.

PCP category	Country of purchase	Sample Code	μBs' concentration (items/mg product)	Size (μm)			Granular determination methods	Color	Shape	Plastics	Polymer determination methods	Reference
				Range	Mean	Median						
Facial cleanser												
	USA	1	0.09^{*}	100 - 500	272	240	Dissecting scope	N/A	Spherical	Polyethylene (as listed in ingredient list)	N/A	[a]
		2	0.08^{*}	120 - 520	271	260						
		3	0.09^{*}	80 - 400	183	200						
		4	0.08^{*}	60 - 420	215	200						
		5	0.08^{*}	200 - 800	317	300						
		6	0.10^{*}	60 - 540	265	260						
		7	0.09^{*}	120 - 600	274	250						
		8	0.08^{*}	100 - 600	286	240						
		9	0.08^{*}	180 - 600	293	270						
	China	A	7.15×10^{-3}	7 - 1020	315	N/A	Dissecting scope	White, transparent	Irregular	Polyethylene	Raman FTIR	[b]
		B	12.64×10^{-3}	N/A	272	N/A		White, transparent	Irregular	Polyethylene		
		C	11.06×10^{-3}	N/A	200	N/A		White, transparent	Irregular	Polyethylene		
		D	6.84×10^{-3}	N/A	380	N/A		White, transparent	Irregular	Polyethylene		
		E	4.92×10^{-3}	N/A	366	N/A		White, transparent	Irregular	Polyethylene		
		F	3.73×10^{-3}	N/A	380	N/A		White, transparent	Irregular	Polyethylene		
		G	9.86×10^{-3}	N/A	255	N/A		White, transparent	Irregular	Polyethylene		
	N/A		124	80 - 700	183	N/A		N/A	Irregular	Polyethylene	FTIR	[c]
	New Zealand	A	N/A	10 - 1075	N/A	197	Microscope	Transparent (?)	Irregular (granular, ellipses, threads)	Polyethylene		[d]
		B	N/A	53 - 848	N/A	375		Transparent (?)	Granular	Polyethylene		
		C	N/A	4 - 1240	N/A	248		Transparent (?)	Irregular, rounded to thread-like	Polyethylene		
		D	N/A	32 - 418	N/A	197		Transparent (?)	Elliptical Slightly granular	Polyethylene		
	N/A	PCP2	3.10×10^{8}	N/A	3	N/A	SP-ICP-MS		Irregular	Polystyrene	ATR-FTIR	[e]
	N/A	PCP3	4.40×10^{7}	N/A	2	N/A			Irregular			
	The UK	A	$1.89 \times 10^{4**}$	8 - 56	164	N/A	Laser particle sizer	White, pink	Smooth	Polyethylene (?)	FTIR	[f]
		B	N/A^{**}	10 - >2000	327	N/A		White	Smooth			
		C	919^{**}	N/A	318	N/A		White	Smooth			

PCP category	Country of purchase	Sample Code	µBs' concentration (items/mg product)	Size (µm)			Granular determination methods	Color	Shape	Plastics	Polymer determination methods	Reference
				Range	Mean	Median						
		D	N/A**	N/A	289	N/A		White, light blue	Smooth			
		E	6.43 x 10³**	N/A	290	N/A		White, dark blue	Smooth			
		F	1.89 x 10⁴**	N/A	293	N/A		White, dark blue	Rough			
	China	A	9.93	24 - 832	126	96	Microscope	Blue Light blue, Colorless	Spherical Irregular, mainly granular	Wax LDPE	ATR- FTIR	[g]
		B	28.90	24 - 988	148	104		Blue, Colourless	Spherical, Irregular, mainly granular	LDPE		
		C	34.24	24 - 517	111	101		Colourless Orange	Irregular, mainly granular Irregular	LDPE Polyethylene		
		D	5.22	24 - 829	142	112		Colourless	Irregular, mainly granular	LDPE		
		E	12.36	24 - 620	161	167		Colourless White	Granular Mainly spherical	Polyethylene PVC		
		F	14.42	24 - 485	85	72		Colourless	Irregular, mainly granular	Polyethylene		
		G	25.10	24 - 813	88	63		Blue Colourless	Irregular, mainly granular Spherical	LDPE Wax		
		H	50.39	24 - 440	92	76		Colourless	Irregular	LDPE		
		I	7.19	24 - 790	186	190		Colourless Colourless	Irregular Spherical	LDPE Wax		
	Slovenia	A	3108	N/A	38	N/A	Scanning electron		Irregular	Polyethylene	FTIR	[h]
		C	2185	N/A	56	N/A	Microscope		Irregular	Polyethylene		
	China		9.91	363 - 945	663	652	Microscope	Blue	Spherical	Polyethylene		[i]
	Spain	A	0.53	112 - 1588	419	N/A	Malvern	White	Irregular	Polyethylene	FTIR	[j]
		B	3.73	12 - 1445	264	N/A	Mastersizer	White	Irregular	Polyethylene		
		C	1.60	9 – 2188	424	N/A	2000LF laser	White	Irregular	Polyethylene		
	The UAE	A2	3.60	18 - 253	107	N/A	Stereoscope	Colorless	Irregular	Polyethylene	FTIR	[k]
		A8	1.90	13 - 184	79	N/A		Green/White	Irregular	Polyethylene		
		A10	0.53	15 - 128	60	N/A		Colorless	Irregular	Polyethylene		
		A12	6.89	15 - 213	76	N/A		White, Red	Irregular	Polyethylene		
		A13	0.25	16 - 274	116	N/A		White	Irregular	Polyethylene		
		A16	1.57	14 - 150	54	N/A		Red, White/Brown	Spherical, Granular	Polyethylene, Polyacrylate copolymer		
		A17	7.62	15 - 191	60	N/A		White	Irregular	Polyethylene		
		A18	7.02	29 - 146	87	N/A		White/Brown	Irregular	Polyethylene		
		A23	1.85	29 - 132	63	N/A		Blue,	Spherical	Polyethylene		
		A24	0.90	29 - 185	86	N/A		Blue, White	Spherical	Polyethylene		
		A27	1.30	15 - 204	65	N/A		White	Irregular	Polyethylene		

PCP category	Country of purchase	Sample Code	µBs' concentration (items/mg product)	Size (µm)			Granular determination methods	Color	Shape	Plastics	Polymer determination methods	Reference
				Range	Mean	Median						
Body scrub												
	Slovenia	B	853	N/A	71	N/A	Scanning electron microscope		Irregular	Polyethylene	FTIR	[h]
		D	625	N/A	96	N/A			Irregular	Polyethylene		
		E	1186	N/A	75	N/A			Irregular	Polyethylene		
	Spain	D	0.12	10 - 2188	520	N/A	Malvern Mastersizer 2000LF laser	White		Polyethylene	FTIR	[j]
		E	0.56	12 - 2188	465	N/A		White	Irregular	Polyethylene		
		F	2.14	5 - 2188	400	N/A		White	Spherical	Polyethylene		
		G	0.86	15 - 2188	551	N/A		White	Spherical	Polyethylene		
		H	2.32	23 - 1260	406	N/A		White	Irregular	Polyethylene		
		I	0.30	15 - 1260	409	N/A		White	Irregular	Polyethylene		
		J	1.00	20 - 832	297	N/A		White	Irregular	Polyethylene		
Shower gel												
	China	H	4.77×10^{-3}	110 - 970	458	N/A	Dissecting scope	White, transparent	Irregular	Polyethylene	Raman/ FTIR	[b]
		I	1.65×10^{-3}	N/A	348	N/A		White, transparent	Irregular	Polyethylene		
		J	1.27×10^{-3}	N/A	461	N/A		White, transparent	Irregular	Polyethylene		
Toothpaste												
	Malaysia	G	48.99×10^{-3}	13 – 110	N/A	N/A	Microscope couple	Blue	Granular	LDPE	FTIR	[l]
	Turkey	2	N/A	4 - <20	N/A	N/A	Microscope	Opaque	Irregular	Polyethylene	FTIR	[m]
		3	N/A	4 - <20	N/A	N/A		Opaque, transparent	Irregular	Polyethylene		
		12	N/A	4 - <20	N/A	N/A		Opaque, transparent	Irregular	Polyethylene		
		20	N/A	4 - <20	N/A	N/A		Opaque, transparent	Irregular	Polyethylene		
	UK	1	1.18^{**}	50 - 590	248	N/A	Scanning electron microscope	Blue	Irregular	Polyethylene	FTIR	[n]

(*): g/mL, (**): particles/mL, LDPE: low-density polyethylene, N/A: not available, FTIR: Fourier-transform infrared spectroscopy

References

[a] Chang M 2015 Reducing microplastics from facial exfoliating cleansers in wastewater through treatment versus consumer product decisions *Mar. Pollut. Bull.* **101(1)** 330–333 https://doi.org/10.1016/j.marpolbul.2015.10.074

[b] Lei K, Qiao F, Liu Q, Wei Z, Qi H, Cui S, Yue X, Deng Y, An L 2017 Microplastics releasing from personal care and cosmetic products in China. *Mar. Pollut. Bull.* **123(1-2)** 122–126 https://doi.org/10.1016/j.marpolbul.2017.09.016

[c] Kokalj JA, Horvat P, Skalar T and Kržan A 2018 Plastic bag and facial cleanser derived microplastic do not affect feeding behaviour and energy reserves of terrestrial isopods *Sci Total Environ.* **615** 761–766 https://doi.org/10.1016/j.scitotenv.2017.10.020

[d] Fendall LS and Sewell MA 2009 Contributing to marine pollution by washing your face: Microplastics in facial cleansers *Mar. Pollut. Bull.* **58(8)** 1225–1228 https://doi.org/10.1016/j.marpolbul.2009.04.025

[e] Laborda F, Trujillo C and Lobinski R 2021 Analysis of microplastics in consumer products by single particle-inductively coupled plasma mass spectrometry using the carbon-13 isotope. *Talanta* **221** 121486 https://doi.org/10.1016/j.talanta.2020.121486

[f] Napper IE, Bakir A, Rowland SJ and Thompson RC 2015 Characterisation, quantity and sorptive properties of microplastics extracted from cosmetics. *Mar. Pollut. Bull.* **99(1-2)** 178–185 https://doi.org/10.1016/j.marpolbul.2015.07.029

[g] Cheung P K and Fok L 2017 Characterisation of plastic microbeads in facial scrubs and their estimated emissions in Mainland China *Water Res.* **122** 53–61 https://doi.org/10.1016/j.watres.2017.05.053

[h] Kalčíková G, Alič B, Skalar T, Bundschuh M and Gotvajn AŽ 2017 Wastewater treatment plant effluents as source of cosmetic polyethylene microbeads to freshwater *Chemosphere* **188** 25–31 https://doi.org/10.1016/j.chemosphere.2017.08.131

[i] Cheung P K and Fok L 2016 Evidence of microbeads from personal care product contaminating the sea. *Mar. Pollut. Bull.* **109(1)** 582–585 https://doi.org/10.1016/j.marpolbul.2016.05.046

[j] Godoy V, Martín-Lara MA, Calero M and Blázquez G 2019 Physical-chemical characterization of microplastics present in some exfoliating products from Spain *Mar. Pollut. Bull.* **139** 91–99. https://doi.org/10.1016/j.marpolbul.2018.12.026

[k] Habib RZ, Salim Abdoon MM, Al Meqbaali RM, Ghebremedhin F, Elkashlan M, Kittaneh WF, Cherupurakal N, Mourad A-H I, Thiemann T, Kindi RA 2020 Analysis of microbeads in cosmetic products in the United Arab Emirate *Environ. Pollut.* **258** 113831. https://doi.org/10.1016/j.envpol.2019.113831

[l] Praveena SM, Shaifuddin SNM and Akizuki S 2018 Exploration of microplastics from personal care and cosmetic products and its estimated emissions to marine environment: An evidence from Malaysia. *Mar. Pollut. Bull.* **136** 135–140. https://doi.org/10.1016/j.marpolbul.2018.09.012

[m] Ustabasi G.S and Baysal A 2019 Occurrence and risk assessment of microplastics from various toothpastes. Environ. Monit. Assess. **191(7)** 438 https://doi.org/10.1007/s10661-019-7574-1

[n] Bråte ILN, Blázquez M, Brooks S J and Thomas KV 2018 Weathering impacts the uptake of polyethylene microparticles from toothpaste in Mediterranean mussels (*M. galloprovincialis*). *Sci Total Environ.* **626** 1310–1318 https://doi.org/10.1016/j.scitotenv.2018.01.141

Appendix C. The microbeads containing PCPs sold in Ho Chi Minh City, the plastic microbeads are shown in bold.

Category	PCPs	Price	Country	Use	Ingredients	Link
Facial cleanser/ scrub						
	Himalaya herbals clear complexion whitening face scrub[a]	4.2$	India	Scrubbing away dead skin cells, blackheads, and deeply embedded impurities	H_2O, Stearic acid, Caprylic / Capric Triglycerice, Cetyl alcohol, Sorbitan Stearate & Sucrose Cocoate, **Polyethylene**, Prunus americana seed Powder, Carbomer, Glyceryl stearate SE, Phenoxyethanol, Perfume, Sodium Hydroxide, Benzyl alcohol, Methylchloroisothiazolinone & Methylisothiazolinone, Tocopheryl Acetate, Glycyrrhiza glabra root extract, Disodium EDTA, Vateria indica bark extract, Cl 14700.	1
	Himalaya herbals gentle exfoliating daily face wash[a]	5.2$	India	Deep cleaning, removing dead skin cells	H_2O, Ammonium Lauryl Sulfate, Cocamidopropyl Betaine, Sodium Cocoyl Glutamate & Disodium Cocoyl Glutamate, **Polyethylene**, Aloe Barbadensis Leaf Extract, Glycerin, **Acrylates/C10-30 Alkyl Acrylate Crosspolymer**, Citrus Medica Limonum Peel Extract, Melia Azadirachta Leaf Extract, Sodium Hydroxide, Perfume, Phenoxyethanol, Prunus Americana Seed Powder, Lens Esculenta Fruit Extract, Methylchloroisothiazolinone & Methylisothiazolinone, Carnauba (Copernicia cerifera) Wax D & C Red No.30(Cl 73360), Glycerin, Citric Acid, Disodium EDTA, Tocopheryl Acetate, Sodium Metabisulfite.	2
	ST.Ives energizing coconut & coffee scrub[a]	5.4$	USA	Deep cleaning, whitening skin, removing dead skin cells	Coffee Powder, Benzyl, Parfum, Ceteareth-20, Triethanolamine, Carbomer, Polysorbate 60, Disodium EDTA, Benzoic Acid, Acetylated Lanolin Alcohol, Sorbic Acid, Caffeine, Cocos Nucifera.	3
	ST.Ives blackhead clearing green tea scrub[a]	5.4$	USA	Unclogs pores and calms redness for bump-free skin	Water (Aqua), Juglans Regia (Walnut) Shell Powder, Glycerin, Glyceryl Stearate Se, Cetearyl Alcohol, Cocamidopropyl Betaine, Sodium Lauryl Sulfoacetate, Cetyl Alcohol, Triethanolamine, Salicylic Acid, Cetyl Acetate, Blue 1 (Cl 42090).	4
	ST.Ives acne control apricot scrub[a]	5.4$	USA	Reduce breakouts and make your skin glow	Salicylic Acid 2%, Juglans Regia (Walnut) Shell Powder, Glyceryl Stearate SE, Glycerin, Zea Mays (Corn) Kernel Meal, Glyceryl Stearate, Cetyl Acetate, Prunus Armeniaca (Apricot) Fruit Extract.	5
	E100 Curcuma Facial Cleanser with Massage[a]	1.5$	Vietnam	Peeling away dead-cells, bright and smooth skin	Polyethylene glycol, Sodium lauryl ether sulfate, Cocamido betaine, Stearuc acid, Myristic acid, Fragrance, Lauric acid, Ceteareth 20, Potassium Hydroxide, Curcuma extract, Sensidin DO, Glyceryl Monostearate, Titanium Dioxide, Phenoxyethanol, Butylcarbamate, Distilled Water.	6
	Oxy Perfect Wash[a,b]	3.2$	Japan	Perfect cool	Water, Glycerin, Palmitic Acid, Stearic Acid, Lauric Acid, Myristic Acid, Potassium Hydroxide, Butylene Glycol, Sorbitol, Cocamide Methyl Mea, Cocamidopropyl Betaine, Ethanol Glyceryl Stearate, Sea Water, Algae extract, Saliva Officinalis Leaf Extract, Isopropylmethylphenol, Niacinamide, Dipotassium Glycyrrhizate, Hydroxypropyl Methylcellulose, Methol, Propylene glycol, Polyquaternium-7, BHT, Disodium EDTA, Phenoxyethanol, Methylparaben, Propylparaben, Fragrance.	7
	Oxy Deep Wash Scrub Formula[a,b]	3.4$	Japan	Deep wash	Water, Palmitic Acid, Glycerin, Stearic Acid, Sorbitol, Lauric Acid, Myristic Acid, Potassium Hydroxide, Hydrated Silica, Cocamide Mythyl MEA, Cocamidopropyl Betaine, Ethanol, Glyceryl Stearate, Algae Extract Salvia Officinalis (Sage) Leaf Extract, Salicylic Acid, Polylactic Acid, Lacide, Lactic Acid, Isopropylmethylphenol, Charcoal Powder, Dipotassium, Glyrhizate, Menthol, Butylene Glycol, Hydroxypropyl Methylcellulose, Polyquatemium-7, Polyglyceryl-10 Stearate, Polyglycerin-10, Polyglyceryl-10 Mristate, Disodium EDTA, BHT, Citric Acid, Sodium Dehydroacetate, Fragrance.	8

Category	PCPs	Price	Country	Use	Ingredients	Link
	Oxy Total Anti-Acne[a,b]	3.4$	Japan	Extra strength for acne	Water, Glycerin, Sodium Cocoyl Glycinate, Butylene Glycol, Hydroxypropyl Starch, Phosphate, Cocamidopropyl Betaine Stearate, Sodium Lauroyl Aspartate, Sodium Stearoyl Glutamate, Camellia Sinensis Leaf Powder, Isopropylmethylphenol, Salicylic Acid, Niacinamide, Citric Acid, Copolyler, Lauric Acid, Stearic Acid, PEG-32, Polyquaternium-7, Hydroxypropyl Methylcellulose, Polyquaternium-52, Sodium Dodoxynol-40 Sulfate, Sorbic Acid, Charcoal Powder, Polyglycenin-10, Polyglyceryl-10 Myristate, Polyglyceryl-10 Stearate, Methylparaben, Propylparaben, BHT, Disodium EDTA, Fragrance, Cl 19140, Cl 42090.	9
	Men's Biore Oil Clear[a,b]	2.8$	Japan	Oil Clear	Water, Glycerin, Palmitic acid, Stearic Acid, Palmitic Acid, Laureth-6 carboxylic Acid, Propylene Glycol, Myristic Acid, Potassium Hydroxide, Cellulose, Zea mays Starch, Sorbitol, Lauric Acid, PEG-150, Fragrance, Menthol, Magnesium Potassium Fluorosilicate, Decyl Glucoside, Sodium A/Vinyl Alcohol Copolymer, Disodium EDTA, PEG-45m, Kaolin, Magnesium/Potassium/Silicon/Fluoride/ Hydroxide/Oxide, Alcohol, BHT, Camellia Sinensis Leaf Extract, Arginine, Bentonite, Betaine, Caprylic/Capric Triglyceride, O-cymen-5-ol, Cl 11680, Cl 77120, Cl 77499.	10
	Men's Biore Deep Action[a,b]	2.8$	Japan	Deep action - Blackhead	Water, Glycerin, Palmitic Acid, Stearic Acid, Laureth-6 Carboxylic Acid, Myristic Acid, Potassium Hydroxide, Propylene Glycol, Cellulose, Zea Mays (corn) Starch, Sorbitol, Lauric Aicd, PEG-150, Magnesium Potassium Fluorosilicate, Fragrance, Decyl Glucoside, Sodium MA/Vinyl alcohol Copolymer, Disodium EDTA, Menthol, PEG-45M, PEG-6, Magnesium/Potassium/Silicon/ Fluoride/Hydroxide/Oxide, BHT, Caprylic/Capric Triglyceride, Arginine, Bentonite, Betaine, 0-cymen-5-ol, Cl 77499, Cl 77891.	11
	Men's Biore Deep Action Cool[a,b]	1.7$	Japan	Deep action – Extra cool	Water, Glycerin, Palmicte Acid, Stearic Acid, Laureth-6 Carboxylic Acid, Propylene Glycol, Myristic Acid, Potassium Hydroxide, Cellulose, Zea Mays (Corn) Starch, Sorbitol, Lauric Acid, PEG-150, Fragrance, Methyl, Magnesium, Alcohol Copolymer, Disodium EDTA, PEG-45M, Magnesium/ Potassium/Silicon/Fluoride/ Hydroxide/Oxide, BHT, Caprylic/Capric Triglyceride, Arginine, Bentonite, Betaine, 0-Cymen-5-ol, Cl 17200, Cl 19140, Cl 42053, Cl 42090, Cl 77499.	12
	Biore Mild & Smooth[a,b]	2.1$	Japan	Mild & Smooth	Water, Sorbitol, Myristic Acid, Laureth-4 Laureth-6 Carboxylic Acid, Lauric Acid, **Acrylates/C10-30 Alkyl Acrylate Crosspolymer**, Ethylhexylglycerin, Palmitic Acid, Fragrance (Parfum), Polyquaternium-39, Disodium EDTA, PEG-45M, BHT, Phenoxyethanol, Methylparaben, Propylparaben, Cl 77891.	13
	Himalaya Oil Clear Lemon Face Wash[a,b]	5.2$	Indian	Lemon essence	Water, Ammonium Lauryl sulfate, Citrus medica limonum fruit extract, Cocamidopropyl betaine, Sodium cocoyl glutamate & Disodium cocoyl glutamate, Glycerin, Honey (Low HMF), Perfume, Sodium Hydroxide, Citric Acid, Sodium metabisulphite, Methylchloroisothiazolinone & Methylisothiazolinone, Disodium EDTA, Phenoxyethanol, Hydrogenated Jojoba Oil, **Acrylates/C10-30 Alkyl Acrylate Crosspolymer**.	14
	Neutrogena Deep Clean Blackhead Eliminating[a,b]	5.6$	USA	Deep Clean Blackhead Eliminating	Water, Sodium Methyl Cocoyl Taurate, Cocamidopropyl Betaine, **Polyethylene**, Disodium Lauroamphodiacetate, Citric Acid, Lauryl Methyl Gluceth- 10 Hydroxypropyldimonium Chloride, Methyl Gluceth-20, Acrylates/Aminoacrylates/ C10-30 Alkyl PEG-20 Itaconate Copolymer, **Acrylates Copolymer**, PEG-3 Distearate, Glycol Stearate, Salicylic Acid, Sodium Laureth Sulfate, Fragrance, Neopentyl Glycol Dicaprylate/ Dicaprate, Iron Oxide, Mica, Cl 73360, titanium Dioxide, Agar, Propylene Glycol, Methylchloroisothiazolinone, Methylisothiazolinone, Cedrus Atlantica Bark Extract.	15
	Acnes Gel Control Oil[a,b]	2.3$	Japan	Control Oil	Water, Sodium Laureth Sulfate, Glycerin, Sodium Cocoyl Glycinate, Potassium Laureth Phosphate, Potassium Cocoyl Glycinate, **Acrylates Copolymer**, Sorbitol, Niacinamide, Sodium Chloride, Maltooligosyl Glucoside, PEG-200 Hydrogenated Glyceryl Palmate, Potassium Cocoate, Sodium Hyaluronate, Zinc PCA, Hydrogenated Starch Hydrolysate, PEG-7 Glyceryl Cocoate, Cera, Microcristallina, Disodium EDTA, Dipotassium Glycyrrhizate,	16

Category	PCPs	Price	Country	Use	Ingredients	Link
					Isopropylmethylphenol (O-cymen-5-ol), OxybenzoneBenzotriazolyl Dodecyl p-Cresol, Sodium Lactate, Lactic Acid, Fragrance, Propylparaben, Methylparaben.	
	Calliderm gommage apaísant a l'aloe vera[c]	7.8$	France	Remove dead skin cells, deep cleaning, whitening, and softening skin	Water, Kaolin, Glycerin, Cetearyl Alcohol, Bentonite, Potassium Sorbate, Silica, Sodium Hydroxide, Levulinic Acid, Sodium Levulinic, Cl 77289, Hydroxide Green.	17
	Vaseline men anti dullness face scrub[c]	3.1$	Indonesia	Deep cleaning, removing dead skin cells, whitening skin	Myristic Acid, Glycerinwater/Aquapropylene, Glycolpotassium, Hydroxidestearic Acidglycol Distearate, Lauric Acid, Decyl Glucoside, Glyceryl, Stearateperfume/ Fragranceterpineol-phenoxyethanolhydrated, Silicasalicylic, Acidthymolpolyquaternium-7, Disodium EDTA, PEG-4.	18
	Rosetle gommage[c]	4.3$	Japan	Remove dead skin cells, rejuvenating and strengthen-ing the skin healthy, mois-turizing	Water (Purified Water), Ethanol, Glycerin, Steartrimonium Chloride, Carbomer, Allantoin, Lactic Acid, Glucose, Citric Acid, Malic Acid, Green Tea Extract, Methylparaben.	19
	Rosetle gommage moist[c]	4.3$	Japan	Remove dead skin cells, rejuvenating and strengthen-ing the skin healthy, moist-ure balance, improve skin elasticity	Water, Glycerol, Sutearutorimoniumukurorid, Ethanol, Carbomer, Allantoin, Lactate, Glucose, Acid Citric, Acid Malic, Methylparaben.	20
	Calliderm gommage corporel a L'huile D'argan Bio[c]	7.8$	France	Remove dead skin cells, deep cleaning, whitening and softening skin	Water, Kaolin, Cetearyl Alcohol, Glycerin, Bentonite, Ceteareth-33, Citric Acid, Cl 14700 Red 4, Sodium Hydroxide.	21
	Moist Cream Cleanser[d]	19.1$	Korean	Deep cleaning, moisturizing. Suitable for dry skin	Water/Aqua/Eau, Limnanthes Alba (Meadowfoam) Seed Oil, Myristic Acid, Lauryl Hydroxysultaine, Lauric Acid, Potassium Hydroxide, Glycerin, Sodium Chloride, Stearic Acid, **Acrylates/C10-30 Alkyl Acrylate Crosspolymer**, Fragrance / Parfum, Polyquaternium-7, PEG-14m, Disodium EDTA, Helianthus Annuus (Sunflower) Seed Oil, Sodium Benzoate, Tocopherol, Silica, Bht.	22
	Multi Deep-Clean Cleanser[d]	18.7$	Korean	Remove makeup, sunscreen. Suitable for oily and oily combination skin	Water/Aqua/Eau, Glycerin, Myristic Acid, Stearic Acid, PEG-32, Potassium Hydroxide, Palmitic Acid, Lauric Acid, Cocamidopropyl Betaine, Microcrystalline Cellulose, Glyceryl Stearate, PEG-100 Stearate, Fragrance/Parfum, Sodium Chloride, Disodium EDTA, Sodium Benzoate, Polycaprolactone, Propanediol, Lactic Acid, PEG-75, Rubus Idaeus (Raspberry) Fruit Extract, Vaccinium Angustifolium (Blueberry) Fruit Extract, Coffea Arabica (Coffee) Seed Extract, Lycium Chinense Fruit Extract, Sapindus Mukorossi Fruit Extract, Vaccinium Macrocarpon (Cranberry) Fruit Extract, Fragaria Chiloensis (Strawberry) Fruit Extract, Rubus Chamaemorus Seed Extract, Papain, Phenoxyethanol, Capric Acid.	23
	Soy Milk The Facial Foam[d]	4.6$	Japan	Cleaning. Removes dirt, sebum, and other impurities from the skin. Moisturizing	Water, Glycerin, Myristic acid, PEG-8, Stearic Acid, Potassium Hydroxide, Lauric Acid, Glycol Distearate, Palmitic Acid, Cocamide DEA, Cocamidopropyl Betaine, Lactobacillus/soy milk Ferment Filtrate Glyceryl Stearate, Glyceryl Stearate SE, Cetyl Alcohol, Polyquaternium-7, Tetrasodium EDTA, methylparaben, Fragrance.	24

Category	PCPs	Price	Country	Use	Ingredients	Link
	Himalaya Fresh Start Oil Clear Face Wash-Blueberry[d]	6$	India	Removes excess oil, dirt and impurities. Deep cleaning	Aqua, Ammonium Lauryl Sulfate, Sorbitol, Cocamidopropyl, Betaine, PEG-7 Glyceryl Cocoate, Glycerin, **Acrylates/C10-30 Alkyl Acrylate Crosspolymer**, Sodium Cocoyl Glycinate, Fragrance, Vaccinium Angustifolium (Blueberry) Fruit Extract, Sodium Benzoate, Phenoxyethanol, Polysorbate 20, Microcrystalline Cellulose, Titanium Dioxide, Sodium Hydroxide, Salicylic Acid, Citric Acid, Disodium EDTA, Guar Hydroxypropyltrimonium Chloride, Sodium Metabisulfite, Benzophenone-4, PEG-45M, Emblica Officinalis Fruit Juice Powder, Cl 77491, Cl 77492, Cl 77499, Cl 74160.	25
	Himalaya Fresh Start Oil Clear Face Wash-Peach[d]	6$	India	Deep cleaning. Removes excess oil, dirt, and impurities.	PEG-7 Glyceryl Cocoate, Glycerin, **Acrylates/C10-30 Alkyl Acrylate Crosspolymer**, Sodium Cocoyl Glycinate, Fragrance, Sodium Benzoate, Phenoxyethanol, Polysorbate 20, Microcrystalline Cellulose Titanium Dioxide, Sodium Hydroxide, Salicylic Acid, Citric Acid, Disodium EDTA Sodium Metabisulfite, Benzophenone-4. PEG-45M. Guar Hydroxypropyltrimonium Chloride, Emblica Officinalis Fruit Juice Powder, Prunus Persica (Peach) Kamelot Cl 77491, Cl 77492, Cl 77499, Cl 14700.	26
	La Roche Posay Ultra Fine Scrub Sensitive Skin (50ml)[c]	18.5$	France	Purifies and stimulates new cell regeneration. Removes excess oil, dirt, and impur-ities. pH balance of the skin	Aqua / Water; Butylene Glycol; Glycerin; Hydroxyethy piperazine Ethane Sulfonic Acid; Perlite; Ammonium Polyacryloyldimethyl Taurate; Sodium Hydroxide; Pumice; Poloxamer 184; Disodium EDTA; Caprylyl Glycol; Xanthan Gum; T-butyl Alcohol; Polyquaternium-47; BHT; Sodium Benzoate; Phenoxyethanol; Myrtrimonium Bromide; Parfum/Fragrance.	27
	Natureine Aqua Peel Moisture Peeling Gel 300ml[e]	21.7$	Japan	Clean the skin and excess sebum, maintain bright skin. Moisturizing	Water, Glycerin, Dicocodimonium Chloride, Steartrimonium Bromide, Butylene Glycol, Lauryl Betaine, Isopropyl Alcohol, Phenethyl Alcohol, Sodium Hyaluronate, Squalane, Pectin, Royal Jelly Extract, Hydrolyzed Collagen, Tilia Cordata Flower Extract, Althaea Officinalis Root Extract, Arnica Montana Flower Extract, Paeonia Suffruticosa Root Extract, Equisetum Arvense Extract, Humulus Lupulus (Hcps) Flower Extract, Pinus Sylvestris Cone Extract, Citrus Limon (Lemon) Fruit Extract, Rosmarinus Officinalis Leaf Extract, Rose Flower Oil, Sucrose, Alcohol, Rosa.	28
	Eucerin Pro Acne Scrub[e]	15.2$	Germany	Removing dead skin cells, prevent acne, removes dirt and impurities.	Aqua, Coco Glucoside, Coco-Betaine, Microcrystalline Cellulose, **Acrylates/C10-30 Alkyl Acrylate Crosspolymer**, Lactic Acid, Sodium Chloride.	29
Body wash/ scrub						
	Hazeline orange cherry body scrub	3.8$	Vietnam	Removing dead skin cells, promote skin regeneration, keeps skin soft and fresh	Water, Sodium Laureth Sulfate, Cocamidopropyl Betaine, Cocamide MEA, Tetrasodium EDTA, Microcrystalline Wax, **Styrene/Acrylates Copolymer**, Niacinamide, Sodium Lauryl Sulfate, Propylene Glycol, Glycerin, Sodium Sulfate, Citrus Junos Fruit Extract (Cam Yuzu), Prunus Cerasus Juice (Cherry), Trideceth-9, Lactic Acid, Bisabolol, Maltodextrin, Potassium Sorbate, Cl 14700, Cl 73360, Cl 47005, Cl42090.	30
	Enchanteur white creamy body scrub[a]	8.3$	Vietnam	Contains massage beads to remove aging skin cells, dead cells and dark spots. Accelerate new cell regeneration.	Water, Sodium Laureth Sulfate, Cocamidopropyl Betaine, **Acrylates Copolymer**, Glycerin, Potassium Laureth Phosphate, Sodium Cocoyl Isethionate, Sodium Lactate, Guar Hydroxypropyltrimonium chloride, Citri Acid, Titanium Dioxide, Tetrasodium EDTA, Hydrated Silica/Silica (and) Cl 74160, Methylchloroisothiazolinone, Methylsothiazolinone, Lactobacillus/Milk Solids/Glycine Soja (Soybean) Oil Ferment, Cyeclodextrin, Niacinamide, Benzophenone-3.	31
	AN'Sw lavender exfoliating gel[a]	5.5$	Vietnam	Exfoliating, nourishing skin, relaxing body	Water, Glycerin, Propylene Glycol, DMDM Hydantoin, Cetrimonium Chloride, Carbomer, Dipotassium Glycyrrhizate, Olea Europaea Husk Oil, Lavandula Angustifolia (Lavender) Oil.	32

Category	PCPs	Price	Country	Use	Ingredients	Link
	Olay daily exfoliating with sea salts[c]	6.5$	USA	Body cleansing, moisturiz-ing, exfoliating skin smooth bright	Water/eau, Sodium Trideceth Sulfate, Petrolatum, Sodium Chloride, Cocamidopropyl Betaine, Trideceth-3. Fragrance/Parfum, Guar Hydroxypropyltrimonium Chloride, Sodium Benzoate, Xanthan Gum, Citric Acid, Stearyl Stearate, Glyceryl Oleate, Disodium EDTA, Sodium Hydroxide, **Acrylates/C10-30 Alkyl Acrylate Crosspolymer**, Euphorbia Cerifera (Candelilla Wax Cire De Candelilla, Sea Salt/Sel Marin, Jojoba Esters, Methylchloroisothiazolinone, Methylisothiazolinone, Ultramarines.	33
	Felina body scrub apricot kernel scrub calendula[c]	5$	Vietnam	Removing dead skin cells. Moisturizing, whitening, and softening skin	Aqua, Prunus, Armeniaca (Apricot) Kernel Powder, Triethanolamine, Sodium Laureth Sulfate, Cocamidopropyl Betaine, Glycerin, DMDM Hydantoin, Carbomer, Sodium Lactate, PEG-40 Hydrogenated Castor Oil, Tocopheryl Acetate, Calendula Officinalis Flower Extract, Prunus Armeniaca Kernel Oil, Paraffinum Liquidum, Parfum.	34
	Felina body scrub apricot kernel scrub avogado[c]	5$	Vietnam	Removing dead skin cells. Moisturizing, whitening, and softening skin	Aqua, Prunus Armeniaca (Apricot) Kernel Powder, Triethanolamine, Sodium Laureth Sulfate, Cocamidopropyl Betaine, Glycerin, DMDM Hydantoin, Carbomer, Sodium Lactate, PEG-40 Hydrogenated Caster Oil, Tocopheryl Acetate, Persea Gratissima Fruit Extract, Butylene Glycol, Parfum.	35
	Veilment Spa Himalaya Pinksalt[c]	12.1$	Korea	Removing dead skin cells. Moisturizing, whitening, and softening skin	Aqua, Sodium Laureth Sulfate Glycerin, **Acrylates Copolymer,** Microcrystalline Cellulose, Fragrance, Lauryl Hydroxysul-taine Cellulose Acetate, Gocamidopropyl Betaine Alcohol Denat... Sodium Benzoate Caprylyl Glycol, Potassium Byorokide Prunus Armeniaea Seed Powder, Sea Salt, Hexylene Tetrasodium EDTA Rose Extract, Camellia Sinensis Leaf Extract Coix Lacryma Jobi Napyuen Seed Extract, Serine, Butylene Glycol Water (Onsen-sun12hexanediol Allantoin Sodium Chloride, Citric Acid, Wine Extract, Capryucicaprio Triglyceride Hyaluronate, Ceramide np, Rose Flower Oil, Cl 14700, Cl 15985.	36
	Veilment Spa Blackrose[c]	12.1$	Korea	Removing dead skin cells. Moisturizing, whitening, and softening skin	Aqua, Sodium Laureth Sulfate, **Acrylates Copolymer**, Microcrystalline Cellulose, Fragrance, Glycerin, Lauryl Hydroxysultaine, Cellulose Acetate, Cocamidopropyl Betaine, Alcohol Denat, Sodium Benzoate Caprylyl Glycol, Potassium Hydroxide, Sea Salt Hexylene Glycol, Prunus Armeniaca Seed Powder Tetrasodium EDTA, Rose Extract, Camellia Sinensis Leaf Extract, Butylene Glycol, 1,2-Hexanediol, Citric Acid, Punica Granatum Seed Oil, Caprylic Capric Triglyceride, Rose Flower Oil, Cl 14700, Cl 15985.	37
	Veilment Spa Jasmine[c]	12.1$	Korea	Removing dead skin cells. Moisturizing, whitening, and softening skin	Water (Including Onsen-sui), Sodium Laureth Sulfate Glycerin, **Acrylates Copolymer**, Fragrance Microcrystalline Cellulose, Lauryl Hydroxysultaine, Cellulose Acetate Cocamidopropyl Betaine, Alcohol Denat, Phenoxyethanol, Sodium Benzoate Caprylyl Glycolhexylene Glycol, Potassium Hydroxide, Prunus Armeniaca Seed Powder, Sea Salt, Tetrasodium EDTA, Serine, Butylene Glycol Alcohol, Allantoin, Citric Acid, Cananga Odorata Flower Oil, Lavandula Angustifolia Pogostemon Cablin Oilcoix Lacryma-job ma-yuen Seed Extract, Citrus Aurantium Dulcis Flower Oil, Jasminum Sambac Flower Extract Rosa Damascena Flower Oil, Ceramide np, Sodium Hyaluronate.	38
	Veilment Spa Deadsea Salt[c]	12.1$	Korea	Removing dead skin cells. Moisturizing, whitening, and softening skin	Aqua, Sodium Laureth Sulfate, **Acrylates Copolymer**, Glycerin, Lauryl Hydroxysultaine, Microcrystalline Fragrance, Cellulose Acetate. Potassium Cocoyl Glycinate, Benzoate, Alcohol Denat, Potassium Cocoate, Caprylyl Glycol, Sodium Citrate, Prunus Armeniaca Seed Powder, PEG-40 Hydrogenated Gastor Oil, Sea Salt Potassium Hydroxide EDTA Sucrose Chloride, Hexylene Glycol, Citric Acid.	39
	ST.Ives radiant skin pink lemon & mandarin orange exfoliating body wash[c]	7.2$	USA	Removing dead skin cells. Whitening skin	Water (Aqua), Sodium Laureth Sulfate, **Acrylates Copolymer**, Hydrated Silica, Cocamide MEA, Sodium Chloride, Citrus Medica Limonum (Lemon) Fruit Extract, Citrus Nobilis (Mandarin Orange) Peel Extract, Fragrance (Parfum), Glycerin, Citric Acid, Tetrasodium EDTA, PPG-9, Propylene Glycol, Methylchloroisothiazolinone, Methylisothiazolinone, Red 33 (Cl 17200), Yellow 5 (19140).	40

Category	PCPs	Price	Country	Use	Ingredients	Link
	ST.Ives purifying sea salt & pacific kelp exfoliating body wash[c]	7.2$	USA	Removing dead skin cells, cleaning, and softening skin	Water (Aqua), Sodium Laureth Sulfate, **Acrylates Copolymer**, Sodium Chloride, Cocamide MEA, Hydrated Silica, Fragrance (Parfum), Glycerin, Citric Acid, Tetrasodium EDTA, PPG-9, Propylene Glycol, Methylchloroisothiazolinone, Methylisothiazolinone, Sea Salt, Macrocystis Pyrifera (Kelp) Extract, Yellow 5 (Cl 19140), Titanium Dioxide (Cl 77891).	41
	Dove gentle exfoliating[a,b,c]	7.1$	Vietnam	Moisturizing, skin whitening	Water, Cocamidopropyl Betaine, Sodium Laureth Sulfate, Sodium Cocoyl Glycinate, Sodium Chloride, Glycerin, Perfume, Silica, **Acrylates/C10-30 Alkyl Acrylate Crosspolymer**, Sodium Lauroyl Isethionate, Stearic Acid, PPG-9, **Styrene/Acrylates Copolymer**, Phenoxyethanol, Lauric Acid, Sodium Hydroxide, BHT, Sodium Palmitate, Sodium Benzoate, Tetrasodium EDTA, Sodium Isethionate, Citric Acid, Coconut Acid.	42
	A bonne moisturizing shower cream scrub tomato & milk[c]	3.9$	Thailand	Deep Cleaning. Whitening and softening skin	Sodium Chloride, Aqua, Sodium Laureth Sulfate, Cocamide Dea, Cocamidopropyl Betaine, Glycerin, Fragrance, Prunus Amygdalus Dulcis Oil, Allantoin, Hydroyzed Milk Protein, Arbutin, Niacinamide, Solanum Lycopersicum Fruit Extract, Cl 15985, Cl 42090.	43
	A bonne whitening shower cream scrub tamarind & aloe vera[c]	3.9$	Thailand	Whitening, cleaning, and softening skin	Sodium Chloride, Aqua, Sodium Laureth Sulfate, Cocamide Dea, Cocamidopropyl Betaine, Fragrance, Methyl, Prunus Amygdalus Dulcis Oil, Allantoin, Tamarindus Indica Extract, Aloe Barbadensis Leaf Juice, Niacinamide, Arbutin, Cl 15985.	44
	L'affair shower scrub pH 5.5 Whitening[c]	9.3$	Malaysia	Contains body massage beads to fade dark spots	Aqua, Sodium Laureth Sulfate, **Acrylates Copolymer**, Cocamidopropyl Betaine, **Polyethylene**, Glycerin, Triethanolamine, Polyquaternium 7. Fragrance, Polyquaternium 39. Citric Acid, Sodium Chloride, Aloe Barbadensis Leaf Juice, Tetrasodium EDTA, Methyl chloroisothiazolinone & Methylisothiazolinone.	45
	L'affair shower scrub beaute[c]	9.6$	Malaysia	Removing dead skin cells. Moisturizing, whitening, and softening skin	Aqua, Sodium Laureth Sulfate, **Acrylates Copolymer**, **Polyethylene**, Glycerin, Poly Quaterniun 7, Polyquatermiun 39, Fragrance, Trietharnolamine, Citric Acid, Sodium Chloride, Morus Alba (mulberry), Leaf Extract, Tetrasodium EDTA, Methyl Chloroisothiazolinone & Methylisothiazolinone.	46
	Olay Age Defying Vitamin E Body Wash[d]	6.2$	USA	Containing Vitamin E helps prevent skin aging	Sodium Trideceth Sulfate, Petrolatum, Cocamidopropyl Betaine, Niacinamide, Sodium Citrate, Guar Hydroxypropyltrimonium Chloride, Glyceryl, Acrylates/C10-30 Alkyl Acrylate Crosspolymer.	47
	Olay Fresh Outlast White Strawberry & Mint Body Wash[d]	5.7$	USA	Perfect cool	Water, Sodium Laureth Sulfate Cocamidopropyl Betaine, Sodium Chloride, Fragrance, Nachlamide, Panthenol, Sodium Dirate Ollenesulfonate Sou, Benzoate Glycol Distearate Citric Acid, Distaedta Methylchlorosothiazolyone Methylisoehazolinone Red 33 Yellow 5.	48
	Olay Ultra Moisture Shea Butter Body Wash[d]	5.4$	USA	Moistures and protects the skin against UV rays	Sodium Trideceth Sulfate, Petrolatum, Cocamidopropyl Betaine, Niacinamide, Sodium Citrate, Guar Hydroxypropyltrimonium Chloride, Glyceryl, Acrylates/C10-30 Alkyl Acrylate Crosspolymer.	49
	Olay Soothing Orchid & Black Currant Body Wash[d]	5.8$	USA	Hydration and freshness	Water, Sodium Laureth Sulfate, Cocamidopropyl Betaine, Sodium Chloride, Fragrance, Niacinamide, Panthenol, Sodium Citrate, Sodium Xylenesulfonate, Sodium Benzoate, Glycol Distearate, Citric Acid, Disodium EDTA, Methylchloroisothiazolinone, Methylisothiazolinone, Red 33, Blue 1.	50
	Sea Salt & Bergamot Marine Softness Body Wash[d]	5.8$	USA	Moisturizing for soft and fresh skin. Balance skin and provide minerals, soften skin, retain	Glycerin, Lauric Acid, Sodium Methyltaurate, Sodium Benzoate, Stearic Acid, Cocos Nucifera (Coconut) Oil, Citrus Aurantium Bergamia (Bergamot) Fruit Oil, Sea Salt.	51

Category	PCPs	Price	Country	Use	Ingredients	Link
				moisture, bring a feeling of refreshment and relaxation		
	Jojoba Shower Gel[d]	8.2$	Japan	Moisturizing for soft, white and fresh skin preventing aging skin	Sodium PPG-16/PEG-2 Lauryl Ether Sulfate, Acrylates Copolymer, Cocamidopropyl Betaine, Cocamide Dea, Sodium Chloride, Sodium Laureth Sulfate, Sodium Cocoyl Glycinate/PEG-7, Sodium Lauroyl Sarcisinate, Sodium Laurimodipropinate, Sodium Hydroxide, Phenoxyethanol, Disodium EDTA, Isomerized Jojoba Oil.	52
	Caresse Shower Lotion Cotton Flower[d]	3.7$	Spain	Gently cleanses and moistur-izes the skin. Whitening and softening skin. Balance moisture and prevent skin dryness. Protect skin against harmful external factors	Cotton, Aqua (Water), Sodium Chloride, Cocamidopropyl Betaine, Parfum (Fragrance), Sodium Laureth Sulfate,Triethanolamine, Carbomer, Glycerin, **Styrene/Acrylates Copolymer**, Magnesium Nitrate, Gossypium Herbaceum (Cotton) Seed Extract, Methylisothiazolinone, Sodium Benzoate, Potassium Sorbate, Benzyl Salicylate, Coumarin, Geraniol, Hexyl Cinnamal, Linalool.	53
	Caresse Shower Lotion Sweet Almond Milk[d]	4.6$	Spain	Gently cleanses and moistur-izes the skin. Whitening and softening skin. Balance moisture and prevent skin dryness. Protect skin against harmful external factors	Aqua (Water), Sodium Laureth Sulfate, Sodium Chloride, Cocamidopropyl Betaine, Parfum (Fragrance), **Styrene/Acrylates Copolymer**, Triethanolamine, Carbomer, Propylene Glycol, Magnesium Nitrate, Prunus Amygdalus Dulcis (Sweet Almond) Fruit Extract, Magnesium Chloride, Methylchloroisothiazolinone, Methylisothiazolinone, Coumarin, Hexyl Cinnamal, Limonene.	54
	Caresse Shower Lotion Vanilla Flower[d]	3.7$	Spain	Gently cleanses and moistur-izes the skin. Whitening and softening skin. Balance moisture and prevent skin dryness. Protect skin against harmful external factors	Aqua (Water), Sodium Laureth Sulfate, Sodium Chloride, Cocamidopropyl Betaine, Parfum (Fragrance), Glycerin, Laureth-3, **Styrene/Acrylates Copolymer**, Polyquaternium-7, Triethanolamine, Magnesium Nitrate, Prunus Amygdalus Dulcis (Sweet Almond) Oil, Propylene Glycol, Magnesium Chloride, Cetearyl Glucoside, Methylchloroisothiazolinone, Sodium Benzoate, Phenoxyethanol, Glyceryl Stearate SE, Methylisothiazolinone, Xanthan Gum, Sorbitol, Benzoic Acid, Vanilla Planifolia Fruit Extract, Dehydroacetic Acid, Sodium Hydroxide, Ethylhexylglycerin.	55
	Caresse Shower Cream Olive Milk[d]	6$	Spain	Gently cleanses and moist-urizes the skin. Whitening and softening skin. Balance moisture and prevent skin dryness. Protect skin against harmful external factors	Aqua (Water), Sodium Laureth Sulfate, Sodium Chloride, Cocamidopropyl Betaine, Parfum (Fragrance), Laureth-3, Glycerin, **Styrene/Acrylates Copolymer**, Citric Acid, Polyquaternium-7, Magnesium Nitrate, Olea Europaea (Olive) Leaf Extract, Magnesium Chloride, Olea Europaea (Olive) Fruit Oil, Sodium Benzoate, Methylchloroisothiazolinone, Propylene Glycol, PEG-60 Hydrogenated Castor Oil, Glyceryl Stearate, Phenoxyethanol, Methylisothiazolinone, Potassium Sorbate, Rosmarinus Officinalis (Rosemary) Leaf Extract, Helianthus Annuus (Sunflower) Seed Oil, Alpha-Isomethyl Ionone, Butylphenyl Methylpropional, Citronellol, Hexyl Cinnamal, Limonene, Linalool, Cl 19140 (FD&C Yellow No. 5), Cl 42090 (FD&C Blue No. 1).	56
	Caresse Shower Cream Rosehip[d]	6.2$	Spain	Gently cleanses and moist-urizes the skin. Whitening and softening skin. Balance moisture and prevent skin dryness. Protect skin against harmful external factors	Aqua (Water), Sodium Laureth Sulfate, Sodium Chloride, PEG-4 Rapeseedamide, Cocamidopropyl Betaine, **Styrene/Acrylates Copolymer**, PEG-7 Glyceryl Cocoate, Parfum (Fragrance), Coco-Glucoside, Citric Acid, Polyquaternium-7, Magnesium Nitrate, Rosa Moschata (Musk Rose) Seed Oil, Sodium Benzoate, Methylchloroisothiazolinone, Magnesium Chloride, Methylisothiazolin-one, Tocopherol, Alpha-Isomethyl Ionone.	57
	Plaisir Shower Gel Pink Pomegranate[d]	3.8$	Spain	Gently cleanses and moisturizes the skin. White-ning and softening skin. Balance moisture and	Aqua, Sodium Laureth Sulfate, Sodium Chloride, Parfum, Cocamidopropyl Betaine, Laureth-3, Citric Acid, Glycerin, Magnesium Nitrate, Magnesium Chloride, Methylchloroisothiazolinone, Punica Granatum Fruit Extract,	58

IOP Conf. Series: Earth and Environmental Science **964** (2022) 012013 doi:10.1088/1755-1315/964/1/012013

Category	PCPs	Price	Country	Use	Ingredients	Link
				prevent skin dryness. Protect skin against harmful external factors	Methylisothiazolinone, Sodium Benzoate, Potassium Sorbate, Butylphenyl Methylpropional, Citronellol, Geraniol, Limonene, Cl 16255, Cl 60730.	
	Plaisir Shower Gel Argan Oil[d]	3.8$	Spain	Gently cleanses and moist-urizes the skin. Whitening and softening skin. Balance moisture and prevent skin dryness. Protect skin against harmful external factors	Aqua (Water), Sodium Laureth Sulfate, Cocamidopropyl Betaine, Sodium Chloride, Decyl Glucoside, Parfum (Fragrance), Laureth-3, PEG-40 Hydrogenated Castor Oil, Argania Spinosa Kernel Oil, Citric Acid, Polyquaternium-7, Magnesium Nitrate, Magnesium Chloride, Methylchloroisothiazolinone, Methylparaben, Methylisothiazolinone, Propylparaben, Butylphenyl Methylpropional, Cl 19140 (FD&C Yellow No.5), Cl14700 (FD&C Red No. 4).	59
Toothpaste						
	Colgate Max Fresh Bamboo Charcoal[a,b,c]	1.4$	USA	Bright white crystal	Water, Hydrated Silica, Sodium Lauryl Sulfate, Flavor, PEG-12, Tetrasodium Pyrophosphate, Celluose Gum, Cocamudopropyl Betaine, Sodium Saccharin, Sodium Fluoride, Hydroxypropyl Methylcellulose, Bambusa Vulgaris Shoot Extract, Charcoal Powder, Cl 77891, Cl 74260.	60
	Colgate Max Fresh Minty Blash[a,b,c]	1.9$	USA	Cool crystal	Water, Sorbitol, Hydrated Silica, PEG-12, Sodium Lauryl Sulfate, Flavor, Sodium Saccharin, CelluloseGum, TrisodiumPhosphate.	61
	Colgate Max Fresh Green Tea[a,b,c]	1.9$	USA	Cool crystal	Water, Sorbitol, Hydrated Silica, Sodium Lauryl Sulfate, Flavor, CelluloseGum, Cocamidopropyl Betaine, Benzyl Alcohol, Sodium Saccharin, Hydroxypropyl Methylcellulose, Menthol, Cl 47005, Cl 74260.	62
	Colgate Max Fresh Thai (Cooling Crystals)[a,b,c]	1.9$	USA	Cool crystal	Sorbitol, Water, Hydrated Silica, Sodium Lauryl Sulfate, Flavor, Sodium Saccharin, Sodium Fluoride, Menthol, Cl 74260.	63
	Dạlan Mint[a,b,c]	1.3$	Vietnam	Clean bacteria – Fresh cool	Water, Sorbitol, Hydrated Silica,Tetrasodium Pyrophosphate, Flavour Polyethylene Glycol, Cocamidopropyl Betaine, Cellulose Gum, Xanthan Gum, Sodium Fluoride, Cl 42090, Cl 44090.	64
	Dạlan Clean Bacteria[a,b,c]	1.3$	Vietnam	Clean bacteria, fight ging-ivitis	Water, Sorbitol, Hydrated Silica, Calcium Carbonate, Sodium Lauryl Sulfate, Green Tea Flavour, Sodium Benzoate, Sodium Monofluoro Phosphate, Polyethylene Glycol, Tetrasodium Pyrophosphate, Cellulose Gum, Xanthan Gum, Sodium Saccharin.	65
	Dạlan Green Tea[a,b,c]	1.3$	Vietnam	Heat clearing- Fresh cool	Water, Sorbitol, Hydrated Silica, Sodium Lauryl Sulfate, Green Tea Flavor, Tetrasodium phosphate, Polyethylene Glycol, Cellulose Gum, Sodium Benzoate, Sodium Saccharin, Sodium Fluoride, Camellia sinensis (Green Tea) Leaf Extract, Cl 74260.	66
	Close up Mineral Clay Alami& Acai Berry[a,b,c]	3.7$	Vietnam	White attraction x2.5	Sorbitol, Water, Hydrated Silica, Sodium Lauryl Sulfate, PEG-31, Flavors, Cellulose Gum, Sodium Fluoride, Sodium Saccharin, Propylene Glycol, Lactic acid, Charcoal Powder, Cocos Nucifera Fruit Extract, PEG-60 Hydrogenated Castor oil, Mica, Cl 77266, Cl 77891.	67
	Close up Coconut Extract + Bamboo Charcol[a,b,c]	1.8$	Vietnam	Whiter teeth In 2 weeks	Sorbitol, Water, Hydrate Silica, Sodium Lauryl Sulfate, Flavor, PEG-12, Cellulose Gum, Tetrasodium Pyphos-phate, xanthan gum, Cocamidopropyl Betaine, Charcoal Powder.	68

Category	PCPs	Price	Country	Use	Ingredients	Link
	Close up Mandarin Oil Herbal[a,b,c]	2$	Vietnam	Fresh attraction	Sorbitol, Water, Hydrate Silica, Sodium Lauryl Sulfate, PEG-32, Flavor (Mandarin extract, Orange extract, Buchu extract,…) Cellulose Gum, Sodium Fluoride, Sodium Saccharin, Zinc Sulfate, Synthetic, Fluorphlogopite, Sodium Hydroxide, Cl 42090, Cl 47005, Cl 77891.	69
	Close up Minty Matcha[a,b,c]	2$	Vietnam	White attraction x2.5	Sorbitol, Water, Hydrated Silica, Kaolin, Sodium Lauryl Sulfate, Flavor (Mentha Arvensis Leaf Oil), Cellulose Gum, Sodium Fluoride, Sodium Saccharin, Mica, Bentonite, Camellia Sinensis Leaf Extract, Cl 77891, Cl 74260, Cl 16035, Cl 12490.	70
	Close up Fire Feeze[a,b,c]	1.4$	Vietnam	Extreme Cool	Sorbitol, Water, Hydrated Silica, Sodium Lauryl Sulfate, PEG-32, Flavor, Cellulose Gum, Sodium Fluoride, Sodium Saccharin, Zinc Sulfate, Sodium Hydroxide, Glycerin, Cl 12490, Cl 74160, Cl 77019, Cl 77510, Cl 77891.	71
	Close up Snow Crystal[a,b,c]	1.1$	Vietnam	Icy white – Winter blast	Sorbitol, Water, Hydrated Silica, Sodium Lauryl Sulfate, PEG-32, Flavor, Cellulose Gum, Sodium Saccharin, Sodium Fluoride, Perlite, Synthetic Fluor Phlogopite, Cl 77891, Menthol, Cl 74160, Cl 42090.	72
	Close up Minty Deep Action[a,b,c]	1.5$	Vietnam	Fresh breath	Sorbitol, Water, Hydrated Silica, Sodium Lauryl Sulfate, Flavor, Cellulose Gum, Perlite, Sodium Fluoride, Sodium Saccharin, Synthetic Fluorphlogopite, Cl 77891, Cl 74260, Cl 19140, Cl 42090.	73
	Close up Winter Fresh Breath[a,b,c]	1.4$	Vietnam	Ice fresh with Extra white power	Sorbitol, Water, Hydrated Silica, Sodium Lauryl Sulfate, PEG-32, Flavor, Cellulose Gum, Sodium Saccharin, Sodium Fluoride, Perlite, Synthetic Fluorphlogopite, Cl 77891, Menthol, Cl 74160, Cl 42090.	74
	Closeup White Attraction Diamond[a,b,c]	1.3$	Vietnam	White attraction diamond	Sorbitol, Water, Hydrate-d Silica, Sodium Lauryl Sulfate, PEG-32, Flavor, Cellulose Gum, Trisodium Phosphate, Mica, Sodium Fluoride, Sodium Saccharin, PVM/MA Copolymer, Glycerin, Lecithin, Phenylphenol, Cl 74160, Cl 77891.	75
	Close up Lemon & Sea Salt[a,b,c]	2$	Vietnam	Whiter teeth In 2 weeks	Water, Sorbitol, Hydrated Silica, PEG-32, Sodium Lauryl Sulfate, Flavor, Sodium Lauryl Sulfate, Flavor, Cellulose Gum, Sodium Fluoride, Sodium Saccharin, Mica, Sodium chloride, Citrus Limon Juice, Cl 77891, Cl 74260, Cl 47005, Cl 42090.	76

a: CoopXtra/Coopmart; b: Lotte Mart; c: Bach Hoa Xanh; d: Hasaki; e: Guardian

IOP Conf. Series: Earth and Environmental Science **964** (2022) 012014 doi:10.1088/1755-1315/964/1/012014

Impacts of Livelihood Activities on the Environment and Natural Resources of Can Gio Mangrove Forest, Ho Chi Minh City

Lam Vinh Son[1,2*], Nguyen Vu Phong[1], Ta Trung Kien[1], Nguyen Thi Phương Chau[2], Le Thanh Hoa[2,4], Vo Le Phu[3,4*]

[1] HUTECH Institute of Applied Sciences, HUTECH University, Ho Chi Minh City, Vietnam

[2] Faculty of Geography, University of Social Sciences and Humanities – VNU HCM, Ho Chi Minh City, Vietnam

[3] Faculty of Environment and Natural Resources, Ho Chi Minh City University of Technology (HCMUT) 268 Ly Thuong Kiet street, District 10, Ho Chi Minh City, Vietnam

[4] Vietnam National University Ho Chi Minh City, Linh Trung Ward, Thu Duc District, Ho Chi Minh City, Vietnam

*Corresponding authors: lv.son@hutech.edu.vn; volephu@hcmut.edu.vn

Abstract. In recent years, local livelihood activities in the Can Gio district have increased significantly, including aquaculture practices, aquaculture cultivation, salt-works, forest guard, farming, livestock, tourism, and etc. These livelihood activities have contributed remarkably to the local economy and the improvement of living conditions for local people. However, the burgeoning development of these activities have posed negative impacts on the mangrove ecosystem in Can Gio. In which, the use and exploitation of existing human and natural resources are paid inadequate attention. The purpose of this study is to identify livelihoods by conducting a survey with 536 questionnaire samples in both the buffer and transition zones in Can Gio Mangrove Forest. In addition, the quality of surface water sources was examined by collecting 70 surface water and wastewater samples. The results show that the quality of surface water meets the allowable limits of QCVN 08-MT:2015/BTNMT. The extent of the impact from livelihood activities on the environment of mangrove forest is at the medium level. However, the livelihood of aquaculture cultivation practices has the most affect on the environment and natural resources and is posing negative impacts on the quality of water sources in Can Gio Mangrove Forest. Therefore, these findings will be useful information for decision makers of the local authority in building an appropriate planning for aquaculture cultivation area. Further, irrigation and drainage systems for aquaculture practices need to be invested to control and handle wastes from aquaculture cultivation areas.

1. Introduction

Can Gio is a coastal district, located in the Southeastern area of Ho Chi Minh City, consisting of Can Thanh town and 6 villages namely Binh Khanh, An Thoi Dong, Ly Nho, Tam thon Hiep, Long Hoa, and Thanh An [1-3]. Can Gio Mangrove Forest was recognised as a World Biosphere Reserve Area by UNESCO in 2000. It plays a vital role in providing the local communities with specific livelihoods, including aquaculture practices, aquaculture cultivation, salt-works, farming, livestock, tourism, and etc. [4]. Over the past 20 years, the development and expansion of immigrants' livelihoods, associated with supporting policies of the local authority enabled residents to diversify livelihoods. These diverse livelihoods are attached to specific characteristics of the mangrove ecosystems and encompass onshore-offshore fishing, aquaculture cultivation (shrimp, crab, fish, and other aquatic products), salt-works, forest guarding, farming and livestocks, and tourism services [5-6]. However, these livelihoods have had adverse impacts on the environment and natural resources of Can Gio Mangrove Forest (CGMF). Therefore, zoning and assessing the impacts of livelihood activities on the mangrove environment is an imperative requirement to formulate measures for the management, use, and exploitation practices appropriately and sustain the existing natural resources in CGMF.

2. Study Area and Methods

2.1. Study area

The study area is Can Gio Mangrove Forest, Ho Chi Minh City, Vietnam, covers 3 zones that include core zone, buffer zone, and transition zone. This study examined the impacts of the following livelihood activities: aquaculture cultivation (shrimp farming, crab, fish, and other aquatic products), aquaculture practices, salt-works, farming and livestocks. The location of livelihood areas is depicted in Figure 1.

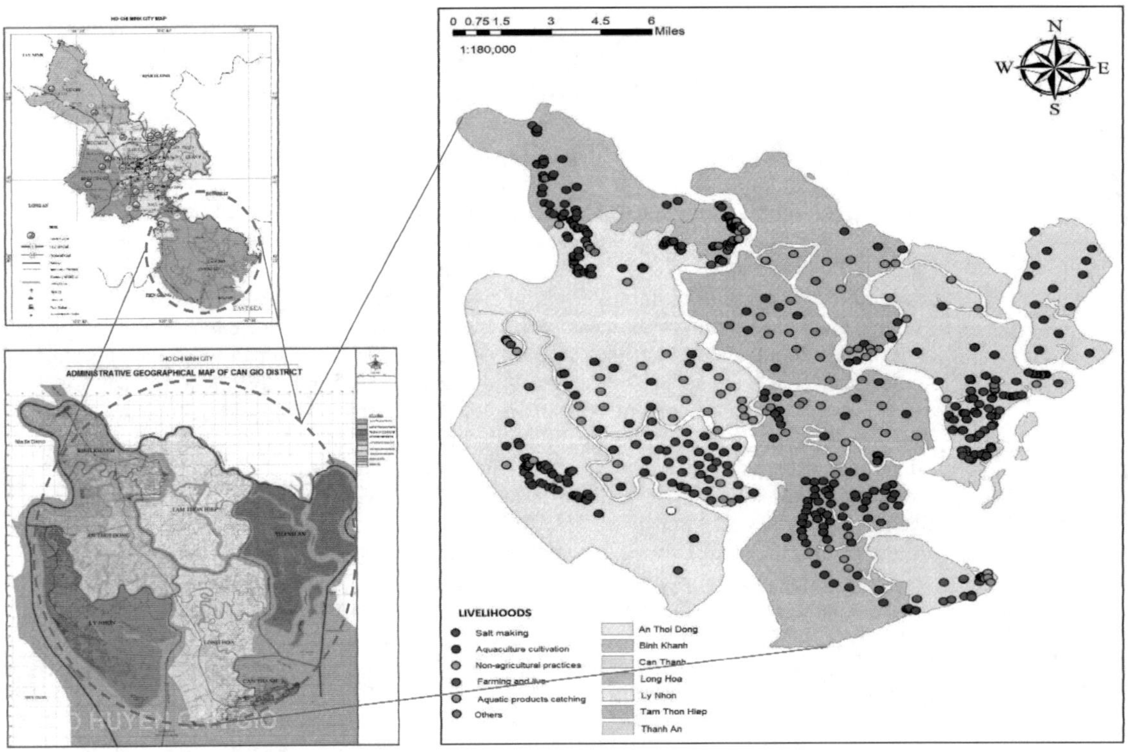

Figure 1. Administrative map and livelihood activities zones of Can Gio District

2.2. Methods

2.2.1. Site Investigation and Survey

There are 9,960 households with above-mentioned livelihoods in CGMF. About 536 out of 9960 households were surveyed and interviewed through a questionnaire set. These households depend on the following livelihoods: aquaculture cultivation (shrimp farming, crab, fish, and other aquatic

IOP Conf. Series: Earth and Environmental Science **964** (2022) 012014 doi:10.1088/1755-1315/964/1/012014

products), aquaculture practices, salt-works, farming and livestocks. Surveyed households were allocated randomly in buffer and transition zones. The number of households surveyed is inequaly distributed denpending on the concentration of these livelihood activities. The survey data was calculated by Yamane formula (1), with standard deviation 0.4% as below.

$$n = \frac{N}{1 + N\,(e^2)} \qquad (1)$$

Where as, n is sampling size, N is total households attached with livelihoods in the study area, e is standard error (%). In addition, 54 households were face-to-face intereviewed as an additional qualitative information on existing practices of local livelihood activities.

2.2.2. Simple addition weight method – radar graphs

A multi-criteria approach was deployed by using the simple addition weight method. Collected data was conversed from qualitative expression (%) with the extent of scale is very poor – poor – average – good – very good to quantitative expression scaling as impact level (extent) from 1 to 5 (corresponding as *Non-impact* – 5 points, *minor impact* – 4 points, *medium impact* – 3 points, *considerable impact* – 2 points, *serious impact* – 1 point). Average point of each impact level is calculated by the below formular (2).

$$\text{Average point} = \frac{\sum m_i \times a_i}{\sum m_i} \qquad (2)$$

In which, m_i is the rate (%) of household's livelihood corresponding to the scale ($m_i = 0 - 100$); a_i is the value corresponding to the extent of scale ($a_i = 1 - 5$). Then, total point of each impact level for each livelihood is calculated and aggregated to livelihood activities and affected entities corresponding impact level (extent) from 1 to 5. Calculated aggregate results are depicted by Radar graph to expressing the level of impact from each livelihood activity.

Further, aggregate data technique is deployed to aggregating similar information and data in the same group (homogeneity).

Table 1. Types of livelihood activites in Can Gio district

	Non-aggregation information	Aggregate information
1	1.1. Shrimp farms/cultivation	Shrimp farming
2	2.1. Fish cultivation 2.2. Floating fish cultivation (Cobia fish) 2.3. Fish farms (Pineapple fish, butterfish) 2.4. Crab cultivation	Fish, crab cultivation
3	3.1. Oyster cultivation 3.2. Clam, shell cultivation 3.3. Shellfish cultivation 3.4. Fishing pond cultivation	Other aquatic products cultivation
4	4.1. Aquaculture exploitation 4.2. Beding net fishing 4.3. Crab burrowing (catching) 4.4. Crabing	Aquaculture exploitation
5	5.1. Salt making	Salt making
6	6.1. Rice cultivation 6.2. Fruit trees cultivation (mango, custard apple) 6.3. Vegetables 6.4. Livestock / fowls 6.5. Swallow nests	Farming and livestocks

2.2.3. Water Sampling and Analysis

A total of 35 wastewater samples and 35 surface water samples at receiving water bodies of wastewater discharging from aquaculture cultivation activities (shrimp, crab, fish, and other aquatic products) were analysed to assess the extent of impacts from livelihood activities on water environment. Water samples were collected in the dry and rainy seasons of 2020. Water sampling and storing were conducted in accordance with TCVN 6631-1:2011 and transported to the lab at HUTECH within 24 hours.

Table 2. Analysis methods for water quality parameters

Parameter	Analysis method/protocol	Permissible limits (B1)	Measurement and analysis instruments
pH	TCVN 6492:2011 Water quality – pH	5.5 – 9	Model Seven Compact S220-K
DO (mg/L)	TCVN 7324:2004 Water quality – DO measurement, Electrode probe.	4	Model Oxi 7310
BOD_5 (mg/L)	TCVN 6001-2:2008 Water quality – Non-dilution protocol	15 50	Model Oxi 7310
COD (mg/L)	TCVN 6491:1999 Water quality – COD analysis	30 150	Elox portable meter
NH_4^+-N (mg/L)	TCVN 6179-1:1996 Water quality – Spectrometer	0.9 10	AmmoniTor
NO_3^--N (mg/L)	TCVN 6180:1996 Water quality – UV Vis	10	UV Vis 730
NO_2^--N (mg/L)	TCVN 6178:1996 Water quality – UV Vis	0.05	UV Vis 730
PO_4^{3-} - P (mg/L)	TCVN 6202:2008 Water quality –UV Vis (amoni molipdate)	0.3	UV Vis 730
Total P	TCVN 6202:2008 Water quality –UV Vis (amoni molipdate)	6	UV Vis 730
TOC (mg/L)	TCVN 6634:2000 Water quality – TOC	4 (A1)	TOC Ultra
Colifrom (MPN)	TCVN 6187-2:1996 Water quality – Most probable number	7500 5000	Most probable number

The quality of surface water was compared with QCVN 02-19:2014/BNNPTNT – National Technical Regulation on Brackish Water Shrimp Culture Farm - Conditions for Veterinary Hygiene, Environmental Protection and Food Safety and QVCN 08-MT:2015/BTNMT – National Technical Regulation on Surface Water Quality. The quality of wastewater was compared with QCVN 40:2011/BTNMT – National Technical Regulation on Industrial Wastewater.

3. Results and discussion

3.1. The Situation of Livelihood Activities

Given the mangrove ecosystem characteristics, local people maintain traditional livelihood activities which play a crucial role in improving their living conditions, including aquaculture practices (shrimp, crab, fish, and other aquatic products cultivation), seafood and aquaculture exploitation, salt-making, farming and livestocks. Each household (corresponding to its livelihood) was surveyed and interviewed by filling one questionnaire set. A total of 536 households was surveyed corresponding to existing livelihood activities, the distribution of different activities surveyed was presented in Table 3.

Table 3. Number and ratio of livelihoods surveyed

	Livelihoods	Number of households surveyed	Percentage of livelihoods surveyed (%)
1	Shrimp farming	124	23.0
2	Fish, crab cultivation	40	7.0
3	Other aquatic products cultivation	118	22.0
4	Aquaculture exploitation	106	20.0
5	Salt making	105	20.0
6	Farming and livestocks	43	8.0
	Total	**536**	**100**

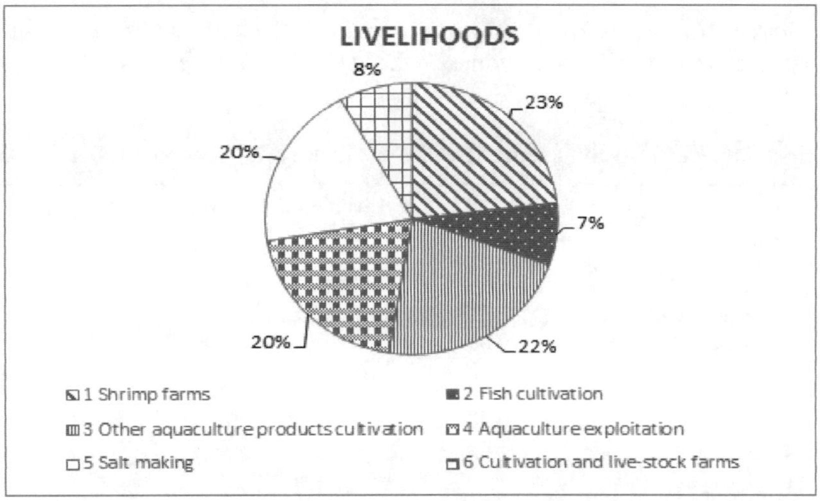

Figure 2. Proportion of Livelihood Activities in Can Gio district

Aquaculture cultivation is the highest livelihood activity, accounting for 52%, followed by salt-making and seafood and aquaculture exploitation, 20% for each. Meanwhile, farming and livestocks is the lowest proportion, contributing 8%.

Figure 3. Aquaculture cultivation in CGMF

Figure 4. Aquaculture exploitation in CGMF

Figure 5. Salt-making in Can Gio

Figure 6. Farming (fruit trees) and livestocks

3.2. Impacts of Livelihood Activities on Water Quality

The extent of impact from each livelihood activity was calculated by addition weight analysis, an aggregation of impacts from livelihood activities on the quality of water sources in CGMF was depicted in Table 4 below.

Table 4. Impacts of livelihood activities on the quality of water sources in Can Gio

Water Quality Issue	**Scale of extent**	**Livelihood Activities**					
		Shrimp farming	Fish, crab cultivation	Other aquactic cultivation	Aquaculture exploitation	Salt making	Farming and livestocks
Water Quality – 10 years ago	Very poor	0.04	0.00	0.00	0.00	0.01	0.10
	Poor	0.67	0.29	0.15	0.51	0.46	0.05
	Medium	0.56	0.50	1.25	1.24	1.54	0.90
	Good	1.45	1.70	1.68	1.65	0.81	1.97
	Very good	0.85	0.81	0.22	0.15	0.17	0.06
Total		**3.57**	**3.30**	**3.30**	**3.55**	**2.90**	**3.08**
Water Quality nước –10 years recntly	Very poor	0.02	0.00	0.00	0.00	0.00	0.06
	Poor	0.51	0.41	0.14	0.34	0.02	0.24
	Medium	0.37	0.79	0.37	0.69	1.00	1.51
	Good	1.55	1.23	2.00	1.50	1.57	1.23
	Very good	1.20	0.69	0.80	0.65	0.44	0.25
Total		**3.65**	**3.12**	**3.31**	**3.18**	**3.03**	**3.29**

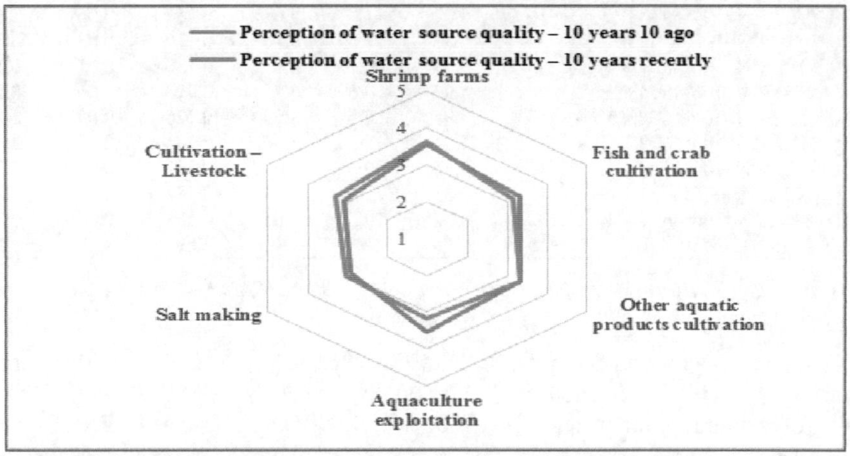

Figure 7. Extent of Impact Level from Livelihood Activities on Water Quality

The analysis showed that the impact of livelihood activities on water environment ranges at medium level (level 3 - medium). However, the impact of aquaculture cultivation practices on the quality of water environment ranges from the level of medium to minor impact (from 3 to 4) compared to other livelihood activities.

Additionally, the quality of surface water at receiving water bodies was also identified by analysing 35 water samples collected at discharging sites of aquaculture cultivation practices (shrimp, crab, fish, and other aquatic products) in CGMF is presented in Table 5.

Table 5. The quality of surface water at receiving water bodies of aquaculture practices

Sample Code*	Coordinates	BOD_5	DO	COD	TOC	NH_4^+-N	PO_4^{3-}	NO_3^--N	NO_2^--N
		mg/L	mg/L	mg/L	mg/L	mg/L	mg/L	mg/L	mg/L
NM1	10°38'09.8"N; 106°48'20.5"E	8	5.21	15.0	6.25	0.184	0.010	3.67	0.012
NM2	10°37'43.2"N; 106°47'11.4"E	9	5.64	10.0	5.19	0.21	0.070	3.15	0.026
NM3	10°37'34.5"N; 106°47'40.5"E	4	5.38	12.0	5.26	0.146	0.060	3.20	0.013
NM4	10°37'08.4"N; 106°47'20.4"E	5	5.61	12.6	2.48	0.245	0.048	2.46	0.024
NM5	10°37'12.8"N; 106°48'35.9"E	10	5.45	14.7	5.26	0.289	0.150	2.35	0.017
NM6	10°36'20.0"N; 106°49'09.7"E	8	5.68	15.9	5.26	0.222	0.052	3.05	0.025
NM7	10°35'04.5"N; 106°49'22.5"E	6	5.92	8.5	3.47	0.192	0.060	2.95	0.012
NM8	10°35'03.2"N; 106°48'23.6"E	7	5.35	16.0	2.21	0.234	0.020	2.45	0.016
NM9	10°35'29.0"N; 106°47'57.6"E	9	6.08	16.8	2.57	0.215	0.057	3.28	0.012
NM10	10°37'06.8"N; 106°50'49.4"E	10	5.42	22.3	6.29	0.111	0.030	3.33	0.019
NM11	10°37'07.0"N; 106°51'05.9"E	9	5.62	15.4	6.24	0.128	0.030	2.24	0.023
NM12	10°36'26.3"N; 106°51'42.9"E	7	6.13	10.0	6.28	0.146	0.050	2.41	0.010
NM13	10°36'08.1"N; 106°51'27.1"E	14	6.08	16.8	6.28	0.134	0.010	2.05	0.021
NM14	10°36'23.0"N; 106°51'14.4"E	10	6.06	12.0	5.23	0.201	0.050	3.16	0.023
NM15	10°35'15.8"N; 106°50'29.4"E	6	5.29	10.6	2.48	0.167	0.013	3.82	0.012
NM16	10°35'14.5"N; 106°50'08.8"E	13	5.35	28.0	8.18	0.149	0.150	2.16	0.012
NM17	10°35'03.9"N; 106°49'42.4"E	8	5.24	14.0	2.18	0.182	0.100	3.05	0.024
NM18	10°33'51.8"N; 106°48'38.1"E	9	5.67	14.6	4.70	0.167	0.200	3.14	0.023
NM19	10°23'50.8"N; 106°54'26.2"E	9	5.88	13.6	2.28	0.165	0.050	3.26	0.031
NM20	10°27'46.0"N; 106°46'24.6"E	5	5.60	19.2	5.1	0.164	0.030	2.16	0.016
NM21	10°28'07.3"N; 106°46'05.7"E	7	5.27	25.6	2.18	0.142	0.020	2.22	0.018
NM22	10°28'46.9"N; 106°45'42.6"E	6	5.40	14.0	5.29	0.136	0.050	2.31	0.023
NM23	10°27'48.3"N; 106°46'49.8"E	4	5.17	8.2	4.64	0.227	0.080	3.61	0.020
NM24	10°28'40.8"N; 106°46'27.2"E	8	6.22	21.5	5.81	0.213	0.060	2.54	0.016
NM25	10°28'51.6"N; 106°46'08.7"E	8	5.64	20.6	2.18	0.152	0.06	2.31	0.02
NM26	10°29'14.8"N; 106°45'57.6"E	15	5.21	27.0	9.00	0.139	0.07	2.31	0.012

NM27	10°29'33.3"N; 106°45'45.0"E	10	6.02	11.0	5.67	0.157	0.06	2.67	0.019
NM28	10°30'10.0"N; 106°45'47.5"E	7	5.88	15.6	5.18	0.126	0.03	2.16	0.017
NM29	10°30'09.8"N; 106°45'47.4"E	6	5.64	20.7	2.23	0.150	0.01	3.16	0.016
NM30	10°30'53.0"N; 106°45'54.4"E	9	5.90	18.6	6.23	0.213	0.06	3.24	0.034
NM31	10°31'22.9"N; 106°46'32.9"E	13	5.91	14.5	2.98	0.120	0.05	3.67	0.016
NM32	10°31'02.1"N; 106°46'32.7"E	12	5.54	17.5	2.50	0.184	0.03	2.15	0.017
NM33	10°32'27.0"N; 106°45'42.1"E	16.5	5.35	29.0	5.67	0.167	0.06	3.26	0.014
NM34	10°33'22.3"N; 106°45'31.1"E	15	5.33	27.0	6.25	0.182	0.45	3.15	0.010
NM35	10°33'06.3"N; 106°46'31.4"E	8	6.27	19.2	3.05	0.138	0.06	3.24	0.016
QCVN 08-MT:2015/BTNMT		**15**	**4**	**30**	**4**	**0.9**	**0.3**	**10**	**0.05**

Notes: NM is the code of surface water samples; QCVN 08-MT:2015/BTNMT – National Technical Regulation on Surface Water Quality.

The results showed that the contents of DO, BOD_5, COD, NH_4^+-N, NO_3^--N, NO_2^--N from most of water samples meet the allowable levels of surface water quality – QCVN 08-MT:2015/BTNMT (Class B1) and the quality of shrimp farming water – QCVN 02-19:2014/BNNPTNT (see Fig. 8 – 9). However, the concentration of total organic carbon (TOC) at the same sampling sites exceeds permissible limits of QCVN 08-MT:2015/BTNMT (> 4 mg/L) (Fig. 8). The content of PO_4^{3-}-P at sampling site NM34 is the highest (0.45 mg/l) exceeding the permitted level of QCVN QCVN 08-MT:2015/BTNMT (with PO_4^{3-}-P < 0.3 mg/L) (Fig. 9). This is caused by the fact that the sampling site at NM34 closes to many intensive and large shrimp farms which discharges a huge volume of organic wastes.

This result is similar to previous study conducted by Nguyen Van Trai and Momtaz (2006) on the assessment of water quality at shrimp farms of three communes, namely Binh Khanh, An Thoi Đong, and Long Hoa. Nguyen Van Trai and Momtaz (2006) claimed that the higher production of shrimp farming the worse water quality, in which COD was 1.51 times higher than the allowable limit of QCVN 08-MT:2015/BTNMT (B1). This caused by the decomposition of organic matter in shrimp ponds which consumed a great deal of oxygen content leading to the depletion of DO level.

Similarly, the content of NO_2^--N is 2.34 folders higher than permissible level of QCVN 08-MT:2015/BTNMT (B1) although the concentration of NH_4^+-N at three sampling sites is lower than the permitted limit (< 0.5 mg/L). This condition of water quality coupled with increasing temperature and pH may poison aquatic ecosystems [7]. However, Swann (1997) suggested that the safe level of NH_4^+-N for aquaculture cultivation should be less than 0.02 mg/L [8].

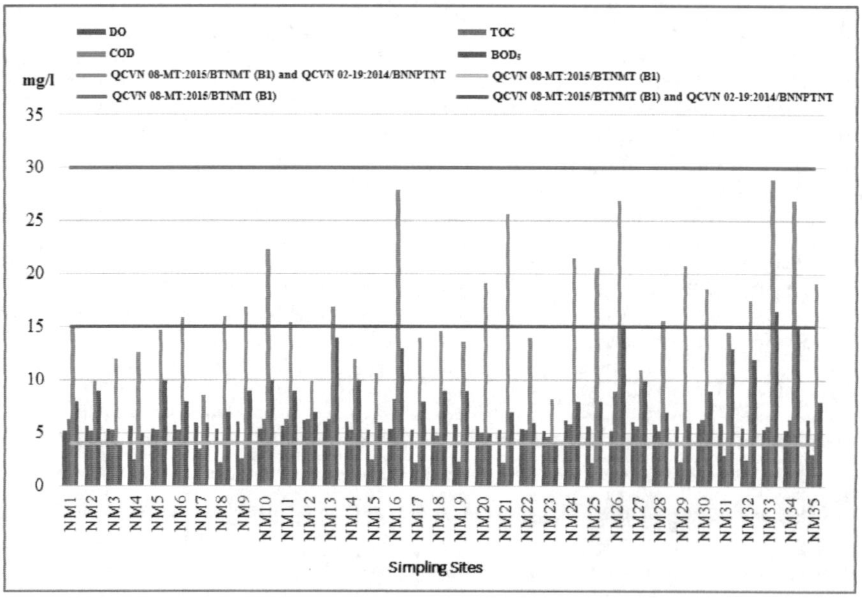

Figure 8. The Content of DO, BOD_5, COD and TOC from Surface Water at Receiving Water Bodies

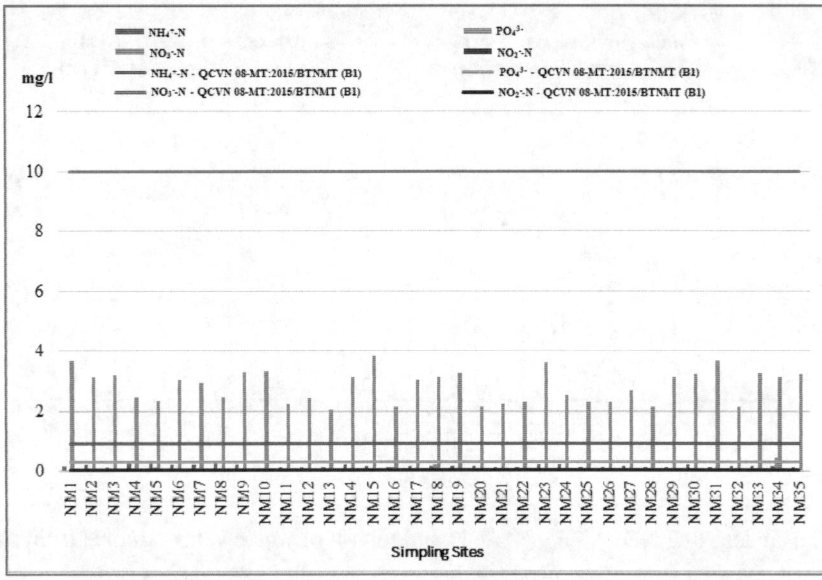

Figure 9. The content of NH_4^+-N, PO_4^{3-}-P, NO_3^--N and NO_2^--N from surface water at receiving water bodies of aquaculture cultivation practices

Importantly, direct and indirect impacts on the mangrove aquatic ecosystems are the causes for the spread of epidemic diseases and the reduction of aquacultural yields and incomes of local people [9-11]. Apart from shrimp farming, other aquacultural practices have been burgeoning recently such as oyster, clam, and shell cultivation, etc. with areas ranging 800 – 1000 ha [12]. However, these kinds of aquaculture cultivation are facing some barriers, including low production yield, low experience and techniques, and unstable incomes. Thus, the extent of impacts from these aquaculture practices is not considerable.

Figure 10. Site visit and investigation of livelihood activities in Can Gio Mangrove Forest

Further, 35 samples of wastewater from shrimp farms were collected to assess the quality of wastewater discharging into the receiving water bodies (Table 6). The results showed that most of samples has the contents of BOD_5, COD, NH_4^+-N, Total P meet the permissible limits of QCVN 40:2011/BTNMT (B1) (Fig.11). However, the content of Coliform in most of samples exceeds the allowable level of QCVN 40:2011/BTNMT (<5,000 MPN/100ml) and QCVN 02-19:2014/BNNPTNT (Fig.12). The sampling site of NT19 has the highest Coliform content (27,384 MPN/100ml) owing to the high contents of organic matter from wastewater.

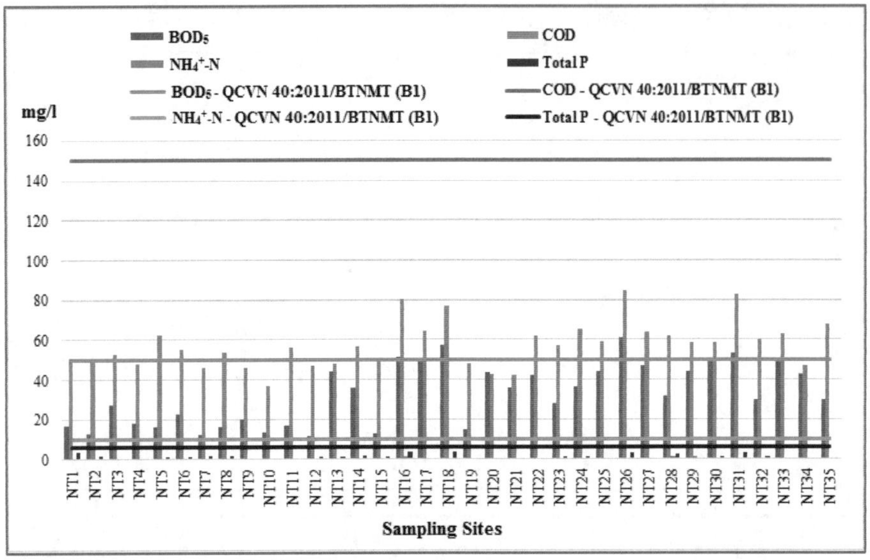

Figure 11. The content of BOD_5, COD, NH_4^+-N and total P of wastewater samples from shrimp farms

The quality of surface water and wastewater from sampling sites suggested that the extent of impact from livelihood activities, in particularly in aquacultural practices, is minor impact. Nevertheless, the quality of water sources is likely threatened by aquacultural cultivation practices if the local government does not make a master plan for livelihood activities properly. In addition, farmers do not usually comply with adequate techniques for cultivation practices as regulated resulting in the degradation of surface water sources, increasing aquatic epidemic diseases, and leading the deterioration of the environments and mangrove ecosystems.

Table 6. The quality of wastewater from shrimp farms

Sample Code*	pH	BOD_5 (mg/L)	COD (mg/L)	NH_4^+-N (mg/L)	Total P (mg/L)	Coliform (MPN/100Ml)
NT1	7.24	16.80	50.60	0.190	3.65	14,436
NT2	6.81	13.25	48.70	0.067	1.78	20,244
NT3	6.97	27.60	52.80	0.150	0.69	24,360
NT4	6.83	18.50	48.13	0.100	0.90	6,702
NT5	7.32	16.01	62.70	0.059	1.04	20,634
NT6	7.36	22.70	55.12	0.250	1.09	15,402
NT7	6.72	12.40	46.20	0.071	2.17	20,463
NT8	7.10	16.60	53.80	0.250	2.00	26,583
NT9	6.83	20.40	46.14	0.106	0.61	27,384
NT10	6.92	13.87	37.00	0.062	0.32	18,753
NT11	6.61	17.06	56.30	0.167	0.21	17,205
NT12	6.35	11.48	47.10	0.055	1.23	13,158
NT13	6.79	44.17	48.20	0.145	0.99	5,256
NT14	6.85	36.01	56.80	0.422	1.67	8,907
NT15	6.68	12.80	50.70	0.146	1.13	6,693
NT16	6.70	51.20	80.00	1.146	4.20	11,334
NT17	6.60	49.23	64.30	0.122	0.77	5,592
NT18	6.54	57.17	77.00	0.263	3.89	23,223
NT19	6.50	15.16	48.00	0.219	0.45	16,353
NT20	6.91	43.80	42.68	0.052	0.78	20,157
NT21	6.70	35.60	42.17	0.122	0.56	17,166

NT22	6.48	42.30	62.01	0.164	0.54	17,244
NT23	5.48	27.80	57.40	0.250	1.01	17,748
NT24	6.65	36.61	65.20	0.066	1.24	19,836
NT25	5.72	44.50	59.30	0.290	0.88	10,284
NT26	6.85	61.01	85.00	0.371	3.56	6,804
NT27	6.68	47.20	64.10	0.868	0.65	10,947
NT28	6.82	31.62	61.80	1.179	2.78	9,801
NT29	5.31	44.60	58.90	1.016	0.67	15,333
NT30	5.43	49.20	58.40	0.860	0.99	15,702
NT31	6.82	53.17	83.00	0.226	3.10	2,883
NT32	6.84	30.08	60.10	0.400	1.47	7,371
NT33	6.90	49.62	63.20	0.195	0.66	25,257
NT34	6.78	42.84	47.20	0.256	0.79	14,283
NT35	6.77	29.70	67.80	0.500	0.72	6,967
QCVN 40:2011/BTNMT	**5.5 – 9**	**50**	**150**	**10**	**6**	**7,500**

Notes: NT is the coding of wastewater samples; QCVN 40:2011/BTNMT – National Technical Regulation on Industrial Wastewater.

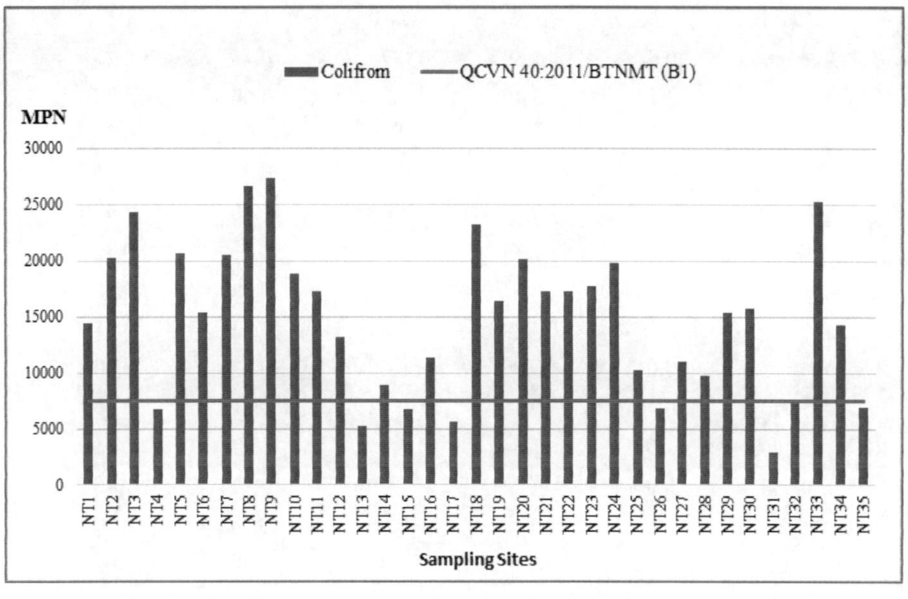

Figure 12. The content of Coliform in wastewater samples from shrimp farms

3.3. Impact of Livelihood Activities on Food Sources and Aquatic Fauna Biodiversity

The extent of impacts from livelihood activities was calculated and analyzed by aggregation of addition weight factor of each livelihood type, an aggregation of impacts from livelihood activities on the diversity of food sources and aquatic fauna in Can Gio Mangrove Forest was presented in Table 7.

Table 7. Impacts of livelihood activities on the diversity of food sources and aquatic fauna

Issues	Scale of extent	Livelihood Activities					
		Shrimp farming	Fish, crab cultivation	Other aquactic cultivation	Aquaculture exploitation	Salt making	Farming and livestocks
Natural food sources – 10 years ago	Very poor	0.04	0.00	0.00	0.00	0.03	0.29
	Poor	0.54	0.34	0.02	0.20	0.25	0.13
	Medium	0.81	0.40	1.42	1.50	2.30	0.50
	Good	1.55	2.00	1.65	1.80	0.75	0.80
	Very good	0.51	0.63	0.31	0.30	0.12	0.77
Total		**3.45**	**3.37**	**3.40**	**3.80**	**3.45**	**2.49**
Natural food sources – 10 years recently	Very poor	0.01	0.00	0.00	0.00	0.01	0.29
	Poor	0.43	0.45	0.20	0.13	0.04	0.19
	Medium	0.6	0.87	0.62	1.04	1.51	1.44
	Good	1.42	1.40	1.60	1.98	1.25	0.71
	Very good	0.64	0.68	0.78	0.45	0.21	0.35
Total		**3.10**	**3.40**	**3.20**	**3.60**	**3.02**	**2.98**
Diversity of aquatic fauna – 10 years ago	Very poor	0.07	0.00	0.00	0.00	0.00	0.27
	Poor	0.37	0.45	0.12	0.37	0.36	0.13
	Medium	1.35	0.78	1.43	1.30	1.97	0.46
	Good	0.85	1.35	1.70	1.71	0.72	0.64
	Very good	0.65	0.62	0.05	0.05	0.36	0.50
Total		**3.29**	**3.20**	**3.30**	**3.43**	**3.41**	**2.00**
Diversity of aquatic fauna – 10 years recently	Very poor	0.01	0.00	0.00	0.04	0.00	0.29
	Poor	0.40	0.45	0.12	0.31	0.06	0.60
	Medium	1.07	1.64	1.05	0.74	1.41	0.82
	Good	0.90	0.7	1.49	2.11	1.43	0.45
	Very good	0.67	0.62	0.66	0.70	0.10	0.14
Total		**3.05**	**3.41**	**3.32**	**3.90**	**3.00**	**2.30**

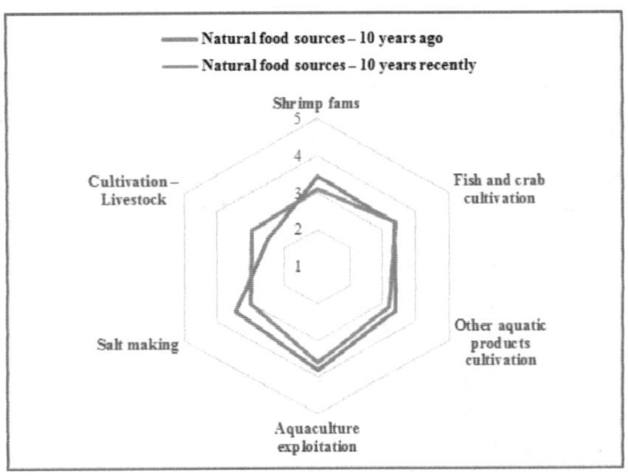

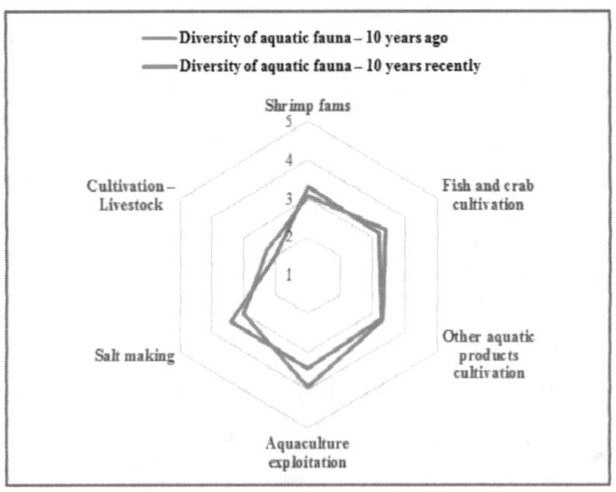

Figure 13. Extent of impacts from livelihood activities on natural food sources

Figure 14. Extent of impacts from livelihood activities on the diversity of aquatic fauna

Livelihood activities also influence the natural food chain and the biodiversity of aquatic fauna at medium level (Fig. 13 and 14). In fact, aquaculture resources are declining not only due to an over-exploitation of crowded local fishing farmers but also fishermen from neighbouring provinces of Long An and Tien Giang at the same fishing ground of Can Gio district [5, 13-14]. This is similar to the same findings reported by previous studies by Pham Van Tuan et al. (2019), in which aquaculture cultivation and fishing imposed negative impacts on the environment and ecosystems of the mangrove forest [14] as summarised in Table 8.

Table 8. Impacts of different fishing exploitation on the aquaculture resources

Livelihoods	Impacts on aquaculture resources	Aquaculture/ Fishing equipment
Net fishing	Negative impacts, harming and exterminating coastal aquaculture resources and damaging beding ecosystems, etc.	Very small mesh size, 15 – 30 mm (catching small fish, accounting 25 – 40% of production)
Pa kua fishing	Exacerbating adjacent fish polulation mining due to over-exploitation of fish at immature stage	Small mesh size, 15 – 20 mm (catching more small and immature fish)
Te xiep fishing	Posing considerable impacts on shrimp, prawn, and aquatic species	Very small mesh size ranging from 10 mm to 15 mm, which attached with small boats or vessels. This kind of fishing equipment is so-called *te xiep* locally.
Beding net fishing	This is a kind of exterminate fishing which was forbidden by Ministry of Agriculture and Rural Development (MARD). This destroys spawns, larvae, and aquatic ecosystems.	

3.4. Impacts of Livelihood Activities on Mangrove Landscape and Alluvial Grounds

The extent of impacts from livelihood activities was calculated and analysed by aggregation of addition weight factor of each livelihood type, an aggregation of impacts from livelihood activities on the landscape of mangrove forest and alluvial grounds in Can Gio Mangrove Forest was depicted in Table 9 below.

Table 9. Impacts of livelihood activities on mangrove landscape and alluvial grounds

Issues	Extent of Scale	Livelihood Activities					
		Shrimp farming	Fish, crab cultivation	Other aquactic cultivation	Aquaculture exploitation	Salt making	Farming and livestocks
Mangrove landscape – 10 years ago	Very poor	0.03	0.05	0.00	0.00	0.01	0.14
	Poor	0.15	0.10	0.00	0.03	0.04	0.05
	Medium	1.50	0.67	1.04	0.75	1.23	0.73
	Good	1.31	2.25	2.19	1.52	2.14	1.23
	Very good	0.81	0.56	0.27	0.90	0.11	0.75
Total		**3.80**	**3.63**	**3.50**	**3.20**	**3.53**	**2.90**
Mangrove landscape	Very poor	0.00	0.00	0.00	0.00	0.01	0.12
	Poor	0.18	0.12	0.02	0.03	0.12	0.23

– 10 years recently	Medium	1.02	1.80	1.60	1.37	1.56	1.02
	Good	1.53	0.88	1.33	1.00	1.39	0.83
	Very good	0.92	0.59	0.40	0.60	0.30	0.90
Total		**3.65**	**3.39**	**3.34**	**3.00**	**3.38**	**3.10**
Alluvial grounds – 10 years ago	Very poor	0.01	0.00	0.00	0.00	0.06	0.29
	Poor	0.26	0.00	0.06	0.19	0.08	0.05
	Medium	1.65	1.85	1.56	0.79	1.66	0.35
	Good	1.08	1.25	1.52	1.37	1.00	1.31
	Very good	0.20	0.00	0.34	1.47	0.00	0.60
Total		**3.20**	**3.10**	**3.48**	**3.82**	**2.80**	**2.60**
Alluvial grounds – 10 year recently	Very poor	0.00	0.00	0.00	0.00	0.00	0.27
	Poor	0.15	0.00	0.09	0.04	0.04	0.30
	Medium	1.11	0.90	1.36	1.01	1.90	1.20
	Good	1.37	2.30	1.75	1.64	0.40	0.33
	Very good	1.05	0.00	0.40	1.15	0.10	0.44
Total		**3.68**	**3.20**	**3.60**	**3.84**	**2.40**	**2.54**

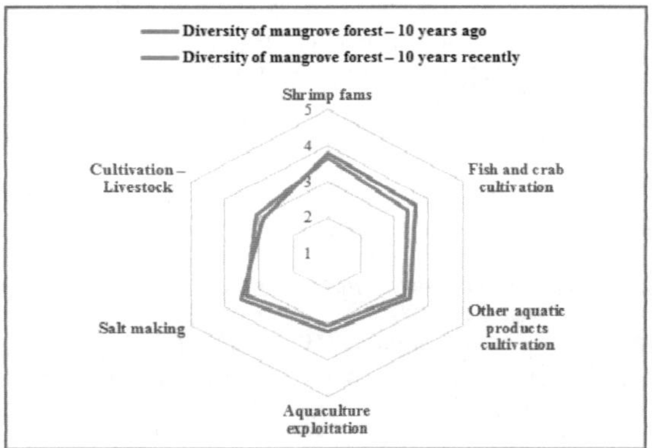

Figure 15. Impacts of livelihood activites on landscape of CGMF

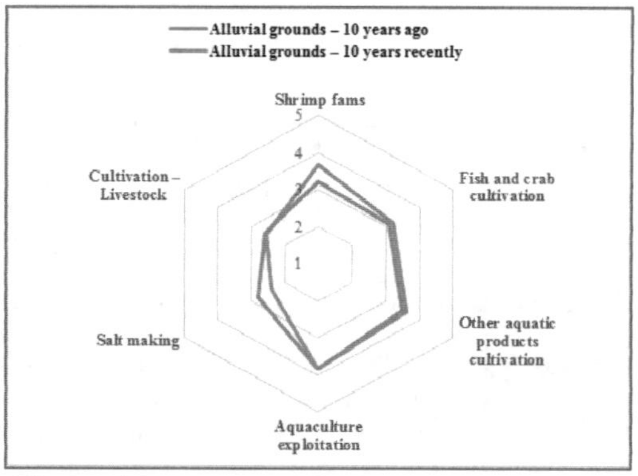

Figure 16. Impacts of livelihood activities on alluvial grounds

The livelihoods of aquaculture cultivation and practices have posed direct impacts on the coverage of mangrove and the landscape of alluvial grounds, but at the minor level (level 4). In parallel with an increase in aquaculture production, the area of aquaculture practices and changes in land use were also expanded. The shift of mangrove areas to aquaculture cultivation areas is a major cause for the diminution of landscape and area of mangrove forest. Phan Nguyen Hong (2000) argued that "the area of mangrove forests in Vietnam was declined sharply from 400,000 ha in 1940 to 150,000 ha at present; just one-third of the original mangrove area, mainly re-forested and secondary mangrove forests" [9]. The loss of mangrove forest was due to land use changes in aquaculture practices, including intensive cultivation, semi-intensive farms, and extensive farming [4, 15]. Aquacultural practices deteriorated aquatic ecosystems and plankton species, and deprived living conditions of reptile, avian [3-4, 16]. These adverse impacts were addressed in previous studies on the construction of mega farms for aquacultural practices along the coastline and estuaries [9-10]. Phan Nguyen Hong (2000) and Nguyen Van Trai and Momtaz (2006) claimed that these engineering aquaculture cultivation practices declined areas for tidal exchange and allocation. Particularly, high tides coupled with northeast winds induce intrusion of seawater into further land causing salt intrusion.

3.5. Aggregated Impacts of Livelihood Activities on the Environments of Can Gio Mangrove Forest

The extent of impacts from livelihood activities was calculated and analysed by aggregation of addition weight factor of each livelihood type, an integrated impacts from livelihood activities on water environment; food sources and biodiversity of aquatic fauna; landscape of mangrove forest and alluvial grounds in Can Gio Mangrove Forest was presented in Table 10.

Table 10. Integrated impacts of livelihood activities on Can Gio Mangrove Forest

Issue	Livelihood Activities					
	Shrimp farming	Fish, crab cultivation	Other aquactic cultivation	Aquaculture exploitation	Salt making	Farming and livestocks
Integrated impacts – 10 years ago	3.464	3.222	3.362	3.523	3.221	2.616
Integrated impacts – 10 years recently	3.469	3.265	3.374	3.471	3.048	2.647

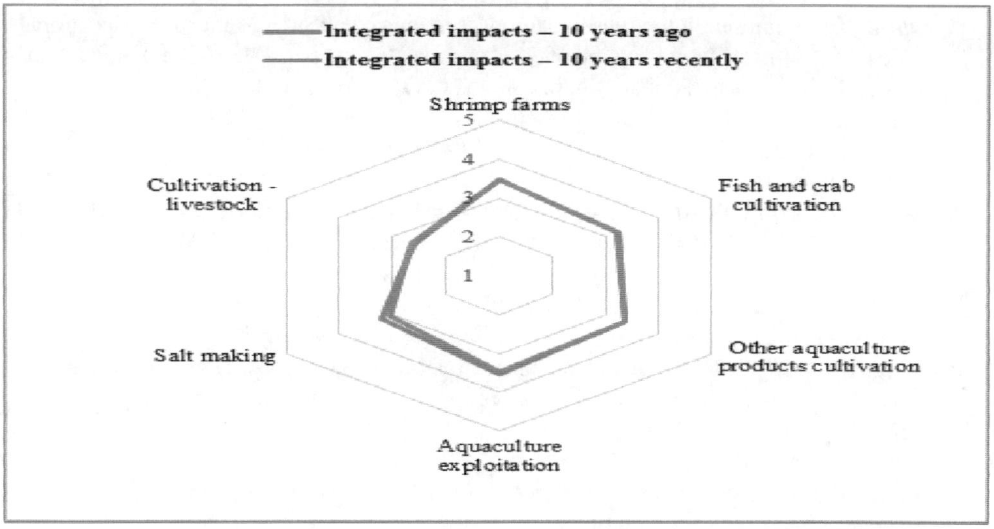

Figure 17. Integrated Impacts of Livelihood Activities on Can Gio Mangrove Forest

Integrated impacts of livelihood activities are described in Figure 17. Generally, the extent of integrated impacts is medium level (ranging from 3 to 4). However, the livelihood of aquaculture cultivation is posing negative impacts on the quality of water sources of Can Gio Mangrove Forest (CGMF).

4. Conclusion

Can Gio Mangrove Forest, a coastal district of Ho Chi Minh City, has been burgeoning over the past 20 years with many aquacultural practices. However, the lack of planning for cultivation areas has resulted in a crowded development of aquacultural cultivation practices (shrimp farming, crab, fish, and other aquatic products) both the scale and area of cultivation. This has posed an unstable condition on water environment and exacerbated the quality of surface water at receiving water bodies. The contents of phosphorus (PO_4^{3-}), total organic carbon (TOC) and Coliform from wastewater exceed the allowable limits of water sources for aquaculture cultivation.

The livelihood activity of aquacultural practices (shrimp farming, crab, fish, and other aquatic products) has a minor impact (level 4), though this is the highest proportion of livelihood activities in Can Gio (over the past 20 years). Nevertheless, this kind of livelihood activity is posing negative impacts on water environment. Therefore, an appropriate planning for aquaculture cultivation area is an imperative. Further, irrigation and drainage systems for aquaculture practices need to be invested to control and handle wastes from aquaculture cultivation areas.

Acknowledgement

We acknowledge the support of time and facilities from Ho Chi Minh City University of Technology (HCMUT), VNU-HCM for this study.

References

[1] UNESCO/MAB 2000 Valuation of the Mangrove Ecosystem in Can Gio Mangrove Biosphere Reserve Vietnam. Final report, Hanoi Vietnam.

[2] Vien Ngoc Nam Le Van Sinh Miyagi T Baba S Chan T H An 2014 Overview of Can Gio District and Mangrove Biosphere Reserve. ISME Mangrove Ecosystem Technical Report No.6.

[3] Vo Quoc Tuan and Kuenzer C 2012 Can Gio Mangrove Biosphere Reserve Evaluation 2012 Final report of UNESCO/MAB Project implemented by the Vietnam MAB National Committee.

[4] Clough B 2013 Continuing the journey amongst mangroves ISME Mangrove Educational Book Series No 1 International Society for Mangrove Ecosystems (ISME) Okinawa Japan and International Tropical Timber Organization (ITTO) Yokohama Japan.

[5] Ngo Thi Phuong Lan 2016 Livelihood Activities in Can Gio, Ho Chi Minh City: An Interaction of Political and Market Factors *Science and Technology Development Journal* **19**(X3) 95-113 (in Vietnamese).

[6] Nguyen Dang Hiep Pho An 2016 Approach to Sustainable Livelihood Framework of DFID in Assessing the Livelihood Activity of Ma ethnic in Cat Tien National Park *Journal of Science Development – Dong Nai University* **2** 101-112 (in Vietnamese).

[7] Boyd C E and Tucker C S 1992 *Water quality and pond soil analysis for aquaculture* 173-176

[8] Swann L 1997 *A fish farmer's guide to understanding water quality*

[9] Phan Nguyen Hong 2000 Effects of mangrove restoration and conservation on the biodiversity and environment in Can Gio District *Mangrove Management and Conservation Workshop 21st May 2000 Okinawa Japan* 111-137

[10] Nguyen Van Trai and Momtaz S 2006 Water pollution concerns in shrimp farming in Vietnam: A case study of Can Gio, Ho Chi Minh *City International Journal of Environmental, Cultural, Economic & Social Sustainability* **3**(2) 129-137

[11] Le Manh Tan 2006 Assessment of impacts on water quality of shrimp farming in Can Gio *Science and Technology Development Journal* **9**(4) 77-84 (in Vietnamese).

[12] Nguyen Van Trai 2008 The Influences of Shrimp Farming and Fishing Practices on Natural Fish Conservation in Can Gio Ho Chi Minh City Vietnam (PhD Thesis) University of Newcastle, Australia.

[13] Nguyen Van Trai 2016 Small-scale of fishing, managing and conserving aquaculture resources in Can gio, Ho Chi Minh City *Journal of Agriculture and Development* **2** 353-361 (in Vietnamese).

[14] Pham Van Tuan 2019 Impacts of some aquaculture practices on aquaculture resources *National Forum on Biology and Sustainable Development* Hai Phong 70 – 79 (in Vietnamese).

[15] White A T and Cruz Trinidad A 1998 Values of Philippine coastal resources: Why protection and management are critical *Coastal Resource Management Project* Cebu city, Philippines, 96 p.

[16] Hamilton S E and Collins S 2013 Livelihood responses to mangrove deforestation in the northern provinces of Ecudor *Bosque* **34**(2) 143-153.

Energy recovery from municipal solid waste landfill for a sustainable circular economy in Danang City, Vietnam

Phong Hoang Nguyen[1,2,*], Quynh Khanh Nguyen Cao[1,2], Long Ta Bui[1,2,*]

[1] Laboratory of Environmental Modelling, Faculty of Environment and Natural Resources, Ho Chi Minh City University of Technology, 268 Ly Thuong Kiet Street, District 10, Ho Chi Minh City, Vietnam
[2] Vietnam National University Ho Chi Minh City, Linh Trung Ward, Thu Duc District, Ho Chi Minh City, Vietnam

[*] Corresponding authors: nhphong@dcselab.edu.vn, and longbt62@hcmut.edu.vn

Abstract. Sustainable development of Danang City in the direction of circular economy (CE) and a zero-waste city is an urgent solution because the impacts of local municipal solid waste (MSW) generation in the city's districts are increasingly causing serious pressure for MSW management and treatment. Segregation of waste at source, reuse, recycling, and energy recovery from landfill gas (LFG) generated is considered as one of the keys to solving the dilemma of sustainable waste management. This study analyzed and evaluated the generation of greenhouse gases (GHGs), mainly CH_4 and CO_2 gases from the Khanh Son landfill based on the application of the EnLandFill software and assessed the potential of energy recovery, clean electricity generation, as well as GHG emission reduction in the period of 2021 – 2050 based on the CE-oriented scenario of the city government. With the potential to recover LFGs in the period of 2021 – 2050 could reach 136.9 million m^3 (with efficiency E = 90%), the total annual potential value of electricity generation is estimated at 420.767 million kWh, equivalent to the total potential for GHGs emission reduction (GWP) about 271.25 thousand tCO_2-eq. At the same time, this will be a baseline study to serve as the basis for extensive assessments and to suggest the most appropriate waste management strategies and policies create a circular economy in the future.

1. Introduction

Danang City is the fourth largest city in Vietnam after Ho Chi Minh City, Hanoi Capital, and Hai Phong City in terms of urbanization and socio-economic development. Located on the Eastern Sea coast with the estuary of the Han River, Danang is one of the Port cities with a strategic location in Central Vietnam and one of five municipalities directly under the Central Government. The total area of the city is 1,284.88 km^2 [1]. The population of Danang City has reached 1,134,310 people in 2019 [1]. This place has a rapid urbanization rate, people's lives are increasingly improved, the demand for material life and resource use is increasing, leading to an increase in the amount of solid waste in general and the amount of municipal solid waste (MSW) in particular is increasing [1]. In this context, currently, Danang City has been aiming to build a zero-waste city with a sustainable circular economy.

The term of CE first appeared in a study by Pearce and Turner [2] referring to the link between the environment and economic activities that have attracted considerable attention from many different target groups worldwide in recent years [3]. However, the boundary of CE definition is not defined, there are variations based on many factors and different perspectives [4]; typically, a variety of principles and recommendations have emerged over the past decades, including "regenerative design" [5], "performance economy" [6], "Cradle-to-Cradle" [7] and "industrial ecology" [8]. But in general, the goal of CE models is towards a better alternative to the traditional economic development model, called "take, make and dispose" [9–11]. Currently, CE has been recognized as a guiding principle for policy in many countries and has approached implementation in a variety of ways [12]. In the European Union (EU), CE has become a central aspect of development strategies and policies and is seen as part of the Circular Economy Action Plan [13]. On a global scale, countries including Germany, Japan, China, and the EU are recognized as having developed legal systems for the implementation of CE principles [4], [14]. China has particularly considered CE as the central goal of the 11th and 12th Five-Year Plans on National Economic and Social Development [14], and also in 2009 officially promulgated the "Circular Economy Promotion Law", in which CE is defined as a general term for actions to reduce, reuse and recycle in production, circulation, and consumption [4]. Especially, China's CE approach is a top-down solution, which is quite different from other countries that have developed CE policies with a bottom-up approach [3]. The basic CE models have contributed to the formation of a new set-up system to optimize the use of materials and bring value throughout lifecycle phases as well as minimize emissions [15–17]. This is also considered as one of the typical management strategies to transit to a low-carbon and less polluting economy [18].

The adaptation of circular solutions in environmental management is worldwide recognition from academic institutions to governmental and non-governmental organizations [19] due to the issues of depletion of natural resources as well as the adverse effects of climate change [18]. Although there is increasing interest in implementing recovery solutions such as reuse and recycling, but it is inefficient waste management solutions that most of the waste is only buried or dumped without applying environmental protection measures [16], [20]. When not being handled properly (especially in open-dumping sites), landfill MSW can form hazards from toxic substances generated from decomposition and direct permeation into soil, mainly as chlorinated solvents, heavy metals, Aromatic Hydrocarbons and polycyclic aromatic or Vinyl Chlorides [21], [22]. On the other hand, when MSW is properly collected and disposed of in sanitary landfills, anaerobic biodegradation also generates significant emissions lead to global warming, most notably, CH_4 contributes up to 15% of the total emissions of GHGs and is also the main contributor to GHG emissions from the waste sector [16], [20].

This study was carried out with the goal of analyzing and evaluating the evolution of LFG emissions (CH_4 and CO_2 gases) for the period of 2021 – 2050 from the Khanh Son landfill site with analysis of uncertainty factors by the EnLandFill software. The study also conducts an assessment of the potential for energy recovery based on CE policies applied to meet various objectives, consisting of energy conversion and electricity generation for urban areas, as well as achieving the national common interest related to GHGs control in order to reduce the impacts of climate change. Based on comparative evaluations with similar policies and solutions from other regions and countries showing the effectiveness of CE-oriented sustainable MSW management, with the ultimate goal of the study is to provide policy makers and relevant agencies/organizations with information on the diversification of energy sources for urban and surrounding areas, along with the benefits achieved from reducing the contribution of GHG emissions from the waste sector and promoting renewable energy production in Danang City, Vietnam.

2. Study area and methods

2.1. Study area

In Danang City, the need to handle the entire amount of waste generated up to the present and in the future is also extremely urgent and requires a timely solution to ensure the complete treatment of generated waste along with the socio-economic development of the city. Currently, most of the MSW in Danang City has been collected, transported, and treated [1]. The main method of domestic waste treatment in the city is still burial, this activity takes place at the Khanh Son landfill (Figure 1). After nearly 20 years of existence, the Khanh Son landfill is facing the problem of overcrowding, which is a dilemma for the city when the amount of MSW is increasing, whilst the alternative is currently in the process of being completed and has not kept pace with the city's waste generation progress, at least in the next few years.

Figure 1 presents description of the study area, specifying the scope of the MSW treatment area in the Khanh Son landfill. The landfill with a total existing area of 32.4 ha located in Hoa Khanh Nam ward, Lien Chieu district, Danang city [1]. The landfill area is mainly low hills located in the West of Danang city, the terrain is relatively rough and steep, the altitude varies from 13.5 m to 19.0 m that the traffic is relatively convenient [1]. Currently, the treated MSW volume has been up to 1,132 tonnes/day in the Khanh Son landfill, specifically domestic waste is 1,056 tonnes/day, non-hazardous industrial waste is 50 tonnes/day, non-hazardous medical solid waste is 17 tonnes/day, and sewage sludge is 9 tonnes/day [1].

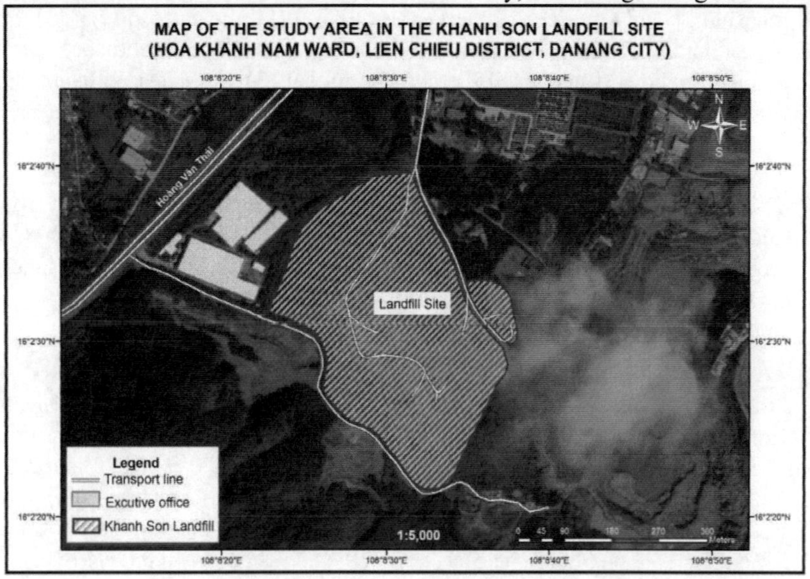

Figure 1. Description of the Khanh Son landfill site in Danang City, Vietnam

The current position of non-hazardous waste burial cells is located to the West – Southwest of the landfill area, including five existing garbage cells (landfill cells 1, 2, 3, 4, 5) and one planned expansion (cell 6) [1]. The total area of these cells is about 13.83 ha, as shown in Table 1. The existing non-hazardous MSW treatment areas have been operated since 2007, particularly the cell 1 received MSW from 2007 with the highest elevation about 38.74m; the cell 2 operated from July 2012 to June 2017 and covered 0.3m of soil layer with the highest elevation about 43.93m. Whilst the cells 4 and 5 landfilled MSW from September 2015 to June 2017 with the highest elevation roughly 43.93m and June 2017 to current with the highest elevation around 40.50m, respectively. Furthermore, the hazardous waste treatment area has a total area of roughly 1 ha that the hazardous waste treatment subdivision has an area of 6,500 m^2 and the treated hazardous waste cell has an area of 5,000 m^2 [1]. In addition, in the Khanh Son Complex Zone area, there are other facilities including an operating office, weighing station, truck wash station, leachate treatment plant, and operating roads.

Table 1. The status of landfill cells in the Khanh Son landfill, Danang City

Order	Subdivision	Area (ha)	Ratio (%)
1	Trash box 1 (Cell 1)	2.27	7.0
2	Trash box 2 (Cell 2)	2.73	8.4
3	Trash box 3 (Cell 3)	2.61	8.1
4	Trash box 4 (Cell 4)	2.85	8.8
5	Trash box 5 (Cell 5)	3.37	10.4
6	Hazardous waste burial box	0.50	1.5
7	Hazardous waste treatment area	0.65	2.0

2.2. Analysis of LFGs emissions contribution

The EnLandFill software (Environmental Information – Model Integrated System for Air Emission and Dispersion Estimation from Landfill) is the name of an integrated system of environmental information and mathematical model was built to allow calculation of potential gas emissions CH_4 (L_0) and determination of the optimal CH_4 (k) gas generation rate constant for a landfill [23], [24]. The model block are considered as an essential component in the EnLandFill with three integrated mathematical models [23], including the model (1) calculates the potential to generate CH_4 (L_0) based on the theory of first-order decomposition (FOD) model of Pipatti et al. (2006) [25] (the equation Eq. [1], Eq. [2] and Eq. [3]); the model (2) estimates emissions of LFGs based on the LandGEM model theory of Alexander et al. (2005) [26] (the equation Eq. [4] and Eq. [5]); and the model (3) simulates pollutant dispersion from landfills. Figure 2 shows the steps flowchart of the implementation for this research.

$$L_0 = DDOC_m \times (16/12) \times F, \tag{1}$$

where (16/12) is the molecular weight ratio of CH_4/C (ratio), F is the fraction of CH_4 by volume in the generated landfill gas (the value range recommended by the IPCC is 0.5 – 0.55), $DDOC_m$ is the mass of the decomposable degradable organic carbon (DOC) deposited into the landfill in year T (tonne/year). $DDOC_m$ can be estimated as shown in Eq. (2) below.

$$DDOC_m = W_T \times DOC \times DOC_F \times MCF, \tag{2}$$

where W_T is the mass of solid waste deposited into the landfill in year T (tonne/year), and $W_T = 1$ is selected to determine the $DDOC_m$ value of 1 tonne of MSW in the landfill; DOC_F is the fraction of DOC that can decompose (fraction) (the value recommended by the IPCC is 0.5); MCF is the CH_4 correction factor for aerobic decomposition in the year of deposition (fraction) (MCF = 0.6) based on actual conditions of landfill [25], DOC is the DOC in the year of deposition (fraction, tonne C/tonne waste). The DOC value represents the volume of organic carbon in waste and depends on the waste composition ratios in various components of the waste stream. The DOC can be estimated as follow Eq. (3):

$$DOC = \sum_{1}^{n} DOC_i \times W_i\ , \tag{3}$$

where DOC_i is the fraction of the DOC in waste type i(%), and W_i is the fraction of waste type i by waste category (%); W_i is the fraction of mass of waste type i and deposition (%). The value L_0 is the input data of model (2) to simulate emissions of LFGs.

$$Q_{CH_4} = \sum_{i=1}^{n} \sum_{j=0.1}^{1} kL_0 \left[\frac{M_i}{10}\right] e^{-kt_{ij}}\ , \tag{4}$$

$$Q_{CO_{2,Yeari}} = Q_{CH_4} \times \left(\frac{100}{F} - 1\right) \times D_{CO_2}\ , \tag{5}$$

where i is the step index in year which ranges from 1 to n (year); n is the difference (year of the calculation) – (initial year of waste acceptance); j is is equal to 0,1 time increment in year; M_i is is the disposed waste mass in year i in tonne; t_{ij} is the age of the j portion of the waste M_i in year i within the operation year (year as a decimal number); L_0 is the optimal CH_4 generation potential (m^3/tonne waste); and k is the optimal rate of CH_4 generation ($year^{-1}$). Moreover, D_{CO2} and D_{CH4} are the value of the specific gravity of the CO_2 and CH_4 gases (tonne/m^3) and F is the percentage of CH_4 in the total generated landfill gas (%). However, considering the circumstances of biogas collection system installation assumed that not all the volume of CH_4 produced in landfills can be captured for electricity generation potential [27]. A fraction of CH_4 generated emission is able to release directly into the atmosphere as a type of tracer gas, oxidation, or accumulation [27]. Therefore, a coefficient of landfill CH_4 collection efficiency (E, %) is added in Eq. (6) below to estimate CH_4 emission from landfills with regard to biogas recovery [23].

$$Q_{CH_{4Yeari}} = (1-E)\times\sum_{i=1}^{n}\sum_{j=0.1}^{1} k.L_0\left[\frac{M_i}{10}\right]e^{-kt_{ij}}\times D_{CH_4} \tag{6}$$

2.3. Analysis of power generation potential from LFGs

The electricity generation potential of MSW landfills depends on the total volume of CH_4 recovered from landfill gas collection systems [28], [29]. The FOD model in the EnLandFill software can be used to determine LFG emissions for each year in this research area. The total generated CH_4 gas volume from landfill captured to produce energy can be estimated as Eq. (7) according to Bui & Nguyen [30]:

$$CAP_{CH_{4Yeari}} = E\times\sum_{i=1}^{n}\sum_{j=0.1}^{1} k.L_0\left[\frac{M_i}{10}\right]e^{-kt_{ij}}\times D_{CH_4}, \tag{7}$$

The electricity generation potential, $EP_{LFG,yeari}$ (unit: kWh/year) from the total captured CH_4 gas volume estimated for each operating year can be obtained as Eq. (2) [30–32]:

$$EP_{LFG,yeari} = \frac{CAP_{CH_{4Yeari}}\times LHV_{CH_4}\times\delta\times\varepsilon}{\phi}, \tag{8}$$

where LHV_{CH_4} is the Lower Heating Value (LHV) of CH_4 gas (unit: MJ/m^3), and the LHV_{CH_4} value is about from 35 MJ/m^3 to 37.2 MJ/m^3 [28], [33], [34]; δ is the capacity factor of the entire recovered CH_4 combustion process to generate energy source, the common δ value is roughly 85% [28], [35]; ε is the electricity generation efficiency of the gas turbine engine, and is given a range of 30 – 35% [32], [36]; ϕ is the conversion factor from MJ to kWh, and ϕ value is taken as 3.6 [28], [29]. The energy plant size from captured CH_4 gas of landfill ($LFGTE_{(size)}$) assuming it is able to operate throughout the year is calculated in kW or MW as Eq. (9) below [9], [23], [30]:

$$LFGTE_{(size)} = \frac{EP_{LFG,yeari}}{D_{hr}\times\gamma}, \tag{9}$$

where D_{hr} is the number of hours in a day (unit: hours), and γ is the number of days that power plant is worked in a year (unit: days).

2.4. Analysis of GHG emissions reduction potential

The Global Warming Potential (GWP) can be understood as a certain amount of GHG, released into the atmosphere causes a warming effect on the Earth [37] over a given period of time (normally 100 years) [38], [39]. The GWP values used to convert the GHG emissions from different unit to homogeneous unit called CO_2 equivalent or CO_2-eq and estimated based on Eq. (10) follows [38], [39]:

$$Emission_{GHGi,CO_2-eq} = Emission_{GHGi}\times GWP_{index,i}, \tag{10}$$

where $Emission_{GHGi,CO_2-eq}$ is the emission of GHG i converted to the unit of CO_2-eq; $Emission_{GHGi}$ is the emission of GHG i estimated in the unit of tonne or kg, and $GWP_{index,i}$ is the Global Warming Potential of GHG i. To calculate the total value of GHG emission reduction potential generated from landfills for each year based on the computing scenario plan having biogas recovery to produce power generation can be shown in Eq. (11) [30], [40], [41].

$$\sum RE_{GHGs,yeari} = Q'_{CH_{4Yeari}}\times GWP_{CH_4} + Q'_{CO_{2Yeari}}\times GWP_{CO_2} \tag{11}$$

where $\sum RE_{GHGs,yeari}$ is the total GHGs emission reduction potential of the year i (unit: tCO_2-eq/year); $Q'_{CH_{4Yeari}}$ and $Q'_{CO_{2Yeari}}$ is the emissions of CH_4 and CO_2 gases generated from landfill in the year i can be decreased; GWP_{CH_4} and GWP_{CO_2} is the GWP values of CH_4 and CO_2 gases.

2.5. Scenario analysis

According to the analysis for the current state of facilities to meet the treatment, the collection and MSW treatment, the potential for MSW generation, as well as the waste management orientations of Danang City. A detailed computing scenario is set up to assess the effectiveness of each planned MSW treatment option (Table 2 and Figure 3), which proposes to implement a systematic separation of waste at the source with all MSW generated are collected and treated at the Complex; in which 10% of waste including paper, plastic, plastic bags, metal is treated at recycling subdivisions; the remaining waste components are buried at Khanh Son landfill (in the expanded treatment area). At the same time, It proceeds to install the LFGs collection system generated from the landfill cells of the Khanh Son landfill to burn and serve for electricity generation (with efficiency, E% between 90% and 95%). The entire amount of electricity generated from the LFGs recovery type will be used for the operation of the Complex and several households in Lien Chieu District, Danang City. Theoretically, the treatment efficiency from the developed scenario contributes to clearly reflect the "Zero Waste City" goal that the city government is aiming for 100% MSW collected and treated; in which about 90 – 95% of the waste is recycled, reused, energy recovered or composted in the period of 2025 – 2050 and the landfill volume is reduced by 15 – 20% [42]. This is also a contribution to the common national goals in the period of 2020 – 2030 aiming to energy recovery, control and the reduction of CH_4 emissions in the waste sector [24].

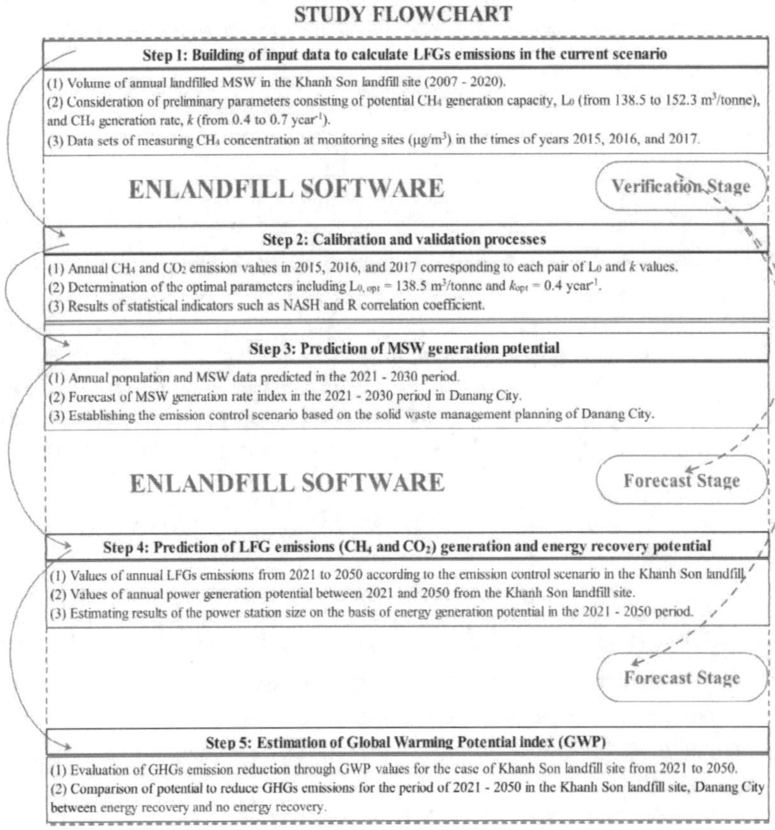

Figure 2. The flowchart presenting the steps of the methodology

Table 2. Summary of the analyzed scenario to forecast the LFGs emission

Period	Approaches of treatment
2007 – 2012	About 90% of MSW in Danang City collected to the Khanh Son landfill; MSW landfilled in trash boxes 1 and 2 (Cell 1 and 2).
2012 – 2015	About 90% of MSW in Danang City collected to the Khanh Son landfill; MSW landfilled the trash box 3 (Cell 3).
2015 – 2017	About 90% of MSW in Danang City collected to the Khanh Son landfill; MSW landfilled in trash box 4 (Cell 4).
2017 – 2020	About 95% of MSW in Danang City collected to the Khanh Son landfill; MSW landfilled in trash box 5 (Cell 5).
2021 – 2050	Total MSW in Danang City collected to the Khanh Son landfill; About 10% of waste including paper, plastic, metal treated in recycling area; From 70 to 75% of MSW used for composting generation; Installing and capturing the LFG emissions to create the clean energy; From 15 to 20% of MSW landfilled in the trash boxes 5 and 6 (Cell 5 and 6).

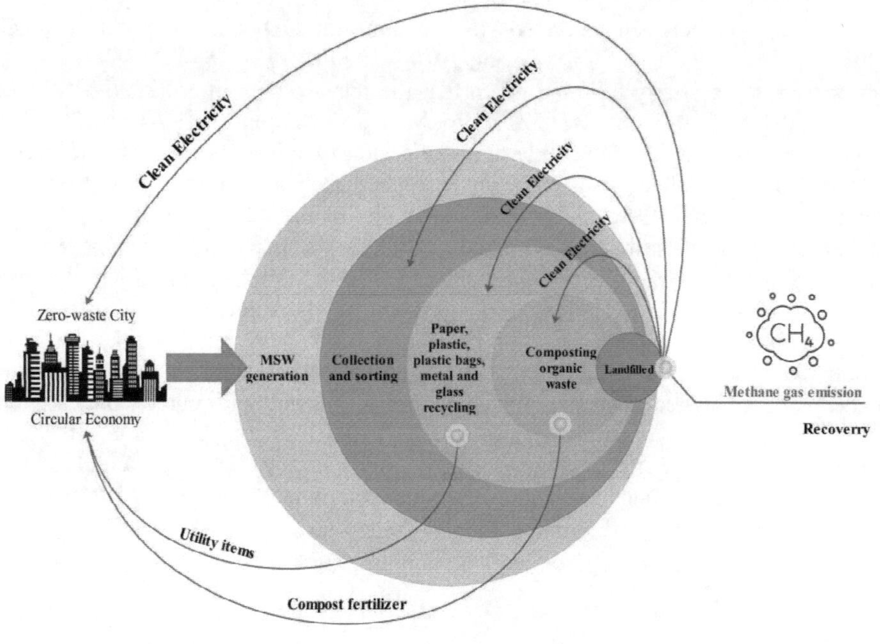

Figure 3. The CE diagram of the Khanh Son landfill in MSW management in Danang City, Vietnam

3. Results and discussion

3.1. Assessment of the MSW management status in Danang City

MSW management has become one of the main challenges for the government of Danang City, mainly due to the rapid pace of urbanization, especially the development of local tourism along with the rapid generation of different types of solid waste in urban areas. This is clearly illustrated in the relationship between Gross domestic product (GDP) per capita value and generated MSW volume in Danang City in the period of 2010 – 2020; in specific, the GDP per capita values rose significantly from 31.21 million VND to 103.61 million VND per person, which was relative with the increase of MSW volume from 279,014.3 tonnes up to 356,813.6 in the 2010 – 2020 period [43] (Figure 4).

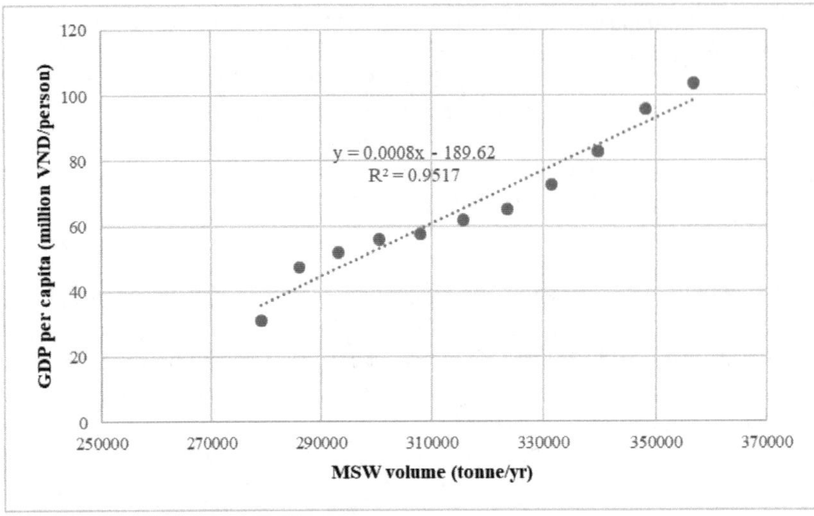

Figure 4. Relationship between generated MSW volume and GDP per capita in Danang Ciy in the period 2010 – 2020

In addition, the results is clearly reflected given that the total volume of collected MSW increased by 17.3% between 2010 and 2013 from 223,521 tonnes/yr to 262,182 tonnes/yr [44]; by 2017 the collected volume reached 314,565 tonnes/yr [42], and estimated to increase of 40.7% and 20.0% respectively for the years 2009, 2013 and by 2019 the statistical volume reached about 401,500 tonnes/yr [43], an increase of 27.6% compared to 2017 with the collection rate of the whole period at 92 – 95% [42]. Figure 5 presents MSW generation levels of all districts in 2019, and Table 3 shows annual MSW volume treated from 2007 to 2020 in the Khanh Son landfill site. Similarly, the city's MSW generation index has also increased year over year from 0.675 $kg.person^{-1}.d^{-1}$ [44] to 0.97 $kg.person^{-1}.d^{-1}$ [43] in the period 2010 – 2019. With all pressure above, the operation of Da Nang Urban Environment Company, public service unit responsible for the collection, transportation and processing waste under the coordination and management of the Provincial People's Committee as well as the Department of Natural Resources and Environment [45] (Figure 6), has arised many problems. When at 6 in 11 transfer stations and some gathering points are always overloaded with waste transshipment capacity only reaching 7% [42], the backlog problems of MSW continously persists, causing bad odors and serious urban environmental pollution [43].

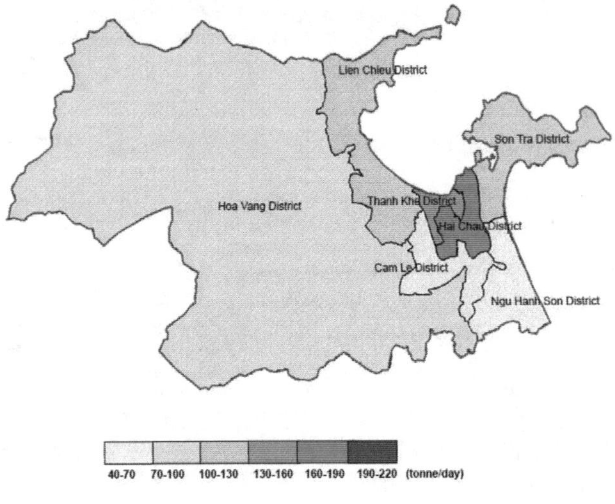

Figure 5. The map of solid waste generation levels of Danang Ciy in the year 2019

IOP Conf. Series: Earth and Environmental Science **964** (2022) 012015 doi:10.1088/1755-1315/964/1/012015

Table 3. Annual treated MSW volume between 2007 and 2020 in the Khanh Son landfill (Unit: kg/day)

Year	Cam Le District	Hoa Vang District	Hai Chau District	Lien Chieu District	Ngu Hanh Son District	Son Tra District	Thanh Khe District	Total
2007	67,389	88,554	153,064	109,652	49,726	100,672	137,632	706,690
2008	70,939	90,857	157,043	112,502	51,019	103,290	141,210	726,861
2009	72,784	93,219	161,127	115,518	52,346	105,976	144,882	745,851
2010	74,596	95,540	165,139	118,394	53,649	108,615	148,490	764,423
2011	76,453	97,919	169,251	121,342	54,985	111,319	152,187	783,457
2012	78,357	100,358	173,465	124,364	56,354	114,091	155,977	802,965
2013	80,308	102,856	177,784	127,460	57,757	116,932	159,861	822,959
2014	82,308	105,418	182,211	130,634	59,195	119,844	163,841	843,451
2015	84,357	108,042	186,748	133,887	60,669	122,828	167,921	864,453
2016	86,458	110,733	191,398	137,220	62,180	125,886	172,102	885,977
2017	88,610	113,490	196,164	140,637	63,728	129,021	176,387	908,038
2018	90,817	116,316	201,049	144,139	65,315	132,233	180,779	930,648
2019	93,078	119,212	206,055	147,728	66,941	135,526	185,281	953,822
2020	95,396	122,180	211,186	151,407	68,608	138,901	189,894	977,572

In order to enhance the recycling rate (mainly for paper, plastic, metal) and waste management efficiency, the action plan No. 1577/KH-UBND vision to 2025 for waste segregation at source has been widely implemented from the ward/commune level [43]. Until June 2018, positive transformation from the results showed that over 80% of residential areas have implemented waste separation at source according to the collection process, and three in six districts of Da Nang have apply 3R activities (Reduce, Reuse and Recycle) [43].

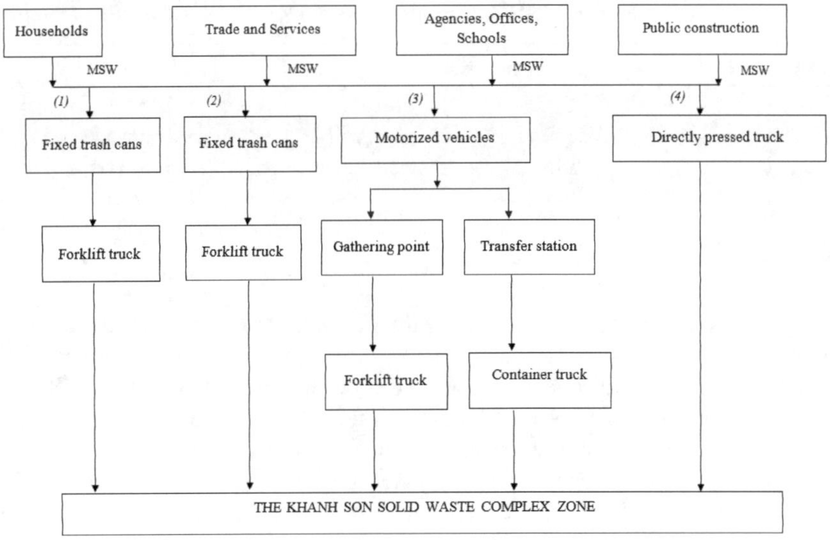

Figure 6. The processes of MSW collection and treatment in Danang City, Vietnam

3.2. Prediction of the GHGs (CH_4 and CO_2) emission potential in the Khanh Son landfill site

The Khanh Son landfill site is located in the area of Danang City – Central Vietnam. The climate here is hot and arid. The CH_4 generation potential (L_0) and CH_4 gas generation rate constant (*k*) mainly depend on the climatic conditions of each area. It can be shown that the values of k and L_0 determine most of the gas emission load at the landfill. In general, the composition of solid wastes in the Khanh Son landfill, unchanged significantly over the years and has continued to remain with an assumption for the next period from 2021 – 2050 (Table 4 and Figure 7).

Along with the coefficient $DOC_f = 0.5$; coefficients $MCF_{KhanhSon} = 1$ [26], F is estimated in the value range of 50 – 55% to create corresponding potential CH_4 generation capacity (L_0) values being 138.5 m^3/tonne, 141.3 m^3/tonne, 144.0 m^3/tonne, 146.8 m^3/tonne, 149.6 m^3/tonne, and 152.3 m^3/tonne respectively.

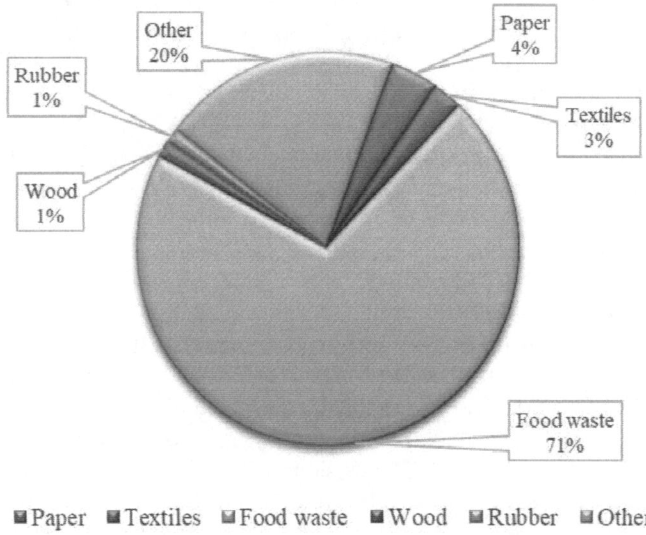

Figure 7. The main MSW components in the Khanh Son landfill, Danang City, Vietnam

Table 4. Synthesis of buried MSW components of the Khanh Son landfill, Danang City

MSW components	W_i mean (%)	Range of DOC_i (%)	DOC_i mean (%)	W_i * DOC_i
Paper/Carton	3.98	40 - 50	42.97172	0.017103
Textiles	2.53	25 - 50	46.98571	0.011887
Organic matter (Food)	70.68	20 - 50	31.43181	0.222160
Wood	1.39	46 - 54	57.24273	0.007957
Rubber	1.18	47	47	0.005546
Other waste	20.24	-	-	-
Total	**100.00**	-	-	**0.264653**

In general, the CH_4 emission load tends to increase gradually from 2008 to 2021 and then tends to decrease rapidly in the period of 2021 – 2050. The CH_4 emission load value peaks in 2021 at around 32,500 tonnes/yr and these cumulative emission values reach about 342,000 tonnes between 2008 and 2021. However, with selecting the value of L_0 through the sensitivity to each k value that this value is determined by the solid waste composition at the landfill and a number of other factors. Solid waste composition changes over time (hourly, day by day), not fixed, it is recommended to test with L_0 which is suitable to the climatic conditions of the area and make comparisons for testing, choose the best value. The optimal input CH_4 (k_{opt}) generated rate constant for the LFG emission load forecasting model is determined by the experimental method based on the initial range of k values. The result of running calculation iterations determines the load, the concentration of the contagant at the measuring locations have been compared and verified to have determined the optimal k_{opt} value for landfill $k_{opt} = 0.4$. It would be the most obvious in the difference in value L_0 of CH_4 over the years. The following Figure 8 is the result of the L_0 values at optimal k ($k_{opt} = 0.4$ year^{-1}).

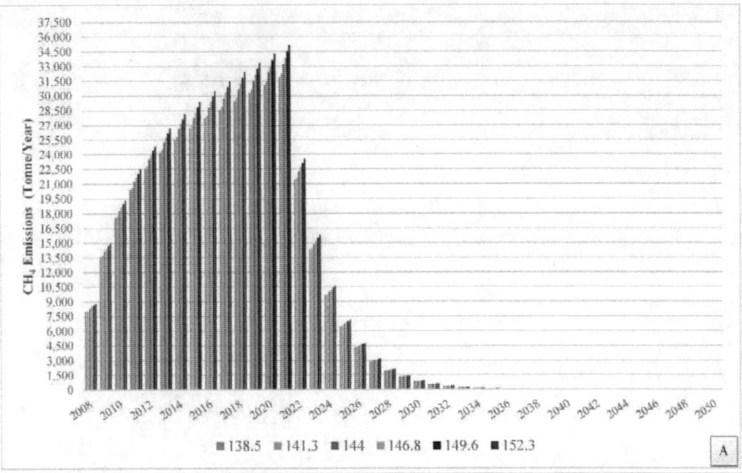

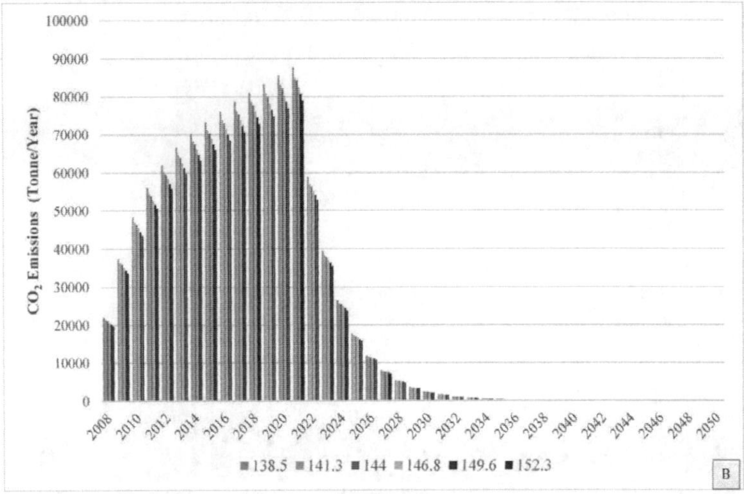

Figure 8. The difference in L_0 values of CH_4 (A) and CO_2 (B) generation with the optimal *k* value

The result estimates CH_4 emissions from 2008 according to the trend of CH_4 (*k*) emission rate constant, specifically looking at the graph, it is possible to give an accurate assessment about the difference between the L_0 values. It means k_{opt} = 0.4 $year^{-1}$ considered as being the most suitable coefficient for the climate conditions in Vietnam in general and the Khanh Son landfill of Danang City in particular, as well as with simulating cases using the EnLandfill software. With selecting value k through sensitivity for each value L_0 fixed that this value commonly depends on the climatic characteristics of the area under consideration. In Vietnam-specific conditions, the value of coefficient *k* is established in the range 0.4 – 0.7 [23]. The emission simulation for each value of each LFG in the period of 2008 – 2050 is conducted to compare and choose the most suitable k for the EnLandfill model. For each L_0 value, the model is run and exported output results with four different k values. The Figure 9 below shows the results respectively with L_0 = 138.5 (m^3/tonne) corresponding to F = 50% (Fig. 9-A1, B1); L_0 = 141.3 (m^3/tonne) corresponding to F = 51% (Fig. 9-A2, B2); L_0 = 144 (m^3) corresponding to F = 52% (Fig. 9-A3, B3); L_0 = 146.8 (m^3/tonne) corresponding to F = 53% (Fig. 9-A4, B4); L_0 = 149.6 (m^3/tonne) corresponding to F = 54% (Fig. 9-A5, B5); and L_0 = 152.3 (m^3/tonne) corresponding to F = 55% (Fig. 9-A6, B6).

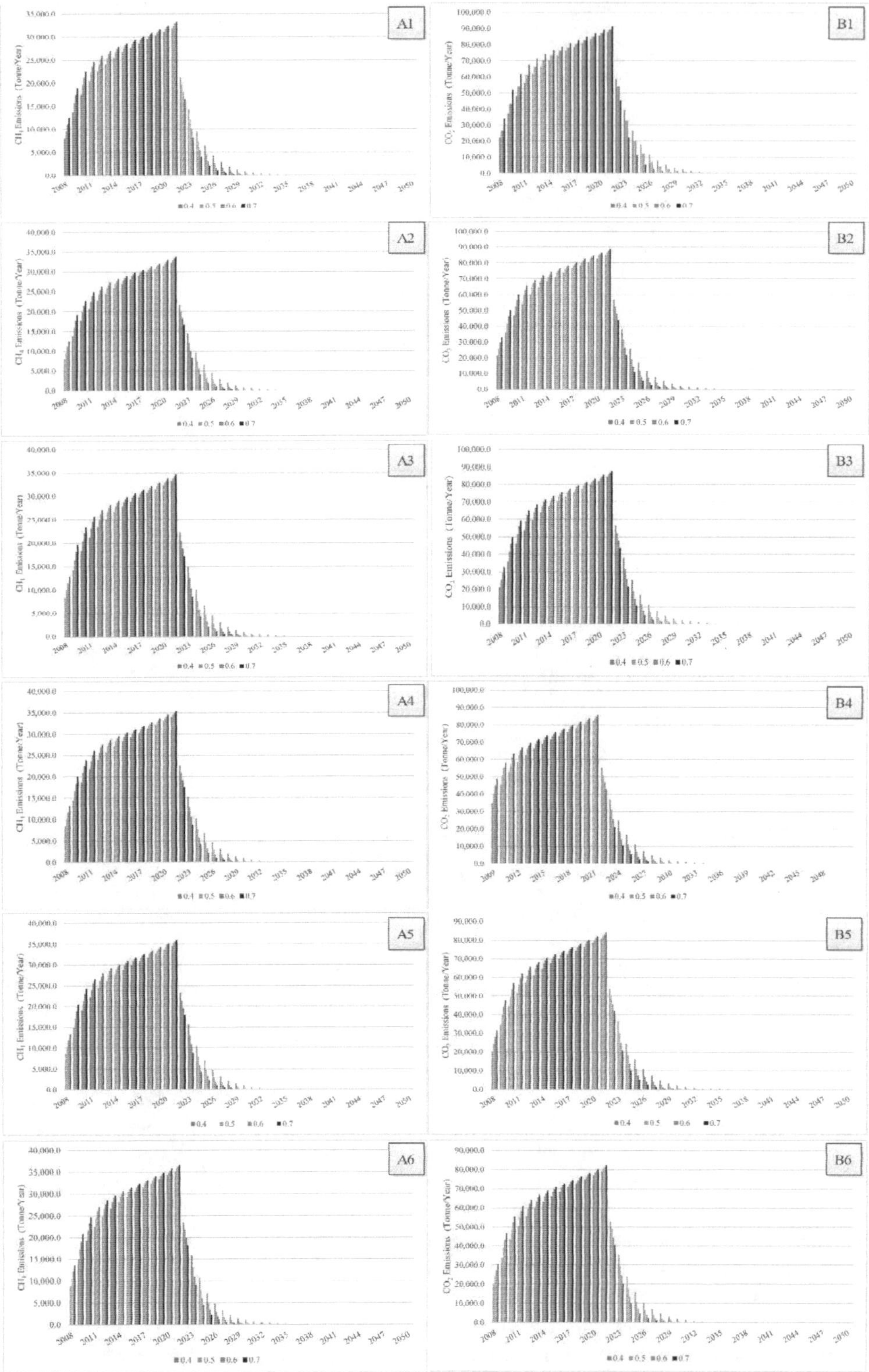

Figure 9. The difference in *k* values of CH_4 (A1, A2, A3, A4, A5, A6) and CO_2 (B1, B2, B3, B4, B5, B6) gases generation with the fixed L_0 values

Whilst visualizing to compare emission values, a couple of coefficients give quite many different values. Because of the complete difference of these L_0 values, it can be seen that there is an error and it is not suitable to run the model. To be able to find an optimal L_0 value, the estimation of pollutant dispersion needs to carry out in order to compare the model and measured substance concentrations with NASH statistical index above 0.75 [23], [24] (Table 5). After calibration and validation, the optimal L_0 value is determined at about 138.5 m^3/tonne and looking at the chart comparing the sensitivity to L_0 of the *k* values, which is not significantly different in CH_4 emissions between the different k values over the years. Hence, the potential of CH_4 gas generation (L_0) is more difficult to determine the emissions of gases than the CH_4 generation rate constant (*k*). Figure 10 below shows the spatial distribution map of average CH_4 emissions in the years 2017, 2018, and 2019, which can be seen that with the optimal $L_{0, opt} = 138.5$ and $k_{opt} = 0.4$ year^{-1}, CH_4 emission load values are quite high. It reached 16,136.43 mg/s; 16,621.66 mg/s; 17,091.47 mg/s respectively in the years 2017, 2018 and 2019. Furthermore, CH_4 emissions generate focus on the Northern areas of 1, 3, 4, 5 borders the existing hazardous waste treatment area.

Table 5. Validation results of CH_4 concentration between observed and simulated values using the EnLandFill software

Time	Type	KK01[(1)]	KK02[(2)]	KK03[(3)]	KK04[(4)]
20/03/2015	Observation	0.005	0.007	0.026	0.800
	Simulation	0.006	0.007	0.070	1.005
16/06/2015	Observation	0.006	0.008	0.028	0.670
	Simulation	0.006	0.100	0.080	0.672
24/09/2015	Observation	0.004	0.007	0.026	0.810
	Simulation	0.005	0.005	0.026	0.816
30/12/2015	Observation	0.005	0.008	0.018	0.820
	Simulation	0.005	0.101	0.100	0.940
07/04/2016	Observation	0.001	0.008	0.001	0.140
	Simulation	0.100	0.006	0.100	0.143
16/06/2016	Observation	0.001	0.091	0.001	0.100
	Simulation	0.003	0.120	0.100	0.237
27/09/2016	Observation	0.350	0.300	0.730	0.420
	Simulation	0.346	0.306	0.800	0.429
07/12/2016	Observation	0.310	0.290	0.690	0.400
	Simulation	0.322	0.292	0.522	0.600
28/03/2017	Observation	0.160	0.190	0.740	0.420
	Simulation	0.140	0.250	0.811	0.422
15/06/2017	Observation	0.100	0.110	0.420	0.310
	Simulation	0.200	0.210	0.560	0.320
12/09/2017	Observation	0.160	0.190	0.740	0.370
	Simulation	0.270	0.191	0.636	0.471
NASH index		**0.808**	**0.764**	**0.918**	**0.815**
R^2		**0.871**	**0.864**	**0.932**	**0.916**

[(1)] KK01 (X = 16.046391°; Y = 108.146622°): at residential area in the head of the main wind direction;
[(2)] KK02 (X = 16.041588°; Y = 108.134399°): at residential area at the end of the main wind direction;
[(3)] KK03 (X = 16.044896°; Y = 108.144302°): at hazardous waste burial trash box;
[(4)] KK04 (X = 16.041863°; Y = 108.141857°): at non-hazardous waste burial trash box.

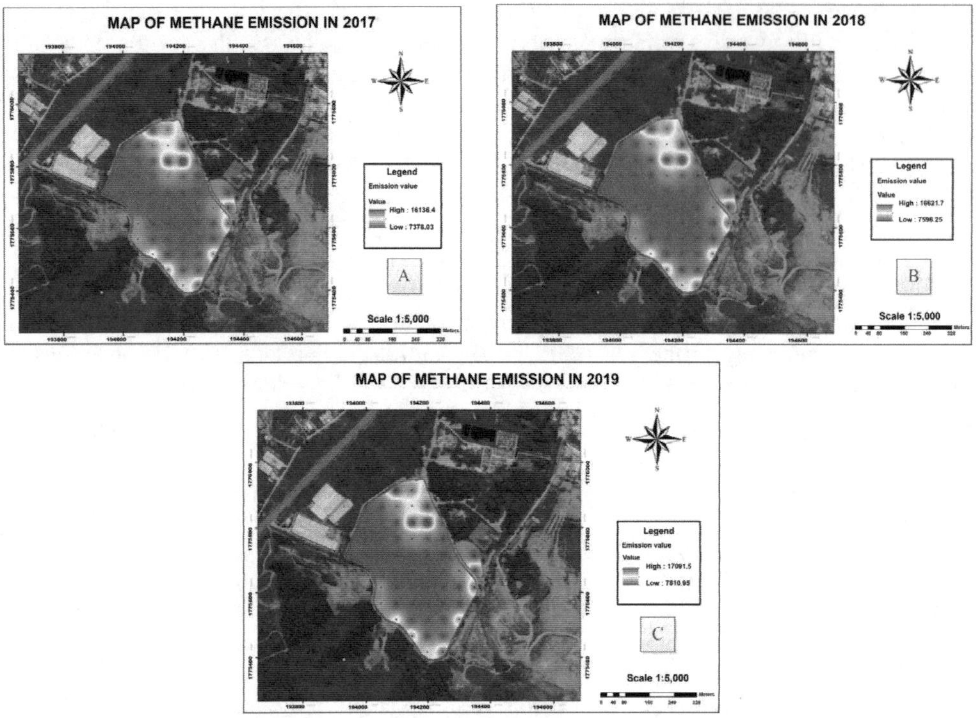

Figure 10. The spatial allocation of annual mean CH_4 emissions in the years 2017 (A), 2018 (B) and 2019 (C) in the Khanh Son landfill, Danang City

3.3. Assessment of the energy recovery potential and GHGs reduction from the Khanh Son landfill

Based on the estimated results of the total amount of CH_4 emissions recovered from the collection system with the recovery efficiency E = 90% in the period of 2021 – 2050 according to the analysis scenarios (according to the optimal parameters of L_0 and k_{opt}), the annual value of electricity generation potential $EP_{LFG,year\ i}$ (kWh/year) has been estimated according to the formula Eq. (8) is shown in Figure 11. From the construction scenario orientation to the analysis, it is found that the potential for annual electricity generation in the period 2021 – 2050 is highest in 2021, with a total estimated value of about 138,717 million kWh/year. In general, the potential for electricity generation tends to decrease gradually during the forecast period from 138,717 million kWh/year to 3.79 million kWh/year in the period of 2021 – 2030 and almost no potential for energy recovery from 2040 to 2050 (approximately 0.001 million kWh/year in 2050). Moreover, it could be seen that the electricity generation potential directly related to the amount of captured CH_4 emissions estimated on the basis of the determined optimal factors $L_{0,\ opt}$, and k_{opt} and increased rapidly from 2008 – 2021, peaked in 2021, and then reduced significantly by 2050. Thus, the recovery potential of CH_4 gas to generate clean energy is merely suitable in the 2021 – 2030 period.

From there, the power station size ($LFGTE_{(size)}$) can be determined according to the formula Eq. (9) from CH_4 gas of the Khanh Son landfill site with the assumption that the generating station is capable of operating throughout the year with D_{hr} = 24 hours/day and the number of days the generating station operates in a year with γ = 365 days. Thus, with the above assumption, in the analysis scenario, the size of the power station will gradually increase from 15,835.26 kW in the year 2021 and then gradually decrease to about 432.68 kW in the year 2030 and by the year 2040, it will almost no longer be able to generate electricity for usage demands (Figure 11).

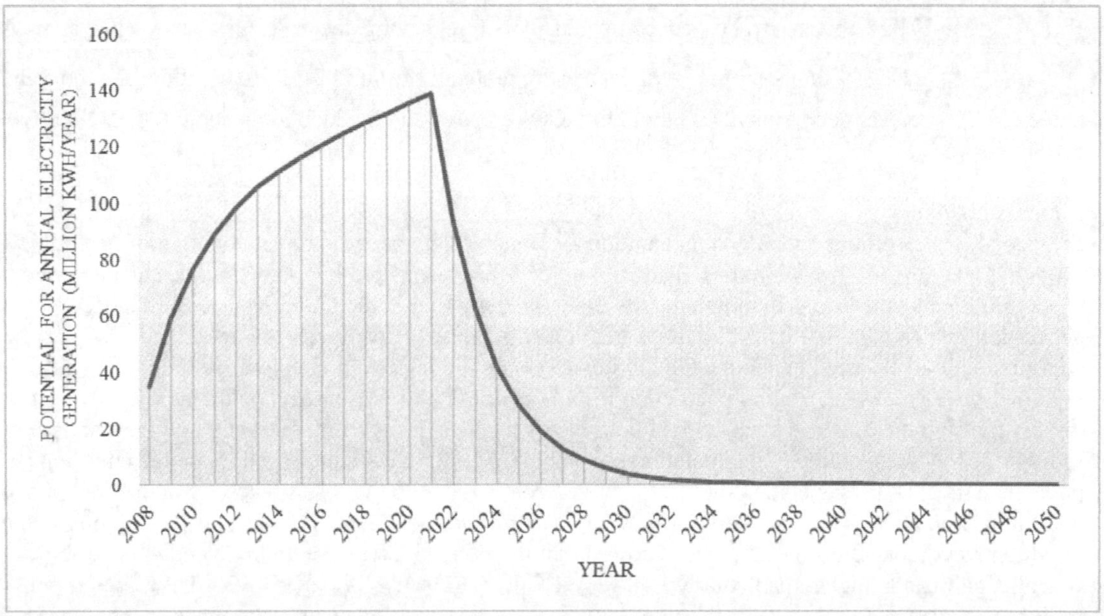

Figure 11. The potential forecast for annual generated electricity changes (million kWh/yr) in Khanh Son landfill

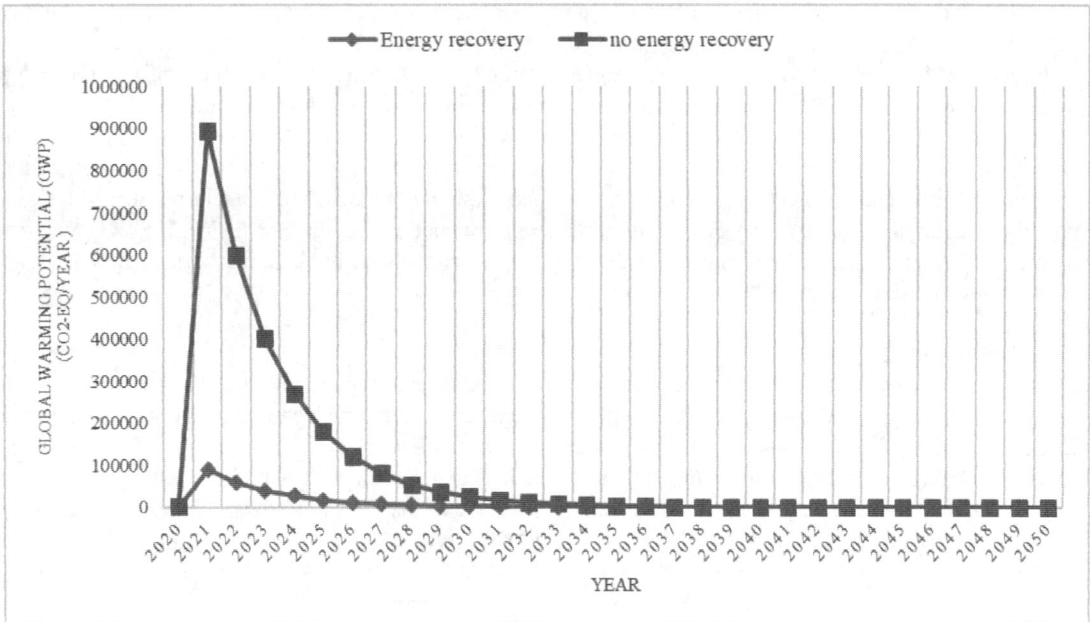

Figure 12. The comparison of potential to reduce GHGs emissions for the period of 2021 – 2050 in the Khanh Son landfill site, Danang City

In order to assess the potential to reduce emissions of GHGs, including CH_4 and CO_2, the GWP index is considered as a general level of conversion that has been calculated in CO_2-eq units based on based on the results of the "5th Assessment Report" (AR5) of IPCC 2014 [39] according to the formulas Eq. (10) and Eq. (11). The estimated results are detailed in Figure 12. With the analysis scenario, the potential to reduce GHGs emissions is calculated for the period 2021 – 2050 with the recovery efficiency E = 90% due to the installation of the generated LFGs recovery system. The total estimated mitigation potential

$\sum RE_{GHGs,re}$ = 271.25 thousand tCO_2-eq compared to before having the LFG gas recovery system is $\sum RE_{GHGs,no-re}$ = 2,712.49 thousand tCO_2-eq, decreasing gradually from 2021 – 2030 with 89.42 thousand tonnes of CO_2-eq/year decreasing 24.43 thousand tCO_2-eq/year; and by 2050, only about 0.8 tCO_2-eq will remain per year.

3.4. Discussion

In general, the coefficient of MSW generation in Danang City ranges from 0.8 – 0.9 kg/person/day, averaging about 0.838 kg/person/day. The volume of MSW generated by districts varies, depending on economic development and living standards. In particular, Lien Chieu District has the highest MSW generation rate (around 0.9 kg/person/day) because it is home to many entertainment and food service areas; many markets are sold and people's living standards are improved. Whilst Hoa Vang is the district with the lowest MSW generation coefficient of only about 0.8 kg/person/day, owing to it is a suburban district far from the center; the majority of households work in the agricultural sector and the standard of living is low. Moreover, the CH_4 generation potential (L_0) is more difficult to determine the emissions of gases than the CH_4 (k) generation rate factor. When compared with the case studies at South Binh Duong landfill in Binh Duong province [46] and Tan Lap 1 landfill in Tien Giang province [30], belonging to the Southern Key Economic Zone, Vietnam showed that the optimal CH_4 generation potential value, $L_{0,\ opt(x)}$ was all significantly higher, particularly $L_{0,\ opt\text{-}SouthBinhDuong} = 81 < L_{0,\ opt\text{-}TanLap1} = 110 < L_{0,\ opt\text{-}KhanhSon} = 138$ m^3/tonne. Nevertheless, this optimal value, $L_{0,\ opt\text{-}KhanhSon}$ was still lower than the recommended default values of IPCC [25], [40]. This can be explained by the fact that the MSW receiving sources from the Tan Lap 1 landfill have been smaller in scale and had much less diverse waste components than in the Khanh Son landfill, and in the case of the South Binh Duong landfill, organic matter components have commonly accounted for only 45%, since roughly 55% of the organic wastes have sorted and treated at the composting areas [46]. Furthermore, according to the trend of CH_4 emission rate constant, k in the range 0.4 – 0.5 $year^{-1}$ is the most suitable for the climate conditions in the Khanh Son landfill site. Similarly, the optimal CH_4 gas generation rate constant value (k_{opt}) in all three areas had certain similarities with $k_{opt\text{-}SouthBinhDuong}$ = 0.355 $years^{-1}$, $k_{opt\text{-}TanLap1}$ = 0.23 $years^{-1}$, and $k_{opt\text{-}KhanhSon}$ = 0.40 $years^{-1}$ reflecting the considerable influence of general characteristics of geographical location and climate on decomposition conditions at landfill sites, and were also within the recommended value range of IPCC [25], [40] with k = 0.15 – 0.70 $years^{-1}$ for medium and rapid decomposition landfills in hot, humid tropical regions with average annual rainfall over 1,000 mm.

4. Conclusions

This study had evaluated the effectiveness of solid waste management planning in Danang City following the direction of a CE with a focus on analyzing the potential of creating clean energy from the recovered LFGs and reducing the contribution of GHG emissions, which is considered one of the highlights in the research. The assessment was carried out at Khanh Son landfill area with the goal of forecasting the amount of CH_4 and CO_2 gases generated by the EnLandFill software in the period of 2021 – 2050, as a basis for setting up an analysis scenario for estimation of energy recovery and power generation from LFGs following the trend of building a sustainable CE and zero-waste city orientation of Danang City, Vietnam. The results of estimating the potential value of CH_4 gas generation is from 8,015.7 – 11,9672.8 tonnes/year and that of CO_2 gas generation are from 21,993.3 – 87,706.98 tonnes/year from 2008 to 2021 with an optimal L_0 = 138.5 m^3/tonne of waste. The optimal input CH_4 (k_{opt}) generated rate constant for the LFG emission load value forecasting model is determined by the experimental method based on the initial range of k values. The result of running calculation iterations determines the load, the concentration of the contagion at the measuring locations have been compared and verified to has determined the optimal k_{opt} value for landfill is k_{opt} = 0.4 $year^{-1}$. The trend of CH_4 and CO_2 generation has increased over the period from 2008 to 2020. It has reached the peak of 119,672.8 tonnes/year (CH_4) and 87,706.98 tonnes/year (CO_2) in 2021. The potential value of CH_4 and CO_2 gas generation has a tendency to plummet in the next 29 years to 2050. It will have been 0.293 tonnes/year (CH_4) and 0.804 tonnes/year (CO_2) in 2050. At the same time, the obtained results also reported that in the period from 2021 – 2050, the total potential for energy recovery from CH_4 gas of the Khanh Son landfill is estimated at about 420,767 million kWh, equivalent to the total potential for GHGs emission reduction (GWP) about 271.25 thousand tCO_2-eq. Therefore, the exploitation of energy recovery from landfill gas can not only help to reduce GHGs emissions, but also provide a locally satisfying energy source for operating subdivisions in the Complex or other neighborhood areas.

Author contribution statement: P.H.N: Methodology, Models, Formal analysis, Writing – Original draft, Writing – Review & Editing. Q.K.N.C: Data curation, Data Formal analysis, Formal analysis, Validation, GIS. L.T.B: Conceptualization, Funding acquisition, Investigation, Project administration, Resources, and Supervision.

Acknowledgements: This research was funded by the Viet Nam National University Ho Chi Minh City (VNU–HCM), grant No: B2019–20B–01. The authors would also like to thank the members of the Laboratory for Environmental Modelling, Ho Chi Minh City University of Technology for discussions that improved the quality of the publication.

Competing interest statement: The authors declare no conflict of interest.

References

[1] Da Nang City People's Committee, "Report on environmental impact assessment of the project of upgrading and renovating several items at Khanh Son landfill in 2019," Danang City, 2019.

[2] D. W. Pearce and R. K. Turner, "Economics of Natural Resources and the Environment," *Balt. MD Johns Hopkins Univ. Press*, p. 378, 1990, doi: https://doi.org/10.2307/1242904.

[3] P. Ghisellini, C. Cialani, and S. Ulgiati, "A review on circular economy: The expected transition to a balanced interplay of environmental and economic systems," *J. Clean. Prod.* **114**, 11–32, 2016, doi: 10.1016/j.jclepro.2015.09.007.

[4] R. Merli, M. Preziosi, and A. Acampora, "How do scholars approach the circular economy? A systematic literature review," *J. Clean. Prod.* **178**, 703–722, 2018, doi: 10.1016/j.jclepro.2017.12.112.

[5] J. Motloch, "Regenerative design for sustainable development," *Landsc. Urban Plan.* **32**, 3, 198–201, 1995, doi: 10.1016/0169-2046(95)90009-8.

[6] W. R. Stahel, "The Performance Economy: Business Models for the Functional Service Economy," in *Handbook of Performability Engineering*, K. B. Misra, Ed. London: Springer London, 2008, pp. 127–138.

[7] M. Braungart, W. McDonough, and A. Bollinger, "Cradle-to-cradle design: creating healthy emissions – a strategy for eco-effective product and system design," *J. Clean. Prod.* **15**, 13, 1337–1348, 2007, doi: https://doi.org/10.1016/j.jclepro.2006.08.003.

[8] S. Erkman, "Industrial ecology: An historical view," *J. Clean. Prod.* **5**, 1, 1–10, 1997, doi: https://doi.org/10.1016/S0959-6526(97)00003-6.

[9] D. Ness, "Sustainable urban infrastructure in China: Towards a Factor 10 improvement in resource productivity through integrated infrastructure systems," *Int. J. Sustain. Dev. \& World Ecol.* **15**, 4, 288–301, 2008, doi: 10.3843/SusDev.15.4:2a.

[10] Y. Geng and B. Doberstein, "Developing the circular economy in China: Challenges and opportunities for achieving 'leapfrog development,'" *Int. J. Sustain. Dev. \& World Ecol.* **15**, 3, 231–239, 2008, doi: 10.3843/SusDev.15.3:6.

[11] H. Salmenperä, K. Pitkänen, P. Kautto, and L. Saikku, "Critical factors for enhancing the circular economy in waste management," *J. Clean. Prod.*, 2**80**, 2021, doi: 10.1016/j.jclepro.2020.124339.

[12] D. A. R. George, B. C. Lin, and Y. Chen, "A circular economy model of economic growth," *Environ. Model. Softw.* **73**, 60–63, 2015, doi: https://doi.org/10.1016/j.envsoft.2015.06.014.

[13] R. Hughes, "The EU Circular Economy Package - Life Cycle Thinking to Life Cycle Law?," *Procedia CIRP* **61**, 10–16, 2017, doi: 10.1016/j.procir.2016.12.006.

[14] B. Su, A. Heshmati, Y. Geng, and X. Yu, "A review of the circular economy in China: Moving

from rhetoric to implementation," *J. Clean. Prod.* **42**, 215–227, 2013, doi: 10.1016/j.jclepro.2012.11.020.

[15] N. M. P. Bocken, I. de Pauw, C. Bakker, and B. van der Grinten, "Product design and business model strategies for a circular economy," *J. Ind. Prod. Eng.* **33**, 5, 308–320, 2016, doi: 10.1080/21681015.2016.1172124.

[16] M. R. Esa, A. Halog, and L. Rigamonti, "Developing strategies for managing construction and demolition wastes in Malaysia based on the concept of circular economy," *J. Mater. Cycles Waste Manag.* **19**, 3, 1144–1154, 2017, doi: 10.1007/s10163-016-0516-x.

[17] P. Brown, N. Bocken, and R. Balkenende, "Why do companies pursue collaborative circular oriented innovation?," *Sustain.* **11**, 3, 1–23, 2019, doi: 10.3390/su11030635.

[18] P. Priyadarshini and P. C. Abhilash, "Circular economy practices within energy and waste management sectors of India: A meta-analysis," *Bioresour. Technol.* **304**, November 2019, p. 123018, 2020, doi: 10.1016/j.biortech.2020.123018.

[19] L. A. López Ruiz, X. Roca Ramón, and S. Gassó Domingo, "The circular economy in the construction and demolition waste sector – A review and an integrative model approach," *J. Clean. Prod.* **248**, 2020, doi: 10.1016/j.jclepro.2019.119238.

[20] S. Suárez, X. Roca, and S. Gasso, "Product-specific life cycle assessment of recycled gypsum as a replacement for natural gypsum in ordinary Portland cement: application to the Spanish context," *J. Clean. Prod.* **117**, 150–159, 2016, doi: https://doi.org/10.1016/j.jclepro.2016.01.044.

[21] C. Ludwig, S. Hellweg, and S. Stucki, *Municipal Solid Waste Management*. Villigen PSI, Switzerland, 2003.

[22] S. M. Ali, A. Pervaiz, B. Afzal, N. Hamid, and A. Yasmin, "Open dumping of municipal solid waste and its hazardous impacts on soil and vegetation diversity at waste dumping sites of Islamabad city," *J. King Saud Univ. - Sci.* **26**, 1, 59–65, 2014, doi: 10.1016/j.jksus.2013.08.003.

[23] B. T. Long and N. H. Phong, "Integrated model for methane emission and dispersion assessment from landfills: A case study of Ho Chi Minh City, Vietnam," *Sci. Total Environ.* **738**, 139865, 2020, doi: 10.1016/j.scitotenv.2020.139865.

[24] P. H. Nguyen, D. C. M. Nguyen, and L. T. Bui, "Potential of methane emission reduction and energy recovery from the municipal solid waste landfills - a case study of Binh Duong province, Vietnam," in *The National Conference on Geography Information System 2020 (GIS Conference) - GIS for smart cities towars sustainable development*, 2020, pp. 381–393.

[25] R. Pipatti *et al.*, "Chapter 3: Solid Waste Disposal," in *2006 IPCC Guidelines for National Greenhouse Gas Inventories*, IPCC, Ed. 2006, pp. 1–40.

[26] A. Alexander, C. Burklin, and A. Singleton, "Landfill Gas Emissions Model (LandGEM) Version 3.02 User's Guide," no. May, p. 56, 2005.

[27] J. Bogner and K. Spokas, "Landfill CH4: Rates, fates, and role in global carbon cycle," *Chemosphere* **26**, 1–4, 369–386, 1993, doi: 10.1016/0045-6535(93)90432-5.

[28] D. Cudjoe and M. S. Han, "Economic and environmental assessment of landfill gas electricity generation in urban districts of Beijing municipality," *Sustain. Prod. Consum.* **23**, 128–137, 2020, doi: 10.1016/j.spc.2020.04.010.

[29] D. Cudjoe, M. S. Han, and W. Chen, "Power generation from municipal solid waste landfilled in the Beijing-Tianjin-Hebei region," *Energy* **217**, 2021, doi: 10.1016/j.energy.2020.119393.

[30] L. T. Bui and P. H. Nguyen, "Prediction of potential for greenhouse gas mitigation and power recovery from a municipal solid waste landfill case in Tien Giang province , Vietnam," *Vietnam J. Hydrometeorol.* 2021, **7**, 32–52, 2021, doi: 10.36335/VNJHM.2021(7).32-52.

[31] D. Cudjoe, M. S. Han, and A. P. Nandiwardhana, "Electricity generation using biogas from organic fraction of municipal solid waste generated in provinces of China: Techno-economic and environmental impact analysis," *Fuel Process. Technol.* **203**, March, 106381, 2020, doi: 10.1016/j.fuproc.2020.106381.

[32] P. Ghosh *et al.*, "Assessment of methane emissions and energy recovery potential from the municipal solid waste landfills of Delhi, India," *Bioresour. Technol.*, 611–615, 2019, doi:

10.1016/j.biortech.2018.10.069.

[33] N. Scarlat, V. Motola, J. F. Dallemand, F. Monforti-Ferrario, and L. Mofor, "Evaluation of energy potential of Municipal Solid Waste from African urban areas," *Renew. Sustain. Energy Rev.* **50**, 1269–1286, 2015, doi: 10.1016/j.rser.2015.05.067.

[34] T. R. Ayodele, A. S. O. Ogunjuyigbe, and E. E. Ekoh, "Outlook of Agricultural Sector in the Face of Changing Global Climate: The Case of Nigeria," *Agric. Res. Technol. Open Access J.* **5**, 3, 68–71, 2017, doi: 10.19080/artoaj.2017.05.555664.

[35] A. S. O. Ogunjuyigbe, T. R. Ayodele, and M. A. Alao, "Electricity generation from municipal solid waste in some selected cities of Nigeria: An assessment of feasibility, potential and technologies," *Renew. Sustain. Energy Rev.* **80**, May, 149–162, 2017, doi: 10.1016/j.rser.2017.05.177.

[36] T. R. Ayodele, A. S. O. Ogunjuyigbe, and M. A. Alao, "Economic and environmental assessment of electricity generation using biogas from organic fraction of municipal solid waste for the city of Ibadan, Nigeria," *J. Clean. Prod.* **203**, 718–735, 2018, doi: 10.1016/j.jclepro.2018.08.282.

[37] Climate Leadership Group - C40 Cities, "Workshop on developing city-level greenhouse gas inventory activities - Technical manual for participating experts," Ho Chi Minh City, 2018.

[38] M. Brander, "Greenhouse Gases, CO_2, CO_2e, and Carbon: What Do All These Terms Mean?," *Ecometrica*, no. August, p. 3, 2012.

[39] Green House Protocol, "Global Warming Potential Values," 2015.

[40] IPCC, "CH4 Emissions from Solid Waste Disposal," *IPCC Good Pract. Guid. Uncertain. Manag. Natl. Greenh. Gas Invent.*, pp. 419–439, 2006.

[41] C. Yaman, I. Anil, and O. Alagha, "Potential for greenhouse gas reduction and energy recovery from MSW through different waste management technologies," *J. Clean. Prod.* **264**, 121432, 2020, doi: 10.1016/j.jclepro.2020.121432.

[42] Da Nang City People's Committee, "Report on domestic solid waste management in Da Nang city," Danang City, 2018.

[43] MONRE, "Vietnam National Environmental Report 2019: Waste Management," Ha Noi Capital, 2020.

[44] Da Nang City Environmental Protection Department, "Assessment of solid waste management status in Danang City," Danang City, 2019.

[45] MONRE, "Vietnam National Environmental Report 2011: Waste Management," Ha Noi Capital, 2012.

[46] L. H. Anh *et al.*, "Site-specific determination of methane generation potential and estimation of landfill gas emissions from municipal solid waste landfill: a case study in Nam Binh Duong, Vietnam," *Biomass Convers. Biorefinery*, March, 2021, doi: 10.1007/s13399-020-01192-0.

Impact of weather factors on hospital admission of hand, foot and mouth disease under climate change scenarios in Ho Chi Minh City.

Phung Duc Nhat[1*], Vo Le Phu[2,3*], Đặng Văn Chính[4], Duong Thi Minh Tam[4], Mai Tien Thanh[4]

[1] University of Medicine Pham Ngoc Thach, 2 Quang Trung street, District 10, Ho Chi Minh City, Vietnam

[2] Faculty of Environment and Natural Resources, Ho Chi Minh City University of Technology (HCMUT), 268 Ly Thuong Kiet street, District 10, Ho Chi Minh City, Vietnam

[3] Vietnam National University Ho Chi Minh City, Linh Trung ward, Thu Duc District, Ho Chi Minh City, Vietnam

[4] Institute of Public Health, 159. Hung Phu street, District 8, Ho Chi Minh City, Vietnam

Corresponding authors: nhatphd@pnt.edu.vn; volephu@hcmut.edu.vn

Abstract. Hand, foot, and mouth disease (HFMD) is one of the most common communicable diseases in Vietnam. The present study aims to examine the association between weather factors and HFMD in association with hospitalisation. Daily and weekly weather and HFMD data from 2013 to 2018 in Ho Chi Minh City were deployed. Poisson regression model combined with a distributed lag non-linear model (DLNM) was applied to examine the relationship between weather factors and HFMD. The forecasting model for HFMD was performed by using the Global Climate Model (GCM) and Yasushi Honda model. The result showed that the average daily temperature induces an increase in the risk of HFDM hospitalisation was 26°C- 30.1°C. The average daily humidity also caused increasing the risk of hospitalisation of HFMD was 75% - 85%. However, the average daily humidity <60% reduced the risk of getting HFMD. The study provides quantitative evidence that the incidence of HFMD cases was associated with meteorological variables including average daily temperature and daily humidity in Ho Chi Minh City. This findings implies that there is a need for building a public health policy for eliminating and mitigating climate change impact on community health in a resilient approach.

1. Introduction

Climate change is one of the biggest challenges in the 21st century in which a wide range of socio-economic sectors has been affected ubiquitously. The Intergovernmental Panel on Climate Change (IPCC) found out that the average global surface temperature on earth increased 0.850C during period 1880-2012 [1]. Hand, foot, and mouth disease (HFMD) is a common infectious disease spread from human to human, easy to become an outbreak caused by an intestinal virus. The most common pathogens are Coxsackie virus A16 and Enterovirus 71 (EV71). In recent decades, Asian countries endured frequent and expanded outbreaks of HFMD with a high rate of mortality mostly occurred in children. However, there is no effective and specific treatment approach nor vaccine to this disease [2]. In Viet Nam, there are few studies on the relation of climate change and HFMD. According to Ebi and Hess (2017), only one-third of studies on the impact of climate change on health were conducted in developing countries [3]. This study aims to build a model to predict the impact of climate change on community health based on updated scenarios of climate change and to provide the City authority with evidence to integrate into economic and social development plan of the City. Further, this study findings implies the need for building a public health policy for eliminating and mitigating the impact of climate change on community health in HCM City in a resilient way.

2. Materials and methods

2.1. Data collection

Daily counts of HFMD cases in 24 districts of Ho Chi Minh City from 1 January 2013 till 31 December 2018 were obtained from Center for Diseases Control of Ho Chi Minh City (HCDC). Since 2008, HFMD has been categorized into an infectious disease, local hospitals and clinics were required to report HFMD cases to HCDC communicable disease surveillance system within 24 hours All HFMD cases were first diagnosed symptomatically. Then, the clinical diagnoses of HFMD were strictly examined and verified through laboratory evidence. Extracted data included age, sex, date of admission, date of discharge, residential district, and the International Classification Diseases 10th revision (ICD10) code. To avoid exposure misclassification, patients from locations other than HCMC were excluded. Therefore, data of HFMD cases were reliable and fully reflected changes in HFMD cases in HCMC. Overall, the reporting was representative and accuracy of HFMD in Ho Chi Minh City.

There are no meteorological monitoring stations within each districts in Ho Chi Minh City, data from the station in adjacent cities closest to the city center were used. Therefore, we obtained representative weather data from 7 monitoring stations managed by Southern Regional Hydro meteorological Center to cover for the whole city. Daily weather data from 1 January 2013 to 31 December 2018 were collected in the same period with hospitalization data. The necessary information included temperature (minimum, maximum, and average), temperature range, humidity, wind speed, and rainfalls. These daily data were then converted to the mean of weekly average temperature (oC), wind speed (m/s), relative humidity (%), and cumulative rainfall (mm) for analysis.

2.2. Data analysis

Firstly, meteorological data and HFMD cases were linked by date and analyzed using a time-series design. A distributed lag non-linear model (DLNM) was constructed to represent a modeling framework to flexibly describe associations showing potentially non-linear and delayed effects in time-series data. In present study, DLNM combined with a quasi-Poisson distribution was performed to quantify the relationship between daily temperature and HFMD counts adjusted with other meteorological variables. The model was specified as follows:

$$E(Yi) = \exp\{\alpha + f(Ti|\beta 1) + (\text{specific covariates}) + (\text{smooth function of time}) \qquad (1)$$

F (Ti/β1): The function of temperature

Specific covariates: Confounding variable

Smooth function of time: The function of the time variable, to adjust for the long-term trend and seasonality.

In addition, to account for the potential non-linear relation between temperatures and health outcomes, we applied a distributed lag non-linear model (DLNM) using a cross-basis function of multiple lag-day temperatures [4]. DLNM can help determine the non-linear and lag effects of the relationship between weather factors and health, simultaneously. The optimal model and parameters were selected and validated using post estimation residuals and the Akaike's information criterion (AIC) in which the model with the least AIC explain better and be chosen.

IOP Conf. Series: Earth and Environmental Science **964** (2022) 012016 doi:10.1088/1755-1315/964/1/012016

To forecasting hospital admissions incremental and rising temperatures, our study assumes factors: population, health services, and treatment remain unchanged, and forecasts to 2025, 2030, 2050 by performing analysis in 2 stages:

The first step used Global Climate Model (GCM) PRECIS developed by Hadley Meteorological Centre, the UK to forecast weather factors including max, min, average temperature of HCMC in 2025, 2030, 2050 on 3 scenarios of climate change RCP4.5 (medium), RCP6.0 (high medium), and RCP8.5 (high). This study applied the GFDL-CM3 and CNRM-CM5 of the PRECIS model and downscaled at 25 km spatial resolution to forecast the climate of HCMC. In previous studies, the PRECIS model had better results than other models in predicting Vietnam's climate.

In the second step, we applied method developed by Yasushi Honda et al. (2014) in a previous study to calculate incremental number of admitted cases to hospitals due to hot temperature events with forecasted weather data in the years 2020, 2025, 2030, and 2050 [5].

3. Results and Discussion

3.1. Descriptive analysis

A retrospective statistical data on admission cases of HFMD disease in Ho Chi Minh City from 2013 to 2018 is depicted in Table 1 below.

Table 1. Weekly admission cases and weather factors in HCMC during 2013 – 2018

Variables	Mean	SD	Min	Median	Max
HFMD cases	46.98	67	2.00	13	618.00
Temperature (° C)	28.42	1.28	24.1	28.5	32.1
T. range (° C)	8.38	1.007	4.8	8.3	13
Humidity (%)	73.1	5.97	58	73.1	86
Wind velocity (m/s)	2.76	0.70	1.6	2.7	5.2
Rainfall (mm)	36.50	49.03	0	15.7	279.8

Weekly mean value of temperature in HCMC was 28.42 °C during 2013 – 2018, in which the lowest and highest temperature were 24.1°C and 32.1°C, respectively. Temperature range weekly was 8.38°C, with the lowest and highest values of 4.8°C and 13.5°C. Relative weekly humidity was 73.1%, with the lowest and highest values of 58.0% and 86%. Weekly wind velocity was 2.76 m/s, with the lowest and highest values of 1.6 m/s and 5.2 m/s, respectively. The weekly average rainfall was 36.5 mm.

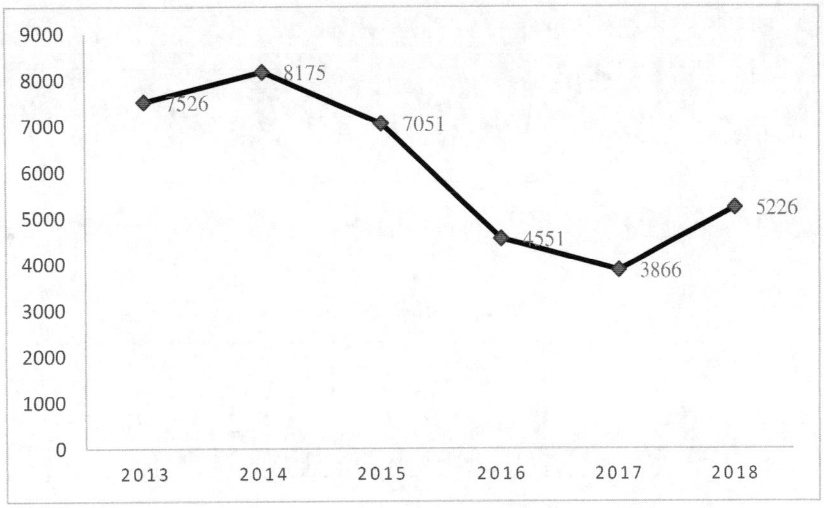

Figure 1. Weekly number of admitted HFMD cases 2013 - 2018

In recent years, the situation of HFMD infection remains severe in Ho Chi Minh City. Figure 1 shows that annual admission cases were highest in 2014 (8,175 cases) . Afterwards, the trend declined since 2015-2018. The two years of 2013 and 2014 were HFMD outbreak events with high number of cases, then there were a declined trend (2014-2017) and after that the cases raised in 2018. Recent data from the World Health Organisation (WHO) showed that HFMD cases increased on a seasonal trend. Until April, 2014 accumulative cases of HFMD in China, Macau, and Singapore increased 1.1-1.5 times compared to the same time period [2]. Increased trend of HFMD cases could be observed in other countries. Some studies also revealed that this increased trend was observed in many Asian countries, including Linsuwanon in Thailand [6], Qingdao, China [7], Edmond Ma in Hongkong, China [8].

Weekly number of HFMD case over time

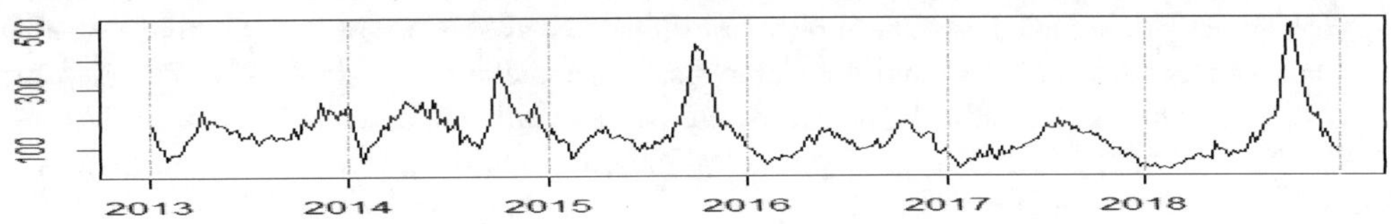

Weekly mean temperature over time

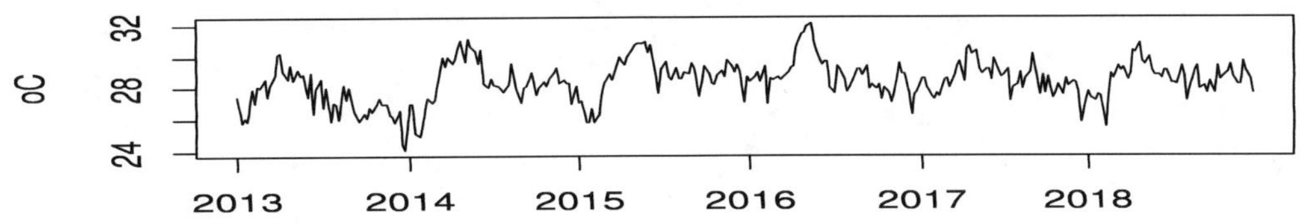

Weekly humidity over time

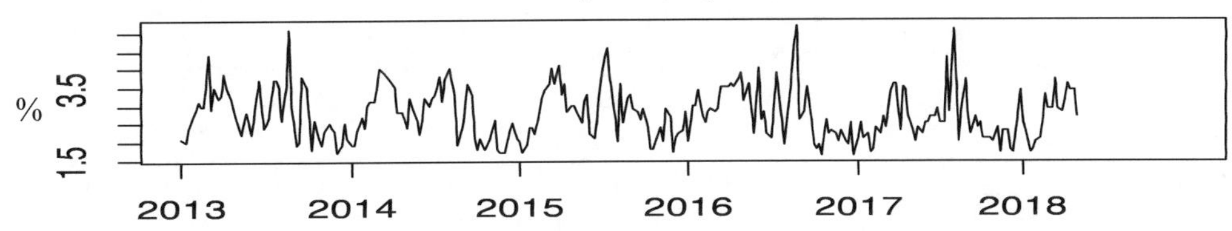

IOP Conf. Series: Earth and Environmental Science **964** (2022) 012016 doi:10.1088/1755-1315/964/1/012016

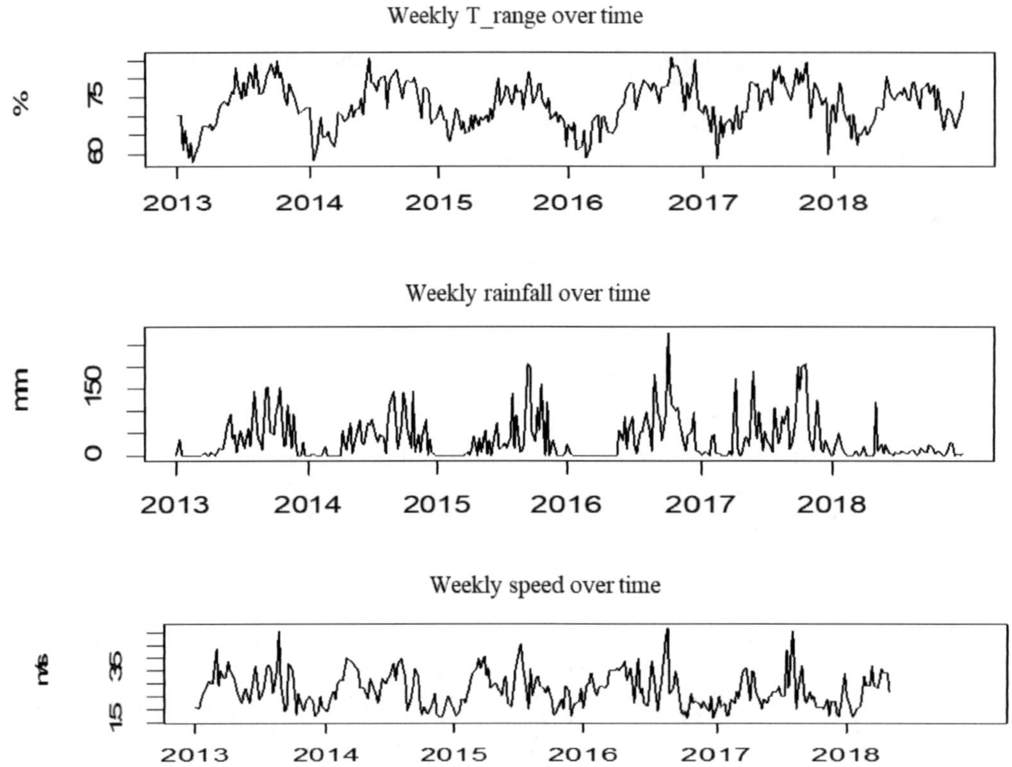

Figure 2. Time series of HFMD cases and weather factors by week, 2009 – 2018

Since 2011, HFMD is officially reported in Vietnam by Circular 48/2010/TT-BYT, issued on December 31st, 2010 by Ministry of Health. After that HFDM cases increased 6 to 7 times yearly compared to previous year and registered as one of the top ten mortality diseases in Vietnam. A study by Huynh Nguyen Phuong on "relation between communicable diseases and climate change in Can Tho city 2014 – 2018" proved that admission of HFMD increased year by year with 510 cases in 2014 and 1,270 cases in 2018 [9]. In Southern region of Viet Nam, an investigation of HFMD showed an abrupt increase of HFMD cases (an epidemic outbreak) in 2011 by 6 – 24 times compared to period 2008 - 2010 [10]. Further, an analysed data from sentinel surveillance system of Northern region of Viet Nam during the period of June 2016 - December 2018 also revealed this increased trend [11].

3.2. Relation between daily HFMD cases and weather factors

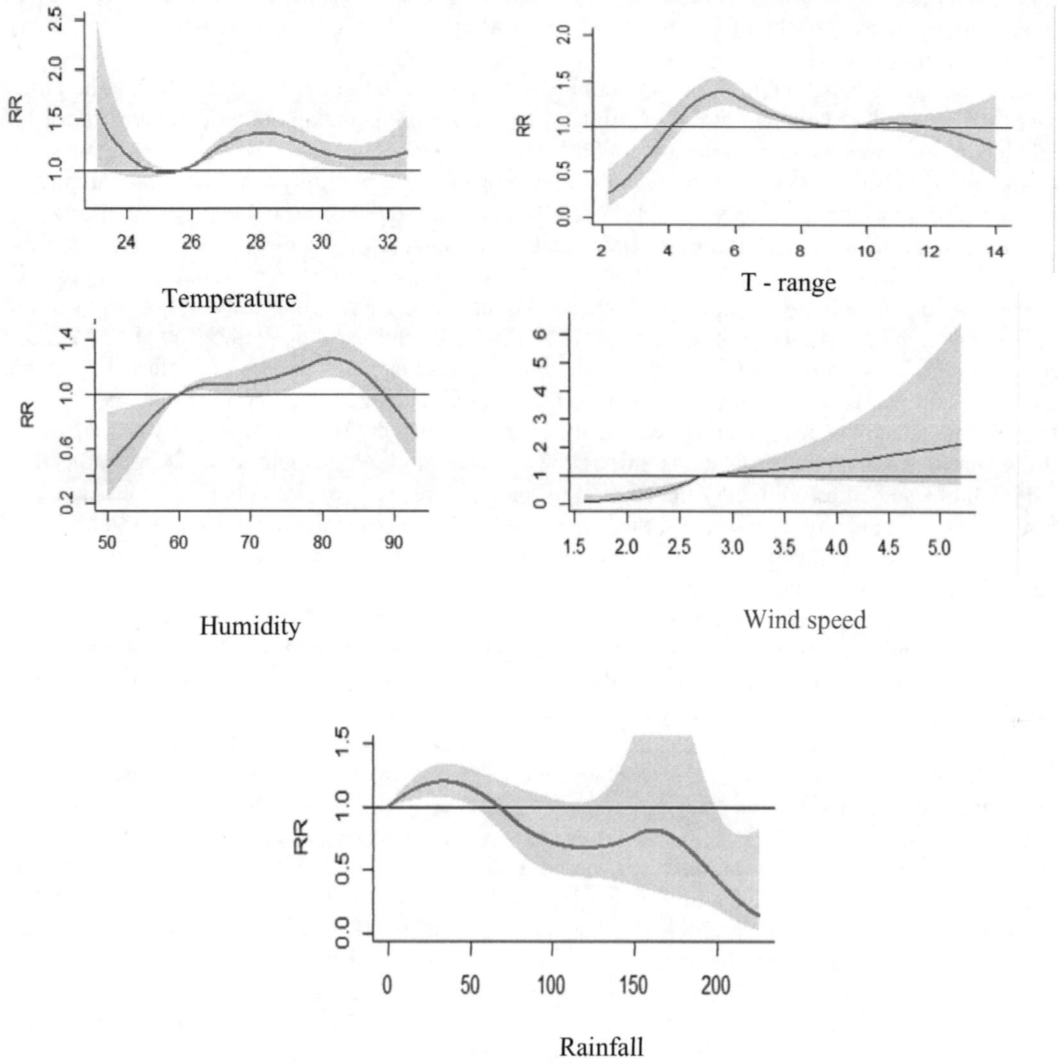

Figure 3. Relation between daily HFMD cases and weather factors

Many studies have proved that meteorological factors contribute to the epidemic of HFMD, including temperature, relative humidity, rainfall, wind speed. This study's results showed that an increase in admission risk of HFMD by daily average temperature in the range of 26°C- 30.1°C. The findings of this study were generally consistent with findings from different regions of the world [12, 13]. A study by Chen (2014) in Canton city, Guangtong province, China indicated a positive correlation between HFMD cases with a rise in temperature with a lag of 3-7 days [14]. On temperature range, admission risk increased in the range of 4.5°C to 7.5°C, however, temperature range of less than 4°C would decrease admission cases significantly. It was possible that temperature could affect the survival and transmission of the HFMD virus in the environment, as well as the behavior and activities of the hosts, thereby influencing the dynamics of the infection transmission [15]. HFMD is transmitted through direct contact with nose discharge, saliva, and fluid from the blisters. Additionally, it can spread through food and water which have been contaminated with fecal droplets or liquid from the mouth or nose of infected persons. Weather variables may affect the transmission of HFMD in both directly and indirectly way. The rate of replication of viruses and the survival of viruses in the environment may directly be affected by the surrounding temperature. Climate changes may affect the human behavior and hence influence the dynamics of HFMD.

Similarly, daily average humidity of 75% to 85% increased admission risk of HFMD and humidity of less than 60% reduced admission risk. A study in Japan also found out an increase in humidity of 1% would increase the weekly admission cases of HFMD by 4.7% (95%CI: 2.4-7.2) [16]. In addition, study in China in 2014 also revealed a similar result, an increase of 1% in humidity would increase admission cases of HFMD by a range of 0.51% to 1.42% [14]. Previous studies on the pathway of transmission showed that toys and other stuffs sharing contributed to the spreading of HFMD in children, especially in kindergartens. HFMD can also transmitted through pulmonary droplet or by faecal-oral route [17]. Another investigation of 2,912 Chinese districts in 2012 on the relation of humidity and HFMD showed a positive relation in only 54.4% of all districts [18]. First, enteroviruses, the pathogens of HFMD, can survive in the environment for a period of time, increasing the chance to infect vulnerable individuals[19]. Relative humidity may impact the survival time and reproductive capacity of enteroviruses, thereby increasing or decreasing the chance of contact [20]. Previous study has suggested that the sharing of toys and other items among the children might contribute to virus infections, and HFMD could also be transmitted by the fecal–oral route and respiratory droplets [17]. It was possible that, during the humid and rainfall days, the virus could be easily attached to the small particles in the air or the toys and then facilitated the disease transmission [21]. However, this possible mechanism needs to be confirmed in future studies.

Compared with most studies investigating the effects of temperature and humidity on HFMD incidence using monthly or weekly data, our findings may be more accurate. Analyses of the effects of weather factors on HFMD incidence using daily data could be more suitable for HFMD control and prevention. The ability to forecast HFMD on the basis of day-to-day weather variation could permit preventive improvements on public health infrastructure, including access to health care resources, scientific knowledge, active disease surveillance, and targeted more specific control to reduce the risk of the increased transmission.

A few limitations should be considered in this study. First, our study design was ecological in nature which did not allow us to explore individual-based association. Second, our analysis was preliminary and exploratory; we could not exclude the possibility of a spurious finding or unmeasured on founding factors that may be associated with both weather variables and HFMD occurrence. However, since this infection is a common disease in the study area, it is not difficult for the local doctors to make a correct diagnosis, so this concern should not have affected the result to a great extent. Thirdly, this was an ecological study, and the meteorological data were obtained from fixed monitoring stations rather than the real exposure of individuals. Therefore, measurement error is inevitable and may be randomly distributed, consequently underestimating the weather factors-HFMD association.

3.3. Predicted number of increased cases attributable to high temperature in children group of 4-5 years old for future decades.

Table 2. The number of increased cases attributable to high temperature in children group of 4-5 years old

No.	Decades	RCP4.5		RCP8.5	
		Number of cases	Relative cases (%)	Number of cases	Relative cases (%)

1	2010-2019 (present)	45	1,0	43	1,0
2	2020-29	213	4,8	183	4,1
3	2030-39	375	8,4	421	9,4
4	2040-49	195	4,4	264	5,9
Total (2020 – 2049)		**783**	**5,9**	**868**	**6,5**

The number of cases admitted to hospitals in children group of 4-5 years old are attributable to high temperature was showed in Table 2. In two different scenarios of RCP, the incremental cases for the period of 2020 – 2049 in scenario RCP4.5 will be 783 cases, less than the cases of scenario RCP8.5, which are 868 cases.

In the children group of 4-5 years old, there were only 2,688 cases (6%) admitted during the period of 2013 – 2018. However, by speculation with the risk function this age group is the main group to be affected by high temperature in future.

With an assumption that all other groups will be affected by high temperature in a similar weather pattern like in children group, there will be 783 cases (5.9%) admitted to hospitals in the period of 2020 – 2049 by RCP4.5 scenario and 868 cases (6.5%) by RCP8.5 scenario. However, resisting strength of HFMD of different ages will be different, therefore, impact of high temperature will be different by different age groups.

Furthermore, in the future period of 2020 – 2049, children group of 4-5 years old will be affected by high temperature and will increase their risk of admission by 5.9% and 6.5% with the RCP4.5 scenario and RCP8.5 scenario, respectively.

4. Conclusion

Climate change has posed a wide range of impacts on socio-economic sectors around the world. HFMD is one of the most common communicable diseases in Viet Nam and is closely related with weather-related events and climate risks. The outbreaks of HFMD in recent years indicated the disease interacts with both weather and seasonal factors. This study indicated that admission risk of HFMD increase in the range of daily average temperature from 26°C to 30.1°C. Meanwhile, daily average humidity range of 75% to 85% caused an increase in the admission risk of HFMD but reduced the admission risk with the range of humidity of less than 60%. Daily rainfall ranging from 5 mm to 55 mm also contributed to an increase in admission risk; on the contrary, daily rainfall equal or higher than 200 mm will reduce admission risk of HFMD. It is assumed that all age groups get the same effect by high temperature as group 4-5 years old, the period 2020 – 2049 will get an accumulative admission frequency of 783 cases (5,9%) and 868 cases (6,5%) by RCP4.5 and RCP8.5 scenarios, respectively. Most importantly, these results imply that there is the need for building up a public health policy in Ho Chi Minh City for mitigating the impacts of climate change on community health in a resilient approach.

Acknowledgements: We would like to thank to Ho Chi Minh City Department of Science and Technology for its funding to this research under the contract No. 43/2018/HĐ-SKHCN. The authors also acknowledged the support of time and facilities from Ho Chi Minh City (HCMUT), VNU-HCM for this study.

References

[1] IPCC 2013 Climate change 2013. The Physical Science Basis *New York*. Cambridge University Press p. 1032.
[2] World Health Organization. Hand, Foot and Mouth Disease Situation Update, 13 July 2012.
[3] Ebi KL and Hess JJ 2017 The past and future in understanding the health risks of and responses to climate variability and change *International journal of biometeorology* **61** 71-80.
[4] Gasparrini A 2010 Distributed lag non-linear models *Stat Med.* **29** 2224-2234.
[5] Honda Y 2014 Heat-related mortality risk model for climate change impact projection *Environmental health and preventive medicine* **19** 56-63.
[6] Linsuwanon P 2014 Epidemiology and seroepidemiology of human enterovirus 71 among Thai populations *Journal of biomedical science* **21** 16.
[7] Jiang FC and Yang F and Chen L 2016 Meteorological factors affect the hand, foot, and mouth disease epidemic in Qingdao, China, 2007-2014 *Epidemiology and infection.* **8** 1-9.
[8] Ma E 2010 Changing epidemiology of hand, foot, and mouth disease in Hong Kong, 2001-2009 *63.* **6** 422-426.
[9] Quang H and Hai LH 2016 The relationship between communucable diseases and climate change in Can Tho City 2014 - 2018 2016 *Journal of Preventive Medicine.* **3** 13.
[10] Hữu TN 2012 Epidemic characteristics of hand, foot and mouth disease in 2o Southern provinces of Vietnam during 2005 - 2011 *Ho Chi Minh City Journal of Medicine* **16** 8.
[11] Nghĩa ND 2018 Epidemiology characterics of hand foot and mouth disease in sentinel surveillace in the Northern Vietnam in 2012- 2014 *Journal of Preventive Medicine* **16** 7.
[12] Chen C 2020 Influence of temperature and humidity on hand, foot, and mouth disease in Guangzhou, 2013-2017 *J Int Med Res.* **48** 56.
[13] Wang JF 2011 Hand, foot and mouth disease: spatiotemporal transmission and climate *Int J Health Geogr.* **10** 25.
[14] Chen C and Lin H and Li X 2014 Short-term effects of meteorological factors on children hand, foot and mouth disease in Guangzhou, China *International journal of biometeorology.* **58** 1605-1614.
[15] Lin H and Zou H 2013 Short-term effect of El Nino-Southern Oscillation on pediatric hand, foot and mouth disease in Shenzhen, China *PLoS One* **8** 65-55.
[16] Onozuka D 2011 The influence of temperature and humidity on the incidence of hand, foot, and mouth disease in Japan *Sci Total Environ.* **410-411** 119-125.
[17] Chang LY 2002 Risk factors of enterovirus 71 infection and associated hand, foot, and mouth disease/herpangina in children during an epidemic in Taiwan *Pediatrics* **109** e88–e88.
[18] Maogui H 2012 Determinants of the Incidence of Hand, Foot and Mouth Disease in China Using Geographically Weighted Regression Models *PLoS* **8**123.
[19] Pirtle EC 1991 Virus survival in the environment *Rev Sci Tech.* **10** 733-748.
[20] Sánchez G. Survival of Enteric Viruses in the Environment and Food. In: Viruses in Foods. 2016 pp. 367-392.
[21] Luo C 2020 The burden of childhood hand-foot-mouth disease morbidity attributable to relative humidity: a multicity study in the Sichuan Basin, China *Sci Rep.* **10** 19394.

Marine litter pollution along sandy beaches of Can Gio coast, Ho Chi Minh City, Vietnam

Thanh-Khiet L. Bui[1,2*], Quoc-Khanh Pham[2], Nhu-Thuy Doan[3], Thanh-Ban Nguyen[4], Van-Nghia Nguyen[4], Kieu Lan Phuong Nguyen[1,5], Hong Hanh Nguyen[6], Hong Quan Nguyen[1]

[1*]Institute for Circular Economy Development, Vietnam National University, Ho Chi Minh City, Vietnam

[2]Institute for Environment and Resources, Vietnam National University, Ho Chi Minh City, Vietnam

[3]Faculty of Environment, University of Science, Vietnam National University, Ho Chi Minh City, Vietnam

[4]Institute of Environmental Science, Engineering and Management, Industrial University of Ho Chi Minh City

[5]Faculty of Environmental and Food Engineering, Nguyen Tat Thanh University, Ho Chi Minh City, Vietnam

[6]Leibniz Institute of Freshwater Ecology and Inland Fisheries, Berlin, Germany

*Email: blt.khiet@iced.org.vn

Abstract. Marine debris is a significant threat to the marine environment, human health and the economy in Can Gio island, Vietnam. In this study, we conducted beach litter surveys to quantify and characterize marine litter from six beach transects in this region for the first time using the OSPAR beach litter monitoring guideline. A total of 29,456 items weighting 529,432 g was recorded from 12 surveys in two monitoring campaigns in December 2019 (dry season) and in June 2020 (rainy season). Plastic was the most abundant type of litter in terms of quantity (a total of 26,662 items) and weight (325,606 g), followed by paper and cardboard, cloth, wood, metal, glass and ceramics, rubber and other items. Meanwhile, ropes (less than 1 cm in diameter) and plastic bags accounted for the major parts of sampled plastic items, i.e., 20.18% and 14.46%, respectively. In general, a higher percentage of marine litter, particularly the fishing related items, was found in the rainy season than in the dry season, possibly due to increased fishing and aquaculture activities. In contrast, a reduction pattern of the single use plastics in the rainy season might be due to the decreased tourism activities during the Covid pandemic period. This study not only showed the magnitude of litter pollution, but also provided valuable information that could help decision making to better control and reduce marine litter in the region. In addition, insights from this study indicate that there is an urgent need to design collection, reuse and recycling programs in the area. The collected recyclables, specially plastics, will be a great source of materials for recyclers in a circular economy achieving sustainable development goals.

Keywords: **marine litter, OSPAR, Can Gio coast, plastics.**

1. Introduction

Marine debris (or marine litter, ML) is defined as "any persistent, manufactured or processed solid material discarded, disposed of or abandoned in the marine and coastal environment" [1]. For decades, ML is increasingly accumulated in various marine environment, such as sea surface, sea floor, water column, and beaches [2], [3], [4]. It could take hundreds of years to degrade and remove ML from the natural environment. Most studies have shown that plastics are the most common source of ML, accounting for 75% of global average ML pollutants [2], [5]. Furthermore, plastic issues have currently received increasing attention since they can break down into small pieces and thereafter into microplastics that have potential impacts on marine habitat [6]. Until now, ML are found all over the world, making it a global problem. Education and raising public awareness could be one of the most effective ways to reduce ML; however, monitoring ML in order to improve our knowledge is also of utmost necessity for a more effective pollution control.

Asian countries with long coastline and wide water areas are major waste contributors globally. Vietnam is ranked the fourth most polluting country after China, Indonesia and the Philippines. In 2010, Vietnam produced 1.83 million tons of plastic wastes, of which 0.28 – 0.73 million tons ended up in the ocean [7]. In order to decrease the pollution pressure of marine plastic litter, the Prime Minister has issued a National Action Plan (Resolution 36-NQ/TW dated October 22, 2018) for marine plastic waste management until 2030 with the following key contents: (i) implement effectively Vietnam's initiatives and commitments to the world to address plastic waste issues, specially ocean plastic wastes, ensure the prevention of plastic waste sources, wastes on land and activities at sea, and target to make Vietnam a pioneer in the region in reducing marine plastic wastes; (ii) contribute to promoting the recovery, recycling, and reuse of plastic wastes in Vietnam, and (iii) raise the awareness, behavior and habits of the community and society of using disposable plastic products and non-biodegradable plastic bags. There are some studies on macroplastic in Saigon river, upstream of Can Gio region [8-10]. However, the scientific data regarding the ML pollution in Vietnam are still scarce, but some investigations carried out by NGO show that marine plastic litters are the most abundant artificial polymer materials along beaches in Vietnam [11], [12].

The composition, quantities and distribution of ML in the coastal regions of Vietnam is still not well defined. The main reasons for this are: i) lack of standard monitoring and assessment methods, ii) no appropriate legal framework to support researchers as well as managers for ML pollution investigations. In order to understand the current state of ML pollution in Vietnamese coastline, a case study was conducted in Can Gio beaches. The objective of this study was to conduct a beach litter survey to quantify the litter in a specific sea area in order to evaluate the potential ML pollution threats to the coastal environment of Vietnam.

2. Study area and methods

2.1. Study area

Can Gio is a one of the five main suburban districts located in southeast of Ho Chi Minh City (HCMC), with more than 20 km of coastline running in the southwest - northeast direction. There are large estuaries at Can Gio such as Long Tau, Cai Mep, Go Gia, Thi Vai, Soai Rap, and Dong Tranh Rivers. Can Gio is the only district of the city which is bordered by the sea, i.e., an island separated from the surrounding with a favorable geographical position. The location of Can Gio District was from 106°46'12" to 107°00'50" East longitude and from 10°22'14" to 10°40'00" North latitude [13], [14]. The North area is separated from Nha Be District by Soai Rap River, the South border on the East Sea, the West border on Soai Rap River, the Northeast is separated from Nhon Trach district, Dong Nai province by Long Tau river, and the Southeast border on Tan Thanh district, Ba Ria-Vung Tau bordered by Thi Vai river.

The climate of Can Gio is typically monsoonal with two distinct seasons [13], [14]. The dry season is from December to April of the next year and the rainy season is from May to November. Can Gio is influenced by two main wind directions and mainly the West - Southwest and North - Northeast monsoons. The west-southwest wind from the Indian Ocean blows in the rainy season, from June to October, with an average speed of 3.6m/s and the strongest wind in August, with an average speed of 4.5m/s. The North-Northeast wind blows from the East Sea in the dry season, from November to February, with an average speed of 2.4 m/s. In addition, there is a wind gust, from the south to the southeast, from March to May with an average speed of 3.7 m/s.

2.2. Methods

The study was conducted on six sandy beaches of Can Gio, which are located in recreational and aquaculture areas (bivalve culture) (Figure 1). Specific details of sampling area and characteristics of the six sampling locations are shown in Table 1.

Table 1. Beach locations surveyed in Can Gio, including number of surveys conducted and a brief description of the locations.

Site	Coded site	Replicas	Description
Can Thanh Town	VT1	2	Near park area, tourists and local resident use areas
Can Thanh Town	VT2	2	Used to be an aquaculture area
Can Thanh Town	VT3	2	Aquaculture area
Long Hoa Commune	VT4	2	Recreational area
Long Hoa Commune	VT5	2	Aquaculture area
Long Hoa Commune	VT6	2	Aquaculture area

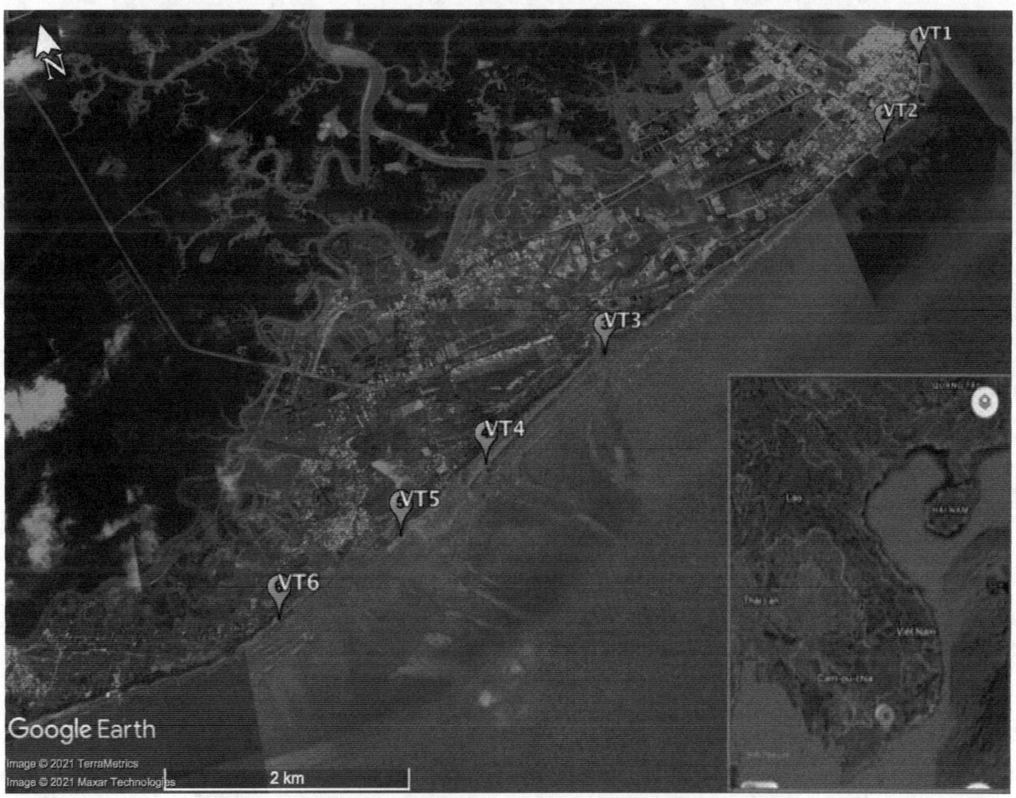

Figure 1. Study area and locations for beach litter sampling

According to the guidance of OSPAR [15], the selection of coastline sampling locations was based on the following criteria:

- Sand or gravel beaches
- No breakwater or jetty affecting local traffic, accumulating or inhibiting debris deposition
- Minimum length of 100 m parallel to the water (i.e., measured along the water's edge)
- No regular cleaning activities.

These locations were surveyed in typical months of the dry season (28th and 29th of December) in 2019 and the rainy season (6th and 7th of June) in 2020, on these days, it was clear, sunny and windy. Each survey consisted of a transect of 100 m length parallel to seawater line and 25 m width extended from the upper beach limit (bushes or rock). In each transect, all non-natural anthropogenic items larger than 0.5 cm on the surface were collected into bags, kept at ambient temperature and transferred to the lab for further processing. In the lab, each bag was emptied on a clear surface, then all of the items were cleaned from the sand and each item was assigned to a litter category according to OSPAR: (i) Recognizable item: whole items that match the category description and/or fragments that are unmistakably from things that fit the category description; (ii) Other items: items clearly recognizable, but lack of designated litter categories; and (iii) Fragments: unidentified broken items classified by material, plastic and polystyrene fragments were sorted into size categories. In addition, similar items were also grouped into single use plastics (SUP), fishing related items (FR) and plastic/polystyrene pieces (PP) according to Hanke et al. [16]. After sorting, the number and dry weight of items per category were recorded.

2.3. Data analysis

We surveyed parameters including density of items (number of items per m^2 and weight of items per m^2), the Clean Coast Index (CCI) [17] and the percentage of single use plastics (%SUP), fishing related items (%FR) and plastic/polystyrene pieces (%PP) were calculated based on the number of plastic items collected. Item density was calculated as follow:

$$D = \frac{N}{w \times l} \quad (1)$$

where D is the density of items, N is the total number/weight of items on surveyed transect, w is the width of the transect (25 m), and l is the length of the transect (100 m). As for beach cleanliness, we applied the CCI ranking using the following equation:

$$CCI = \frac{TL}{TA} * K \quad (2)$$

where TL is the total items counted on a surveyed transect, TA is the total area of a transect, and K is a constant factor which was assigned a value of 20. Beaches were classified from very clean to extremely dirty based on the litter condition on the beach (Table 2).

Table 2. Clean Coast Index ranking

Value	Coast index	Visual assessment
0 – 2	Very clean	no litter is seen
2 – 5	Clean	no litter is seen over a large area
5 – 10	Moderate	a few pieces of litter can be detected
10 – 20	Dirty	a lot of debris on the shore
20+	Extremely dirty	most of the beach is covered with plastic debris

Source: [17]

In order to determine whether number of items varied significantly in different (dry and rainy) seasons and locations, one-way ANOVA tests were conducted. Prior to the statistical analyses, log transformation was applied to stabilize the variances. Normal distribution and variance homogeneity were checked before performing one-way ANOVA tests. When differences were detected in ANOVA tests, post-hoc Tukey's tests were applied. All statistical data analyses were performed using R Statistical Software (version 3.6.1) (R Foundation for Statistical Computing, Vienna, Austria).

3. Results

3.1. Marine litter abundance and composition

A total of 29,456 items weighting 529,432 g was collected along six beach transects during the two sampling periods, and were classified into 8 major groups. Litter composition could be found on Supplementary Table S1 and abundance on each beach at Can Gio is shown in Table 3, meanwhile percentage of SUP, FR and plastic/polystyrene pieces (PP) are listed in Table 4. The highest percentage of SUP was found at VT1 in the dry season (85.14%). Highest percentage of PP was also recorded in VT1 but in the rainy season (45.11%), of which plastic/polystyrene pieces of size 2.5 – 50 cm was absolutely predominant. The highest percentage of FR (42.93%) was recorded at VT2 in the rainy season.

Overall, the number of items ranged from 294 to 6,779 items per transect, with a mean of 2,455 ± 1,681 items per transect. In term of the weight, the amount of beach litter varied from 8,294 g to 110,345 g per transect, with a mean of 44,177.67 ± 29,161.44 g per transect (Table 4). The high standard deviation values show that beach litter abundance varied substantially in both space and time at each location. The number of items on the least polluted beaches (294 at VT1 in rainy season) and the most polluted ones (6779 at VT4 in rainy season, Table 3) differed by an order of magnitude. Plastic was the most abundant group of ML found on the beaches, representing 89.92% (ranging from 75.41 to 96.34%) of sampled items. Paper and cardboard were the second abundant group (3.74%), followed by cloth (3.01%), wood (1.57%), metal (0.81%), glass and ceramics (0.45%), rubber (0.40%) and other items (0.02%).

IOP Conf. Series: Earth and Environmental Science **964** (2022) 012017 doi:10.1088/1755-1315/964/1/012017

Table 3. Abundance (items per transect) and weight (g per transect) of litter collected from six beaches during two sampling periods (on December 2019 and June 2020)

	VT1		VT2		VT3		VT4		VT5		VT6		Mean	SD
	Dry season	Rainy season	Dry season	Rainy season	Dry season	Rainy season	Dry season	Rainy season	Dry season	Rainy season	Dry season	Rainy season		
Abundance (number of items per transect)														
Artificial polymer materials	1193	266	878	3712	1135	3131	1960	6021	2880	2003	798	2685	2221.83	1531.43
Rubber	5	0	4	1	8	2	17	9	19	24	6	2	8.08	7.49
Cloth/textile	263	3	9	28	34	16	77	169	75	70	21	23	65.67	73.99
Paper/Cardboard	43	11	29	112	39	40	103	443	184	91	41	73	100.75	112.80
Processed/worked wood	1	7	15	46	28	45	28	123	47	21	23	57	36.75	30.73
Metal	75	4	8	5	6	3	16	6	9	2	5	11	12.50	19.20
Glass/ceramics	2	3	7	12	3	13	10	8	7	14	2	27	9.00	6.79
Other (unidentified)	0	0	0	0	0	0	5	0	0	0	0	0	0.42	1.38
Total	**1582**	**294**	**950**	**3916**	**1253**	**3250**	**2216**	**6779**	**3221**	**2225**	**896**	**2878**	**2455.00**	**1681.06**
% Plastics	75.41	90.48	92.42	94.79	90.58	96.34	88.45	88.82	89.41	90.02	89.06	93.29	89.92	
Weight (g per transect)														
Artificial polymer materials	10650	6994	13502	37270	15034	34580	27287	56000	30499	36775	8650	48365	27133.83	16230.50
Rubber	600	0	680	300	1110	50	1055	3150	4102	4825	1650	300	1485.17	1641.32
Cloth/textile	6352	10	375	1475	1505	1950	5650	11310	4750	10195	1655	4650	4156.42	3718.78
Paper/Cardboard	1370	55	150	1300	155	575	360	6560	675	1520	300	1250	1189.17	1772.50
Processed/worked wood	300	1000	1112	6055	3905	13375	3655	31350	8450	4200	3500	15670	7714.33	8852.43
Metal	3380	35	475	200	755	300	850	625	125	125	150	700	643.33	906.62
Glass/ceramics	1850	200	1360	1450	125	3500	655	1350	1550	2550	250	3825	1555.42	1224.67
Other	0	0	0	0	0	0	3600	0	0	0	0	0	300.00	1039.23
Total	**24502**	**8294**	**17654**	**48050**	**22589**	**54330**	**43112**	**110345**	**50151**	**60190**	**16155**	**74760**	**44177.67**	**29161.44**

Note: SD is the standard deviation (n = 12)

Table 4. Beach litter densities (items/m^2 and g/m^2), %SUP, %FR, %PP and CCI in Can Gio beaches

Locations	Dry season						Rainy season					
	items/m^2	g/m^2	%SUP	%FR	%PP	CCI	items/m^2	g/m^2	%SUP	%FR	%PP	CCI
VT1	0.63	9.80	85.14%	1.84	3.01%	12,66 (d)	0.12	3.32	42.48%	4.89	45.11	2.35 (c)
VT2	0.38	7.06	50.34%	14.17	18.14	7,6 (m)	1.57	19.22	37.49%	42.93	12.44	31.33 (ed)
VT3	0.50	9.04	62.73%	6.56	16.01	10.02 (d)	1.3	21.73	46.41%	36.20	11.55	26.00(ed)
VT4	0.89	17.24	44.26%	4.60	30.96	17.73 (d)	2.71	44.14	50.68%	28.84	14.66	54.23 (ed)
VT5	1.29	20.06	70.85%	10.11	7.28	25.77 (ed)	0.89	24.08	46.87%	28.86	13.91	17.80 (d)
VT6	0.36	6.18	62.63%	10.25	7.00	7.14 (m)	1.15	29.90	63.7575%	16.49	11.65	23.02 (ed)

Note: CCI abbreviations given in parenthesis refer to(vc) for very clean, (c) for clean, (m) for moderate, (d) for dirty and (ed) for extremely dirty

3.2. Comparisons of ML compositions, quantities and distribution among locations

A distinctive variation pattern in the ML quantity was found among sites and seasons (Table 3, Table 4). The highest amount of ML was observed at the VT4 location whereas the lowest was observed at the VT1 during the rainy season (Table 3). The mean ML count density was twice higher during the rainy season than during the dry season. Particularly, the litter count density fluctuated in the range from 0.36 to 1.29 items/m^2 (mean of 0.67 items/m^2) and from 0.12 to 2.71 items/m^2 (mean of 1.29 items/m^2) in the dry and rainy seasons, respectively. Beach litter by weight was 11.56 g/m^2 in the dry season, ranging from 6.18 g/m^2 in the VT6 to 20.06 g/m^2 in the VT5, and 23.73 g/m^2 in the rainy season, ranging from 3.32 g/m^2 in the VT1 to 44.14 g/m^2 in the VT4 (Table 4). Except for the VT1 location, an increasing pattern was observed in most of other sampling sites in both beach litter count density and beach litter by weight in the rainy season. However, in the VT5 location, there was a decrease in the count density but an increase in weight of ML during the rainy season, which was due to the increase in size of litter that also lead to the increase in the overall weight. The other four areas reported an increase in both of the count density and weight in the rainy season compared to the dry season. According to the CCI, only the VT1 in rainy season was found to be clean, meanwhile VT2 and VT6 in the dry season were defined to be moderately dirty, and the remaining locations were classified either dirty or extremely dirty in both seasons (Table 4). Regarding the SUP quantities, the highest percentage was recorded at the VT1 in the dry season (85.14%). This location is closed to a park, with a lot of recreational activities from the local community and tourists, and it is also close to the river mouth, which creates a close pathway of pollution transports of inland wastes. Most of the SUP was plastic bags and crisps packets/sweets wrappers. In general, except VT4 and VT6, SUP decreased in the remaining locations in the rainy season compared to the dry season, which may be due to the decreased number of tourists visiting Can Gio during the Covid-19 pandemic period. The highest percentage of FR items (42.93%) was recorded at the VT2 in the rainy season and most of items were strings and cords with a diameter of less than 1 cm that were buried in the sand and were deteriorating. The highest percentage of PP was found at the VT1 (45.11%) in the rainy season, where the abundant pieces of sizes 2.5–50 cm accounted for the majority and were originating from Styrofoam food containers. Overall, the SUP decreasing trend in the rainy season could be attributed to fewer tourists due to the Covid-19 epidemic. However, the pattern of FR wastes tends to be the opposite, i.e., increasing in the rainy season, possibly due to increased fishing and aquaculture activities during the previous months of dry season.

There was no significant difference in the collected data during the dry season between locations (ANOVA, $F_{5,276} = 1.38$, $p = 0.23$). However, in the rainy season, there was a significant difference in sampled items (ANOVA, $F_{5,250} = 6.06$, $p < 0.05$), and ML densities varied significantly among the location VT1 with the rest five locations (post-hoc Tukey's test, $p < 0.05$). Particularly, there was no significant difference between the results during both the dry and the rainy seasons in some locations, including VT3 (ANOVA, $F_{1,84} = 1.71$, $p = 0.19$), VT5 (ANOVA, $F_{1,} = 0.03$, $p = 0.85$), and VT6 (ANOVA, $F_{1,80} = 2.30$, $p = 0.13$); however, significant differences in collected ML items were found at locations VT1 (ANOVA, $F_{1,75} = 4.95$, $p = 0.03$), VT2 (ANOVA, $F_{1,92} = 4.55$, $p = 0.04$) and VT4 (ANOVA, $F_{1,102} = 7.51$, $p = 0.01$) between the two seasons. High abundance of ML was observed in all sites for both seasons, except VT1 in the rainy season. At the VT1, the amount of litter in rainy season decreased sharply despite the fact that it is located in a close proximity to the river mouths, and rivers are considered as land-based transports of plastic litter. One explanation to this might be due to the fact that this location is close to a park and most of litter sources came from recreation activities; however, due to Covid-19, the park was closed until the time of the survey and such kind of litters were greatly reduced. There is an increase in litter during the rainy season in most of the locations, most likely due to an increase in precipitation and/or wind leading to increase number of items linked to increased discharge rate from rivers and also increase in fish and aquaculture activities in these locations.

3.3. Top ten ML items

The top ten items for Can Gio beaches are shown in Table 5, all of them are plastics and accounted for 77.9% of found items. The top item, string and cord (diameter less than 1 cm), was mainly small pieces rather than whole items, they were found mostly in sand under a deteriorated state and accounts for 1/5 of the total number of collected items. Seven items out of the top ten in Can Gio beaches are single-use plastics.

Table 5. Top ten items found in Can Gio beaches in twelve surveys

Rank	Item	Percentage
1	String and cord (diameter less than 1 cm)	20.18
2	Shopping Bags incl. pieces	14.46
3	Plastic/polystyrene pieces 2.5 cm > < 50 cm	11.76
4	Crisps packets/sweets wrappers	10.55
5	Cups and cup lids	4.70
6	Small plastic bags, e.g. freezer bags incl. pieces	4.51
7	Cutlery and trays	3.59
8	Plastic caps/lids drinks	3.22
9	Straws and stirrers	2.51
10	Other plastic/polystyrene items (identifiable)	2.42
Total percentage		77.9

4. Discussion

The results obtained from this study showed that plastic varied between 75.41% - 96.34%% by count, which was higher than the global average (75%) [2], [5] and 43.46% - 84.32% by weight. These results are in line with the findings obtained from other surveys around the world [18], [19], [20]. The presence of plastics could have great potential impacts on ecosystems since they are persistent in the environment and could break into smaller pieces (microplastics), causing far greater harm than the debris itself. This means plastics could be transported to further distance and more into the food chain; in addition, additives compounds, plasticizers and adsorbing-desorbing hazardous chemicals could leach out and be absorbed by victim animals [21], [22], [23]. Therefore, an urgent preventive measure is needed to reduce plastics in the environment.

The present findings also showed that there was an increasing trend in the amount of ML in the rainy season compared to the dry season, which could be due to the weather and/or climate conditions [24]. During the rainy season, high intensity of wind and rain causes flooding in urban areas, which washes down more plastic litters and other debris into rivers [25], [26]. There are evidences that rivers transport waste from land to the ocean [27], [28], [29] and Can Gio is located close to river mouths;

therefore, the amount of litters increased in the rainy season is justifiable. In addition, high cloud coverage and cooler ambient temperature in Can Gio lead to slow degradation of ML. On the other hand, during the time between February and April, there was Covid epidemic that resulted in the low frequency of human appearance (waste pickers, beach clean-up activities, etc.) on beaches, which might also explain the high number of litters during this period.

ML results also suggested that both land-based activities and sea-based activities are major contributors since one-fifth of collected litters were string and cord, mainly from aquaculture and fishing activities in the area; unlike some studies which showed that land-based activities and foreign litters were major contributors [30], [31] [32], [33], [34]. In this study, most of remaining litters were domestic products or routinely found items in shops.

ML in Can Gio beaches were derived from various sources and a majority of them were SUP (plastic bag, crisp/sweet packets, disposable cutlery, etc.). For example, string and cord were extensively used by local fishermen and seafood farmers; plastic bags could come from daily use or were discarded directly on the beaches by visitors. Food-related SUPs (cups, crisp/sweet packets, disposable cutlery, straw) mostly came from beach visitors. Plastic/polystyrene pieces found on beaches were primarily originated from Styrofoam food containers, which are used extensively in inland and beach visitors, and part of them were from blocks of Styrofoam foam which are used by local fishermen as float for their nets.

ML abundance was compared with previous studies from Small Island Developing States and other islands in Pacific region (Table 6). ML was calculated as number of items per square meter of studied areas. Although ML value was higher in Can Gio, it was not substantially different from those of French Polynesia, and compared to other sites in Pacific region, litter abundance in Can Gio was still low. This may due to low population density of the area when compared to other sites of studies like Indonesia, where high amount of waste was mismanaged [7]. However, the abundance of ML is within the range of values for SIDS in Pacific region [30], [35]. However, there were differences in the types of litter in different studies, of which plastic was still the most dominant type of ML in all studies. When comparing to other studies in Vietnam [11], [12], we have difficulties since these studies only focused on beach clean-up aspect and they did not publish data on litter items as well as amount of item collected.

Table 6. Abundance of marine litter in different studies. Data were adapted from [30]

Area	Items/m^2	Study
Can Gio beach	0.98 (0.12 – 2.79)	This study
Solomon Islands (Guadalcanal Island)	2.5 (0.38 – 6.2)	
Vanuatu (Efate Island)	1.25 (0.38 – 3.54)	
French Polynesia Mo'Orea Island	0.9 (0.05 – 2.03)	
French Polynesia Tahiti Island	0.9 (0.22 – 1.85)	
Bootless Bay, Papua New Guinea	4.5 (1.2 – 11.8)	
Cocos (Keeling) Islands, Australia	21.6 (4.72 – 55.67)	
Henderson Islands	239 (max 671.6)	
Tasmania Island	0.016 – 2.03	
Indonesia	41.6	

Top ten items were compared between different beaches (Table 7) in the Pacific Ocean region, and the results reflected the characteristics of areas. Top ten items were very similar to those found in Vanuatu and Solomon Islands [30] and French Polynesia [35]. The similarity of top ten items in these regions may be due to the fact that the main source of ML is from land-based activities. Moreover, this study also showed that a large amount of ML originated from sea-based activities, which is different from the SIDS in Pacific region where most ML came from land-based activities. In this study, the top ten items were all plastic groups and belonged to fish related items as well as single use plastics, while in other studies, there were products belonging to groups other than plastic (e.g., glass, metal). Overall, the results in this study showed the similarity of ML sources with other studies in the Asia Pacific region; these sources were mostly related to land-based activities.

Table 7. Comparisons of top items found in this study and in Vanuatu and Solomon Islands and in French Polynesia. Items are ranked by abundance

Rank	This study	Solomon Island	Vanuatu Island	French Polynesia
1	String and cord (diameter less than 1cm)	Foam food packs	Glass broken	Glass beverage bottle and fragments
2	Shopping Bags incl. pieces	Polysterene pieces 0 – 2.5 cm	Plastic pieces 0 – 2.5 cm	Hard plastic fragments
3	Plastic/polystyrene pieces 2.5 cm > < 50 cm	Drink bottles, containers, drum	Plastic pieces 2.5 – 50 cm	Sheet plastic fragments
4	Crisps packets/sweets wrappers	Cigarette butts	Polystyrene pieces 2.5 – 50 cm	Cigarette butts and filters
5	Cups and cup lids	Drink cans	Crisp and sweet packets	Tile fragments
6	Small plastic bags, e.g. freezer bags incl. pieces	Glass broken	Polystyrene pieces 0 – 2.5 cm	Sheet plastic food and confectionary wrappers
7	Cutlery and trays	Plastic pieces 0 – 2.5 cm	Cigarette butts	Polystyrene fragments
8	Plastic caps/lids drinks	Crisp and sweet packets	Caps and lids	Hard plastic beverage bottle caps
9	Straws and stirrers	Ice lolly plastic containers	String and cord (diameter < 1 cm)	Rubber strapping/inner tubes
10	Other plastic/polystyrene items (identifiable)	Caps and lids	Lolly sticks	Metal beverage caps

The results of this study could be useful for addressing preventive measures of ML. Since a large amount of found litters was SUP; a ban on plastic bags, food packaging especially Styrofoam food container is suggested. This measure could help to reduce 3% of ML in Vanuatu and 17% of ML in Solomon Islands [30]. SUP is known as an easily break-down plastics in the environment [36], especially in coastal area [37]. The release of large amounts of SUP microplastics into the environment of Can Gio beaches might affect the bivalve culture activities in this region, as it makes easier to introduce microplastics through ML degraded wastes into the human food chain. Therefore, strong preventive measures of SUP in Can Gio are needed. This study also showed that waste mismanagement coupled with low awareness of people (both tourists and residents) are the main contributors of ML in Can Gio.

The effectiveness of a preventive measure should be monitored overtime; therefore, it is necessary to apply a standardized method for beach litter monitoring and data collection and comparison with other data in the world. This study used OSPAR guideline and showed the applicability of this method for the future applications; however, it is important to modify the protocol and expand the list of litter categories to include local-specific items, adding more value for waste management as well as providing preventive measures. In addition, understanding how the new country-specific ML categories fit within the original OSPAR list is critical to maintain comparability in the assessment approach.

5. Conclusion

This research is the first to document the abundance and composition of beach litter along Can Gio coast. For the first time, an OSPAR protocol was used in this region in order to gather data of beach litter to support the development of marine litter action program.

Through twelve surveys along sandy beaches of Can Gio coast, litters were classified into eight major groups (plastic, rubber, fabric, paper, wood, metal, glass, combination and other). Our study showed that plastics was the most abundant group and accounted for 89.9% of found items, in which rope (diameter less than 1 cm) and plastic bag contributed the most. The rest of plastic litter was mostly single use plastics. Results of this study demonstrated that both sea-based activities and land-based activities are significant contributors to beach litter in Can Gio. Therefore, an active program to control and reduce litters from these sources is necessary.

This study also showed that it is possible to adopt an international protocol to region in order to collect data on marine litter, and sources of marine litter could be identified for future legislations in this region. However, additional items should be added to the OSPAR category list to allow global comparisons and to measure the effectiveness of active programs in terms of reducing marine litter in future assessments.

6. Acknowledgement

This research is funded by Vietnam National University Ho Chi Minh City (VNU-HCM) under grant number C2020-24-03.

7. References

[1] UNEP 2009 Marine litter: a global challenge. (Nairobi: UNEP p 232 pp

[2] Galgani F, Hanke G and Maes T 2015 *Marine Anthropogenic Litter,* ed M Bergmann*, et al.* (Cham %@ 978-3-319-16510-3: Springer International Publishing) pp 29-56

[3] Watts A J R, Porter A, Hembrow N, Sharpe J, Galloway T S and Lewis C 2017 Through the sands of time: Beach litter trends from nine cleaned north cornish beaches *Environmental Pollution* **228** 416-24

[4] Ostle C, Thompson R C, Broughton D, Gregory L, Wootton M and Johns D G 2019 The rise in ocean plastics evidenced from a 60-year time series *Nature Communications* **10** 1622

[5] Hanke G, Galgani F, Werner S, Osterbaan L, Nilsson P, Fleet D, Kinsey S, Thompson R, Palatinus A, Van Franerker J A, Vlachogianni T, Scoullos M, Veiga J M, Matiddi M, Alcaro L, Maes T, Korpinen S, Budziak A, Leslie H, Gago J and Liebezeit G 2013 Guidance on Monitoring of Marine Litter in European Seas . EUR 26113. (Luxembourg (Luxembourg): Publications Office of the European Union; 2013. JRC83985

[6] Gall S C and Thompson R C 2015 The impact of debris on marine life *Marine Pollution Bulletin* **92** 170-9

[7] Jambeck J R, Geyer R, Wilcox C, Siegler T R, Perryman M, Andrady A, Narayan R and Law K L 2015 Plastic waste inputs from land into the ocean *Science* **347** 768

[8] Lahens L, Strady E, Kieu-Le T-C, Dris R, Boukerma K, Rinnert E, Gasperi J and Tassin B 2018 Macroplastic and microplastic contamination assessment of a tropical river (Saigon River, Vietnam) transversed by a developing megacity *Environmental Pollution* **236** 661-71

[9] van Emmerik T, Kieu-Le T-C, Loozen M, van Oeveren K, Strady E, Bui X-T, Egger M, Gasperi J, Lebreton L, Nguyen P-D, Schwarz A, Slat B and Tassin B 2018 A Methodology to Characterize Riverine Macroplastic Emission Into the Ocean *Frontiers in Marine Science* **5** 372

[10] van Emmerik T, Strady E, Kieu-Le T-C, Nguyen L and Gratiot N 2019 Seasonality of riverine macroplastic transport *Scientific Reports* **9** 13549

[11] Nguyen T T 2018 Coastal Clean Ups in Vietnam – Looking Backwards and Facing Forward. In: *6th International Marine Debris Conferences,* (San Diego

[12] Pham N A and Hoang A 2018 Coastal management in Vietnam: Experience and sharing from MCD. In: *6th International Marine Debris Conference,* (San Diego

[13] UNESCO/MAB 2000 Valuation of the Mangrove Ecosystem in Can Gio Biosphere Reserve, Vietnam. Final Report of UNESCO/MAB Project Implemented by the Vietnam MAB National Committee)

[14] Tuan V Q and Kuenzer C 2012 Can Gio Mangrove Biosphere Reserve Evaluation 2012. (Hanoi, Vietnam p 102

[15] OSPAR 2010 Guideline for Monitoring Marine Liter on the Beaches in the OSPAR Maritime Area.
[16] Hanke G, Walvoort D, Van Loon W, Addamo A, Brosich A, Del Mar Chaves Montero M, Molina Jack M, Vinci M and Giorgetti A 2020 EU Marine Beach Litter Baselines, EUR 30022 EN. (Luxembourg
[17] Alkalay R, Pasternak G and Zask A 2007 Clean-coast index—A new approach for beach cleanliness assessment *Ocean & Coastal Management* **50** 352-62
[18] Terzi Y and Seyhan K 2017 Seasonal and spatial variations of marine litter on the south-eastern Black Sea coast *Marine Pollution Bulletin* **120** 154-8
[19] Peraš I, Divanović M, Pešić A, Joksimović A, Marković O, Đurović M and Mandić M 2017 Composition and abundance of beach litter in Montenegro (South Adriatic Sea) *Studia Marina* **30** 17-27
[20] Pasternak G, Zviely D, Ribic C, Ariel A and Spanier E 2017 Sources, composition and spatial distribution of marine debris along the Mediterranean coast of Israel *Marine Pollution Bulletin* **114** 1036-45
[21] Engler R E 2012 The Complex Interaction between Marine Debris and Toxic Chemicals in the Ocean *Environmental Science & Technology* **46** 12302-15
[22] Gallo F, Fossi C, Weber R, Santillo D, Sousa J, Ingram I, Nadal A and Romano D 2018 Marine litter plastics and microplastics and their toxic chemicals components: the need for urgent preventive measures *Environmental Sciences Europe* **30** 13
[23] Galgani L, Beiras R, Galgani F, Panti C and Borja A 2019 Editorial: Impacts of Marine Litter *Frontiers in Marine Science* **6** 208
[24] Browne M A, Galloway T S and Thompson R C 2010 Spatial Patterns of Plastic Debris along Estuarine Shorelines *Environmental Science & Technology* **44** 3404-9
[25] Roebroek C T J, Harrigan S, van Emmerik T H M, Baugh C, Eilander D, Prudhomme C and Pappenberger F 2021 Plastic in global rivers: are floods making it worse? *Environmental Research Letters* **16** 025003
[26] van Emmerik T, Tramoy R, van Calcar C, Alligant S, Treilles R, Tassin B and Gasperi J 2019 Seine Plastic Debris Transport Tenfolded During Increased River Discharge *Frontiers in Marine Science* **6** 642
[27] Rech S, Macaya-Caquilpán V, Pantoja J F, Rivadeneira M M, Jofre Madariaga D and Thiel M 2014 Rivers as a source of marine litter – A study from the SE Pacific *Marine Pollution Bulletin* **82** 66-75
[28] Schwarz A E, Ligthart T N, Boukris E and van Harmelen T 2019 Sources, transport, and accumulation of different types of plastic litter in aquatic environments: A review study *Marine Pollution Bulletin* **143** 92-100
[29] Meijer L J J, van Emmerik T, van der Ent R, Schmidt C and Lebreton L 2021 More than 1000 rivers account for 80% of global riverine plastic emissions into the ocean *Science Advances* **7** eaaz5803
[30] Binetti U, Silburn B, Russell J, van Hoytema N, Meakins B, Kohler P, Desender M, Preston-Whyte F, Fa'abasu E, Maniel M and Maes T 2020 First marine litter survey on beaches in Solomon Islands and Vanuatu, South Pacific: Using OSPAR protocol to inform the development of national action plans to tackle land-based solid waste pollution *Marine Pollution Bulletin* **161** 111827
[31] Lavers J L, Dicks L, Dicks M R and Finger A 2019 Significant plastic accumulation on the Cocos (Keeling) Islands, Australia *Scientific Reports* **9** 7102
[32] Smith S D A, Banister K, Fraser N and Edgar R J 2018 Tracing the source of marine debris on the beaches of northern New South Wales, Australia: The Bottles on Beaches program *Marine Pollution Bulletin* **126** 304-7

[33] Ryan P G, Dilley B J, Ronconi R A and Connan M 2019 Rapid increase in Asian bottles in the South Atlantic Ocean indicates major debris inputs from ships *Proceedings of the National Academy of Sciences* **116** 20892
[34] Ryan P G 2020 Land or sea? What bottles tell us about the origins of beach litter in Kenya *Waste Management* **116** 49-57
[35] Verlis K M and Wilson S P 2020 Paradise Trashed: Sources and solutions to marine litter in a small island developing state *Waste Management* **103** 128-36
[36] Jahnke A, Arp H P H, Escher B I, Gewert B, Gorokhova E, Kühnel D, Ogonowski M, Potthoff A, Rummel C, Schmitt-Jansen M, Toorman E and MacLeod M 2017 Reducing Uncertainty and Confronting Ignorance about the Possible Impacts of Weathering Plastic in the Marine Environment *Environmental Science & Technology Letters* **4** 85-90
[37] Chubarenko I, Efimova I, Bagaeva M, Bagaev A and Isachenko I 2020 On mechanical fragmentation of single-use plastics in the sea swash zone with different types of bottom sediments: Insights from laboratory experiments *Marine Pollution Bulletin* **150** 110726

IOP Conf. Series: Earth and Environmental Science **964** (2022) 012017 doi:10.1088/1755-1315/964/1/012017

Supplementary Table S1. Macro litter classification for Can Gio beach. Empty spaces represent zero values.

	VT1		VT2		VT3		VT4		VT5		VT6	
	Dry season	**Rainy season**	**Dry season**	**Rainy season**	**Dry season**	**Rainy season**	**Dry season**	**Rainy season**	**Dry season**	**Rainy season**	**Dry season**	**Rainy season**
Plastics												
Shopping Bags incl. pieces	338	39	121	291	193	724	197	942	466	231	42	665
Small plastic bags, *e.g.* freezer bags incl. pieces	93	1	16	75	63	145	86	339	265	66	4	173
Drink bottles <= 0.5l	10	5	12	14	8	25	34	37	10	21	22	33
Drink bottles > 0.5l	1		5			2	5	2		1		2
Cleaner bottles and containers				2		2				7		3
Food containers incl. fast food containers	139	6	8	78		17	17	117	50	75	17	50
Beach use related cosmetic bottles and	32	2	14	23	21	10	45	11	7	17		13
Jerry cans (square plastic containers with handle)	1		1								5	
Car parts	3											
Plastic caps/lids drinks	77	8	55	210	43	31	66	197	89	46	29	95
Plastic caps/lids chemicals, detergents	9						3					
Cigarette lighters	4	2	5	26	5	12	13	44	4	14	4	18
Cigarette butts and filters			1		44	7	2	15	3		44	5
Pens and pen lids	3		1	10	1	3	7	10	5	4	1	5

Combs/hair brushes/sunglasses	1						1					
Crisps packets/sweets wrappers	210	11	90	259	180	262	141	641	574	230	149	355
Toys and party poppers	18	3	35	25	15	10	35	4	63	26	24	14
Cups and cup lids		26	55	217	41	127	102	282	170	154	62	146
Cutlery and trays	67	13	59	188	107	88	147	387				
Straws and stirrers									388	82	121	148
Fertiliser/animal feed bags											9	
Mesh vegetable bags			10	27	6	22	77	54	12	26	13	39
Gloves (washing up)	3	2		7	2	2	16	18	9	13	11	17
Gloves (industrial/professional	19			2	13	7	10	14	2	4	13	
String and cord (diameter less than 1cm)	21	11	100	1566	58	1122	73	1681	268	540	75	416
Nets and pieces of net		1	22	11	12	10	11	42	18	28	6	9
Floats for fishing nets	1	1	3	17	5	2	7	16	7	17	1	18
Strapping bands		2	10	50	26	47	19	86	66	18	54	31
Hard hats/Helmets		1		1	1							1
Shoes/sandals	4		13	20	6	12	41	22	24		5	27
Foam sponge	12				1		1					

IOP Conf. Series: Earth and Environmental Science **964** (2022) 012017 doi:10.1088/1755-1315/964/1/012017

Plastic/polystyrene pieces 0 - 2.5 cm			15	10	122				3			
Plastic/polystyrene pieces 2.5 cm > < 50cm	35	120	136	450		362	612	884	207	282	56	313
Plastic/polystyrene pieces > 50 cm	1		9	2					1			
Masking tape		1	7	5		11	4	10	17	4		21
Telephone (incl. parts)	2						2					
Plastic flower pots						2	2	1	4			2
Cable ties										2		
Diapers/nappies	40		3	9	9	2	15	42	27	6	7	7
Syringes/needles	2			3	1		4	4		1		1
Medical/Pharmaceuticals containers/tubes	42	2	2	6	11	3	18	16	19	34	10	
Dog faeces bag												
Flip-flops		3	2	25	2	8	6	18	4	13		17
Other plastic/polystyrene items (identifiable)	5	6	68	83	78	54	141	85	98	41	10	41
Total	1193	266	878	3712	1135	3131	1960	6021	2880	2003	794	2685
Rubber												
Balloons and balloon sticks					3		2		1			
Inner-tubes and rubber sheet	1											

Wheels	2		1	1	4		4	5	11	12	3	2
Rubber bands (small, for kitchen/household/post	2		3		1	2	11	4	7	12	3	
Total	5	0	4	1	8	2	17	9	19	24	6	2
Cloth/textile												
Clothing / rags (clothing, hats, towels)	162		6	10	22	9	41	60	36	16	14	23
Shoes and sandals (*e.g.* Leather, cloth)				2			1		1	37	1	
Backpacks and bags					1							
Carpet and Furnishing	1				3		14		3			
Tampons and tampon applicators	10	2	1	2	3		2	4	4		4	
Other textiles (incl. rags)	90	1	2	14	5	7	19	105	31	17		
Total	263	3	9	28	34	16	77	169	75	70	21	23
Paper/Cardboard												
Cardboard (boxes and fragments)										6		
Cartons/Tetrapack Milk	5	9	15	91	31	37	79	80	96	65	24	65
Cartons/Tetrapack (others)	1							34				
Cigarette packets		2	3	4		3	4	21	17	10	10	2
Newspapers and magazines							6					

IOP Conf. Series: Earth and Environmental Science **964** (2022) 012017 doi:10.1088/1755-1315/964/1/012017

Paper fragments				13			8	141	35	7	7	
Other paper items	37		11	4	8		6	167	36	3		6
Total	43	11	29	112	39	40	103	443	184	91	41	73
Processed/worked wood												
Ice-cream sticks, chip forks, chopsticks,			3	10	7	5	2	13	14			4
Paint brushes				3			3	3		6		3
Other woods < 50 cm	1	7	12	33	21	40	23	98	33	15	23	46
Other woods > 50 cm								9				4
Total	1	7	15	46	28	45	28	123	47	21	23	57
Metal												
Aerosol/Spray cans industry									1			
Cans (beverage)	11	1	4	4	4	2	14	5	5	2	5	11
Cans (food)				1								
Bottle caps, lids and pull tabs	20	2	1									
Appliances (refrigerators, washers,	1								2			
Other cans (< 4 L)			2					1				
Gas bottles, drums and buckets (> 4 L)		1										

Wire, wire mesh, barbed wire	43				1							
Other			1		1	1	2		1			
Total	75	4	8	5	6	3	16	6	9	2	5	11
Glass/ceramics												
Bottles incl. pieces	1		5	2	1	8		7	6	7		7
Jars incl. pieces				4			6		1		1	2
Light bulbs		3	1	6	2	5	1	1		6		18
Construction material (brick, cement, and	1											
Fluorescent light tubes										1		
Glass or ceramic fragments >2.5cm			1									
Other glass items							3				1	
Total	2	3	7	12	3	13	10	8	7	14	2	27
Other (glass, metal, tar) <5mm							5					

IOP Conf. Series: Earth and Environmental Science **964** (2022) 012018 doi:10.1088/1755-1315/964/1/012018

The association between high temperature and hospitalizations for cardiovascular diseases under climate change scenarios in Ho Chi Minh City

Phung Duc Nhat[1*], Duong Thi Minh Tam[2*], Dang Van Chinh[2], Vo Le Phu[3,4], Tran Ngoc Dang[5], Mai Tien Thanh[2]

[1] University of Medicine Pham Ngoc Thach, 2 Quang Trung street, District 10, Ho Chi Minh City, Vietnam
[2] Institute of Public Health, 159 Hung Phu street, District 8, Ho Chi Minh City, Vietnam
[3] Faculty of Environment and Natural Resources, Ho Chi Minh City University of Technology (HCMUT), 268 Ly Thuong Kiet street, District 10, Ho Chi Minh City, Vietnam
[4]Vietnam National University Ho Chi Minh City, Linh Trung ward, Thu Duc District, Ho Chi Minh City, Vietnam
[5] University of Medicine Ho Chi Minh City, 217 Hong Bang street, District 5, Ho Chi Minh City, Vietnam
Corresponding authors: nhatphd@pnt.edu.vn duongtam81@gmail.com

Abstract. Global warming is anticipated to induce an increase in the frequency and intensity of hot days and heatwaves, which ultimately have effects on public health. The study aimed to identify the impacts of high temperature and developing climate forecasting projections focused on cardiovascular causes in Ho Chi Minh city (HCMC). The projections were built up based on updated climate scenarios in HCMC. Poisson regression model combined with a distributed lag non-linear model (DLNM) was applied. The forecasting model for cardiovascular causes was performed by using the Global Climate Model (GCM) and Yasushi Honda model. Result showed that the average daily temperature induces an increase in the risk of hospitalisation, in which temperature below 25.7^{0}C reduced number of patients due to cardiovascular disease, meanwhile temperature above 25.7^{0}C has increased hospitalisations. Heat waves over 31^{0}C had the strongest impact on the > 60 years old elderly people after 5 days lag and decreased its impact consecutively to age groups of 41 to 60, 16 to 40, and less than 15 years old. The incremental prediction for the hospitalised cardiovascular disease cases based on the RCP4.5 scenario was 79,713 cases and based on the RCP8.5 scenario was 81,362 cases, respectively.

IOP Conf. Series: Earth and Environmental Science **964** (2022) 012018 doi:10.1088/1755-1315/964/1/012018

1. INTRODUCTION

Climate change is one of the most challenges in the 21st century. The Intergovernmental Panel on Climate change (IPCC) found that global average temperature on earth increased 0.85^{0}C in period 1880-2012 [1]. Many studies observed an increase in number of hospitalisation in cases and deaths of cardiac diseases due to extreme weather events. A study in California in 2006 revealed that on a heat wave the number of hospitalisation caused by heart diseases and other climate-related diseases increased significantly [2]. A retrospective study in Taiwan on temperature, cardiovascular diseases death during the period 1991-2006 was carried out to measure effect of temperature change on heart diseases' death. Results showed that the proportion of death increase 0.226% when temperature fluctuation increases by 1%. The death cases of cardiac diseases increased from 1.2% to 4.1% when other environmental factors were included into the predict model. In Viet Nam, there are few studies on relation between climate change and cardiac diseases. Meanwhile, Ebi K. (2017) found out that only about one-third of studies on climate change were conducted in developing countries [3]. This study aims to build up a model to predict the impact of climate change on community health by updated scenarios on global warming and increase of sea level to supply the policy makers with evidence to integrate into socio-economic development plan of Ho Chi Minh City (HCMC) to eliminate and to increase climate change resilience of community health of Ho Chi Minh City.

2. MATERIALS AND METHODS

2.1. Data collection

We obtained daily admission data for the period from 1 January 1997 to 31 October 2018 from the hospital records of health centres and hospitals in 24 districts of HCMC through The Health Department. Data on hospital admissions included daily counts for arterial hypertension, myocarditis. Data extracted from the admission records included those on primary and discharge diagnoses (ICD-10 codes), date of admission, date of discharge, age, sex, and the district of residence of individual patients. To avoid exposure misclassification, patients from locations other than HCMC were excluded.

There are no meteorological monitoring stations within each districts in Ho Chi Minh City, data from the station in adjacent areas closest to the city center were used. Therefore, we obtained representative weather data from 7 monitoring stations managed by Southern Regional Hydro meteorological Center to cover for the whole city. Daily weather data from 1 January 1997 to 31 December 2018 were collected in the same period with hospitalization data. The necessary information included temperature (minimum, maximum, and average), temperature range. These daily data were then converted to the mean of weekly average temperature (°C) for analysis

IOP Conf. Series: Earth and Environmental Science **964** (2022) 012018 doi:10.1088/1755-1315/964/1/012018

2.2. Data analysis

Firstly, meteorological data and cardiovascular hospitalisation cases were linked by date and analyzed using a time-series design. A distributed lag non-linear model (DLNM) was constructed to represent a modeling framework to flexibly describe associations showing potentially non-linear and delayed effects in time-series data. In present study, DLNM combined with a quasi-Poisson distribution was performed to quantify the relationship between daily temperature and cardiovascular hospitalisation counts adjusted with other meteorological variables. The model was specified as follows:

$$E(Yi) = \exp\{\alpha + f(Ti|\beta 1) + (\text{specific covariates}) + (\text{smooth function of time}) \quad (1)$$

F (Ti/β1): The function of temperature

Specific covariates: Confounding variable

Smooth function of time: The function of the time variable, to adjust for the long-term trend and seasonality.

In addition, to account for the potential non-linear relation between temperatures and health outcomes, we applied a distributed lag non-linear model (DLNM) using a cross-basis function of multiple lag-day temperature [4]. DLNM can help determine the non-linear and lag effects of the relationship between temperature and health, simultaneously. The optimal model and parameters were selected and validated using post estimation residuals and the Akaike's information criterion (AIC) in which the model with the least AIC was chosen due to its better explanation.

To forecasting hospital admissions incremental and rising temperatures, our study assumes factors: population, health services, and treatment remain unchanged, and forecasts to 2025, 2030, 2050 by performing analysis in 2 stages:

The first step used Global Climate Model (GCM) PRECIS developed by Hadley Meteorological Centre, the UK to forecast temperature including max, min, average temperature of HCMC in 2025, 2030, 2050 on 3 scenarios of climate change RCP4.5 (medium) and RCP8.5 (high). This study applied the GFDL-CM3 and CNRM-CM5 of the PRECIS model and downscaled at 25 km spatial resolution to forecast the climate of HCMC. In previous studies, the PRECIS model had better results than other models in predicting Vietnam's climate.

In the second step, we applied method developed by Yasushi Honda (2014) in a previous study to calculate incremental number of admitted cases to hospitals due to hot temperature events with forecasted weather data in the years 2020, 2025, 2030, and 2050 [5].

IOP Conf. Series: Earth and Environmental Science **964** (2022) 012018 doi:10.1088/1755-1315/964/1/012018

3. RESULTS AND DISCUSSION

3.1. Cardiovascular diseases in HCMC by day and week during 1997 - 2018

Table 1. Cardiovascular diseases in HCMC by day during 1997-2018

Data	Mean	SD	Min	Percentile			Max
				25%	50%	75%	
Weather							
Max. temperature (°C)	33.7	1.9	25.0	32.5	34.0	35.0	38.5
Mean temperature (°C)	28.4	1.5	23.1	27.5	28.5	29.4	32.6
Min. temperature (°C)	25.3	1.7	18.0	24.3	25.5	26.4	30.0
Cardiovascular hospitalisation cases							
Arterial hypertension	76.4	24.0	19	58	74	93	170
Acute coronary syndrome	4.6	2.4	0	3	4	6	15
Myocarditis	0.1	0.3	0	0	0	0	2
Cardiovascular cases by age,	81.1	24.8	22	62	79	98	179
Gender							
Male	26.8	9.6	0	20	26	33	74
Female	49.0	15.2	10	37	48	60	100
0 – 15 years old	0.4	0.8	0	0	0	1	8
16 – 40	3.6	3.5	0	2	3	4	26
41 – 60	22.7	7.7	5	17	22	28	51
> 60	54.4	17.6	13	41	53	66	120

Average mean temperature was 28.4°C, min and max temperature consecutively were 23.1°C and 32.6°C, respectively. In cardiovascular diseases group, admission was highest in arterial hypertensions (76.4%), in which the number of female cases was higher than male, and by age group the highest group was older than 60 years old group.

3.2. *Distribution of hospitalisation by group of cardiovascular diseases by year (arterial hypertension, acute coronal syndrome, myocarditis).*

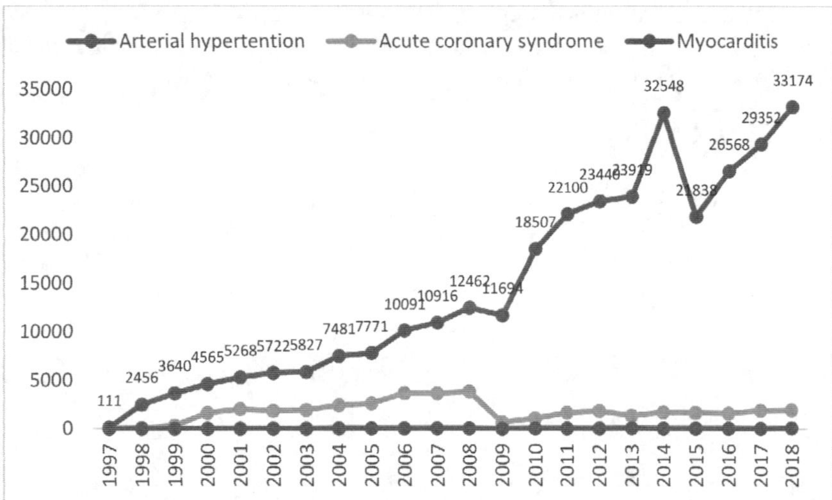

Figure 1. Hospitalisation of cardiovascular diseases by year in HCMC period 1997 – 2018

IOP Conf. Series: Earth and Environmental Science **964** (2022) 012018 doi:10.1088/1755-1315/964/1/012018

The result showed that hospital admission of cardiovascular diseases had an increase trend by year. During period of 1997 to 2018, a total of 319,450 cases of arterial hypertensions admitted to hospitals was reported in HCMC. With cardiovascular diseases group, during 2006 – 2018 hospitalisation of arterial hypertensions increased by year. Meanwhile, acute coronary syndrome fluctuated slightly during 1997 to 2018. In Viet Nam, cardiovascular diseases caused 31% total of deaths annually. On a national survey on risk factor of non-communicable diseases (NCDs) in 2015, percentage of hypertensions in adults from 18-69 years of age was 18.9%, however only 13.6% of them were reported to be managed in a health settings [6]. The proportion of communicable diseases was in a decrease trend, but NCDs proportion was in an increase trend. In 2000, worldwide there were 972 million arterial hypertension cases (26.4% of total world population), in which 639 million people were in developing countries. It is projected that there will be 1.56 billion of cases with arterial hypertensions in 2025 (equivalent to 29.2% of total world population) [7].

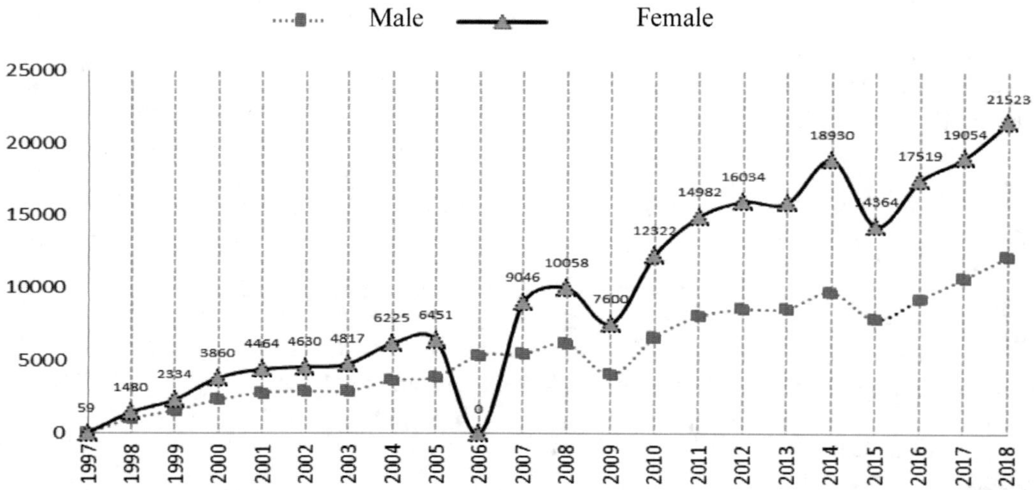

Figure 2. Cardiovascular diseases hospitalisation yearly by gender in HCMC during 1997 –2018

During the study period, a total of 358,995 cases of cardiovascular diseases was hospitalised in HCMC. The result showed that hospitalisation cases increased by year. There was only 113 cases in 1997, and was 35,211 cases in 2018. Further, hospitalisations also increased in both male and female, in which number of female outweigh over that of male.

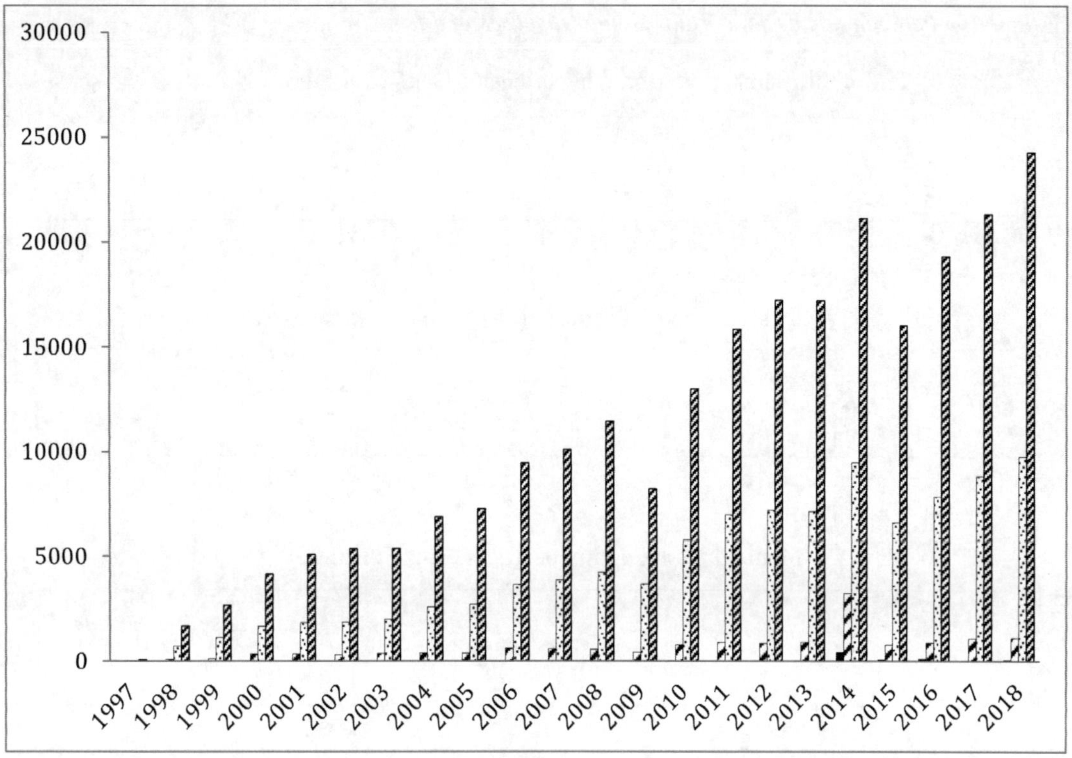

Figure 3. Cardiovascular diseases hospitalisation yearly by age group in HCMC during 1997 –2018

Hospitalisation caused by cardiovascular diseases was highest in group over 41 year of age, in which the highest one was over 60 years old group. Controversy, there were very rare cases under 15 years old, less than 100 cases each year. Over 60 years old was always the group with highest hospitalised cases. The risk of cardiac event increased by age. Epidemiological studies found out age as one of the most predictable factor for heart disease. In group older than 65 years old, they comprised more than half of heart stroke and four-fifth deaths by stroke. Increase of these types of diseases was in relation with lifestyle, behavior, and habits. This findings is similar to that found in the USA in which coronary disease was the top reason of deaths in the US with 82% of total death was in over 65 year old group in 2005 [8]. Further, the cases of coronary were estimated to be about 80 millions of American, and half of them were from 60 years old [9].

3.3. Relation between number of cardiovascular diseases and temperature

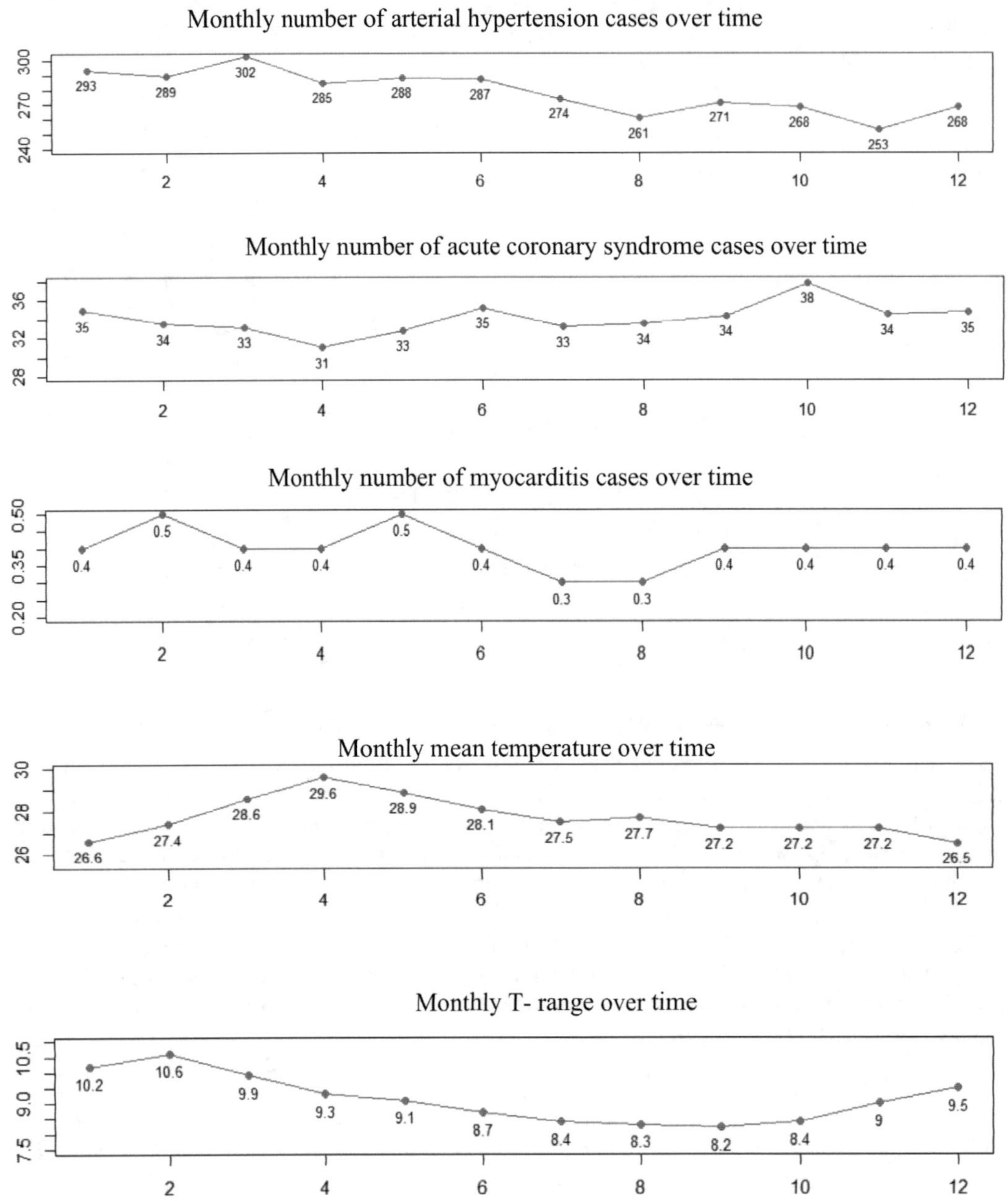

Figure 4. Time series cardiovascular diseases and temperature by month during 1997 – 2018

Monthly average temperature in HCMC period 1997 – 2018 was highest during March to May (28.6 °C – 28.9 °C); lowest in January (26.6 °C) and December (26.5 °C). Monthly average temperature range was highest in January and February (10.2 °C – 10.6 °C), then decreased from March to September. The result shows that when temperature increased, the number of arterial hypertension hospitalisation also increased.

3.4. Relation between number of cardiovascular diseases and daily temperature

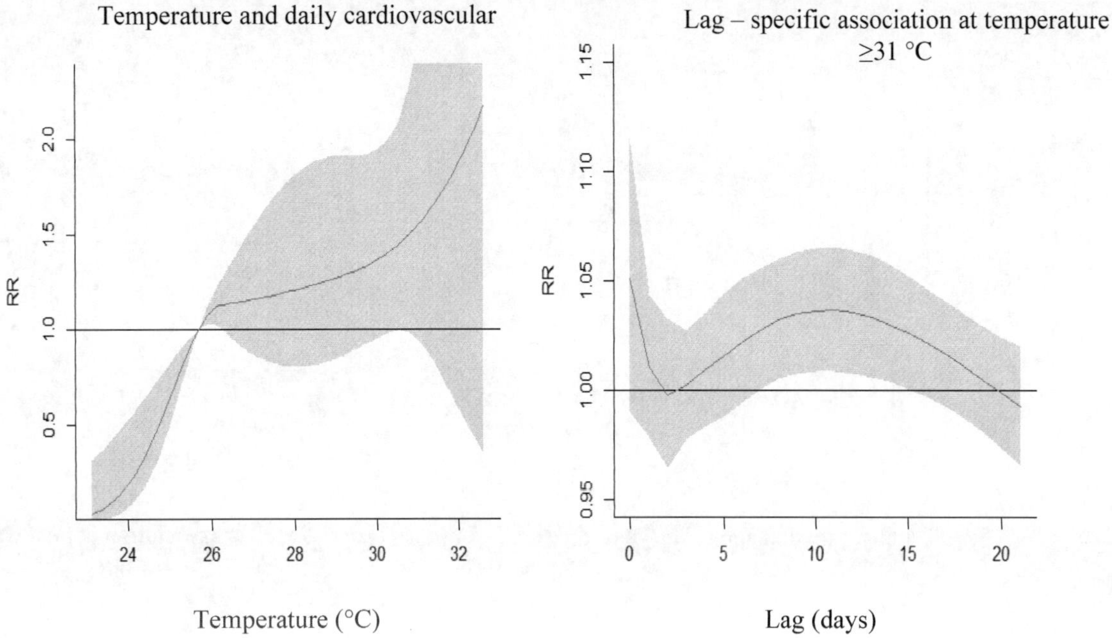

Figure 5. Short term relation between temperature and hospitalisation of cardiovascular cases in HCMC during 2013-2018

Days with temperature under 25.7°C decreased the number of hospitalisation by cardiovascular diseases and temperature above 25.7°C increased cardiovascular cases. However, when identifying the effect of temperature ≥ 31°C in relation with daily hospitalisation caused by cardiovascular diseases, we could see an increase trend after 0 to 2 days, and a decrease trend after 3 days but the number increases again from 4 to 19 days later.

Previous studies have documented that people with chronic diseases, especially cardiovascular diseases, are more susceptible to high temperature. The mechanism of heat effect on cardiovascular includes changes in thermoregulation response, blood viscosity, surface blood circulation, and endothelial cell damage [10]. As the temperature increases, the capacity for dry heat exchange is reduced because of a reduction in the temperature gradient between the skin surface and the ambient air. As a results, losing water and salt from sweating during exposure to high temperatures. This finding implies that patients with cardiovascular disease should be taken care of during both hot and cold periods [11].

IOP Conf. Series: Earth and Environmental Science **964** (2022) 012018 doi:10.1088/1755-1315/964/1/012018

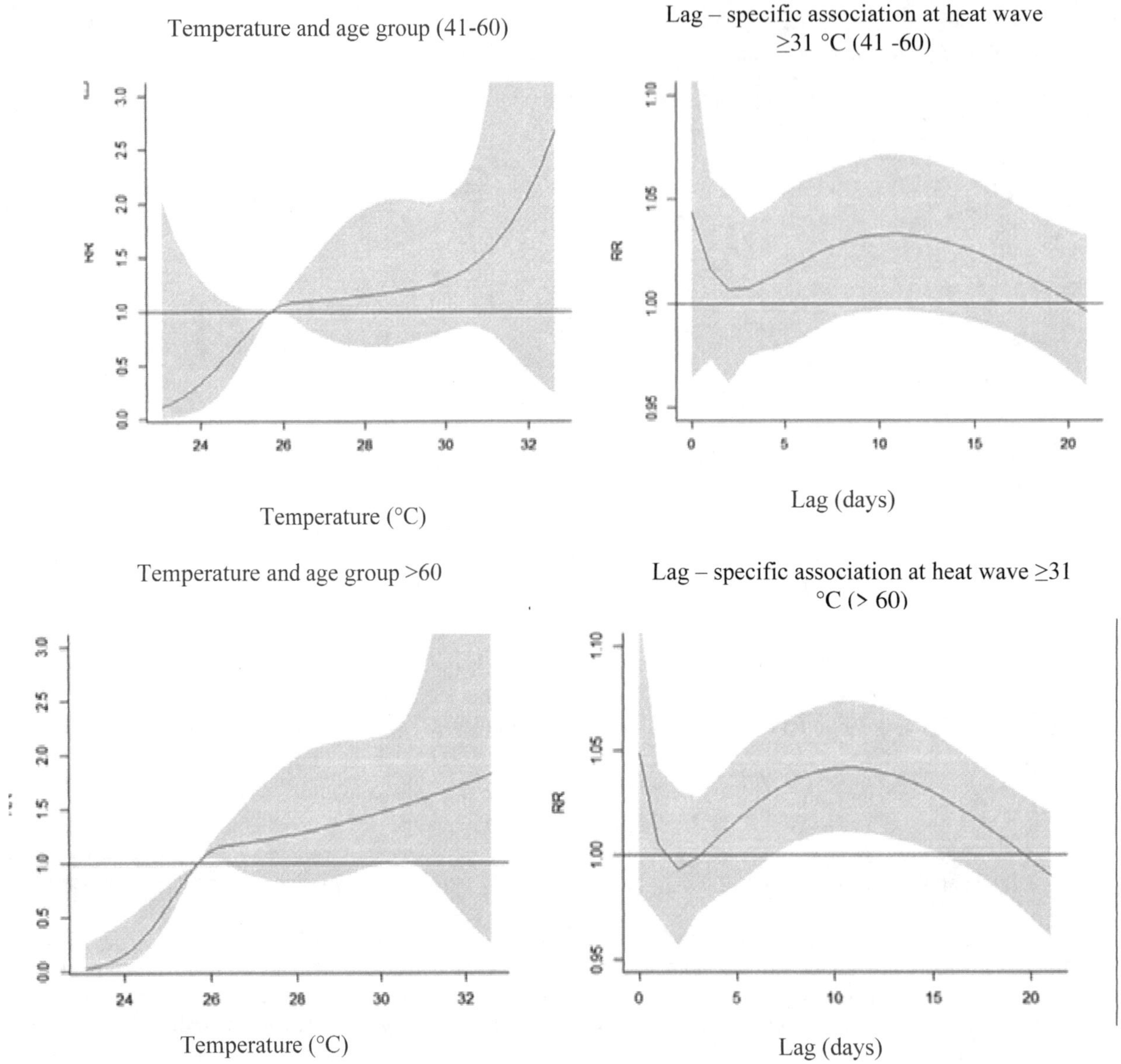

Figure 6. Influence of heat waves on cardiovascular hospitalisation by age groups (Age groups from 41-60 years old and > 60 years old)

Optimal temperature for the age group of < 15 was 30.3°C, 16 – 40 years of age was 29,5°C and > 40 years of age was 25.7°C, respectively. There was a significant statistical difference between 40 - 60 years old group and > 60 years old group on optimal temperature. Heat wave ≥ 31°C had a strong relation in older people group with relative risk (RR) on 5 days-lag of age group above 60 years higher than group 41 – 60 years old and group less than 15 years old.

Population aging has been viewed as an important factor that could substantially exacerbate the health burden of future rising temperatures globally. Older adults are particularly sensitive to climate change[12]. Numerous studies have provided similar evidence that the elderly population is among the most vulnerable groups [13, 14]. The vulnerability of the elderly is associated with physiological and social factors, such as living alone, pre-existing chronic diseases, reduced physiological function. The vulnerability of the elderly to heat exposure is mainly due to a combination of impaired thermoregulatory capacity to heat, and a high prevalence of comorbidities involving the cardiovascular,

IOP Conf. Series: Earth and Environmental Science **964** (2022) 012018 doi:10.1088/1755-1315/964/1/012018

respiratory, endocrine, and renal systems [10]. Aging induces a decrease in thermoregulatory abilities, together with the increased the prevalence of chronic diseases, which are likely to contribute to vulnerability to temperature effects in elderly people [15].

b) Effect of temperature and heat wave on gender

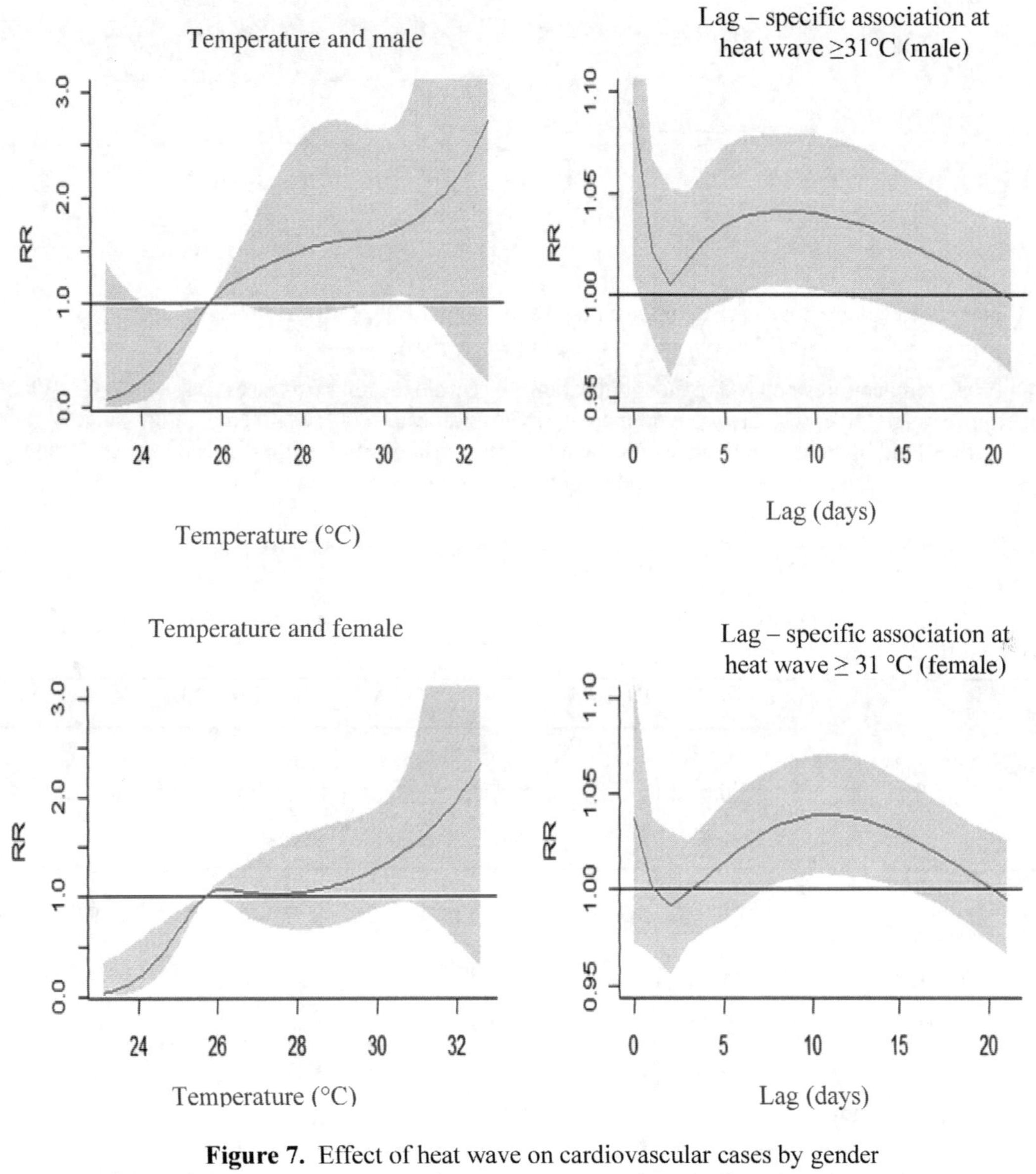

Figure 7. Effect of heat wave on cardiovascular cases by gender

Both male and female were affected by the same temperature threshold of 25.7°C, however this threshold affected statistically significance only in female group. In consideration of gender factor, we could see that in the first 5 days of heat wave period, male was affected more than female but after 5 days female was affected more than male.

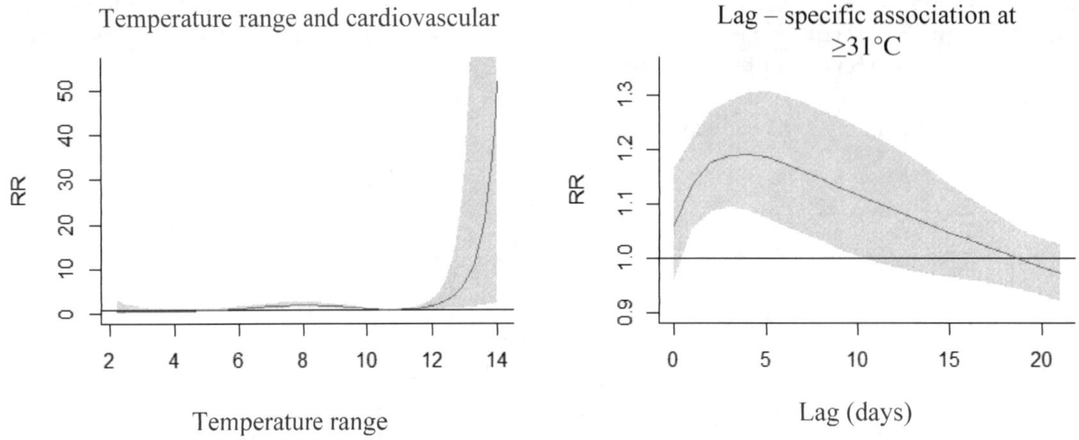

Figure 8. Relation between temperature range and hospitalisation of cardiovascular diseases in HCMC during 2013-2018

Optimal temperature range for hospitalisation of cardiovascular diseases was 10.7°C. When temperature range ≤ 10.7°C there were no increase in hospitalisation but when temperature range higher than the threshold of > 10.7°C there would be an increase in hospitalisation. Consider of threshold ≥ 13°C, number of hospitalisation cases increased dramatically on 9 days lag and decreased after 10 days.

3.5. Anticipation of number of hospitalisation cases caused by cardiovascular diseases in the coming decades

Table 2. Number of hospitalisation by cardiovascular diseases due to high temperature during 2010 – 2049 by different climate change scenarios (RCP)

No.	Decade	RCP4.5		RCP8.5	
		Number of hospitalisation cases	Percentage of hospitalisation (%)	Number of hospitalisation cases	Percentage of hospitalisation (%)
1	2010-2019 (present)	13,169	4.4	13,760	4.6
2	2020-29	24,005	8.1	20,979	7.1
3	2030-39	29,646	10.0	29,579	10.0
4	2040-49	26,062	8.8	30,804	10.4
Total (2020 – 2049)		**79,713**	**9.0**	**81,362**	**9.2**

The number of hospitalisation cases caused by the increase of temperature during 2020 – 2049 based on scenario RCP4.5 was 79,713, which was less than the number of hospitalisation cases based on scenario RCP8.5, 81,362 cases. These hospitalisation cases were appropriate with the prediction model of this study when temperature increased 1°C then after a lag from 1 to 4 weeks hospitalisation of cardiovascular diseases increased 8%.

4. Conclusion:

Global climate change induced an increase in the frequency and intensity of hot days and heatwaves, which ultimately have effects on human health. Exposure to high temperature can lead to an increase in hospital admissions for cardiovascular causes. The results of this study showed that the total hospitalisation of cardiovascular cases was in an increase trend in HCMC. Female cases outnumbered male cases. The highest cases were found in people over 40 years old. The mean average temperature weekly was 28.4°C, minimum 23.1°C and maximum 32.6°C, highest during March to May (28,6 °C – 28,9 °C); lowest in January (26.6 °C) and December (26.5 °C). When considering the relation between hospitalisation of cardiovascular cases and daily weather factors, the temperature of less than 25.7°C decreased number of hospitalisation cases, meanwhile the temperature of higher than 25.7°C resulted in increased hospitalisation cases. Heat wave $\geq$ 31°C had a higher impact on higher age groups, with relative risk (RR) on a 5 day- lag of age group above 60 years old which is higher than group 41 – 60 years old and group less than 15 years old.

Acknowledgements: We thank to Ho Chi Minh City Department of Science and Technology for its funding to this research under the contract No. 43/2018/HĐ-SKHCN in which data and information were deployed to enable this paper. The authors also acknowledged the support of time and facilities from Ho Chi Minh City University of Technology (HCMUT), VNU-HCM for this study.

References

[1] IPCC 2013 The Physical Science Basis *New York*. **Cambridge University Press** p. 1032.
[2] Knowlton K 2009 The 2006 California heat wave: impacts on hospitalizations and emergency department visits *Environ Health Perspect* **117** 61–67.
[3] Ebi K and Hess J 2017 The past and future in understanding the health risks of and responses to climate variability and change *International journal of biometeorology* **61** 71-80.
[4] Gasparrini A 2010 Distributed lag non-linear models *Stat Med.* **29** 2224-2234.
[5] Honda Y 2014 Heat-related mortality risk model for climate change impact projection *Environmental health and preventive medicine* **19** 56-63.
[6] World Health Organization Viet Nam National STEPS Survey 2015. In: 2015.
[7] Whelton P 2004 Epidemiology and the Prevention of Hypertension *J Hypertens*.
[8] Yazdanyar A 2009 The burden of cardiovascular disease in the elderly: morbidity, mortality, and costs *Clin Geriatr Med.* **25** 563-vii.
[9] Lloyd-Jones D 2009 Heart disease and stroke statistics--2009 update: a report from the American Heart Association Statistics Committee and Stroke Statistics Subcommittee *Circulation* **119** 181.
[10] Kenny GP 2010 Heat stress in older individuals and patients with common chronic diseases *CMAJ* **182** 1053-1060.
[11] Carder M 2005 The lagged effect of cold temperature and wind chill on cardiorespiratory mortality in Scotland *Occup Environ Med.* **62** 702-710.
[12] Bunker A 2016 Effects of Air Temperature on Climate-Sensitive Mortality and Morbidity Outcomes in the Elderly; a Systematic Review and Meta-analysis of Epidemiological Evidence *EBioMedicine*. **6** 258-268.
[13] Hajat S 2007 Heat-related and cold-related deaths in England and Wales: who is at risk? *Occup Environ Med.* **64** 93-100.
[14] Yang J 2012 Daily temperature and mortality: a study of distributed lag non-linear effect and effect modification in Guangzhou *Environmental Health* **11**.
[15] Gasparrini A, Armstrong B 2012 The effect of high temperatures on cause-specific mortality in England and Wales *Occupational and Environmental Medicine* **69** 56-61.

IOP Conf. Series: Earth and Environmental Science **964** (2022) 012019 doi:10.1088/1755-1315/964/1/012019

Adaptation to saline intrusion for agriculture farming transformation in the coastal Ben Tre province, Vietnam

Nguyen Thi Hong Diep[1], Phan Ky Trung[1], Dinh Thi Cam Nhung[1], Nguyen Thi Thanh Huong[2], Pham Thanh Vu[1] and Le Anh Tuan[3]

[1] College of Environment and Natural Resources, Can Tho University, Campus II, 3/2 street, Ninh Kieu district, Can Tho city 900000, Vietnam
[2] Faculty of Agriculture and Forestry, Tay Nguyen University, 567 Le Duan, Buon Ma Thuot, Dak Lak city 630000, Vietnam
[3] DRAGON-Mekong Institute, Can Tho University, Campus II, 3/2 street, Ninh Kieu district, Can Tho city 900000, Vietnam

nthdiep@ctu.edu.vn, pktrung@ctu.edu.vn, pkdiem@ctu.edu.vn, dtcnhung20@gmail.com, huongthanh.frem@gmail.com, ptvu@ctu.edu.vn, latuan@ctu.edu.vn

Abstract. This study aimed to evaluate the adaptive capacity (AC) of agriculture farming transformation to improve for the adaptability by saline intrusion of those culture systems in the coastal Ben Tre province. The primary data was collected by questionnaires in three coastal districts of Ba Tri, Thanh Phu and Binh Dai with 178 households to distribute on agriculture land use transformations. An Object-Based Image Analysis (OBIA) and multi-temporal image analysis approach was developed to detect how LULC changes during 2010-2020 in the coastal Ben Tre province using Landsat TM and OLI data. The MODIS (MOD09 8-day reflectance) data was processed into monthly NDVI maps with the Time Series Product Tool software package and then used to classify regionally common rice crops LULC types using unsupervised classification by ISODATA algorithm. Based on primary data collection, evaluating adaptation measures was assessed the compilation of modification options based on a detailed description and criteria including human, society, infrastructure and natural conditions by standardized a weight ranges of adaptive indicators in the agriculture land use changes. The results detected five agriculture land use transformation including rice-shrimp crop rotation to aquaculture, single rice crop to aquaculture, double rice crop to perennial plant/orchards, double rice crop to rice-shrimp crop rotation, and triple rice crop to double rice crop. The adaptability of these transformations conducted the shrimp-rice crop rotation to aquaculture in Binh Dai of strongly high adaptability. Next is followed by double rice crops to orchards and double rice crops to rice-shrimp crop rotation with high adaptability while the rice-shrimp crop rotation to aquaculture is only moderate in Thanh Phu. The low adaptability levels were identified on single rice crop to aquaculture and triple rice crop to double rice crop transformations due to tolerant condition of saltwater for rice crop cultivation. The studies results can further contribute to support adaptation planning in these sectors by using, developing and streamlining this framework to additional and different socio-ecological contexts.

1. Introduction

The Vietnamese Mekong Delta (MKD) is located at the end of the Mekong River, one of the 10th largest rivers in the world and flows through six countries (China, Myanmar, Thailand, Laos, Cambodia and Vietnam) [1]. The Mekong Delta is a key agricultural and aquaculture production area of the country. The Mekong Delta contributes three major export products, which are rice, fruit and seafood. Annually, the Mekong Delta provides more than 50% of the national rice production, 90% of export rice production, 70% fruit, 40% of caught fish and 74% of the country's aquaculture [2]. However, agricultural production and people's lives here also face many difficulties, especially in coastal areas. Every year, in the dry season, the upstream water flows less and the rainfall decreases, so saline water intrusion the field, making more than 2 million hectares of saline soil, affecting crops, livestock and community activities, especially the poor [3,4]. In the Mekong Delta, the problem of saltwater intrusion and water scarcity in dry season has been complicated and tend to become more serious, especially in the context of global climate change and over-exploitation Mekong River's water source in the upstream of the country and seriously affects the production as well as the daily life of people in the Mekong Delta provinces.

Faced with the situation like the ones from above, Ben Tre province is the most affected locality, especially in coastal rural areas where the quality of water for crop irrigation is vital to the economy. Currently, in Ben Tre province, saline water continues to penetrate deep into major rivers: Co Chien, Ham Luong, and Cua Dai rivers, affecting agricultural production and people's daily life on a large scale [5]. Three coastal districts of Ben Tre province, in which Ba Tri district has more than 26,000 hectares of agricultural production on land, accounting for 2/3 of the natural land area. Every year, saltwater intrusion goes deep into the field in the dry seasons, when the Mekong River's water flow decreases [6]. According to the report of the Provincial People's Committee, during the last salt season of 2016, Ben Tre province suffered heavy damage due to the early appearance of saline water. The damaged rice area is about 20,356 ha, about 458.31 ha of crops, 5,240.48 ha of orchards, 151,357 seedlings. The rate of damage was about 60 - 70%, in addition, the area of industrial crops also lost 1,302.51 ha, and 1,380,115 flowers and ornamental plants died due to salt water.

According to the National Climate Change Scenario, Ben Tre province with sea level rise of 12 cm in 2020, Binh Dai district will be affected the most, having the highest percentage of flooded area; next is Thanh Phu district, whose boundary line is adjacent to Ham Luong, Co Chien rivers and has a long coastline, thus the flooding rate is quite high; Ba Tri and Giong Trom districts also have relatively high salinity areas. At a sea level rise of 30cm by 2050, Binh Dai and Ba Tri will be most affected.

Adaptability to salinity intrusion is the capacity of human, natural, infrastructural and social systems to resist adverse conditions caused by salinity intrusion, which is the opposite of vulnerability, adaptability is an activity aimed at reducing adverse effects, reducing harm or taking advantage of opportunities. Adaptation to salinity intrusion is the adjustment of natural, social, human systems to changing circumstances or environments to reduce vulnerability to existing or potential saltwater intrusion and to utilize and take advantage of the opportunities it presents. Assessing the current adaptive capacity is an important scientific basis for building effective adaptation strategies to salinity intrusion. Assessment of adaptability to salinity intrusion aims to review current development activities, plans and adaptation options to the risks posed by salinity intrusion.

The study "Assessment of adaptability of transformative farming models in coastal areas of Ben Tre province in the period 2010-2020" was carried out to assess the adaptability of transformative farming models to intrusive salinity intrusion, as a basic premise in land use planning, thereby improving the adaptability of farming models.

2. Study area and methods

2.1. Study area

Ben Tre is located in the Vietnamese Mekong Delta (VMD), which is about 85 km Southwest of Ho Chi Minh City. The province lies between two tributaries of the Mekong branch (normally known as Tien River), one of two main branches of the Mekong River at the VMD. Smaller inland rivers and canals created complex waterway networks, leading the province to be major agricultural production for the region, especially rice farming systems. Ben Tre province consisted of three islands including An Hoa, Bao and Minh among Mekong branch in the North, Co Chien River in the South and Ham Luong River straight down from Ben Tre city center. The discharge flows through those channels before eventually running off to the sea via three estuaries (Figure 1).

Ben Tre was considered as one of the most provinces seriously affected by climate change impacts in the VMD. According to the provincial Department of Agriculture and Rural Development, serious saline water intrusion had periodically occurred further not only in inland rice farming systems but also affected households in terms of water storage and purification, hygiene promotion, nutritional support and livelihood recovery in each four to five years before 2000. However, the frequency of serious saline water intrusion has intended to be more regular since 2000, in two-year or even annually period. In the years of 2000, 2002, 2004, 2005, 2007, 2009, 2010, the water salinity were reported at 4‰ at some times during dry seasons, in which the monitoring in 2004, 2005, 2010 indicated the furthest inland saline water intrusion ever recorded in Vam Mon, being about 60 km from Ham Luong estuary. In those years, salinity increased by 1‰ to cover almost entirely Ben Tre province. Moreover, salinity variability was reported as one of the main factors of climate change dramatically impacting the coastal provinces based on sea-level rise underflow discharge degradation from the upstream part of the VMD. These changes have accumulatively led to serious consequences in agricultural production. In 1995 - 2008, drought and saline water intrusion caused damage of VND 672,305 billion. Additionally, the drought-induced by freshwater scarcity damaged and reduced the yield of 1,575 hectares of rice, 4,500 hectares unproductive, damaged and reduced the yield of 10,162 hectares of orchards in 2010, which total damage is estimated at 198 billion VND.

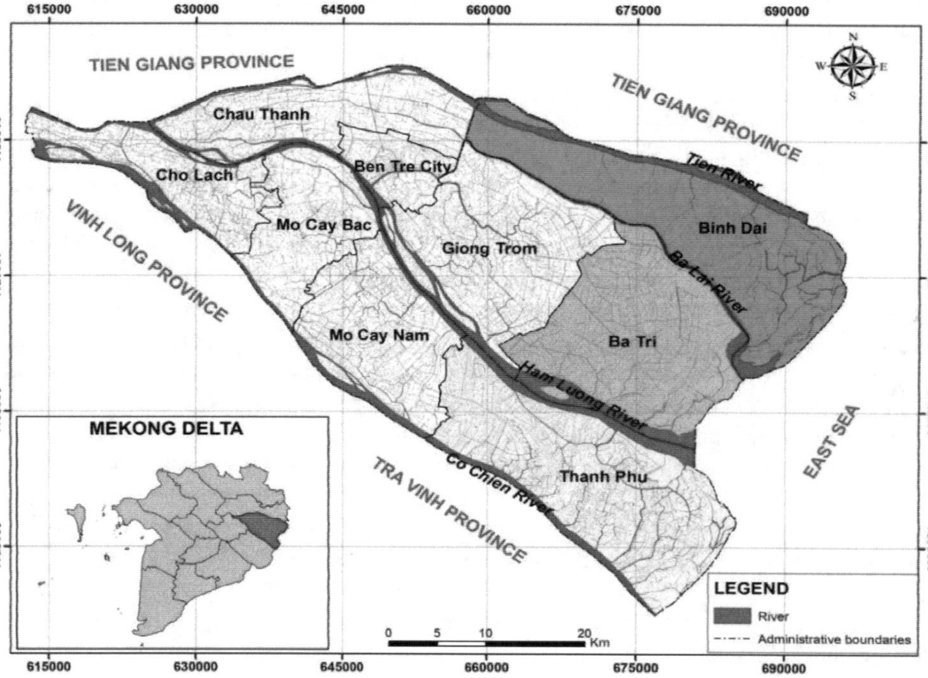

Figure 1. Study area location map.

2.2. Method

2.2.1. Data collection

LANDSAT data: the images including Landsat 4-5 TM in 2010 and Landsat 8 OLI in 2020 at 30-meter spatial resolution and 16 days of temporal resolution were collected through downloading from http://earthexplorer.usgs.gov/ of the United States Geological Survey (USGS). Images were orthorectified and georeferenced to UTM projection, zone 48 to provide users with standard Level-1 product.

MODIS data: MODIS/Terra data (MOD09Q1) acquired from the U.S. National Aeronautics and Space Administration for two years of 2005 and 2020 from January to December each year. This data product has two spectral bands (red and near infrared) and a spatial resolution of 250 m. Each pixel of the data product contains the best possible L2G observation during an 8-day period, selected based on high observation coverage, low viewing angle, absence of clouds, and aerosol loading. The data have been geometrically and radiometrically corrected for these two spectral bands [7].

Secondary data: Three types of primary data including spatial maps, annual reports and related research in the study area were collected to provide the general information and baseline for the research. Detail of collected primary data was illustrated in Table 1.

Table 1. List of primary data collection.

No.	Data content	Periods	Data sources
1	Land-use map in Ben Tre province	2010 - 2020	Ben Tre Provincial Department of Natural Resources and Environment
2	- Water-based risk annual report - Water infrastructure system	2020	Ben Tre Provincial Agency of Irrigation
3	Relevant studies on saltwater intrusion in the Vietnamese Mekong Delta	2010 - 2020	- International Journals (ISI, Scopus, SCI) - Recognized Vietnamese Journals

Primary data:

- 178 household surveys were conducted in three coastal districts in Ben Tre province, consisting of Thanh Phu, Ba Tri and Binh Dai through directive interview approach to understand possible impacts of saltwater intrusion on local livelihoods and the relationship between land use practices and saltwater intrusion-induced vulnerability.
- The sample size of 178 was chosen according [8], where the size from 30 and above being possibly used for multiple research criteria and the size from 100 and above can work for most of social survey-based research at the local scale. This research then implemented 30 household surveys per each changing land use type based on the random sampling technique. Indicators for structuring the survey were framed aiming at identifying possible impacts-induced vulnerability of water hazards on local agricultural land use types, which this study paid attention to on saltwater intrusion in 2010 – 2020 [9]. The questionnaire was developed depending on four baselines (Table 2) regarding human-being (1); society (2); infrastructure (3); and natural system (4).

Table 2. Criteria for building the questionnaire.

Serial	Criteria	Description
1	People	Household demography, perception, community communication, social learning and behavior change.
2	Society	Technical training and governmental financial loan provision.
3	Infrastructure	Water irrigation, water infrastructure improvement and natural disaster protection.
4	Natural system	Cultivation areas affected by saltwater intrusion and areas potentially switched to another land use practice type.

2.2.2. LANDSAT data processing.

Atmospheric Correction: the products were then atmospherically corrected using Fast Line-of-sight Atmospheric Analysis of Hypercubes (FLAASH) tool based on MODTRAN radiative transfer model in ENVI software to obtain surface Reflectance images [10].

Defining class features: Five main classes were defined for this study: Agriculture, Orchards, Forest, Build-up area, Aquaculture and River. These classes and their subsequent inheritance relationships were established in the class hierarchical window of eCognition. The GeoTIFF file was modified a study area of Tra Vinh province and reduce the processing time. All of the necessary classes defined in the sample selection were present in the study area. Class features were defined by image indices and brightness images in Table 3.

Table 3. The class features used in this study.

No.	Image Indices	Equation	Author
1	Normalized Difference Vegetation Index (NDVI)	$NDVI = \frac{NIR - Red}{NIR + Red}$ (1)	Khan et al (2010) [11]
2	Normalized Difference Water Index (NDWI)	$\text{NDWI} = \frac{GREEN - NIR}{GREEN + NIR}$ (2)	McFeeters (1996) [12]
3	Normalized Difference Index (NDBI)	$\text{NDBI} = \frac{\text{MIR - NIR}}{\text{MIR + NIR}}$ (3)	Zha et al (2005) [13]
4	Brightness value	μ =(Red+Green+Blue)/3 (4)	

Segmentation: Image segmentation was carried out on the pan-sharpened imagery. Multi-resolution segmentation (MRS), as implemented in Trimble eCognition software, was used for generating the image objects. MRS is a region-merging pair-wise algorithm in which each data point is associated with another data point, where the delineation algorithm will predict which of the two is "more" relevant according to the input parameters. It segments on pixel or object level based on user-defined scale, shape, and compactness parameters [14]. The multi-resolution function was applied for the extraction of image objects using modifiable scale parameters, single layer weights, and homogeneity concerning color and shape [15]. Within the shape setting, smoothness and compactness can also be weighted from 0 to 1. In our study, segmentation parameter used for analysis applying with shape, compactness and scale at 15, 0.5, 0.5, respectively.

Rule-Based Classification: we applied the Membership Function to obtain the suitable feature values from Sample Editor for each class. We selected the features which have less overlap from one another. Range of the feature's value were selected and also changed based on the features statistical value displayed in Sample Editor window. The suitable values to specify each class based on the statistical analysis were selected for the development of Membership functions. For the main classes, we use the Classification process in Process tree with the Active Class which needed to classify for five main classes in Tra Vinh province.

2.2.3. MODIS data processing

MODIS NDVI Time-Series Data: these data have an effective measurement of vegetation with high temporal resolution from MODIS, NDVI data facilitate meaningful comparisons of seasonal and inter-annual changes in vegetation growth and activity. The strength of NDVI lies in its rationing concept, which reduces many forms of noise (illumination differences, cloud shadows, atmospheric attenuation, and certain topographic variations) present in multiple bands. In addition, a comparison of sensitivity to the dynamics of vegetative cover shows that NDVI data are more sensitive because they enable more accurate change detection as a result of NDVI data presenting higher variability and abrupt changes in the growing season [16,17]. Reflectance values in band 1 and band 2 in the MOD09A1 product are used to calculate NDVI, as shown in Equation 1 above [12].

IOP Conf. Series: Earth and Environmental Science **964** (2022) 012019 doi:10.1088/1755-1315/964/1/012019

Rice crop phenology detection: We estimated sowing and harvesting dates from phonological events of rice plant computed from MODIS NDVI time-series data (Fig. 2). Rice fields were often irrigated before sowing; NDVI responses from rice fields were relatively low during this period. The rice plants grow quickly after sowing, quickly increasing NDVI values; thus, the sowing date can be estimated using the greenup onset or inflection point (i.e., the point at which the concavity changes from negative to positive). The point earlier than this inflection point was determined as the sowing date. Likewise, NDVI values begin leveling off after harvesting, the harvesting date can thus be estimated from the inflection point (i.e., the point at which the concavity changes from positive to negative). The point after this inflection point was determined as harvesting date. In this research, we detected rice crop phenology due to the fluctuation of NDVI in year (Figure 2).

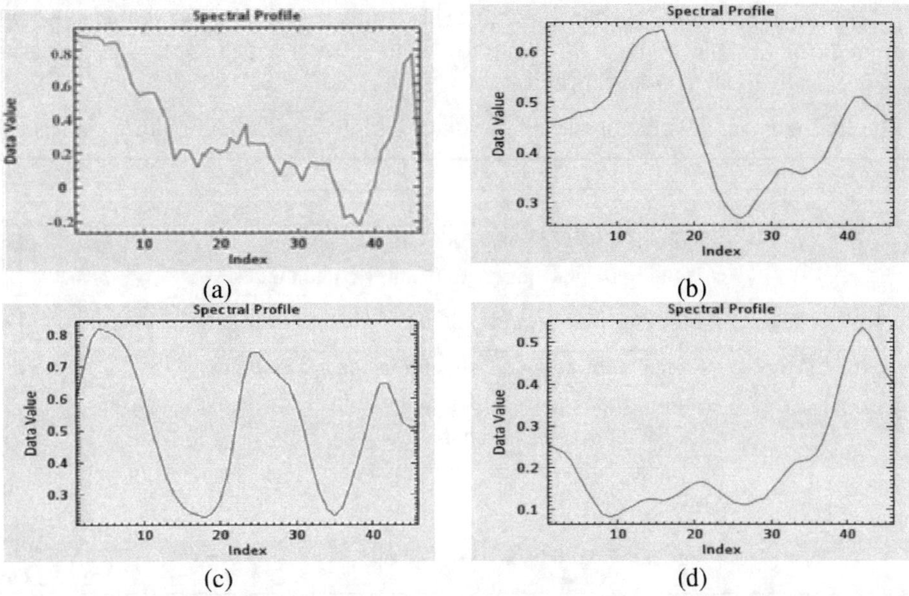

Figure 2. NDVI value to determine phenological events of rice plant during a cropping cycle (a) single rice crop, (b) double rice crop, (c) triple rice crop, (d) rice-shirmp rotation.

Rice crop classification: Rice crop classification was subsequently performed using information for sowing, heading, and harvesting dates. We defined the heading date as the date at which the rice plant is completely emerged, leading to the maximum NDVI value. This date can be identified from the smooth NDVI profile using the local maximum algorithm, and it must be between the maturity onset (i.e., the date when plant green leaf area reaches the maximum NDVI value) and the senescence onset (i.e., the date when photosynthetic activity and green leaf area begin to decrease). Similarly, the greenup and senescence dates can also be estimated by the transition dates corresponding to the times when the rate change in the curvature in the NDVI data exhibit local minima or maximums. In this study, a threshold of 0.6 was set based on the analysis of NDVI rice profiles to remove unrealistic cropping patterns. Moreover, the duration from sowing to harvesting should not be longer than 120 days, given the length of rice cycle in the study region is approximately 90-100 days [18].

2.2.4. Accuracy assessment

Overall accuracy (OA) is easily interpreted as it represents the percentage of classified pixels or objects that corresponds to errors of commission and omission [14,19-24]. While Kappa index (K) can be used to assess statistical differences between classifications. The estimate of K is based on the difference between the actual agreement (the major diagonal) and the chance agreement indicated by the row and column totals. This value is computed for each error matrix and is a measure of how well the produced classification agrees with the reference data, values range from -1 to +1 [14]. The higher the value of kappa, the better the classification performance. As values in the off-diagonal increase, the value of kappa decreases.

2.2.5. GIS technique

The advantage of GIS techniques it's not only linked to exploitation of database capabilities, but also to the ability to manage different LULC maps by means of typical vector format like "intersect" and "union", in order to easily evaluate the amount of change [25].

2.2.6. Adaptive Capacity (AC) indicators

The adaptability index was selected according to four main indicators which are also the criteria for developing [26] to the questionnaire illustrated in Table 2. Each indicator was defined and calculated via a set of predictor variables to be divided into four main aspects which are further downscaled to a group level and finally to an item level (Table 4).

Table 4. The Indicator s and Predictor variables for adaptability index of agriculture land use changes.

Serial	Indicators	Predictor variables	Unit
1	Human-being (AC1)	- Individual perceptions about saltwater intrusion impacts;	Level
		- Training participatory ratio of study communities;	Level
		- Social learning towards saltwater intrusion knowledge sharing;	Level
		- Formal saltwater intrusion information accessibility;	Level
		- Areas of changing agricultural-based land use types towards saltwater intrusion adaptation;	Hectare (ha)
2	Society (AC2)	- Quantity of saltwater intrusion impacts and solution annual training;	Number
		- Agricultural production-based loan;	Amount of money
3	Infrastructure (AC3)	- Frequency of agricultural irrigation system improvement;	Level
		- Agricultural farming sufficiently irrigated;	Level
		- Quantity of water infrastructure for natural disaster prevention and protection;	Level
4	Natural condition (AC4)	- Agricultural farming areas affected by saltwater intrusion;	ha
		- Areas of changing agricultural-based land use types possibly adapting with future saltwater intrusion;	ha

2.2.7. Estimating Adaptive Criteria

The predictor variables were estimated on average by the 30 interviewed households in each predictor variable of each indicator (AC) for agricultural land use changes. Then, the adaptability was estimated by the average of each indicator (AC). Qualitative predictor variables were coded into number values ranging from 1 to 5 according to their levels before normalization. Those predictor variables were either binary or continuous variables that ranges and units were diverse and the normalization was utilized to put all variables to an optimal range towards preventing potential bias. The normalization values due to the variable components and sub-variables are used to evaluate the variable structures in each transformation of land use types. The highest values indicate either those variables presenting or adaptive capacity levels (Table 5).

Table 5. Normalization data for each component variable.

Indicators	Predictor variables	Average values: Predictor variables	Average values: Indicators
People (AC1)	- Individual perceptions about saltwater intrusion impacts (AC11),	0.814	0.53
	- Training participatory ratio of study communities (AC12),	0.0	
	- Social learning towards saltwater intrusion knowledge sharing (AC13),	0.627	
	- Formal saltwater intrusion information accessibility (AC14),	1.000	
	- Areas of changing agricultural based on land use types towards saltwater intrusion adaptation (AC15).	0.229	
Society (AC2)	- Quantity of saltwater intrusion impacts and solution annual training (AC21),	1.000	0.5
	- Agricultural production based on loan (AC22).	0.000	
Infrastructure (AC3)	- Frequency of agricultural irrigation system improvement (AC31),	1.000	0.46
	- Agricultural farming sufficiently irrigated (AC32),	0.000	
	- Quantity of water infrastructure for natural disaster prevention and protection (AC33).	0.372	
Natural condition (AC4)	- Agricultural farming areas affected by saltwater intrusion (AC41),	1.000	0.5
	- Areas of changing agricultural based on land use types possibly adapting with future saltwater intrusion (AC42).	0.0	

2.2.8. Standardization Adaptive Capacity levels

The Standardization Adaptive Capacity levels was estimated the range [0,1] to normalize predictor variables following the formula (5) according [27].

$$x_{ij} = \frac{x_{ij}(t) - Min\{x_{ij}(t)\}}{Max\{x_{ij}(t)\} - Min\{x_{ij}(t)\}} \quad (5)$$

Where: xij, yij were normalized values of variable i in the study district/commune j; xij(t) were raw values of variable i in the study district/commune j; Max and Min respectively were maximum and minimum values of each predictor variable;

Table 6. Standardization Adaptive Capacity levels on the agriculture land use changes.

Adaptive Capacity (AC) levels	Land use changes: Double rice crop - Orchards	Single rice crop - Rice-shrimp	Double rice crop - Rice shrimp	Triple rice crop - Double rice crop	Rice shrimp - Aquaculture: Thanh Phu	Rice shrimp - Aquaculture: Binh Dai
AC1	2.38	1.50	2.26	1.29	2.01	2.47
AC2	0.38	0.38	0.18	0.28	0.24	0.32
AC3	1.41	1.36	1.24	1.57	1.19	1.47
AC4	0	0.08	0.43	0	0.33	0.41
AC	**0.67**	**0.12**	**0.63**	**0**	**0.40**	**1.00**

The adaptation levels will be divided into 4 levels: low level (0<AC≤0.25), moderate level (0.25<AC≤ 0.5), high level (0.5< AC≤0.75) and the highest level (0.75<AC≤1.0) [28]. The adaptability in five types of agricultural land use changes including single-rice crop to rice-shrimp, rice-shrimp to aquaculture, double-rice crop to orchards, double-rice crop to rice-shrimp and triple-rice crop to double rice crop were assessed the adaptive capacity levels in each district location in the coastal area of Thanh Phu, Ba Tri, Binh Dai in Ben Tre province.

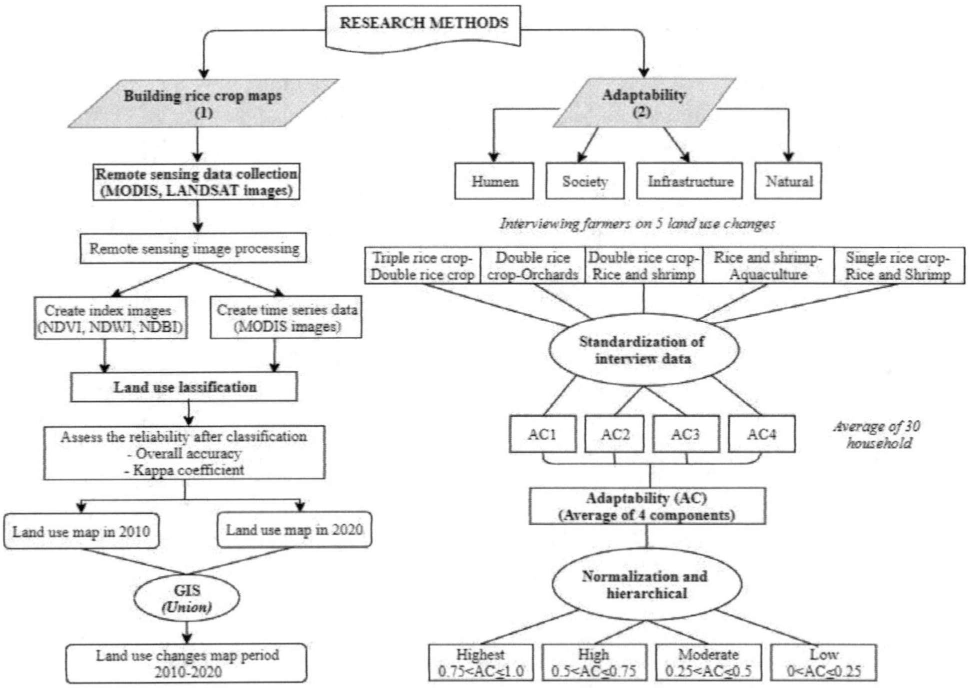

Figure 3. The flowchart of the methods.

3. Results and discussion

3.1. Accuracy assessment

To assessment the accuracy of image interpretation results conduct sampling and scoring surveys by Google Earth platform. Based on the formula for estimating the survey points of Corchan (1977) [29], the research has sampled a total of 385 sampling, the survey points were distributed in different land use types depends on the percentage of the area to be analyzed in each land use on the map (Table 7).

Table 7. Survey points to accuracy assessment after classification.

Land use types	Survey samlings	
	2010	2020
Perennial plant /Orchard	207	151
Rive	71	67
Rice crop	60	38
Aquaculture	22	61
Forest-shrimp	16	19
Urban area	5	41
Forest	4	5
Rice-shrimp	-	5
Total	**385**	**385**

The results overall accuracy of the classification results as 88.7% and the Kappa coefficient = 0.77 (in 2010); and overall accuracy 86.7% and the Kappa coefficient = 0.73 (in 2020). From this result, it is shown that the accuracy of classification results has statistical significant (no less than 75%) and the classification ability indecates a good reliability (Kappa ≥ 0.70) (Table 8).

Table 8. Accuracy assessment results after classification.

Parameters	2010	2020
Overall accuracy	88.7%	86.7%
Kappa	0.77	0.73

3.2. Land use/Land cover in 2010 and 2020 in the coastal area in Ben Tre province

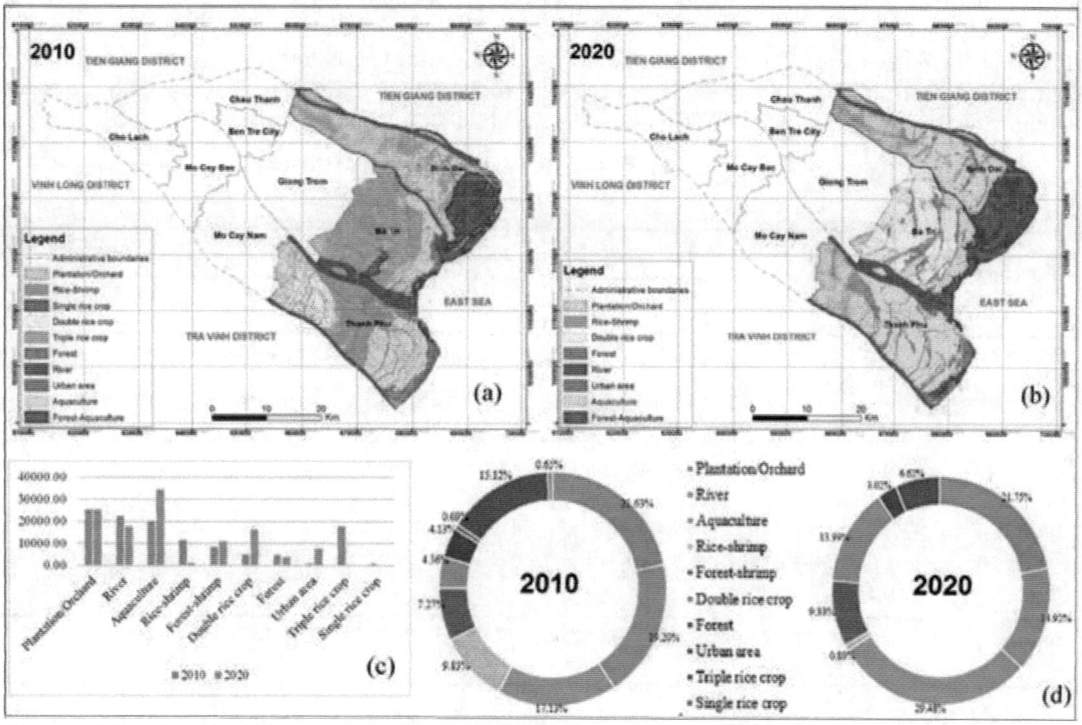

Figure 4. Land use/Land cover in 2010 and 2020 in the coastal area in Ben Tre province, (a) Land use area distribution in 2010, (b) Land use area distribution in 2020 (c) Land use area in 2010 and 2020, (d) Land use percentage in 2010 and 2020.

In 2010, there are ten land use types to be identified in the coastal area in three districts of Thanh Phu, Ba Tri and Binh Dai in Ben Tre province (Figure 4a). The perennial plants/orchards cover with the highest area at 21.63 % of total area. The next land use types of river, aquaculture and triple-rice crop are high distribution at 19.20%, 17.13% and 15.12%, respectively. The rest land use types (rice-shrimp rotation, double rice crop, forest and forest-shrimp) were distributed lower than 10% of total area. The last land use types are single rice crop and urban area with the distribution area at 0.65% and 0.69%, respectively. In 2020, there are eight land use types covering in the study area which were disappeared two land use types including single-rice crop and triple-rice crop (Figure 4b). The increasing of aquaculture, build-up area and double-rice crop were transformed by single and triple-rice crops (Figure 4c). The highest areas increasing identified on aquaculture (17.13 to 29.48%), double-rice crop (4.36 to 13.99%) and urban area (0.69 to 6.52%) and the decreasing areas on rice-shrimp rotation (9.83 to 0.89). The triple -rice crop and single-rice crop are dramatically decreasing areas at 15.12% and 0.65%, respectively (Figure 4d).

The areas of three districts of Thanh Phu, Binh Dai and Ba Tri distributed with 35.31%, 34.37% and 30.31% of total area, respectively (Tab. 9) in which Ba Tri district has the lowest than the remaining provinces. In Binh Dai district, perennial plant /orchard has the highest distribution area at 31.17% with two times higher than the others land use types. In Ba Tri district, triple rice crop has the highest area at 40.46% with the nearest 3 times higher than the others land use types distribution. Lastly, in Thanh Phu district, there are four land use types of rice-shrimp, aquaculture, river and orchards having the highest area from 23.26% to 16.2% approximately 0.5 time area distribution comparing to the others land use types (Table 9). There have no rice-shrimp and double rice crop that could be culture in Ba Tri district, single and double rice crops in Binh Dai district and single rice crop in Thanh Phu district in 2010. In 2020, double rice crop covers the highest area in Ba Tri at 46.17%, both districts of Binh Dai and Thanh Phu have the highest are of aquaculture at 27.83% and 42.48%, respectively. Next to perennial plant/orchards is distribution a similar area in both Binh Dai and Thanh Phu at 25.16% and 26.35%, respectively. There have no double rice crop distribution in Binh Dai and Thanh Phu districts and rice-shirmp in Ba Tri and Binh Dai districts. The lowest of land use type is forest in Ba Tri and Binh Dai districts with only at 1.73% and 2.23%, respectively; forest-shrimp culture system is a land use types with the lowest area to distribute in Thanh Phu district at 2.07% (Table 9).

Table 9. Land use area distribution in the coastal area in 2010 and 2020 in Ben Tre province.

Unit: ha

Land use types	2010			2020			Total		Percentage (%)	
	Ba Tri	Binh Dai	Thanh Phu	Ba Tri	Binh Dai	Thanh Phu	2010	2020	2010	2020
Perennial plant	6,047.1	12,476.2	6,664.7	4,420.8	10,072.8	10,837.1	25,188.0	25,330.7	21.6	21.8
Double rice crop	-	-	5,076.9	16,296.9	-	-	5,076.9	16,296.9	4.4	14.0
Forest	987.0	817.0	3,008.9	611.3	891.9	2,016.9	4,812.9	3,520.1	4.1	3.0
River	6,220.8	8,183.2	7,952.0	3,741.6	7,217.2	6,420.0	22,355.9	17,378.8	19.2	14.9
Urban area	428.4	213.2	157.4	3,013.1	2,195.0	2,501.0	798.9	7,709.1	0.7	6.6
Aquaculture	4,192.6	6,192.2	9,568.7	5,715.3	11,140.2	17,472.3	19,953.5	34,327.9	17.1	29.5
Rice-shrimp	-	3,252.1	8,196.9	-	-	1,032.1	11,449.0	1,032.1	9.8	0.9
Forest-shrimp	474.4	7,673.4	315.1	1,499.5	8,512.4	852.1	8,462.9	10,864.0	7.3	9.3
Total	**35,298.60**	**40,029.43**	**41,131.48**	**35,298.6**	**40,029.4**	**41,131.5**	**116,459.5**	**116,459.5**	100.0	100.0
%	30.31	34.37	35.31	30.31	34.37	35.31	**100**	**100**		

A disappearance of land use types in 2020 comparison in 2010 were identified in single rice crop and triple rice crop that farmer could not culture those cultivation system because of the impact of saline water intrusion, people are not proactive about water sources, causing damage to crops especially for rice that is not adapted to current conditions. From there, changing the farming model suitable for saline water intrusion conditions is shown in the change of land use status.

3.3. Land use changes in the coastal area in Ben Tre province

From 2010 to 2020, there are five types of agriculture systems transformation in the coastal area of Ben Tre province. In Thanh Phu district, there are three land use types transformation including shrimp-rice crop rotation to aquaculture, double-rice crop to perennial crops/orchards and double-rice crop to shrimp-rice crop rotation. In Ba Tri district, there are two types of agriculture transformation which is single-rice crop to aquaculture and triple-rice crop to double-rice crop. In Binh Dai district, it has only one farming model transformation from shrimp-rice crop rotation into aquaculture (Figure 5).

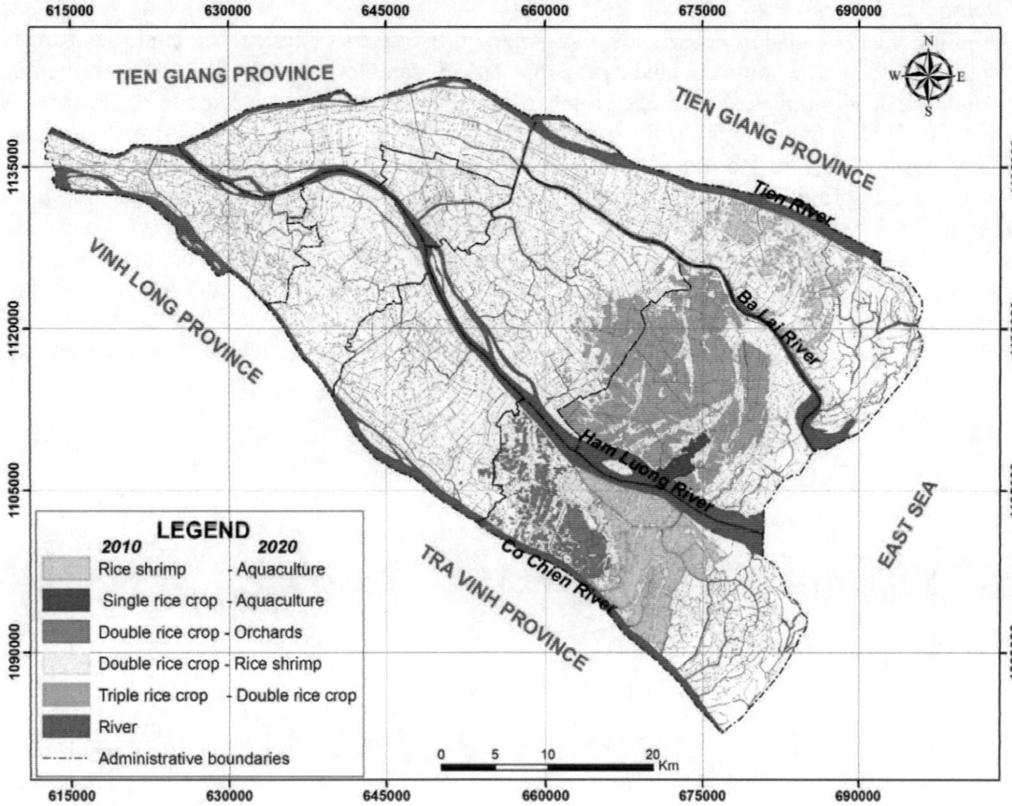

Figure 5. Land use changes map in the coastal area in Ben Tre province in the period of 2010 to 2020.

The five types of agriculture land use has a total area of 30,315 ha in the period from 2010 to 2020 in which Thanh Phu district has three agriculture types changes on shrimp-rice crop rotation to aquaculture which has an area of 7,646 ha (15.13%); double-rice crop to perennial plants/orchards which has an area of 3,971 ha (13.10%), and double-rice crop to shrimp rice crop which has an area of 705 ha (2.33%). In Ba Tri district, there are two types of land use change consists of triple-rice crop to double-rice crop with an area of 14,588 ha (48.12%), single-rice crop to shrimp-rice crop of 658 ha (2.17%), and the rest changing from shrimp-rice crop to aquaculture in Binh Dai district with an area of 2,744 ha (9.05%). The largest change of agriculture land use was detected on triple-rice crop to double-rice crop and the lowest changes were focused on single-rice crop to shrimp-rice crop in this area (Table 10).

Table 10. Area of land use changes in the coast area in Ben Tre province.

No.	Land use changes		Districts	Area (ha)	Percentage (%)
	From	To			
1	Triple-rice crop	Double-rice crop	Ba Tri	14,588	48.12
2	Rice-shrimp	Aquaculture	Thanh phu	7,646	15.13
3	Double-rice crop	Orchards	Thanh Phu	3,972	13.10
4	Rice-shrimp	Aquaculture	Binh Dai	2,744	9.05
5	Double-rice crop	Rice-shrimp	Thanh Phu	705	2.33
6	Single rice crop	Rice-shrimp	Ba Tri	658	2.17
		Total		**30,315**	**100**

3.4. Household interviews

The sampling interviewed 178 households distributed on five types of agriculture land use changes in which the triple-rice crop to double-rice crop at 35 samples, double-rice crop to perennial plants/orchards at 34 samples, shrimp-rice crop to aquaculture at 30 samples. In addition, there are some areas with small scale transformation of agricultural land use with spontaneous transformation scattered distributed in single-rice crop to aquaculture at 26 samples, double-rice crop to shrimp-rice crop rotation at 26 samples and shrimp-rice crop to aquaculture at 27 samples (Figure 6). The adaptability indicators were selected the developing criteria by the questionnaire illustrated according to four main criteria in Table 2 and the indicators and predictor variables in Table 4. For computing the strength of the adaptive capacity, the score must be given according to its properties based on the agriculture land unit is used as an analysis unit which integrated all input data involved indices. Each indicator has specific measurements, causing difficulties in integration and computation. The sequential method is applied for measuring the score by using a relative level for each indicator, denoting that the score of each item represents the relative level of adaptability.

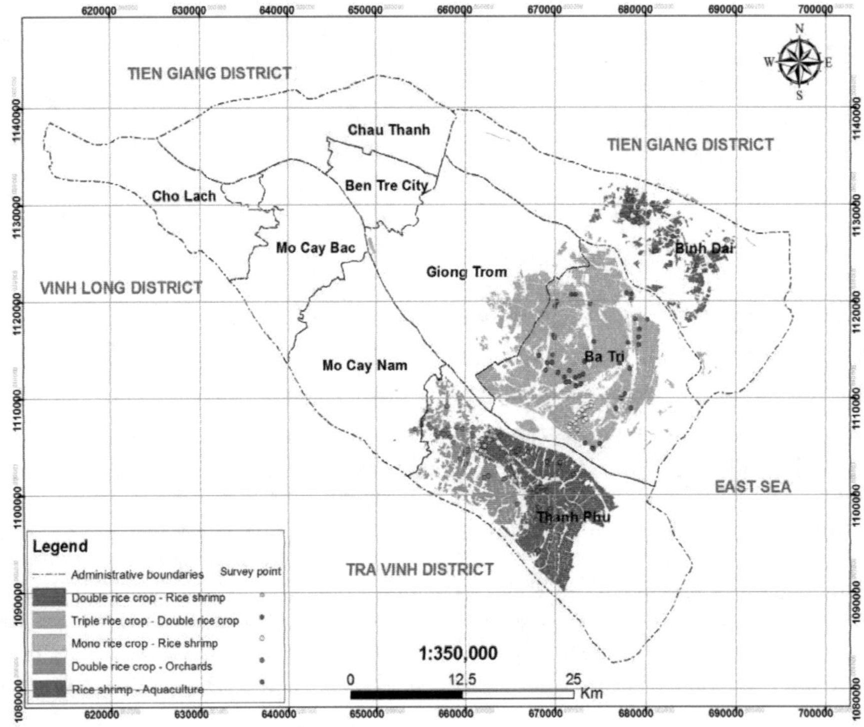

Figure 6. Location of household survey samplings in the coastal area of Ben Tre province.

3.5. Adaptive Capacity Assessment for the transformation of land use area

3.5.1. The Normalization adaptive capacity variables

There are four adaptable component parameters selecting for the changing of land use types: people, infrastructure, society and natural condition in which people are the highest variable effects on the land use changes at the average value of 0.53; next to the natural condition and society at 0.5 and the lowest variables effect of infrastructure at 0.46. It shows that people are one of four the most important adapting factors due to saline water intrusion in coastal area (Figure 7). The people indicator (AC1) shows the highest adaptability identifies on predictor variables of saltwater intrusion information accessibility (AC14) with the maximum value at 1.0. Moreover, the social variable (AC2) found that the predictor variable of quantity of saltwater intrusion impacts and solution annual training (AC21), the natural variable (AC4) with agricultural farming areas affected by saltwater intrusion (AC41) contribute the highest of adaptability at 1.0. In contract, the remaining of data set shows low adaptability contribution because of variable values nearly 0.0.

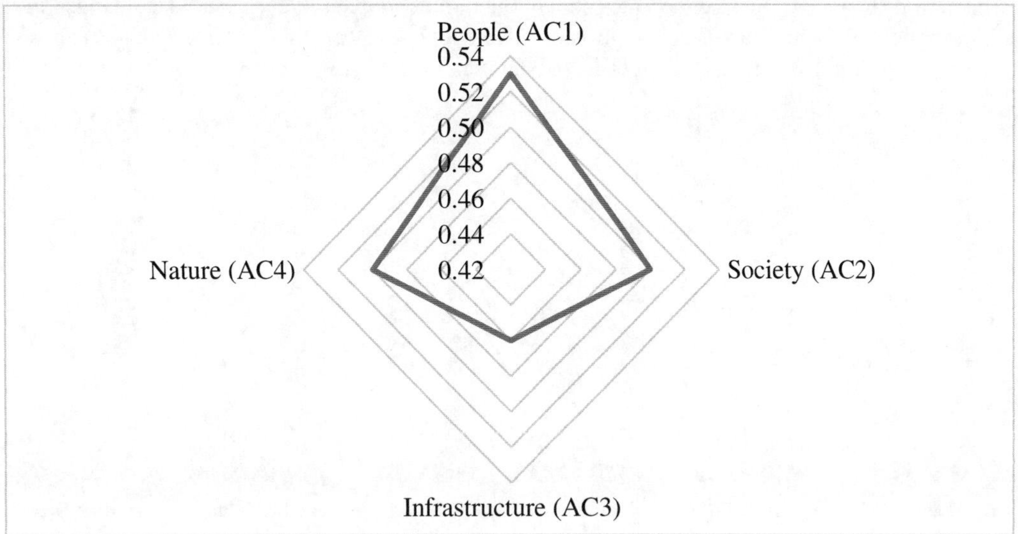

Figure 7. Normalization data for each component indicator.

In Figure 8 identifies the adaptability parameters on each land use type transformation. Society factors highly adapted due to the transformation of single-rice crop to aquaculture and triple-rice crop to double-rice crop in Ba Tri; double-rice crop to perennial plants/orchards in Thanh Phu. Natural factors greatly contributed on the double-rice crop to shrimp-rice crop in Thanh Phu. Also, infrastructure factors have a high influence on the transformation of shrimp-rice crop to aquaculture in Thanh Phu area. The lastly, people indicator have a mainly adaptable on the transformation on shrimp-rice crop to aquaculture in Binh Dai district.

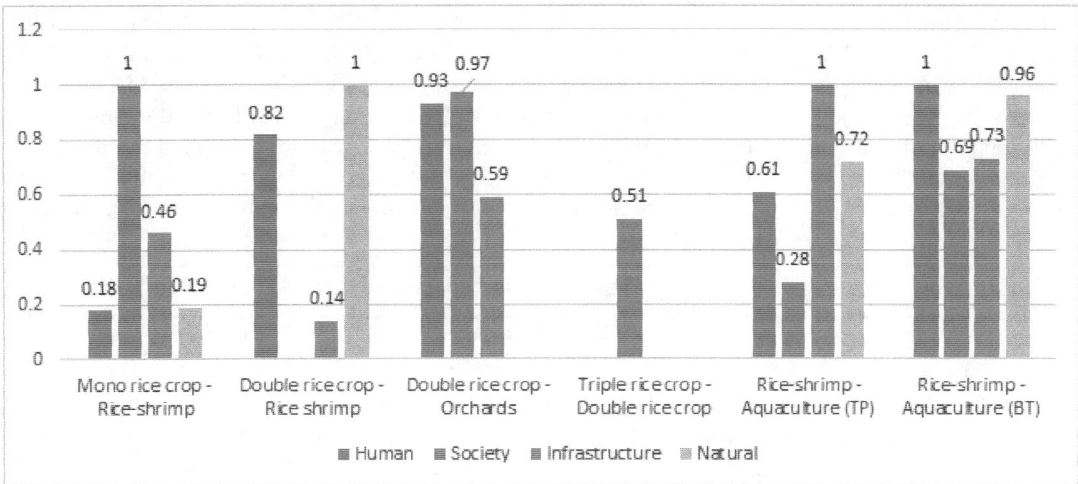

Figure 8. The impact levels in each indicator for whole the land use changes.

3.5.2. Standardization Adaptive Capacity (AC) levels

The adaptability of the agriculture with the transformation from shrimp-rice crop to aquaculture in Binh Dai district has a higher adaptability than other land use transformation, while the converting of triple-rice crop to double-rice crop has lowest adaptability compared with the remaining agriculture land use transformation (Figure 9). People indicator shows the highest impact in each land use changes in three coastal provinces, infrastructure is next to impact and the lowest effects on Adaptive Capacity are society and natural condition for land use changes in coastal area in Ben Tre province.

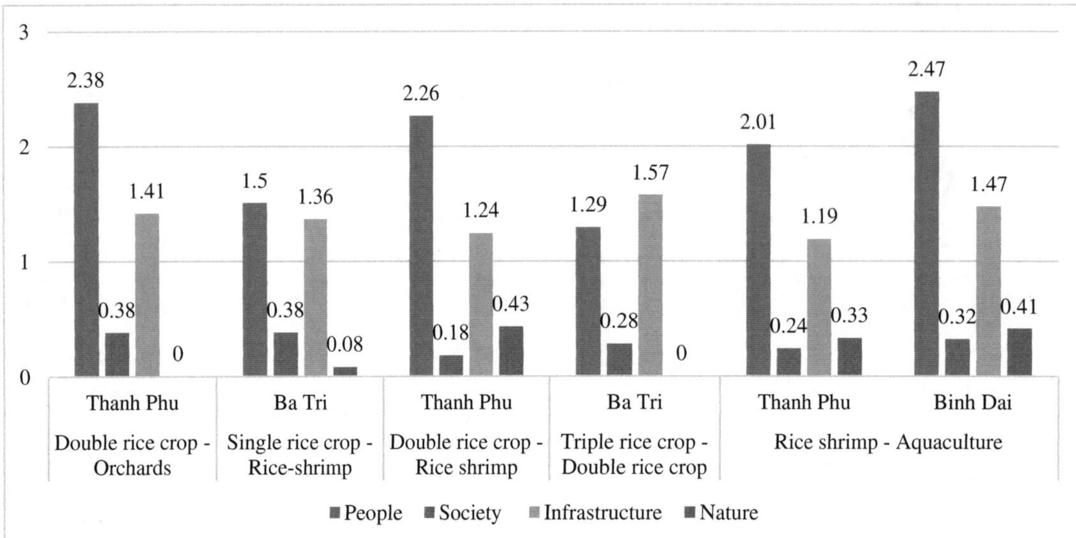

Figure 9. Average Adaptive Capacity (AC) levels for the land use changes.

3.5.3. The adaptability for the changing of land use transformation

The indicator values were converted into a ratio ranges from 0 to 1 and divided into 4 group levels: low, medium, high and very high for adaptability areas. From the collected data and analysis results, it shows that the adaptability of the land use transformation from shrimp-rice crop rotation to aquaculture in Binh Dai area is the highest adaptability to saline water intrusion because the aquaculture cultivation (the industrial aquaculture) is suitable for irrigation and ecological conditions in the saline water intrusion conditions. Moreover, there are two types of land use transformation which have high adaptability to saline water intrusion identifying on double-rice crop to perennial plants/orchards and double-rice crop to shrimp-rice crop in Thanh Phu. Also, the land use transformation from shrimp-rice crop rotation to aquaculture in Thanh Phu district has medium adaptability to saltwater ecosystem. Finally, the triple-rice crop to double-rice crop transformation is a culture system with low adaptability to coastal area due to rice is a poor salt-tolerant crop (Table 11).

Table 11. The value of adaptability index to salinity intrusion in the coastal area in Ben Tre province.

STT	Land use changes	Ecological zone	District	AC (0-1)	Level
1	Rice shrimp - Aquaculture	Saline	Binh Dai	1.00	Strongly high
2	Double rice crop - Orchards	Fresh brackish	Thanh Phu	0.67	High
3	Double rice crop - Rice shrimp	Saline brackish	Thanh Phu	0.63	High
4	Rice shrimp - Aquaculture	Saline	Thanh Phu	0.40	Moderate
5	Mono rice crop - Aquaculture	Saline brackish	Ba Tri	0.12	Low
6	Triple rice crop - Double rice crop	Fresh brackish	Ba Tri	0.00	Low

The adaptable levels for each agriculture land use transformation as well as each district are not similar due to different location. Binh Dai is a district with very high adaptability; high and medium adaptability in Thanh Phu district and the lowest adaptation level is in Ba Tri district (Figure 10).

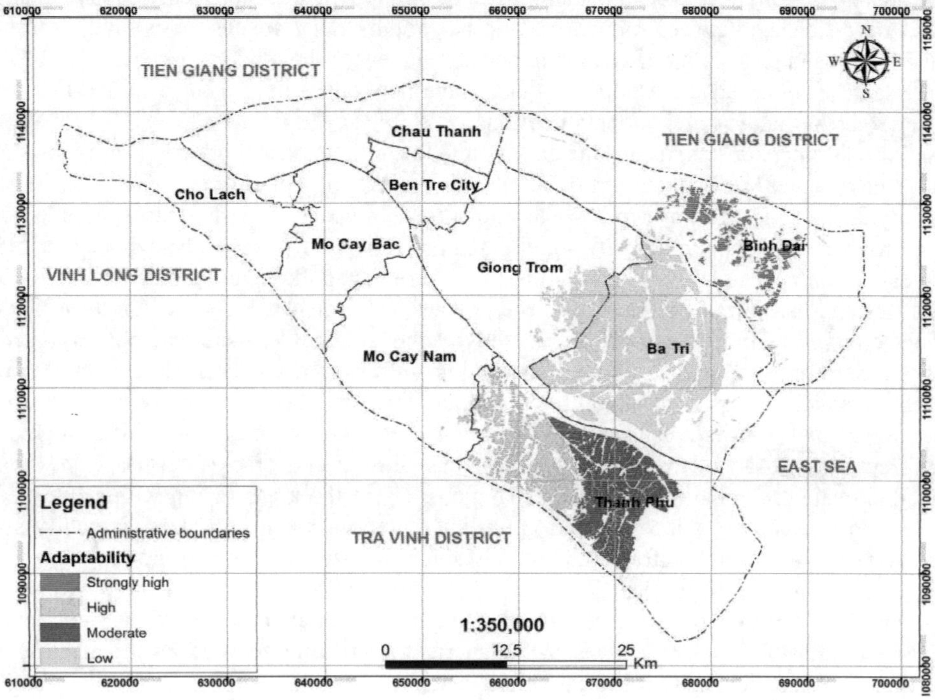

Figure 10. Adaptation map on transition status in the area coastal in Ben Tre province.

4. Discussion

The adaptation options showed the expectation of agriculture farming households which provided useful information for policy makers. Based on the research findings, there were many factors affecting household's decisions responding to different levels of adaptive capacity to saline intrusion. In order to develop adaptation policies and strategy options, agriculture farming households should be involved in balance of economic, social and environmental aspects, especially household's adaptive capacity and government policy supports. The adaptation options at regional/national level should consider regional links and transfer of adaptation experience of saline intrusion between each land use changes models in study area to supervise the implementations of saline intrusion prevention and mitigation in the context of future saline intrusion. According to the results of the adaptive capacity assessment of agricultural land, the entire study region can be selected three land use change types including Rice shrimp – Aquaculture in Binh Dai; Double rice crop – orchards and double rice crop - rice shrimp rotation in Thanh Phu that are low effect in coastal saline intrusion of the adaptation aspect.

In terms of the adaptive capacity assessment framework, this study involved many approaches to moderate data and integrate computation results. For instance, the assessment matrices were employed to establish rules for combining consequences from different sectors including four indicators of people, society, infrastructure and natural conditions. These procedures were limited by the data format, uncertainties in the relationships between criteria, and a lack of information; however, despite these limitations, through the conceptual integration mechanism, the matrices could still effectively simulate the relative locations of adaptive capacity agricultural land, and subsequently develop follow-up strategies.

Agricultural land is one of the essential elements of agriculture. Emphasizing the multiple values of the nature of agricultural land will help to enhance the possibility and application of policy inputs. The main purpose of the application of adaptive capacity assessment is not to explicitly describe the status of agricultural land to climate change, but to help local government and farmers to identify the critical area and to discussing the appropriated adaptive policies and agriculture future. The consequence of adaptive capacity is a relative score and it simplifies the complex values for local people. Local people can feel a real difference between individuals and neighbours. That creates the possibility to connect local government and local farmer and to generate consequences. In addition, by the characteristics of agriculture land location, adaptive policies would have new opportunities to integrate, to efficiently benefit, and to create a new development direction.

Agriculture farming households had the opportunity to continue their production activities under the interest of government investment into the planning stage of agriculture production, especially irrigation, dyke and sluice gate systems which mitigated the impacts of saline intrusion on production activities. Provinces had supported many policies and projections in order to deal with saline intrusion. Nevertheless, according to surveyed households, the government policies and projections had two-sided effects on rice and fish farming households. The policies and projections focused on the efficiency of production activities, including stability of productivity and sustainable livelihood, but less on adaptive capacity. Leaders of provinces confirmed that saline intrusion adaptation must ensure provincial development goals focusing on the strength and adaptability. The adaptation options recommended by provinces were as (1) in the areas facing with severe freshwater shortage, provinces should adjust production activities to adapt to saltwater ecosystems. Shrimp and brackish fish species should be encouraged as main cropping systems in these regions; (2) in the areas facing with salinity-effected freshwater in dry season, provinces should apply integrated production models or modify seasonal calendar to one rice or fish crop in freshwater months-one brackish fish or shrimp crop in saline intrusion duration; (3) in the areas where freshwater was slightly affected by saline intrusion, provinces should encourage households to continue their production activities normally. However, the improvement of farming practices and the development of salt-tolerant rice varieties and fish species should be developed as adaptation options under future saline intrusion impacts.

5. Conclusions

The study identified five agricultural land use transformation in the coastal district of Ben Tre province including shrimp-rice crop rotation to aquaculture in Thanh Phu and Binh Dai, single-rice crop to aquaculture in Ba Tri, double-rice crop to orchards in Ba Tri, double rice crop to rice-shrimp in Thanh Phu and triple- rice crop to double-rice crop in Ba Tri. A set of indicators of Adaptive Capacity (AC) was created due to the effect of salinity intrusion on the land use changes consists of four main variables of people, society, infrastructure, nature conditions and twelve sub-variable components. The results show that the assessment of adaptability levels in saltwater ecosystem on agricultural in the coastal area of Ben Tre province in the transformation on rice-shrimp crop rotation to aquaculture in Binh Dai district has the highest level adaptation and the lowest adaptation on triple-rice crop to double-rice crop transformation. Assessing adaptability is important for defining the risks posed and provides information for identifying measures to adapt to climate change especially in saline intrusion impacts. It enables practitioners and decision-makers to identify the most adaptable areas, sectors and social groups. In turn, this means climate change adaptation options targeted at specified contexts can be developed and implemented for agriculture land use changes in saltwater ecosystem area.

Acknowledgements

The authors thank the survey participants from three districts of Ba Tri, Binh Dai and Thanh Phu provinces, Ben Tre province. We are also in great debt to the Department of Science and Technology and Department of Environment and Natural Resources in Ben Tre province which have been very helpful in providing valuable materials and supporting us during the survey. We highly appreciate a practical project funded in part by the Technological Cooperation Project “Building capacity for Can Tho University to be an excellent institution of education, scientific research and technology transfer” of JICA. Our gratitude extends to the Can Tho University Improvement Project VN14-P6, which was funded by a Japanese ODA loan. Lastly, but of equal importance, we owe an outstanding debt to the project management board and the colleagues at the College of Environment and Natural Resources to support us during this project accomplishment.

References

[1] MRC 2010 *State of the Basin Report* (Vientiane, Laos: Mekong River Commission) p 166
[2] Tran Ngoc 2017 Remove the infrastructure "knots" for the Mekong Delta economy to take off Accessed 12/07/2021 Address https://vov.vn/kinh-te/go-nut-that-ha-tang-de-kinh-te-dbscl-cat-canh-634541.vov
[3] Le Sam 2006 *Salinity Intrusion in The Mekong Delta* (Ho Chi Minh: Agricultural)
[4] Nguyen Thanh Binh 2010 *Evaluation of Vulnerability due to Salinity Intrusion in The Mekong Delta* (Vietnam: Can Tho University)
[5] Tran Thi Phuong Dung 2013 *Application of Modis Remote Sensing Images to Zoning Salinity Intrusion in Ben Tre Province in 2012* (Vietnam: Ho Chi Minh City University of Agriculture and Forestry)
[6] Luong Thi My Hang 2013 *Survey on Some Characteristics Related to Soil Salinity in the Freshwater and Brackish Ecological Zone in Ba Tri District, Ben Tre Province* (Vietnam: Can Tho University)
[7] Vermote E, Kotchenova S and Ray J 2011 *MODIS Surface Reflectance User's Guide* (NASA: MODIS Land Surface Reflectance Science Computing Facility) version 1 p 35
[8] Hutchinson T P 1993 *A Review of "Essentials of Statistical Methods"* (Sydney: Rumsby Scientific) p 41
[9] Nguyen Cao Van, Nguyen Le Tuan, Nguyen Thuc Anh and Pham Van Hieu 2020 Assessment of vulnerability and adaptability due to saltwater intrusion for agricultural production in the context of climate change in the coastal estuary area of Nam Dinh province *Journal of Hydrometeorology* **716** 63–78
[10] FLAASH-Module 2009 *Atmospheric Correction Module: QUAC and FLAASH User's Guide* (Boulder, CO: ITT Visual Information Solutions) Version 4.7 p 41
[11] Khan M R, de Bie C A J M, van Keulen H, Smaling E M A and Real R 2010 Disaggregating and mapping crop statistics using hypertemporal *Remote sensing Int. J. Appl. Earth Obs. Geoinf* **12** 36-46
[12] McFeeters S K 1996 The use of normalized difference water index (NDWI) in the delineation of open water features *International Journal of Remote Sensing* **17** 1425-1432
[13] Zha Y, Gao J and Ni S 2005 Use of normalized difference built-up index in automatically mapping urban areas from TM imagery *International Journal of Remote Sensing* **24** 583-594
[14] Kane J, Kemp J, and Niekerk A V 2016 Effect of pan-sharpening multi-temporal Landsat 8 imagery for crop type differentiation using different classification techniques *Computers and Electronics in Agriculture* **134** 151-159
[15] Yuan F and Bauer M E 2006 *Mapping Impervious Surface Area Using High Resolution Imagery: A Comparison of Object-Based and Per Pixel Classification* (Reno, Nevada: American Society For Photogrammetry And Remote Sensing Annual Conference Proceedings)
[16] Huete A, Didan K, Miura T, Rodriguez E P, Gao X and Ferreira L G 2002 Overview of the radiometric and biophysical performance of the MODIS vegetation indices *Remote Sensing of Environment* **83** 195-213
[17] Lhermitte S, Verbesselt J, Verstraeten W W and Coppin P A 2011 Comparison of time series similarity measures for classification and change detection of ecosystem dynamics *Remote sensing of environment* **115** 129-3152
[18] Bolstad P V and Lillesand T M 1991 Rapid maximum likelihood classification *Photogrammetric Engineering & Remote Sensing* **57** 67- 74
[19] Land Development Department 2016 Sukhothai province land use data 2009 and Sukhothai's land use map 2016 Accessed 12/07/2021 Address http://www.ldd.go.th/

IOP Conf. Series: Earth and Environmental Science **964** (2022) 012019 doi:10.1088/1755-1315/964/1/012019

[20] Lewinski S, Bochenek Z 2009 *Rule-Based Classification of SPOT Imagery Using Object-Oriented Approach for Detailed Land Cover Mapping* (Istanbul, Turkey: Proceedings of the 28th EARSeL Symposium Remote Sensing for a Changing Europe)

[21] Murai H and Omatu S 1997 Remote Sensing Image Analysis using a neural network and knowledge-based processing *International Journal of Remote Sensing* **18** 811-828

[22] Novotny T E, Bialous S A, Burt L, Curtis C, da Costa V L, Iqtidar S U and Tursan d'Espaignet E 2015 The environmental and health impacts of tobacco agriculture *Cigarette manufacture and consumption Bulletin of the World Health Organization* **93** 877–80

[23] Office of Natural Resourse and Enviroment Policy and Planning Thailand 2014 A Summary of Action Plan for the Problem of Tobacco Farmers in Sukhothai Province

[24] Radoux J and Bogaert P 2017 Good practices for object-based accuracy assessment *Remote Sensing* **9** 646

[25] Petit C, Scudder T and Lambin E 2001 Quantifying Processes of Land-Cover Change by Remote Sensing: Resettlement and Rapid Land-Cover Changes in South-Eastern Zambia *International Journal of Remote Sensing* **22** 3435-3456

[26] IPCC 2007 *Climate change 2007: Impact, Adaptation and Vulnerability* (New York: Cambridge University Press)

[27] IPCC 2001 *Climate change 2001: The scientific basis* (New York: Cambridge University Press)

[28] Nguyen Ngoc Truc, Truong Van Thinh, Nguyen Van Thuong and Nguyen Thao Ly 2017 Current status and vulnerability to salinization in the context of climate change in Da Nang city *VNU Science Journal: Earth and Environmental Sciences* **33** 90-107

[29] Cochran W G 1977 *Sampling Techniques* (New York: John Wiley & Sons)

IOP Conf. Series: Earth and Environmental Science **964** (2022) 012020 doi:10.1088/1755-1315/964/1/012020

Effects of sediment deposit on the hydraulic performance of the urban stormwater drainage system

C N X Quang [1, 3], N N H Giang [1, 3], H V Hoa [1, 3], P Q Hung [2, 3]

[1] Department of Hydrology and Water Resources, Institute for Environment and Resources, Vietnam National University Ho Chi Minh City, Nguyen Du Street, VNU-HCM Campus, Di An City, Binh Duong Province, Vietnam
[2] Department of Water Resources Engineering & Management, Faculty of Civil Engineering, University of Technology, Vietnam National University Ho Chi Minh City, 268 Ly Thuong Kiet Street, Ward 14, District 10, Ho Chi Minh City, Vietnam
[3] Vietnam National University Ho Chi Minh City, Linh Trung Ward, Thu Duc District, Ho Chi Minh City 700000, Vietnam

E-mail: cnxquang@hcmier.edu.vn

Abstract. Sediment deposit is a critical factor strongly affecting the drainage capacity of the conduits due to its cross-section area narrowing and roughness increasing. In this study, a numerical model was applied to investigate the influences of sediment deposit on the hydraulic performance of the drainage conduits. The Nhieu Loc - Thi Nghe (NL-TN) basin, located in Ho Chi Minh City, Vietnam, was selected as a case study. The drainage network of the NL-TN basin was simulated by using the EPA-SWMM model. The effects of sediment deposit were assessed by a non-dimensional comparison of the simulated peak flows of the sediment-deposited conduits and the clean ones. The results indicated that the sediment deposit significantly affects the flow capacity of the conduits and could cause severe inundation. Narrowing of the cross-section area has more impact on the hydraulic performance of the conduit than the increase in the roughness. A 40% increase in the Manning coefficient could decline the peak flow to approximately 80-90%, while a 40% increase in sediment thickness depth could degrade the peak flow by about 60-70% to compare with the peak flow of the clean conduit. The findings could support decision makings on the operation and maintenance of the sewer system and adaptation to extreme rainfall events.

1. Introduction

Sediment deposit is one of the most challenging issues in the operation and maintenance of the urban drainage system. Sediment thickness is the major cause of narrowing the cross-section and increasing the Manning coefficient (due to increasing the resistance) of the conduits that lead to disrupting the flow. Problems that arise include blockage, early surcharge, environmental pollution, and costly removal.

Many previous studies have considered the sediment transport process and its effects on the hydraulic performance of the drainage system. Impact of sediment deposit in the combined sewer and indicated many severe effects of sediment on urban drainage infrastructure was studied by Ashley et al. [1]. The impact of sediment deposit on the hydraulic performance of a pipe in the laboratory scale was evaluated by R. Banasiak et al. [2]. It concluded that sediment deposit strongly impacts the hydraulic capacity of the pipe. Recently, the sedimentation transport in urban drainage conduits by using a mathematical model for two-phase fluid was predicted by Y. H. Song et al. [3]. This research was also conducted in a laboratory scale. Due to the lack of data available, the evaluation of the effects of sediment deposit on hydraulic performance for a specific urban drainage system by using a numerical model has not been conducted yet.

In this study, a numerical model-based virtual experiment was applied to assess the hydraulic performance of the urban drainage system with multi sediment deposit scenarios. The research was applied for the NL-TN basin, located in Ho Chi Minh City, Vietnam, as a case study. The EPA Storm Water Management Model (EPA-SWMM) was used to simulate the drainage system of the NL-TN basin. A non-dimensional comparison between the peak flow of the drainage pipes with and without sediment deposit was carried out to quantify the effects of sediment deposit.

2. Study area

The NL-TN basin located in the central part of Ho Chi Minh City (HCMC) covers approximately 33.93 km^2 and includes portions of seven districts of HCMC including District 1, 3, 10, and Phu Nhuan, Tan Binh, Go Vap District. The population of this catchment in 2018 is about 2.356 million people. The topography of the basin is variable, with the north and northwest sections being up to 8.0 m above sea level, while the southern part of the basin averages only 1.3 m above sea level (**Figure 1**). The elevation-area classification is summarized in Table 1.

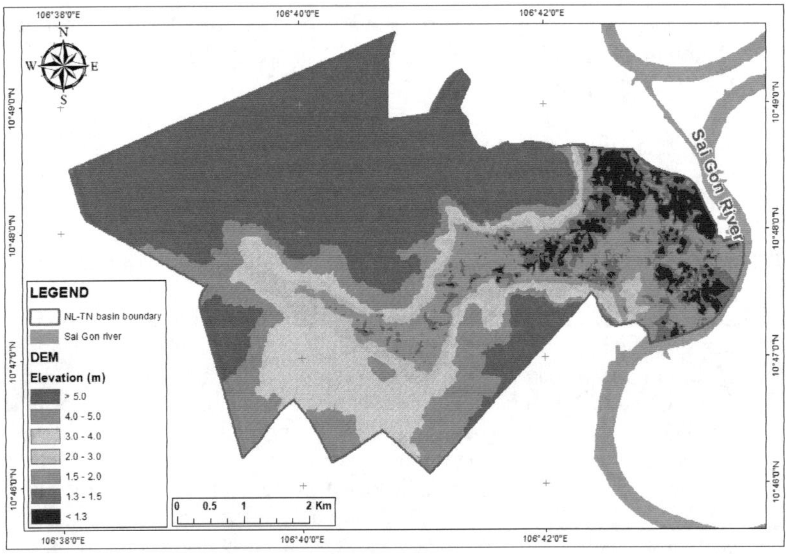

Figure 1. Boundary and DEM of Nhieu Loc – Thi Nghe basin

Table 1: Elevation – area classification in the NL-TN basin

No.	Elevation (m)	Area (ha)	Ratio (%)
1	>5.0	1651.98	48.69
2	4.0 - 5.0	379.31	11.18
3	3.0 - 4.0	504.03	14.86
4	2.0 - 3.0	245.23	7.23
5	1.5 - 2.0	311.33	9.18
6	1.3 - 1.5	215.04	6.34
7	<1.3	86.00	2.53
Total		3392.92	100

Drainage infrastructure in the basin has been fully upgraded under the HCMC Environmental Sanitation (NL-TN basin) project funded by the World Bank and finished in June 2012. However, flooding has recently occurred in some local areas. The flooding's reasons may be caused by the occurrence of rains exceeding the design frequency and the increased impervious surface area beyond the allowable limit. It could be also caused by sediment deposits due to poor maintenance.

3. Data and Methods

3.1. Data description

Data set requires for the rainfall-runoff process simulation including land-use/landcover (LULC), topography, soil properties, and design rainfall hyetograph. LULC data was achieved from Sentinel satellite images in 2020. Topography data was extracted from the Lidar dataset in 2012, and soil properties were obtained from the soil map of HCMC in 2006. The 5-year design rainfall hyetograph at Tan Son Hoa meteorological station was used. To simulate the flow process in the conduit network, the data of the drainage network (conduit size, drainage profile, and drainage network map) were collected from the HCMC Environmental Sanitation (NL-TN basin) project.

3.2. Methods

EPA-SWMM (Environment Protection Agency – Stormwater Management Model) is a comprehensive and widely accepted model for stormwater simulation, which was first developed in 1971 and extended to Ver 5.14 recently [4]. EPA-SWMM conceptualizes a subcatchment as a rectangular surface that has a uniform slope S and a width W that drains to a single outlet channel. Overland flow is generated by modeling the subcatchment as a nonlinear reservoir, as sketched in **Figure 2**.

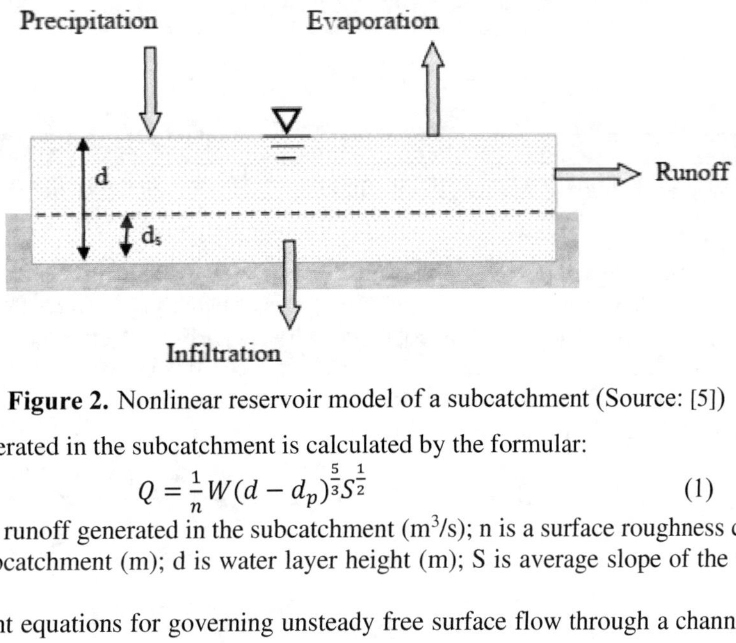

Figure 2. Nonlinear reservoir model of a subcatchment (Source: [5])

The runoff generated in the subcatchment is calculated by the formular:

$$Q = \frac{1}{n} W (d - d_p)^{\frac{5}{3}} S^{\frac{1}{2}} \quad (1)$$

Where: Q is a runoff generated in the subcatchment (m^3/s); n is a surface roughness coefficient; W is width of the subcatchment (m); d is water layer height (m); S is average slope of the subcatchment (m/m).

The St. Venant equations for governing unsteady free surface flow through a channel or pipe can be expressed as [6]:

$$\frac{\partial A}{\partial t} + \frac{\partial Q}{\partial x} = q \quad (2)$$

$$\frac{\partial Q}{\partial t} + \frac{\partial (Q^2/A)}{\partial x} + gA\frac{\partial H}{\partial x} + gAS_f + gAh_L = 0 \quad (3)$$

Where x is distance (m); t is time (sec); A is flow cross-sectional area (m^2); and Q is flow rate

(m^3/s); H is hydraulic head of water in the conduit (Z + Y) (m); Z is conduit invert elevation (m); Y is conduit water depth (m); Sf is friction slope (head loss per unit length); and g is acceleration of gravity (m/sec^2)

The necessary input data were collected and processed to setup the EPA-SWMM to simulate the urban stormwater drainage network in the NL-TN basin. There is total 228 sub-catchments, 333 conduits and 228 junctions were included in the simulation model (**Figure** 3).

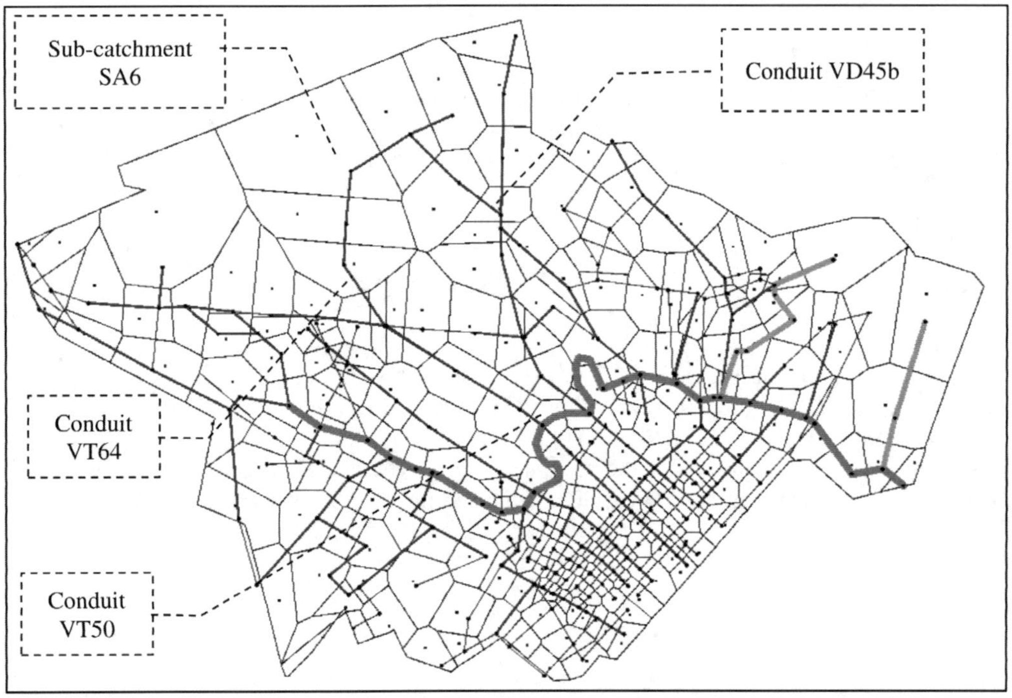

Figure 3. Simulation scheme of urban stormwater drainage network in NL-TN basin

3.3. Development of scenarios

The effects of sediment deposit on the flow capacity were investigated for circular and rectangular cross-section conduits, which are two typical cross-section types of the urban drainage system in the NL-TN basin (**Figure 4**). Three scenarios were developed including (1) Sediment deposit thickness (2) Manning coefficient; (3) Sediment deposit thickness and Manning coefficient. It assumes that the sediment deposit thickness and Manning coefficient of the filled conduit increasing 10, 20, 25, 30, 35, 40, 45, and 50% to compare with the clean one.

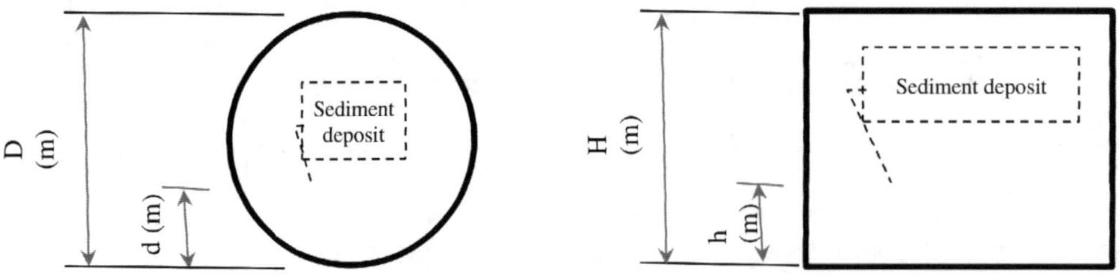

Figure 4. Demonstration of sediment deposit in two typical cross sections of conduits in NL-TN basin

4. Results and Discussion

To overcome the challenge of observed water level and discharge data unavailable, the numerical model was calibrated by expert experiences based on the shape and lag time of runoff hydrograph shape, and runoff coefficient. This approach may obtain an unreliable calibrated model, but this uncertainty could be eliminated by using a non-dimensional comparison approach to evaluate the scenarios based hydraulic performance of the drainage network.

The rainfall - runoff relation of 5-year design rainfall at the outlet of sub-catchment SA46 and conduit VT64 (see **Figure 3.** for the position) presented in **Figure 5**. **Figure 5** showed that the shape of runoff hydrographs seems to be homogenous with the simulated runoff hydrographs from other similar catchments. The lag time and runoff coefficient are 45 minutes and 0.85 respectively that is approximate to the values obtained from the experiment formulas. Thus, the model could be good enough for scenarios analysis in case of observed data for calibrating the model unavailable.

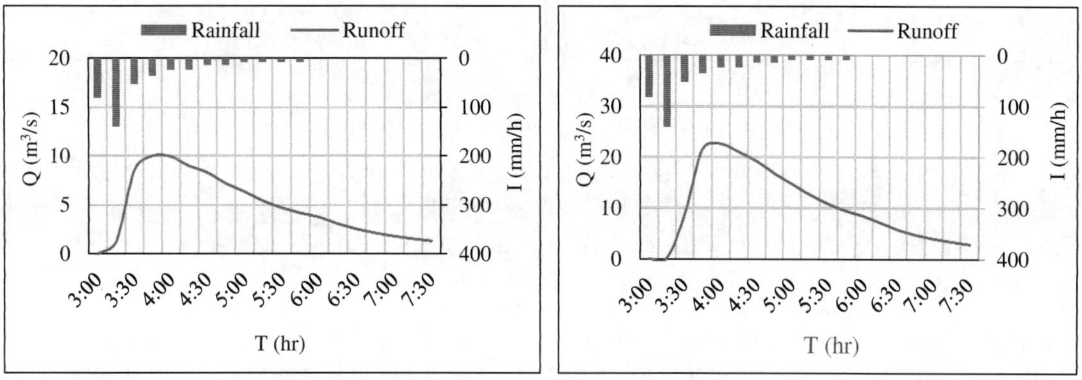

Figure 5. Rainfall - runoff relationship at sub-catchment SA46 outlet and conduit VT64

Sediment deposit causes narrowing of the cross-sectional area and increasing the roughness (Manning coefficient) of the conduits that degrade the hydraulic performance of the conduit system. The calibrated numerical model was applied by replacing their inputs with the scenarios developed in section 3.2. A non-dimensional comparison between the simulated peak flows of the sediment-filled conduits and the clean ones was carried out to quantify the effects of sediment deposit on the flow capacity of the conduits.

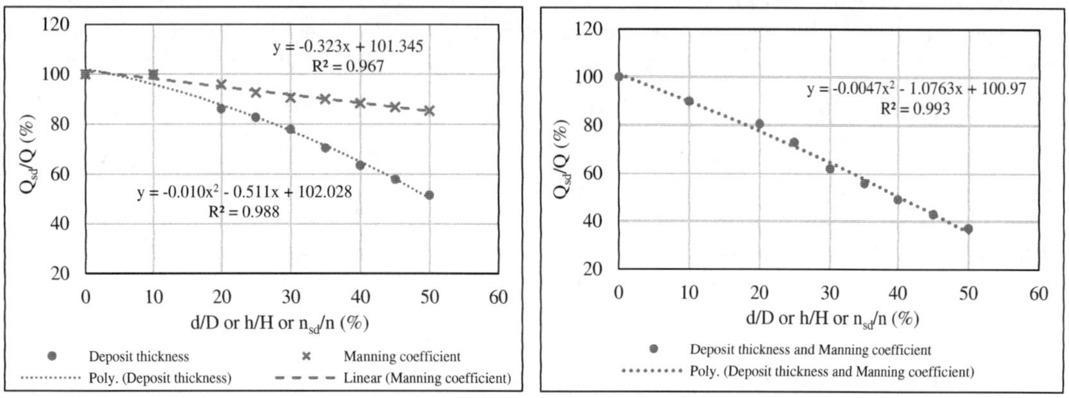

Where: n_{sd}/n (%) is the percentage of the Manning coefficient of the sediment filled and clean conduit; Q_{sd}/Q(%) is the percentage of the peak flow of the sediment filled and clean conduit

Figure 6. Relative reduction of peak flows caused by sediment deposit at conduit VT50 (circular cross-section)

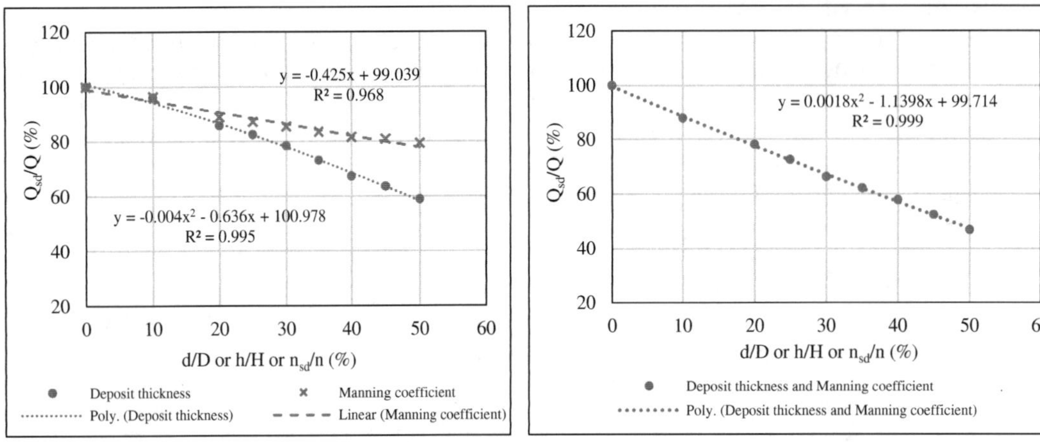

Where: n_{sd}/n is the percentage of the Manning coefficient of the sediment filled and clean conduit;
Q_{sd}/Q is the percentage of the peak flow of the sediment filled and clean conduit

Figure 7. Relative reduction of peak flows caused by sediment deposit at conduit VT45b (rectangular cross-section)

Figure 6 and **Figure 7** plotted the relative reduction of peak flows according to the increase of sediment deposit depth and Manning coefficient at conduits VT50 and VD45b (see **Figure** 3. for positions). The results showed that the increase of sediment thickness declines the flow capacity faster than the increase of the Manning coefficient. A 40% increase in the Manning coefficient and sediment thickness depth, the peak flow degrades to approximately 80-90% and 60-70% of the peak flow of the clean conduit, respectively. The hydraulic performance is rapidly degraded in the case of increasing both sediment deposit thickness and Manning coefficient.

The trend of relative reduction of peak flows could be used to plan the dredging schedule of the urban drainage network. The dredging should start when the sediment deposit has significant effects on hydraulic performance and cause serious flooding. For example, the results showed it may be not necessary to dredge the sediment if the deposit thickness depth is below 10% because it does not significantly affect the hydraulic performance of the conduit. In addition, the hydraulic performance degradation could be used to detect the positions of local blockage due to sedimentation or garbage. The numerical model was also applied to assess the potential impacts of sediment deposit on flood risk caused by extreme rainfall events. Further research is necessary to improve the reliability of the results.

5. Conclusions

The research demonstrated a non-dimensional analysis of the influences of sediment deposit on the flow capacity of sewer pipes. The preliminary results showed the trend of relative reduction of the peak flow due to sediment deposit. The increase of sediment deposit depth has more impacts on hydraulic performance than increases in the roughness of the conduit. A 40% increase in the Manning coefficient, the hydraulic performance degrades to approximately 80-90 %, while a 40% increase in sediment thickness (compared with conduit depth) could degrade the peak flow to about 60-70% compared with the peak flow of the clean conduit. Further research should be carried out to quantify the impacts of the sediment deposit on hydraulic performance with highly accurate and reliable results. The results could support decision makings on the design, operation, and maintenance of the sewer system and adaptation to extreme rainfall events.

Acknowledgments
This research is funded by Vietnam National University Ho Chi Minh City (VNU-HCM) under grant number C2019-48-02.

References

1. R. M. Ashley, A. Fraser, R. Burrows, and J. Blanksby, "The management of sediment in combined sewers," *Urban Water*, vol. 2, no. 4, pp. 263–275, 2000, doi: 10.1016/s1462-0758(01)00010-3.
2. R. Banasiak, "Hydraulic performance of sewer pipes with deposited sediments," *Water Sci. Technol.*, vol. 57, no. 11, pp. 1743–1748, 2008, doi: 10.2166/wst.2008.287.
3. Y. H. Song, R. Yun, E. H. Lee, and J. H. Lee, "Predicting sedimentation in urban sewer conduits," *Water (Switzerland)*, vol. 10, no. 4, pp. 1–16, 2018, doi: 10.3390/w10040462.
4. J. Huber, WC and Dickinson, RE and Roesner, LA and Aldrich, "Storm water management model user's manual, Version 4. Project summary," no. August 1988, 1988.
5. L. A. Rossman and W. H. C., "Storm Water Management Model Reference Manual Volume I – Hydrology (Revised)."
6. L. A. Rossman and Wayne Huber C., "Storm Water Management Model Reference Manual Volume II – Hydraulics," 2017. [Online]. Available: www.epa.gov/water-research.

Strengthen resilience to rivers flooding by the drag reduction technique

Walid BOUCHENAFA[1], Trong DANG-VU[2], Huyen Xuan DANG-VU[3,4]

[1] SNF SA, Rue Adrienne Bolland. 42163. Andrézieux. France
[2] SNF HC. One chemical Plant Road. Riceboro. GA 31323. USA
[3] Ho Chi Minh City University of Technology (HCMUT), Ho Chi Minh City, Vietnam
[4] Vietnam National University, Ho Chi Minh City, Vietnam
Email: wbouchenafa@snf.com

Abstract. Urban agglomerations face the risk of overflowing rivers due to intense urbanization in flood-prone areas and the climate change effects. Despite the important protective measures deployed to reduce the fluvial flooding risk, additional efforts are still needed. This work aims to propose a new complementary non-structural protection measure, used to reduce the river flooding risk. The study is part of the NABRAPOL (NEBRASKA POLYMER) project, which aims to improve knowledge of the drag reduction effect by adding polymers in open-channel flows. The addition of polymers, even in limited concentrations, allows high friction to decrease with the typical Manning coefficient reduced up to 45%. An application case on a real watercourse is presented in this article. Two measurement campaigns are carried out on a river along 30 km. Experimental devices are deployed, and non-intrusive hydraulic measuring instruments are installed at the study field. Surface velocities are evaluated by the Large-Scale Particle Image Velocimetry (LSPIV) technique, and water depth is measured using ultrasonic radar sensors over the river. Measurement results show that the addition of 20 ppm of polymers in the flows results in a marked drag reduction by decreasing the water depth to 18% of its initial depth. The drag reduction technique by addition of small concentrations of polymers can be considered as a new and effective method to reinforce the measures already deployed in the flood risk management strategy since it allows the water depth to be decreased thus avoid overflowing rivers in the extreme flooding event.

Keywords: friction, polymers, flood, boundary layer, LSPIV, flood risk.

1. Introduction

Floods are the most common natural disasters often resulting in considerable economic losses and human and social tragedies [1]. The extreme events in the Caribbean, Bangladesh, and Japan in 2017, India and the Philippines in 2018, show just how devastating the effects of flooding can be. In addition, some mega-cities such as Bangkok, Mumbai, Jakarta, Shanghai, Saigon, Dhaka, London, Paris, and Rotterdam are extremely vulnerable to river flooding due to increased risks associated with urbanization and the effects of climate change [2].

An effective system of protection against the risk of river flooding can significantly improve public safety, reduce social damage and economic losses associated with flooding. Consequently, flood risk management is still a huge challenge for the government and local authorities [3], especially when the adaptation of the riparian population facing the risk of flooding is constantly evolving.

The preferred technique to fight against floods is currently the protection strategy [4]. The drag reduction method by adding polymers can be considered as an effective new technique to reinforce the measures already deployed in the strategy of flood risk management since it makes it possible to reduce the water depth up to 18% of its initial water depth [5] and therefore avoid the overflow of rivers in the flood event.

Discovered by Toms in 1948 [1], the drag reduction by adding polymers is a strategy used to minimize friction in long-distance closed-conduit flow. Industrial applications of polymers began in the 1970s and are growing steadily in different sectors. They are commonly used in pipelines, oil wells, firefighting, irrigation, sewerage systems, water heating and cooling systems, marine systems, and aeronautics [2].

The drag reduction effect allows for dramatic reductions in wall friction and thus linear head losses of up to 80% [6-7]. The origin of this phenomenon lies in the turbulent structures of the flow and the polymer macromolecules added to the fluid [8]. In a fully developed flow over a smooth bed, the polymers modify the velocity profile in the boundary layer. With clear water, the velocity profile in the near-wall regions is composed of a laminar viscous sublayer very close to the wall: $\frac{yu^*}{\upsilon} < 10$, where y is the distance to the wall, $u^*$is the friction velocity $u^* = \sqrt{\tau_w/\rho}$, τ_w is the shear stress at the wall, ρ is the fluid density and υ is the kinematic viscosity of the liquid and the log law slightly further away : $\frac{yu^*}{\upsilon} > 50$. When adding polymers, a so-called elastic zone can be observed in between these two zones for $y^+ > 15$ [9-10], within which the velocity profile follows a logarithmic law with a slope that depends on the polymer concentration. It has been reported that as the extension of this zone increases, the drag reduction efficiency also increases.

Virk [9] also investigated the possibility of using polymers to reduce the drag over rough surfaces. He observed that for walls with a large enough roughness element, the wall roughness tends to decrease the drag reduction process. Using the nondimensional roughness height $K_s^+ = K_s u^*/\upsilon$, where K_s is the equivalent sand roughness height, Virk [9] showed that for $K_s^+ > 50$, drag reduction efficiency decreases compared to a flow over a smooth surface. This holds up to the experimental limit of K_s^+ of approximately 150. Mignot et al. recently confirmed this reduced efficiency of polymers over a rough surface with K_s^+ in the range of 100–135 [11].

Most of the studies reported the use of drag reduction polymers in pressurized pipe or channel flows. However, a few tests have also been performed in open-channel flows, especially within sewer systems [12]. Hart et al. [13] reported the use of such polymers in the sewer network of the city of Whistler, Canada during the 2010 Winter Olympic Games, where the wastewater discharge was strongly increased due to a particularly high population in the area. Including polymer additives in the wastewater decreased the water depth and increased flow velocity, leading to an increased discharge capacity of the sewers. The authorization for releasing polymers in sewer flows was based on a materials safety data sheet (MSDS) that indicated that the chemicals were nontoxic at limited dose concentrations [13].

Mignot [11] works on an open canal flow at the laboratory scale showed that by adding a limited concentration of polymer (approximately 20 ppm) to a steady flow, the Darcy-Weisbach friction coefficient decreased by a factor of 2 in a flow over a smooth surface and by a factor of 1.5 in a flow over a rough surface. The same observation was noted by Janosi et al. [14] in their work on investigating the effect of adding polymers to a dam-break flow experiment on both initially dry and wet beds.

This literature review suggests that this method may have some potential for increasing the bank-full discharge in watercourses. Following this interpretation, the authors hypothesized that similar phenomena also occur at the scale of a river of several kilometers. The aim of the present study was to assess the possibility of decreasing the water depth of a large-scale watercourse by adding polymers.

2. Materials and Methods

2.1. Study Field and Measurement Points

The open channel selected for this study is the Tri-State Canal, maintained by the FIDO (Farmers Irrigation District Office) in Scottsbluff, Nebraska (USA). The length selected for the trial is 40 km. The water depth varies along the canal from 0.85 to 1.50 meters in flow conditions suitable for testing. The canal width varies between 18-21 meters, the small slope is of the order of $S_0 \sim 10^{-4}$. The canal bed contains pebbles, gravel, and alluvium. Strickler's roughness coefficient is 50.

Figure 1. The Tri-State Canal at Scottsbluff and the sediments of its bed.

34 km upstream of the injection site, there is a mobile dam managed by FIDO, allowing to adjust the flow rate and control water depth. The flow of the Tri-State Canal is steady and uniform, the regime is fluvial with an average velocity of 50 cm/s, the flow rate during this measurement campaign is 10 m^3/s. The Manning friction coefficient is about n = 0.017. The outside temperature was about 17°C. The flow in the canal is turbulent, the hydraulic characteristics are summarized in Table 1.

V (m/s)	$\overline{H}$ (m)	Q (m^3/s)	Fr	Re
0.5	1.3	9.65	0.4	5.10^5

Table 1. Flow hydraulic characteristic of the Tri-State channel.

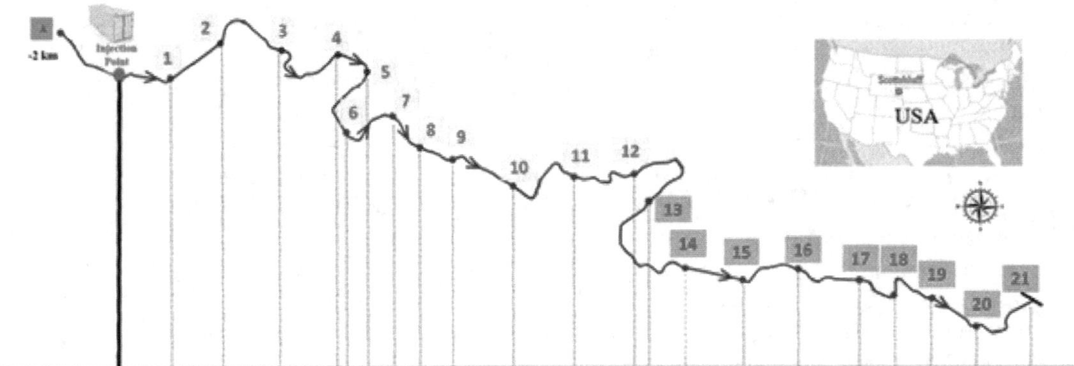

Measuring point	1	2	3	4	5	6	7	8	9	10	11	12	13	14	15	16	17	18	19	20	21
L (km)	1.4	2.9	5.4	7.9	9.0	11.2	13	14.4	15.2	17	19.5	21.5	25.3	28.5	30.2	32.2	33.8	35	36.2	38.8	40
Polymer displace time	47 min	1h + 37 min	3h	4h + 24 min	6h + 13 min	5h	7h + 13 min	8h	8h + 26 min	9h + 26 min	10h + 50 min	12h	14h	15h + 50 min	16h + 45 min	17h + 50 min	18h + 45 min	19h + 225 min	20h	22h + 13 min	22h + 13 min

Figure 2. Study site and identification of monitoring points.

2.2. Polymer Injection and Experimental Devices

The polymers used were anionic polyacrylamides (FLOJET CRP 26), synthesized by copolymerization of AMD and acrylic acid, produced by SNF, in powder form, of very high molecular weight (20 × 10 ^{6}

Da), soluble in water. The polymer concentration of the solution introduced into the canal was set to C = 20 ppm by adjusting the mass rate of polymer powder to the flow rate of the canal: Qm−polymer = C × ρwater × Qwater = 696 kg/h, with ρwate = 1,000 kg/m^3. For an injection lasting 14 consecutive hours. The supplied quantity of water-soluble polymer reached 12 tones. The polymer solution was continuously injected into the Tri-State Canal at the upstream section from 6:00 a.m. until 9:00 p.m.

The installation consists in hydrating the powder polymer before injecting it into the watercourse. The hydration is made using a polymer dispersing device which includes: *1)* a hopper for continuous feeding of the polymer powder, *2)* a metering device for polymer dispersion, *3)* a polymer wetting cone connected to a water inlet circuit taken from the canal using a pump *4)*. The wetting cone is also connected to *5)* a second adjustable flow rate volumetric pump ensuring the transfer of dissolvable polymer under pressure.

The polymer in suspension is injected into the Tri-State Canal continuously using a hose with an injection system containing 10 spray nozzles installed 1 meter above the width of the river to ensure a homogeneous distribution of the injected solution. This set-up provides a good mixture of the polymer with the canal water, and the suspension is quickly prepared before being injected into the canal. The experimental devices are illustrated in Figure 3.

Figure 3. (a) Polymer dispersion device located in a container installed next to the canal; (b) close-up view of the dispersion device, including hopper, metering device, wetting cone, pumping loop, flow rate volumetric pump, (c) injection system containing 10 spray nozzles.

2.3. Measuring Devices

Water Depth Measurement. A set of ten ultrasonic sensors are deployed during the campaign to measure the change in water depth after the polymer injection. Each sensor is fixed to each bridge

IOP Conf. Series: Earth and Environmental Science **964** (2022) 012021 doi:10.1088/1755-1315/964/1/012021

crossing the canal to measure the variation in distance between the position of the sensor and the flow surface level. Wooden planks fixed to the bridges were used for the vertical referencing of the measured levels. The sensor records the measurements automatically every 10 min. Thereafter, the water depth variation is deduced.

Figure 4. Water depth measurement technique.

A time-lapse camera was installed in front of a ruler to film the water depth variation at 15 km from the injection.

Figure 5. Installation of a time-lapse camera Link to the video: https://vimeo.com/382568921

Velocity Measurement. Surface velocity measurements are carried out at each measurement point using Large Scale Particle Image Velocimetry (LSPIV). A widely applied technique for the estimation of hydrometric characteristics in fluvial flow [3]. Each measurement point is equipped with a smartphone containing a video camera filming one-minute videos every 30 minutes.

The various measurement instruments worked satisfactorily throughout the measurement campaign. The measurements carried out thus made it possible to define the change in flow hydraulic characteristics and create a reference datum for characterizing the drag reduction phenomenon.

3. Results and Interpretations

The time evolution of the water depth variation at the ten bridges is depicted in Figure 5. This figure confirms previous experiments in laboratory conditions, that adding polymers to a large-scale watercourse reduces water depth. As the polymer arrived at a cross-section, water depth decreased quite linearly with time and eventually reached a lower plateau of maximum drag reduction with a constant reduced water depth. Measurements showed that the drag reduction phenomenon is reached after 46 min of polymer displacement time. A constant water depth corresponding to constant flow conditions is noted at first. Then, a change in flow regime is observed after 46 min of polymer displacement time. This delay corresponds to the time needed for a complete dissolution of PAM in cold canal water.

Water depth was reduced by up to 33 cm at Bridge #1; this represents 20% of the drag reduction. However, the percentage of drag reduction decreases as a function of the distance from the injection site (Figure 6).

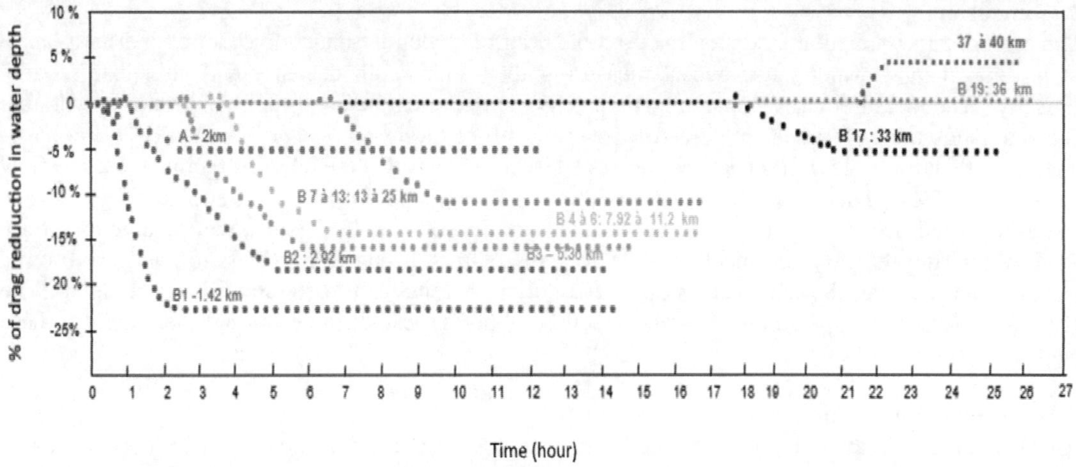

Figure 6. Percentage of drag reduction of water depth variation for monitoring Bridges #1 to #10.

The measurement results also show that the water depth gradually decreases and then stabilizes 2 hours after the arrival of the polymer at each measurement point. This time corresponds to the maximum of DR. These experimental results demonstrate that the decrease in water depth depends on dissolution time and interaction time of polymer with the canal flow, which takes about two hours to obtain the maximum drag reduction effect. However, no variation in water depth was detected at 33 km.

At 37 km, i.e., 21h min of flow from the injection site, an increase in water depth (backwater) to 5% of its initial depth was observed. Indeed, the water height reduction provided by the DRP technique is due to a water flow acceleration. This water then accumulates downstream where the DRP effect is no longer effective. These tests made it possible to define the location of the second injection point to protect all areas, upstream and downstream, against flooding risks.

Velocity Measurement

The flow velocity is influenced by the polymer action on flow dynamics. For the different measurement points, it can be noted that velocity increased by up to 30% which corresponds to a reduction in water depth of 17% as shown in Figure 7. The article by Mignot et al [11], which is also based on the results of tests on an open channel in a laboratory, analyzes the phenomena at the origin of the velocity increase. The percentage increase in flow velocity in the canal also decreases as a function of the distance from the injection point.

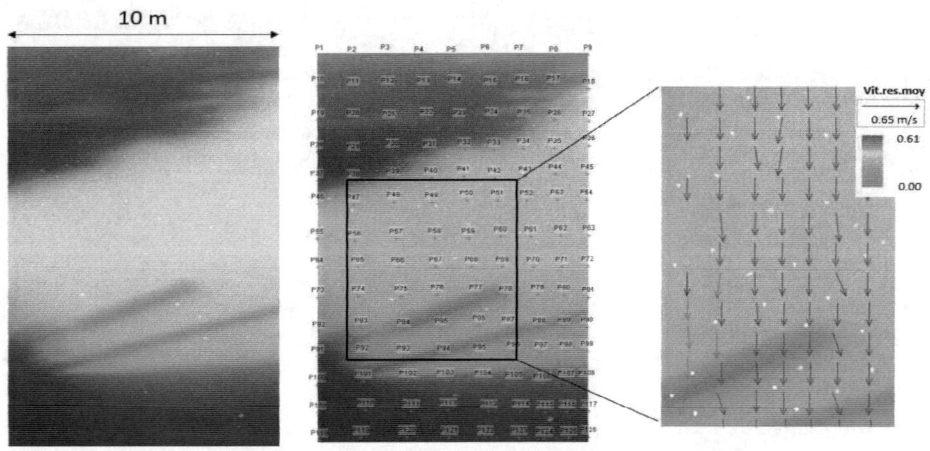

Figure 7. Increase of velocity at bridge #3 at 7.3 km.

4. Conclusion

The present paper aimed at assessing the possible drag reduction by injection of a polymer in a large-scale open-channel watercourse and at measuring the water depth decrease and gain in discharge capacity. The measurement campaign made it possible to define the zones of change of the hydraulic characteristics the experiment presented here was a proof-of-concept of drag reduction by polymer injection at a large scale. It has been demonstrated by these tests that the polymer can increase the flow capacity by 30% and decreases the water depth up to 18% for a polymer concentration of 20 ppm. The analysis carried out in this work showed the non-variation of the physicochemical parameters of the canal water after the polymer injection. However, there are still other parameters to measure. From a technical perspective, the DR mechanics by the polymers remains to be studied in more detail. The processes of the flow change remain to be specified. Another test on the canal will help answer these questions.

5. Data Availability Statement

All data that support the findings of this study are available from the corresponding author upon request.

6. Acknowledgment

This project mobilized more than ten people in the field during the two measurement campaigns. Teams from SNF USA, SNF France, and FIDO Nebraska took turns during the days of the measurement campaign. We thank FIDO, presented by Kevin Adams, for the preparation of the Tri-State Canal for the measurement campaign. Special thanks to Doug Yost, SNF HC in Nebraska.

7. References

[1] Kuang D and Liao K H 2020 *J. Environ. Manage.* **271** 111025
[2] Kundzewicz Z W, Hegger D L T, Matczak P and Driessen P P J 2018 *Proc. National Academy of Sci.* **115** 12321
[3] Driessen P P J, Hegger D L T, Kundzewicz Z W, van Rijswick H F M W, Crabbé A, Larrue C, Matczak P, Pettersson R, Priest S, Suykens C, Raadgever G T and Wiering M 2018 *Water* **10** 1595
[4] Driessen P, Hegger D, Bakker M H N, Rijswick H and Kundzewicz Z 2016 *Ecology Society*, **21** 53
[5] Bouchenafa W, Lefevre A and Quillien B 2018 *Inter. J. Innov. Applied Studies*, **24** 1
[6] Herzhaft B 2000 *Oil Gas Sci. Technol.* **55** 501
[7] White C M, Dubief Y and Klewicki J 2018 *J. Fluid Mech.* **834** 409
[8] Schlichting H 1979 *Boundary layer theory* (New York: McGraw-Hill) p 602
[9] Virk P S 1975 *AIChE J.* **21** 625
[10] Cai W H, Li F C, Zhang H N, Li X B, Yu B, Wei J J, Kawaguchi Y and Hishida K 2009 *Phys. Fluids* **21** 115103
[11] Mignot E, Riviere N, Lefevre A and Quillien B 2019 *J. Hydraul. Eng.* **145** 04019011
[12] Sellin R H J 1988 *Proc., AIAA, ASME, SIAM, and APS, National Fluid Dynamics Congress* (Washington, DC) American Institute of Aeronautics and Astronautics.
[13] Hart J, Cotter L and Jason V 2011 *Proc. Water Environment Federation, Session 88* vol 12 (Victoria, Canada) p 5763
[14] Janosi I M, Jan D, Szabo K G and Tel T 2004 *Exp. Fluids* **37** 219

Proposed Strategy for the Development of a Local Geopark at Mui Ne Red Sand Dunes, Phan Thiet, Vietnam - based on Tourist Perceptions and Experience from Taiwan

V M van Onselen[1], T-Y Lin[1*], L P Vo[2,3*] and Lam T H Nguyen[4]

[1)] Department of Geography, National Taiwan Normal University
162 He Ping East Road, Section 1, Taipei 10610, Taiwan
[2)] Faculty of Environment and Natural Resources, Ho Chi Minh City University of Technology (HCMUT), 268 Ly Thuong Kiet street, Dist. 10, Ho Chi Minh City, Vietnam
[3)] Vietnam National University Ho Chi Minh City, Linh Trung ward, Thu Duc District, Ho Chi Minh City, Vietnam
[4)] Global Health Program, National Taiwan University, 17 Xuzhou Road, Taipei 100, Taiwan
[(*)] Corresponding authors: aeolin@ntnu.edu.tw; volephu@hcmut.edu.vn

Abstract. Vietnam attracts many tourists owing to its unique and diverse geological landscapes. The Vietnam tourism development strategy until 2030 sets a target to make tourism a more sustainable economic sector, in which the advantages of natural resources and development of a variety of tourism products are promoted. Binh Thuan province in central-south Vietnam, hosts many high potential geosites including Mui Ne Red Dunes. Although this is a popular tourist destination, there is no clear tourism strategy for this site and local communities now regulate tourism in an unsustainable way. Online comments from review databases (TripAdvisor and Google reviews) are analysed to find themes in tourism perceptions after a visit to the Red Sand Dunes. Main themes are found that negatively affect tourist perceptions, such as the extended pollution of the dunes, theft reports and disturbance by jeeps, ATV's and motorbikes. Many tourists recommend that this site should be better managed. To conserve this area in the future, the development of an all-inclusive local geopark is suggested, based on the tourism perceptions and previous studies assessing this landscape. This strategy is adapted from the UNESCO geopark framework and an example of a recently established local sand dune geopark in Taiwan is discussed. Implementing a local geopark, following the steps described in this study, can be beneficial for sustainable development and local communities in this area, while also promoting environmental education and enhancing disaster risk reduction.

1. Introduction

The central Vietnam coast is both known for its rapid socio-economic developments as for its attractiveness. This coastal zone hosts large numbers of tourists, is exploited by economic developments and simultaneously suffers from adverse impacts of global climate change [1]. Without adequate management strategies, valuable sites along the coastline might not be conserved in the future. Previous studies clearly demonstrate the value and potential of the high diversity geosites in this area [2-3]. One of these sites is the Mui Ne Red Sand Dunes, in Binh Thuan province, a fascinating landscape with great potential to become a geopark, since it has unique geomorphological, ecological, and aesthetic values [4]. The current situation of community-based tourism in Binh Thuan province is still spontaneous, without a systematic tourism regulation for local communities. The awareness of the authorities at all levels and local communities in tourism development is not adequate [5]. Due to these circumstances, the environmental pressure on this landscape is increasing. Obtaining the status of geopark can promote sustainable development of the region and support other benefits [6]. This site already is a popular tourist destination among foreign and domestic tourists but is still lacking proper management. Online databases are used by tourists to rate and comment their visit to the Red Sand Dunes and these tourism perceptions are used to find themes, specific remarks, and general satisfaction rates. This study will primarily assess tourist perceptions for Mui Ne through these reviews, to explore what tourists point out as positive or negative aspects, find recurring themes in these comments and observe if there is a significant difference between reviews in Vietnamese and reviews in English. For the development of this site, possibly into a local geopark, these comments can give useful insights for future management strategies. A second objective for this study is to propose a successful strategy for the development of a local geopark with local support and discuss if this could be adapted to Mui Ne Red Sand Dunes after evaluating the potentials and issues for this location. This strategy is based on the UNESCO geopark framework, and is designed to support social, economic, and environmental benefits [7]. A recently established local Geopark (Caota Sand Dunes) in Taiwan, will serve as an example [8]. The implementation process to gain the status of a local geopark is discussed in terms of assessment and local community involvement [6].

2. Study area and methods

2.1. Study area

Mui Ne Red Sand Dunes (Doi Hong) currently covers an area of circa 37 hectares and is located on Mui Ne headland, Binh Thuan province, circa 20 km from Phan Thiet city center (Figures 1 and 2). It is an active dune field known for its rusty red color and sunset views (Figure 1). The average dune height varies between 5-10 meters [4]. The sand dunes shift from season to season and are popular amongst tourists for sand surfing, scenic hikes and driving around on quad bikes or in dune buggies. The regional climate is controlled by monsoons, and most tourists visit in the dry season (November-April) as prolonged rains often occur in the summer months from May to October. The sand dunes are mainly driven by the stronger northeastern winter monsoons winds [4].

Figure 1. Field photographs of Mui Ne Red Sand Dunes by T.-Y. Lin, July 2017.

The modern active sand dunes (generally younger than 100 years) are transverse dunes with an NNW-SSE crest orientation, and the sand appears to be derived from Mui Ne beach on the eastern coast [4].

Over time, waves transported sand onshore and once the sand is deposited on the beach, it is picked up by the wind to form sand dunes. The dunes have grown larger and steeper over time and also have the function of a buffer zone, as they can absorb the impact of storms or delay flooding to inland areas. Mui Ne is also known for other tourism destinations, such as the beaches, where recently many resorts have been built [1] the Fairy Stream (where due to erosion sand sculptures have formed along a stream) and the Mui Ne White Sand Dunes. This study will focus on the Mui Ne Red Sand Dunes specifically, and the tourism review analysis is based on tourist's perspectives and comments after visiting this site.

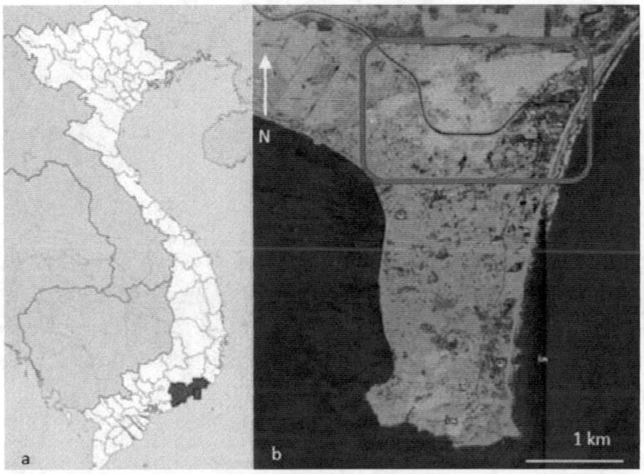

Figure 2- a) Location of study area in Binh Thuan province, Southern Central Vietnam b) The active red sand dune field is located within the red box at Mui Ne headland. Derived from Google Maps and Google Earth.

2.2. Methods

2.2.1. Tourist Perceptions

To assess the tourist perceptions of visitors at the Mui Ne Red Sand Dunes, data from 2 large international review websites (TripAdvisor and Google reviews) are analysed. In these public online databases people can give a rating and/or review comment after they have visited a certain destination. Reviewers can rate up to 5 stars (1 star equals very bad experience, 5 stars equals a wonderful experience). For this study ratings without text are excluded, and although multiple languages were used in the comments, only comments in English and Vietnamese language are selected to avoid possible misinterpretation.

- TripAdvisor July 2021: total 2.037 reviews, rate 3.5 out of 5 stars. A total of 979 reviews in English language and 93 in Vietnamese language, dating back to 2011, are selected for further analysis.
- Google reviews July 2021: total 710 reviews, rate 4.1 out of 5 stars. A total of 49 reviews in English language and 125 in Vietnamese language are selected for further analysis, Google reviews only has data available over the last year.

Thus, in total 1.246 online comments are selected for further analysis. The collected comments reflect opinions of the area, and these reviews are analyzed to explore tourist feelings after visiting the Mui Ne Red Sand Dunes. The textual reviews from the databases are extracted and combined into Excel and the data are analyzed using NVivo 12 to code texts and find themes. The selected comments are assessed one by one, to determine main repetitive themes that come forward, which are labelled to indicate patterns in the online reviews following the Reflexive thematic analysis approach [9].

2.2.2. Local Geopark Framework

Main elements that are addressed in geopark development are pointed out. The policies of tourism development for Binh Thuan province are discussed based on government reports and announcements. The establishment process of the local geopark at Caota Dunes and Taiwan is summarized, based on a review of the current framework for Taiwan Geoparks from reports and publications. The main steps and approaches concerning assessment and local community involvement that are addressed in this framework are described in a 3-step approach: in the initiation phase, establishment phase and evaluation phase.

3. Results and Discussion

3.1. Tourist perceptions

People that use TripAdvisor as a platform for their reviews in general give a lower rating compared to people using Google reviews, but for both platforms there is no significant difference between ratings in Vietnamese and in English (figure 2). Data from TripAdvisor clearly show a decline in English (international) tourism reviews since 2019, while the number of reviews in Vietnamese is increasing since 2019, suggesting more domestic visitors during the time of the global COVID pandemic. Google reviews only date back one year, the number of reviews in Vietnamese (n=126) is more than twice as large as reviews in English (n=50). Over the last year (July 2020 - July 2021) in total 245 people left a review on Google Reviews, and 67 people left a review at Trip Advisor. Although it doesn't represent the actual visitor data (the site is unregulated and there is no data collection on actual numbers), it suggests that is an actively visited site.

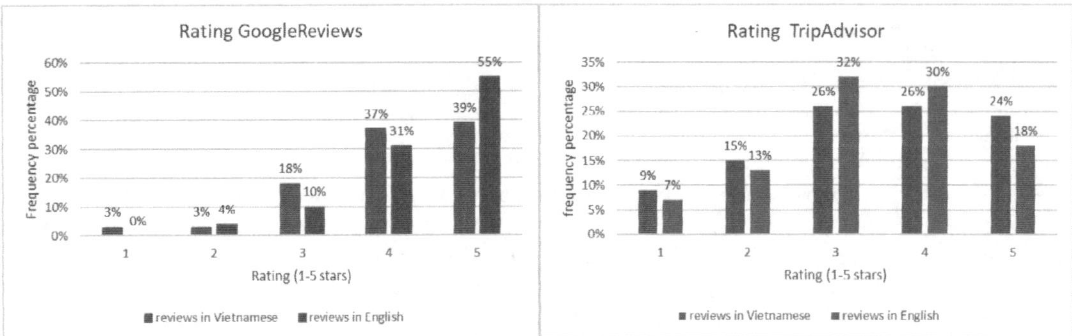

Figure 3 - Reviews on TripAdvisor of the data are on average 3.4 stars, while Google reviews users give an average rating of 4.2 stars. For Google Reviews the average rating of the reviews in Vietnamese is 4 stars, and in English 4.3 stars, while the average rating for TripAdvisor is 3.4 stars for both reviews in Vietnamese and in English.

The Coded themes that were used in NVivo, are related to tourism experiences and tourism perceptions are divided in positive, negative, mixed, and neutral experiences (Figure 4). **Positive** reviews show that visitors have a better experience when visiting during the early morning or late afternoon, since during sunrise and sunset the sand turns more reddish, it's not too hot and they can better enjoy the peaceful landscape. Visitors also highly value the landscape for taking pictures and mention that the views are beautiful. In **Negative** reviews, themes that often occurred were theft

IOP Conf. Series: Earth and Environmental Science **964** (2022) 012022 doi:10.1088/1755-1315/964/1/012022

(pickpocketing) during sand boarding activities and scams by sellers and children that try to rent out sandboards. People felt unsafe because they were being followed, or aggressively harassed by local sellers or children, and this spoiled their visit to the sand dunes. Another theme that frequently occurred in negative reviews is the pollution of the sand dunes; trash and plastic littered over the landscape troubles many people. This is often correlated to how crowded the site is, several comments point out that people go here for picnics, but leave their trash in the dunes. From the comments it seems that most garbage is found near the road and main entrance points, and if you stroll further into the dunes the landscape becomes less polluted. **Mixed** reviews mention both positive and negative feelings after visiting this area. Often people say that the landscape is beautiful, or they had an amazing experience, but there is too much garbage, it's too crowded, or that the local people that are trying to make money out of tourists, somewhat spoil their experience here. Others mention that the site has great potential, but more investment/regulations and or management strategies are needed. Several people also wonder why the government is not paying more attention to environmental regulations of this place and suggest better enforcement and maintenance for this site. Comments classified as **Neutral** often mention that if you are already in Mui Ne, you should not miss this destination; it is interesting enough to visit, but not very special or if you have never seen sand dunes before, you should go.

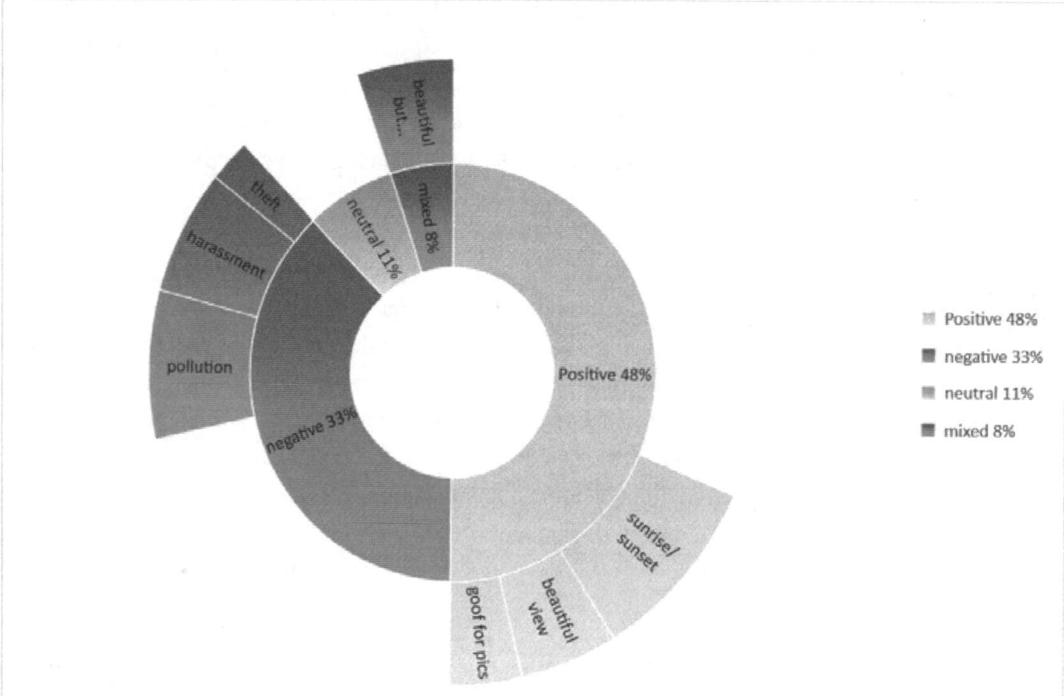

Figure 4 – Main Themes - around 48% of the reviewers left positive comments, often mentioning the amazing sunsets, beautiful landscape and great scenery for taking pictures. Circa 33% of the reviews was mostly negative, pointing out the pollution of the dunes (garbage, trash, plastic), the harassment or rude behaviour of sellers and sledge renters and theft by local people and children. Circa 8% are predominantly mixed reviews and 11 % of the comments was generally neutral.

Other themes correlated to activities that often occur are: '**hot**'(n=61), which is related to visits during mid-day or late morning and early afternoon, when the sun heats up the sand and many people experience their visit as uncomfortable. Some advice not to wear flip-flops or suggest visiting when the sun is less bright. Many people include '**sandboarding/sliding**' into their comments (n=162) and both positive and negative feelings are correlated to this experience, but most agree that it is more an activity for children, since adults can hardly go down on the plastic 'boards' that are provided. It is also mentioned that they use soap and wax to make the board go faster, which leads to more pollution of the environment. Another recurring theme is '**Jeep, motorbike or ATV driving**' on the sand dunes. Only few reviewers (n=19) mention this activity and they also have mixed feelings; they either like it and think it's a fun activity or they think it destroys the peacefulness of the landscape and find it disturbing. Frequently, a '**Comparison to the White Sand Dunes**' is made (n=224), since many tourists visit the Red Sand Dunes as part of a tour package, also visiting the white sand dunes, Fairy Stream and Mui Ne fishing village. In these comments people mainly point out that the white sand dunes are larger, less crowded and less polluted and they are less disturbed by locals and children. But there are more ATV's and motorbikes that drive around in the White Sand Dunes, so people also remark that as a disturbing factor when they compare it to the Red Sand Dunes. Moreover, in these comparisons it is observed that a fee is required to visit the White Sand Dunes, and the Red Sand Dunes can be visited for free.

3.2. Implications for the development of a local geopark

3.2.1. Globally accepted framework for UNESCO geoparks

The human pressure on ecosystems is growing on a global scale and a lack of awareness amongst different stakeholders and insufficient implementation of sustainable development plans will increase this pressure on the environment. This is where geoparks can step in, for touristic areas with a high geoheritage value, such as Mui Ne Red Sand Dunes. Conservation of- and education about the area can help increase awareness and simultaneously conserve a precious environment. A geopark is defined as a site of particular geological importance, which is being sustainably managed to conserve the geoheritage and promote public awareness mainly through tourism [10]. According to a globally accepted framework by UNESCO, geoparks are not solely focussed on geological importance, but are also of value to ecological and cultural aspects and socio-economic development of the surrounding region [11,7,3]. The balance between these different elements, for instance between tourism and environmental conservation, should be considered for each specific area (figure 5). According to this framework, the establishment of a Geopark should be founded on strong community support and local involvement through "bottom-up" processes [11-12]. Fitting management strategies ought to be proposed in consultation with the involved stakeholders and strong support of the (local) government is needed for the provision of necessary financial resources and to make agreements on socio-economic development and conservation strategies [12].

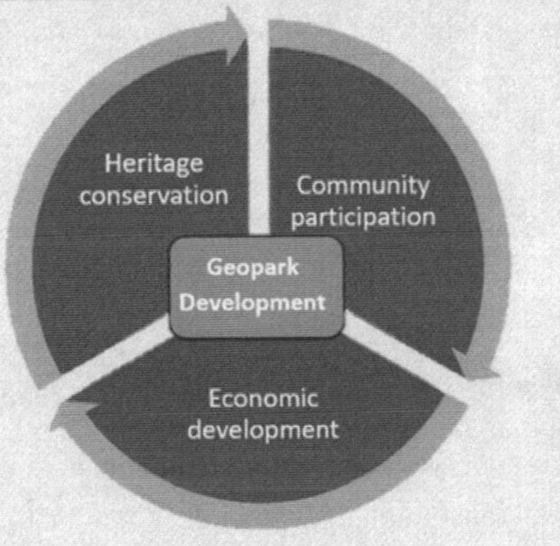

Main elements that are addressed in geopark development, according to UNESCO framework:

- **Heritage conservation**: integrated biological, geological and cultural heritage conservation
- **Community participation**: the community has a leading role in shaping the geopark, to strengthen sense of belonging, cultural promotion and environmental education
- **Economic development:** stimulate economic development through tourism and geopark based products

Figure 5 – Main elements considered in the geopark development process, demonstrating the synergy between heritage conservation, economic development, and community participation, adapted from [12]

3.2.2 Policies for tourism development in Binh Thuan

In 2020, Binh Thuan province issued an implementation plan for a tourism development strategy till 2030, which set the task of developing a plan for the Mui Ne National Tourist Area, promote community-based tourism and create sustainable livelihoods for local people [13]. Therefore, it is fitting to propose a strategy for the development of a Geopark at Mui Ne Red Sand Dunes, to support local communities and manage tourism in a sustainable way. To safeguard cultural and environmental values and to avoid fragmented tourism approaches, the development of a Geopark needs to follow a well-structured plan with cooperation of both the government and local communities. Currently there is still no separate policy for the Mui Ne national tourist area and management is overlapping; the responsibilities between the management unit and the Management Board of Ham Tien, Mui Ne tourist area and Mui Ne ward are unclear. This leads to limitations in management strategies, maintenance and enforcements in tourist sites of Mui Ne lead to corresponding issues such as environmental sanitation is not guaranteed, there are many traffic jams, the sites are overcrowded during holidays, monotonous products are being sold and there is a lack of attractive services for tourists [14]. Therefore, it is essential to have policies to enhance environmental protection, manage tourism, and develop specific products and attractions in line with the unique culture here [13]. Currently, Binh Thuan is implementing an investment project to build Phan Thiet airport, with operational activities for international flights, thereby contributing to strongly promoting socio-economic development of the region and projects to improve infrastructure throughout the province are implemented to further meet the tourism goals.

3.2.3 Example of implementation - local sand dunes geopark in Taiwan

In Taiwan, the TGN (Taiwan Geoparks Network) follows the global UNESCO geopark objectives, and highly prioritizes the participation of local communities by stimulating bottom-up approaches [6]. The Taiwan Geoparks Network developed core values that each geopark in Taiwan needs to fulfill (Figure 6). Because of these values, geoparks can contribute to promoting environmental education, maintain environmentally responsible tourism and stimulate local community involvement.

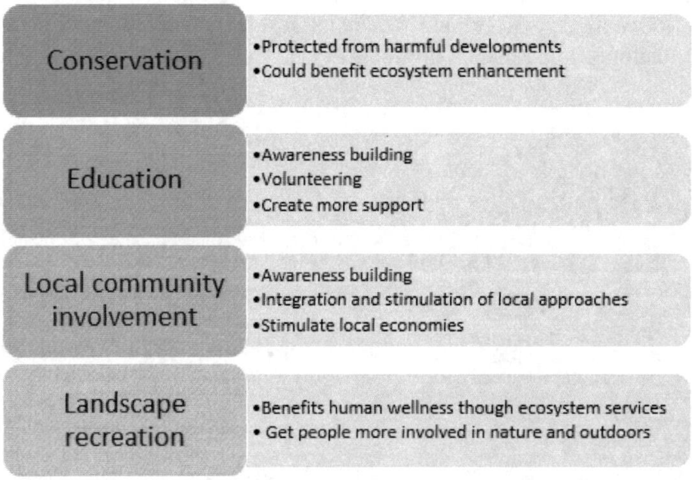

Figure 6 - The four core values for Geoparks in Taiwan

Before bringing a site forward as potential geopark, it is important to gain more knowledge on the system dynamics, human activities, development plans, environmental degradation status and underlying issues [15]. In the case of Mui Ne Red Sand Dunes, most visible environmental issues relate to the rapid development of tourism industries and lack of environmental awareness and regulations for the dune landscape. Main themes found from the review assessment concern waste pollution, the use of plastic boards that are waxed with soap to slide down the dunes or ATV's and jeeps that destroy the peacefulness and leave an environmental footprint. Furthermore, the sand dunes are very dynamic and an issue for local communities in the surrounding area is sand invasion; due to strong winds and flying sand, the sand is invading nearby villages and farmlands, making the land less fertile [16]. This sand invasion is caused by a combination of human factors and natural processes and has local authorities and communities worried. Embankments are built and trees are planted to limit the sand mobility. Improvements in this current situation are needed to work towards a sustainable environment and maintain the characteristic features of the landscape. Local residents are already very much involved in tourism at the Red Sand Dunes, and play a large role in how tourists experience their visit. Therefore, when developing this area into a geopark it is essential to involve local communities in the development plans. Moreover, from an educational perspective, when people are more involved in planning and when public understanding of the environment and dune dynamics is enhanced, they might take more responsibility for the sustainable use of natural resources [17-18]. A newly established local geopark in Taiwan, Caota Sand Dune geopark, conserves the most complete coastal sand dune system in Taiwan [8]. Three general steps were followed to establish this local geopark, the main points are summarized below [19].

- **Initial phase**: This phase starts with an environmental assessment of the area, by listing the values and key points the geopark can promote [2]. A list of involved stakeholders is drawn up, to visualize the network of people related to the geopark site. In this stage, ways to gain support from local communities are inventoried and local communities are involved in the development plans, can observe their local environments and form a network with other geopark communities, on a national or international level, with Global Geopark Network or UNESCO. Through these networks, stakeholders can learn from each other and share experiences [20]. For Caota Dune Geopark, the dune vulnerability is assessed and environmental issues are examined and geological and ecological features are described to form a narrative for educational purposes.

- **Establishment**: This can be done on a local, national or international level. In this phase, coordination with local and/or national governments that are in charge of the tourism development in this area is needed, for arranging funding and drawing up development plans. For example, in Taiwan the local and central government will need to provide a budget for managing geoparks by law, to play a part in the country's sustainable development. In Caota Dunes geopark, bottom-up approaches from within local communities are stimulated and guidelines are drawn for environmental protection in line with the local geopark's core values, protecting the unique dune habitat [21]. Roles which might be fitting for local residents are tour guiding, host activities in the dunes, environmental education, help in conserving and monitoring the landscape or promotion of local goods, for example through geo-products and businesses. Upon assessment of the dune vulnerability, vulnerable zones can be selected where no tourists are allowed, to not disturb natural processes. At Mui Ne, certain areas could be targeted to manage the issue of flying sand, in cooperation with local residents. Construction of designated walking routes, signs, an information centre etc. can all be considered when developing the dune field as a geopark.

- **Evaluation**: This phase entails regular monitoring activities, carried out by scientists or involved stakeholders. To maintain the geopark status regular network meetings should be held on a national and international level, to evaluate maintenance, tourism pressure, observed environmental vulnerability, role of local communities, etc. In the case of Caota Sand Dunes, (local) volunteers are trained to monitor the dune dynamics, but this could also be carried out by local geopark managers. It is important to check if the core values of the geopark are met, by analysing observed benefits for local community, geomorphology and biodiversity. Thus,

IOP Conf. Series: Earth and Environmental Science **964** (2022) 012022 doi:10.1088/1755-1315/964/1/012022

in order to maintain the Geopark, its performance and core values, long-term planning, continuous maintenance and regular evaluation are essential factors [6].

Many elements are open for consideration in these steps and the development plans should fit the local needs and expectations. However, it is essential to all geoparks to stimulate the balanced relationship between humans and the environment.

4. Discussion and further research

4.1. Discussion

Binh Thuan province wants to develop policies to support people to participate in the development of community-based tourism, to improve economic and cultural life and contribute to poverty alleviation for the communities. To work towards these goals, this study suggests developing a local geopark at Mui Ne Red Dunes, with strong local community involvement. This might later be developed in a larger or even UNESCO geopark, together with other geosites at Mui Ne, like the Fairy Stream, White Sand Dunes and Mui Ne fishing village. All combined these geological features are characteristic for this region and its geological history, and many local tour operators already offer a package visiting these 4 sites.

For Mui Ne Red Dunes, many reviewers commented on the large amount of garbage in the sand dunes, stating that this pollution had a very negative impact on their visit. Tourist activities could be managed in a better way to reduce the impact on the environment, by stimulating both visitors and local communities to help keep the dunes clean, through environmental education or providing a fund for clean-up actions. Another element that can be considered for this site is a small entrance fee; this can provide a budget for environmental conservation and provide jobs for local residents in for example monitoring the environment, dune dynamics or visitor numbers. A main objective of a geopark is local community involvement and stimulating the local economy. The local economy can be developed through the sale of products such as food or souvenirs, tourism jobs or jobs related to geopark maintenance and monitoring. Moreover, selling local products can help in promoting the culture and local identity to create a better understanding of the history of the community in the surrounding area. It is hoped that by stimulating the local economy, children that now work at this site for extra income can go to school instead, since another main theme in the tourist reviews was theft and harassment by local children. Additionally, most tourists advise to go here in the early morning or late afternoon, because during the day it is often too hot. Future monitoring can be tailored to this, as these are peak times where it can get very busy

Current tourism activities in the Red Sand Dunes are sandboarding and ATV/ Jeep driving on the dunes. This is controlled by local tour operators and local residents and children, and because of a lack of management rules or environmental regulations, there is no supervision of these activities. These activities damage the sand dunes and can be a threat for the local environment if they are not being monitored. Future management strategies should address these activities to become less of a burden on the environment. Another matter related to the high dynamics of the sand dunes, is desertification of farmlands and villages around the active dune field. In collaboration with local stakeholders, solutions for this issue need to be discussed in terms of the future boundaries of the geopark and management strategies of farmlands surrounding this geosite.

4.2. Limitations and further research

While this study has been conducted very carefully, there are some potential limitations. During the assessment of the reviews, it seemed that not all people understood the rating system; some comments were very positive but only rated 1 star, whilst sometimes very negative reviews were found at a rating of 5 stars. Overall, this was encountered on very few occasions. Additionally, the nationality of reviewers in many cases is unknown, and it is possible that Vietnamese tourists have left comments in English. The general assumption throughout this study has been that English reviews are from international tourists and Vietnamese reviews are from domestic tourists, however no conclusions are made based on this assumption. Lastly, reviews and perceptions can be very personal and people that have already visited many sand dunes or deserts before seeming to be less impressed or give more negative reviews. Nevertheless, the objective of this study was to address main themes that are pointed out in these comments to improve the tourism experience in the future. This study identified main themes related to tourism experiences in the Mui Ne Red Dunes from English and Vietnamese reviews, but further studies can also identify these themes in the remaining reviews in other languages, to gain a broader understanding of the perceptions, or specifically identify themes for tourists from different backgrounds for future information provision.

To develop the Mui Ne Red Sand Dunes into a local geopark, more research is needed on local perceptions and involvement, to stimulate community-driven research. Additionally, the perceptions of the different stakeholders that are involved in this site (e.g., tour operators, policymakers, local government officials, etc.) can be assessed. This data collection can be helpful considering long-term management strategies. Furthermore, to gain a better understanding of the local environmental dynamics, research could focus on biodiversity assessment in the dune ecosystem and on conservation management approaches. Regular monitoring of the system dynamics is required, as dunes are highly dynamic but also vulnerable environments that can quickly change and are sensitive to human impacts.

5. Conclusion

Environmental issues in Mui Ne Red Dunes are pointed out in the tourist reviews, such as waste pollution and destruction of the natural landscape. Other issues that came up were theft and harassment, all having a negative impact on this site. Therefore, future management strategies should address these issues for the sustainable development of this location and find alternative ways to gain profit for the local community. Therefore, when nominating this area as a geopark site, impacts on the natural environment should be evaluated and clear communication with local residents is needed to find an increased social acceptability and local input for future development of the geopark. Implementing a local geopark following the steps described in this study can be beneficial for sustainable development and local communities in this area, and when successful, it can be promoted to develop into a national or even globally recognised UNESCO geopark, including nearby geosites as well. The number of reviews and positive comments point out that Mui Ne Red Sand Dunes is a promising area and attracts many tourists on a yearly basis. By developing this site as a geopark, it can be protected and conserved to ensure that natural resources will be well-maintained and that future generations can also enjoy this remarkable landscape.

Acknowledgement: The authors acknowledged the support of time and facilities from Ho Chi Minh City (HCMUT), VNU-HCM for this study.

References

[1] Vo P Le, Lin T and Nguyen B G 2018 Threat of coastal hazards to socio-economic development in the central coasts of Vietnam *Int. Symp. Lowl. Technol.*
[2] Nguyen-Thuy D, Ta P H, Nguyen-Van H, Van Dinh H, Van Dang B, Dang N H, Do H T T, Nguyen A T K, Tran T D, Van Bui V, Nguyen A N and Hoang T T 2019 Evaluation of Geological Heritage of Geosites for a Potential Geopark in Binh Thuan–Ninh Thuan Coastal Zone, Vietnam *Geoheritage* **11** 689–702
[3] Thi H, Chi P, Nu T, Dan L, Yen C N and Hai H Q 2020 Development of Geo-products - example geosite Suoi Tien – Binh Thuan Province *Int. Conf. Geo-spatial Technol. Earth Resour.*
[4] Tamura T, Ta T K O, Saito Y, Bateman M D, Murray-Wallace C V., Nguyen T M L, Sato T and Nguyen V L 2020 Seasonal control on coastal dune morphostratigraphy under a monsoon climate, Mui Ne dunefield, SE Vietnam *Geomorphology* **370** 107371
[5] Ministry of Sports, Culture and Tourism 2021 Bình Thuận: Phát triển du lịch cộng đồng theo hướng bền vững available from: https://bvhttdl.gov.vn/binh-thuan-phat-trien-du-lich-cong-dong-theo-huong-ben-vung-20210413162434962.htm (in Vietnamese).
[6] Lin, J-C; Su S-J 2019 *Geoparks of Taiwan* Springer, Switzerland
[7] Gabriel R, Moreira H, Alencoão A, Faria A, Silva E and Sá A 2018 An emerging paradigm for the UNESCO global geoparks: The ecosystem's health provision *Geosci.* **8** 1–13
[8] Hetherington C S and W 2020 Taoyuan sand dunes are 'must see,' mayor says available from, *webpage:* https://www.taipeitimes.com/News/taiwan/archives/2020/06/29/2003739042
[9] Braun V, Clarke V 2017 Applied Qualitative Research in Psychology *Appl. Qual. Res. Psychol.* **0887**
[10] Nikolova V and Sinnyovsky D 2019 Geoparks in the legal framework of the EU countries *Tour. Manag. Perspect.* **29** 141–7
[11] UNESCO 2014 United Nations Educational, Scientific and Cultural Organization: Guidelines and Criteria for National Geoparks seeking UNESCO's Assistance to Join the Global Geoparks Network (GGN)
[12] Fauzi N S M and Misni A 2016 Geoheritage Conservation: Indicators Affecting the Condition and Sustainability of Geopark – A Conceptual Review *Procedia - Soc. Behav. Sci.* **222** 676–84
[13] Hoai T 2020 Tầm nhìn chiến lược phát triển du lịch Bình Thuận đến năm 2030 available from: https://trankhangreal.com.vn/tam-nhin-chien-luoc-phat-trien-du-lich-binh-thuan-den-nam-2030/ (in Vietnamese).
[14] Vu N 2019 Gìn giữ, tôn tạo và phát huy vẻ đẹp tự nhiên Đồi cát bay Mũi Né available from: https://bvhttdl.gov.vn/gin-giu-ton-tao-va-phat-huy-ve-dep-tu-nhien-doi-cat-bay-mui-ne-20190428093515474.htm (in Vietnamese).
[15] Lin H M, Shen S M, Lin T-Y, Liaw S C, Chuang Y-C 2011 A study on the industrialized coastal change and degradation of Dayuan- Guanyin in Taoyuan county *Proc. Chinese Geogr. Soc.* 1–13
[16] Yen N 2016 Trồng rừng ở "đồi cát bay, cát nhảy" available from: https://cand.com.vn/Xa-hoi/Trong-rung-de-stop-doi-cat-bay-i387347/ (in Vietnamese).
[17] Gracia A, Rangel-Buitrago N, Oakley J A and Williams A T 2018 Use of ecosystems in coastal erosion management *Ocean Coast. Manag.* **156** 277–89
[18] Dang P T and Aitken S 2014 Conservation of sand dune systems in Canada's capital: A restoration model and case study of the Pinhey Sand Dunes complex *Biodiversity* **15** 148–68
[19] Lin J-C 2018 A strategic plan for managing geopark — an experience of Taiwan *Geophysical Research Abstracts* vol 20 p 5622
[20] Faranasi N T, Coehlo C O A and Costa C M M 2015 Analysis of Network Activities in Geoparks as Geotourism Destinations NEDA *Int. J. Tour. Res.* **113** 101–13
[21] Taoyuan Office of Coast Administration Construction 2021 Caota sand dunes geopark *report*

Modeling impacts of industrial park activity on air quality of surrounding area for identifying isolation distance: A case of Tan Tao Industrial Park, Ho Chi Minh City, Viet Nam

Bang Quoc Ho[1,2*], Hoang Ngoc Khue Vu [1,2], Thoai Tam Nguyen [1,2], Thi-Thu Giang Nguyen[3], Thi-Quynh Nhu Diep[3], Phu Le Vo[2,3] and Thi-Kim Nhung Pham[4*]

[1] Institute for Environment and Resources (IER), Add: 142, To Hien Thanh St, Dist. 10, Ho Chi Minh City 700000, Vietnam
[2] Vietnam National University Ho Chi Minh City, Linh Trung Ward, Thu Duc District, Ho Chi Minh City, Vietnam.
[3] Faculty of Environment and Natural Resources, Ho Chi Minh City University of Technology (HCMUT), Add: 268, Ly Thuong Kiet Street, District 10, Ho Chi Minh City, Vietnam
[4] Branch of National Institute of Occupational Safety and Health in The Southern Vietnam, Add: 124-126, Le Lai St, District 1, Ben Thanh Ward, Ho Chi Minh City, Vietnam.

* Correspondence authors: bangquoc@yahoo.com; cosypham@gmail.com.

Abstract. Industrial factories have been addressed as the main contributor to the amount of air pollution in many urban areas around the world. The emissions of air pollutants from factories, combined with exhausted gases from automobile and domestic cooking activities, have placed enormous adverse effects on human health. Recently, air quality models, which usually figure out for industrial emission with representative indicators such as CO, NO_2, SO_2, and TSP, have triggered an application to identify a suitable isolation distance that could lessen affection on public health. Concerning to develop an air emission inventory for Tan Tao Industrial Park (IP) for three sources of points, line and area sources by using top-down and bottom-up approaches, this study aim to: (i) apply a system model of TAPM – AERMOD model to study the air pollution dispersion from the IP to the surrounding area, and (ii) identify a hygiene isolation distance for sensitive objects around industrial park, especially zones of community, based on their separating demand from contamination. Results show that the point sources are the dominant air emission sources of Tan Tao IP. Total emissions of Tan Tao IP in 2019 estimated at 413.15 tons of TSP/year, 280.9 tons of NO_2/year, 621.99 tons of SO_2/year, and 2720.21 tons of CO/year. Modeling results show that 1-hour maximum concentration of TSP, NO_2, SO_2, and CO in the simulation area is 581µg/m^3, 4.069µg/m^3, 5.478µg/m^3, 40.695µg/m^3, respectively, exceeding the standards. Especially, the pollution levels of NO_2 were 20 times higher than the limit value, similar trends for SO_2 (15 times higher). The hygiene isolation distance was suggested a widely ranged from 2910 meters in the North-West and in the rest directions of Ho Chi Minh City. Some sensitive objects such as residential areas, hospitals and kindergartens recently are inside the affected zone should be reckoned to have suitable solutions that keep their health safe. The method for calculation of hygiene isolation distances from industrial activities has a significant guiding not only for environmental safety action but also for public health protection.

Keywords: Tan Tao Industry park; Atmospheric pollution; Isolation distance; TAPM-AERMOD model; Ho Chi Minh City.

1. Introduction

Industrialization is one of the reasons that worse the quality of life, in the meaning of the air [1]. The air pollutants, releasing into the air from factories, combined with those added by automobiles on the road and cooking activities for workers. So that, demand for distances and additional technical measures should be implemented so as to diminish the risk to people [2]. The air pollution will impact public health such as increasing risk of developing the chronic respiratory disease, lung cancer, heart disease, and many other illnesses, diseases [34]. Whilst a deterministic approach for fixing appropriate distances is done probabilistically or at least using probabilistic elements in some developed countries [2], it just might be pursued in the developing countries.

Ho Chi Minh (HCMC) is the largest commercial industrial city in Viet Nam. Currently, HCMC concentrates 19 industrial zones and numerous small manufactures that could release amount of pollutants affecting on the residential area inside, as simulating results show that the concentration of some classic ambient air (CO, NO_2, O_3 and SO_2) exceeds exposure limits according to national regulation [3]. The running growth of industrialization and the emission scenarios in 2030 represent some drawbacks of implementing the separation distances between industrial activities and residential areas, which base on polluting isolation and health protection demand. These are the lack for regarding on appropriate isolation distances in the planning period and the deficiency of adequate basis on the decision. This boosts the ideas to research finding a technique which could calculate approximately hygiene distances between industrial activities and sensitive objects surrounding. The hygiene distance is an aid that minimizes the impacts of polluting emissions from locating industry to sensitive land uses.

According to the above approaching, there are guidelines to determine the isolation distances applying to stakeholders who are interested in industry development in some countries such as Australia, Germany, Europeans, Malaysia, Singapore and South Africa [4 - 10]. Regarding industrialization, there are three standards to setting up the separation distances depending on the planning policies, and general damaging phenomena in the process plant. In detail, these are distances for an explosion, fire, or release of toxic materials (1); transportation needs (2), and isolated pollution respectively (3) [11]. While determining separation distances in the response of the standard (1) and (2) is applying in European countries, Malaysia and South Africa [4, 5, 8, 10 - 11], the standard (3) have been related by Singapore and Australia [6, 9, 12]. Viet Nam also established the regulation of isolation distance, following the first and second standard, for industrial process in the item of National technical regulation on construction planning (QCVN 01:2009/BXD) [13] and the Decision 3733/2002/QĐ-BYT promulgating 21 Occupational Hygiene Standards, 05 Principles and 07 Occupational Hygiene Measurements [14]. However, these regulations do not be practiced effectively by industrial investors, typically in Ho Chi Minh City (HCMC). Tan Tao IP, being one of the large industrial zones in HCMC, is near the residential areas and consists of different types of processes like electricity and mechanic (18.4%), plastic (15.8%) and textile and garment (14.8%) [15]. There are many sensitive objects surrounding such as hospitals, residential zones, kindergarten and small house might be affected by emission from exhaled sources in IP. This gap provokes a study to find an appropriate distance that could protect sensitive land uses from pollution, which could be supported by model dispersion [11, 16, 17]. The study is aimed to simulate of represented industrial pollutants (NO_2, SO_2, CO and total suspended particular) that are emitted from Tan Tao IP by EMISEN-TAMP-AERMOD system model. The simulating results were used to develop the isolation distances for sensitive land uses surrounding.

2. Materials and Methods

The framework of the study was followed by the Figure 1, including survey steps for collecting data to emission calculation and pollutants dispersion modeling, preparing for determining a suitable distances in the local scope.

Figure 1. Research diagram

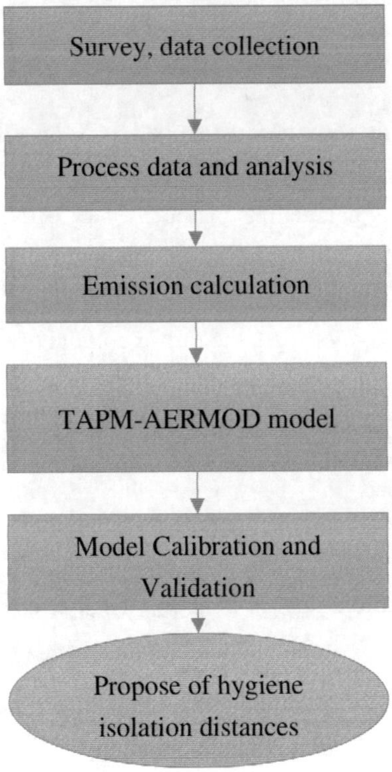

2.1. Data collecting for emission calculation:

Data was collected from three central agencies namely Ho Chi Minh Export Processing Zone and Industrial Park Authority (HEPZA), Tan Tao Investment and Industry Corporation (ITACO), and enterprises of Tan Tao Industrial Park. There are four data collection methods applying on research survey, in details:

i. *Historical method* was used to collect the information on situation of Tan Tao IP, the number of active enterprises, and the distribution of production facilities. All suitable documents were recorded from HEPZA and ITACO.

ii. *Questionnaire method* was applied to get the information about operation process, being supported to the emission calculation process, which includes the source of emissions, the exhaust treatment systems, emission distributions, the amount of raw materials, and the way that fuels were used. These data was collated by survey fulfillments questionnaires at 196 enterprises in Tan Tao IP.

iii. *Observation method* was settled by counting and classifying the types vehicles transmit on 13 main roads representative in the Tan Tao IP. The vehicle type, number of vehicles in circulation, and distance traveled by vehicles supported to the air emissions calculation step.

iv. *Experiment method* was addressed to monitor the ambient air parameters, including TSP, NO_2, SO_2, and CO at four locations in the North, South, West, and East direction of the industrial part. All samples were taken at the same time for later model calibration and validation steps.

2.2. Methodology

To calculate emission for the activities of Tan Tao IP, the research used the emission calculation method based on the emission factor. This method is suitable for Vietnamese conditions [18]. Emission factors for industrial activities in Vietnam are almost nonexistent and not standard. Therefore, the research team applied the emission factor proposed by MONRE in combination with the US EPA AP42 [19] emission factor dataset. Emission calculation formula for production activities in Tan Tao IP is as follows:

$$E = AR \times EF \times \left(\frac{100-ER}{100}\right) \qquad (1)$$

Where: E: emission rate (kilograms/year); AR: operating capacity of the emission source (tons/year) EF: emission factor (kilograms/ton); ER: emission reduction efficiency (%).

2.2.1. EMISENS Model

EMISENS model is used to emission calculate loads due to traffic activities, developed in 2006 – 2010 at the Laboratory of Air and Soil Pollution (LPAS), Lausanne Federal Polytechnic University (EPFL), Switzerland [20]. The theory of EMISENS bases on emissions calculating methods of CORINAIR (European Environment Agency). EMISENS helps users to shorten the calculation time and calculate the errors through the Monte Carlo simulation technique [21]. The input data of the EMISENS model requires: the number of each type of vehicle, the length of each kind of road, the traffic volume of each type of vehicle corresponding to each type of road, the emission factor calculated according to the vehicle system in circulation, etc. Emissions calculation for traffic activities according to the EMISENS model is classified into three types of emissions: hot emissions, E_{hot}; cold emissions, E_{cold}; evaporation emissions, E_{Evap} as follows:

$$E_{Total} = E_{Cold} + E_{Hot} + E_{Evap} \qquad (2)$$

Where: E_{hot}: hot emissions; E_{cold}: cold emissions; E_{Evap}: evaporation emissions; E_{Total}: Total emission. Each type of emission follows a general formula in EMISENS model, which is:

$$E_{ip;ie} = e_{ip;ie} \times A_{ie} \qquad (3)$$

Where: E: total emission; ip: type of pollutant; ie: range of vehicle; e: emission factor; A: traffic activities.

2.2.2. TAPM Model

The TAPM Model is developed by the Commonwealth Scientific and Industrial Research Organization of Australia (CSIRO). This model is used to simulate meteorological conditions and air pollution concentrations in the 3D space environment. This principle function is suitable for supporting meteorological simulation for Tan Tao Industrial Park, where the input data of the model could downscale from the global observed meteorological data, have implemented the detailed meteorological simulations for the whole year 2019. The output meteorological data is the input file for the AERMOD model [22].

IOP Conf. Series: Earth and Environmental Science **964** (2022) 012023 doi:10.1088/1755-1315/964/1/012023

2.2.3. AERMOD Model

The AMS/EPA Regulatory Model (AERMOD Model) is designed to support the management program of the United States Environmental Protection Agency (US-EPA). The model consists of 3 parts: AERMOD (AERMIC Dispersion Model), AERMAP (AERMOD Terrain Tool), and AERMET (AERMOD Meteorological Instrument). Since 1991, The AERMOD model was firstly known in 1991 and officially used on December 9, 2005 after 14 years of research and perfection. AERMOD is the commercial software that uses Gaussian theory to calculate the dispersion of pollutants and is a suitable tool to simulate the dispersion of pollutant concentrations such as TSP, NO_2, SO_2, CO in the atmosphere due to industrial emission activities. Input data, steps, and content to implement the AERMOD model are shown in the following diagram (Figure 2). The meteorological file includes two types: (i) surface meteorological data (*.sam). This file must be the observed data recorded per hour, represented by wind direction, wind speed, air temperature gas, humidity, barometric pressure, precipitation, cloud cover, solar radiation; and (ii) spatial meteorological data (*.ua). This information consists of data, monitoring twice a day at 0 GMT (7:00 LST) and 12 GMT (19:00 LST), and altitude disturbance data. AERMOD would produce simulation results in the form of two-dimensional or three-dimensional images when all data was ready. The grid domain was set 40 km by 40 km with 81 cells x 81 cells, and the grid resolution was 0.5 km x 0.5 km, covering the study areas with Tan Tao IP is in the middle of the simulated topographic map for study area (Figure 2). The output of AERMOD can be imaged on Google Earth platform, making it easily see the impacts of emissions on the survey area [23].

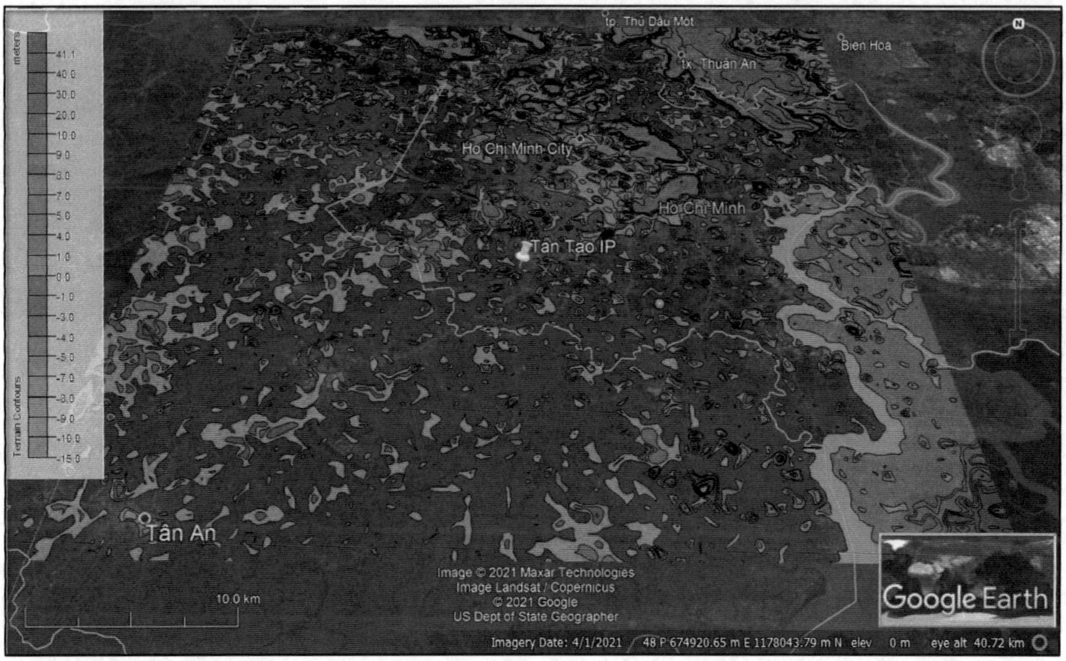

Figure 2. The simulated topographic map

2.2.4. Model calibration and validation

IOP Conf. Series: Earth and Environmental Science **964** (2022) 012023 doi:10.1088/1755-1315/964/1/012023

Meteorological data used to calibrate the model were collected from Da Phuoc meteorological station for 2019. The meteorological data input to the TAPM model is downloaded at the same time from the CSIRO website, which is the global meteorological observation provided by the Australian Meteorological Agency for the world. When TAMP model had been calibrated, it would be run for the area by the AERMOD model. The output of the AERMOD model would be validated by relative error coefficients (MAGE, and R^2) [24],[25] with the relating data of the actual monitoring results at the point (X=674623.69; Y=1187253.90). This is signed by the crossroad of Street No 2 and Street D10 inside IP.

The simulating and adjusting parameters will be calibrated by comparing to monitoring results based on the correlation coefficient and the relative error of the model. The correlation coefficient R^2 (4) and the relative error of the model MNBE (5) will be calculated using the equations as following and must be meet the standard requirement in Table 1 to strengthen the reliability of dispersion results [19].

Correlation coefficient:

$$R = \frac{\sum(Pi - \overline{Pi})(Oi - \overline{Oi})}{\sqrt{\sum(Pi - \overline{Pi})^2 \sum(Oi - \overline{Oi})^2}} \quad (4)$$

Where: Pi: Simulation data, Oi: Monitoring data,

N (n) quantity of data.

Mean Normalized Bias Error (MNBE):

$$MNBE = \frac{1}{N}\sum_{i=1}^{N} \frac{C_{mod}(x_{i,t}) - C_{obs}(x_{i,t})}{C_{obs}(x_{i,t})} \times 100 \quad (5)$$

Where:

$C_{mod}(x_{i,t})$: modeling value at site i, time t

$C_{obs}(x_{i,t})$: observed value at site i, time t

Table 1. Standard of evaluation indicator

Indicator	Standard
Correlation coefficient(R^2)	≥ 0.7
Mean Normalized Bias Error (MNBE)	± 15%

2.3. Steps to determine sanitary buffer/isolation zone

Step 1: Identify sensitive objects around the IP and permissible standards/regulations for concentration of pollutants.

Step 2: After validating the dispersion results of the AERMOD model, the AERMOD model was simulated for 8760 hours/year of 2019. The polluted area was determined based on the air pollution dispersion map of NO_2, SO_2, CO and TSP that exceeds the permissible guideline for ambient air quality. The isolation distance was determined for Tan Tao IP based on Vietnam's technical regulation for ambient air quality limits and the influence of local meteorological parameters. In this study, the allowed threshold guideline for ambient air quality is the National technical regulation in ambient air quality (QCVN 05:2013/BTNMT). The hygiene isolation distance is limited from the border point of the industrial zone boundary to the point, where the polluted area exceeds the QCVN 05:2013/BTNMT determined from the simulation results.

3. Results

Tan Tao IP has two main parks with 232 enterprises that release air emission into the atmosphere. The pointed sources are related to the using of DO oil with total a total consumption volume of above 7 million liters per year. There are 63/196 enterprises with exhaust gas treatment systems at the time of survey and the most commonly technology is the absorption tower. There are also 13 main roads, being inside The Tan Tao IP, are road sources emissions from transportation activities.

3.1. Air emission inventory

Air emissions for point source and area source, according to the formula (1) were calculated by excel tool, and emissions for traffic activities (road source) in the IP were estimated by the EMISEN model. The results show that point sources, being the chimneys, are the main emission source of Tan Tao IP. The emission of each air pollutants from three emission sources is obtained in Table 2. These results are the input data for the AERMOD air pollution dispersion model, in terms of emission loads by gram per second per square meter (usually uses: g/s/m^2).

Table 2: The total air emissions of Tan Tao Industrial Park in 2019

Parameter	TSP (Tons/ year)	NO_2 (Tons/ year)	SO_2 (Tons/ year)	CO (Tons/ year)
Point sources	140.47	262.97	619.28	1918.9
Area sources	270.18	0.53	0.20	0.07
Line sources	2.50	17.40	2.51	801.1

3.2. Developing the hygiene isolation distances from modeling results

3.2.1. Calibration and validation models

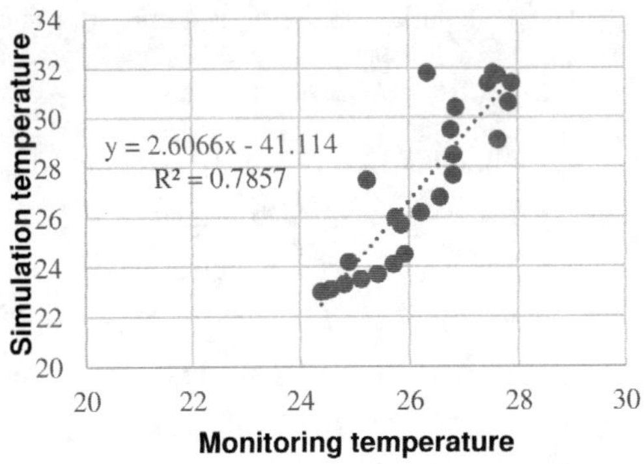

Figure 3. Correlation between simulated temperature and observed temperature on January, 2019

Table 3. The calculating relative error of AERMOD model on April 16th, 2019

Parameter	Results of simulation (µg/m^3)	Results of monitoring (µg/m^3)	MNBE (%)
TSP	35.98	40	+10.1
NO_2	67.3	60	-12.2
SO_2	57.1	50	-14.2
CO	3950	4600	+14.1

The reliability of simulating values is carried out by the model calibration and validation steps. To test the TAMP simulation results, value R^2 was used to validate meteorological TAPM model. The correlation coefficient ($R^2 = 0.7857 > 0.7$) shows a high correlation of TAMP model and monitoring results. This correlation coefficient R^2, being close to 1, means that the results of simulated temperature is very close to results of observed temperature in January 2019 (Figure 3) and the TAPM has good performance/quality to simulate meteorological for the study area. When AERMOD model was validated, in case of a small domain as Tan Tao industrial park and the wind direction in North-West could blow air plumb to South East, only one monitoring result at the *South location* was utilized to validate the model. The relative error between simulated concentrations of CO, SO_2, NO_2, and TSP and their monitoring results on 16th April 2019 is shown in Table 3. The values of MNBE belong to chosen pollutants is from -14.2 % to 10.1% proved that the outcome of the AERMOD model is qualify for this study.

3.2.2. Simulation results for wind directions in study area

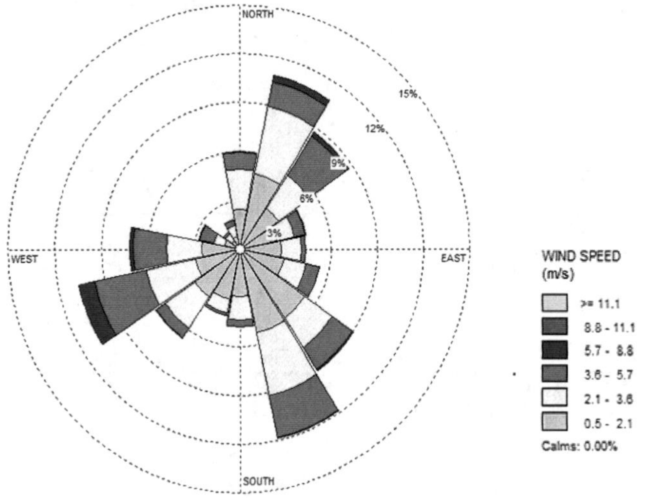

Figure 4: Wind rose in the study area

The calibration values had indicated that the AERMOD model is fully capable of simulating the dispersion of air pollution by the region. So, the model could apply this wind rose to simulate air pollution for different scenarios. The wind rose chart displays the frequency of wind directions and wind speed which has strongly affect dispersion pictures of pollutants in the study area (Figure 4). It is clear to show that the main wind directions of Tan Tao IP are Southwest, Northeast and Southeast.

3.2.3. Simulation results for representing pollutants in Tan Tao Industrial Park

The outcomes of the simulation process are concerning pollutants like CO, SO_2, NO_2, and total suspended particular (TSP), defining in 1-hour maximum, 8-hour mean, 24-hours mean and annual average concentration. Simulation outputs were compared with the air quality in ambient environment standards, regulating on QCVN 05:2013/BTNMT. So that, pollution map results had been extracted from the AERMOD model, including (i) 1-hour maximum concentrations (CO, SO_2, NO_2 and TSP); (ii) 8-hours mean concentration of CO; (iii) 24-hours mean concentrations (SO_2, NO_2 and TSP); and (iv) annual average concentrations (SO_2, NO_2 and TSP). Noticeable outcomes involve 1-hour maximum concentration of NO_2, 8-hours mean concentration of CO, of 24-hours mean concentration and annual average concentration of SO_2, being the highest values, which would be described on Figure 5, 6, 7, and 8.

A quick review of the simulation results (2019) indicated that several pollutants had pollution levels higher than the standards (QCVN 05:2013/BTNMT), but only occurred in the industrial zone areas. This includes CO 8-hours average (12.662 $\mu g/m^3$), NO_2 24-hours average (472.4 $\mu g/m^3$), SO_2 24-hours average (636 $\mu g/m^3$), annual average of NO_2 (80.1 $\mu g/m^3$), and annual of SO_2 (108 $\mu g/m^3$).

Concerning the results on 1-hour maximum type, the average maximum concentration of TSP, NO_2, SO_2, and CO in the simulation area is 581 µg/m^3, 4.069 µg/m^3, 5.478 µg/m^3, 40.695 µg/m^3, respectively, exceeding the standards. Especially, the pollution levels of NO_2 were 20 times higher than the limit value, similar trends for SO_2 (15 times higher). The isolation distances were determined from the center of pollution to the farthest radius where the levels meet the standard. As this circumstance, the suggesting space was 1.75 km for TSP, 2.91 km for NO_2, and 2.15 km for SO_2 toward the northwest direction.

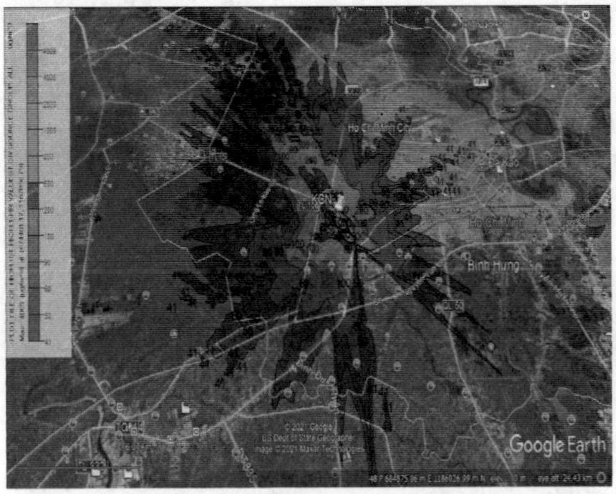

Figure 5. NO_2 dispersion map for 1-hour concentration scenarios

Figure 6. CO dispersion map for 8-hour mean concentration scenarios

Figure 7. SO_2 dispersion map for 24-hour concentration scenarios

Figure 8. SO_2 dispersion map for annual average concentration scenarios

3.2.4. The hygiene isolation distances for Tan Tao Industrial Park

The hygiene isolation distance is determined as the farthest distance, measuring from the edge of the industrial zone to the boundary between the exceeding and permitted standards (QCVN 05:2013/BTNMT). These exposure thresholds are suggested to ensure the health of sensitive objects safely. In the study areas, there are some sensitive objectives such as kindergartens, pagodas, residential areas, clinics and hospitals, which might be impacted by the high levels of pollution from Tan Tao IP. Therefore, the reasonable hygiene isolation distance for Tan Tao Industrial Park should be the maximum length according to the settling measure. According to the concentration levels of considered pollutants, at each time resolution, hygiene isolation distance was calculated as follows QCVN 05:2013/BTNMT and listed as Table 4. At a result, the proposed distance for Tan Tao IP is 2910 m in radius from the center point, to protect public health.

Table 4. Construction of sanitary buffer zone for Tan Tao Industrial Park

Parameter	Scenario base on average time	Values (µg/m³)	QCVN 05:2013/BTNMT (µg/m³)		Exceeding times	Minimum hygiene isolation distance	Main direction
TSP	1h	6 - 581	300	**Exceeded**	194	1750	Northwest
NO_2	1h	41 - 4069	200	**Exceeded**	20	2910	
SO_2	1h	55 - 5478	350	**Exceeded**	157	2150	
CO	1h	407 - 40695	30000	**Exceeded**	14	50	
CO	8h	127 - 12662	10000	**Exceeded**	13	50	Northwest - Southwest
TSP	24h	0.7 – 72.4	200	Lower	-	50	Northwest - Southwest
SO_2	24h	6 - 636	125	**Exceeded**	51	50	
NO_2	24h	4.7 - 272.4	100	**Exceeded**	47	50	Northwest – Northeast & Southeast
TSP	1 year	0.09 - 8.74	100	Lower	-	50	Northwest - Southwest
NO_2	1 year	0.8 - 81.1	40	**Exceeded**	2	50	
SO_2	1 year	1 - 108	50	**Exceeded**	216	50	

4. Discussion

Total emissions of Tan Tao IP in 2019 was estimated at 413.15 tons of TSP/year, 280.9 tons of NO_2/year, 621.99 tons of SO_2/year, and 2720.21 tons of CO/year. This represents that industrial activities emitted a large number of harmful pollutants (TSP, NO_2, SO_2, and CO). The airborne diffusion of pollutants can pose a public health risk, increased risk of developing chronic respiratory disease, lung cancer, heart disease, and many other illnesses, diseases [26 - 27]. It is essential to prepare suitable isolation distance from industrial areas in planning in phases to mitigate health risks for the public surrounding.

Recently, there have been a growing concerns about the air dispersion model that helps users simulate the distribution of pollution in space, for the relating pollutants such as SO_2, NO_2, TSP, VOC_S,

IOP Conf. Series: Earth and Environmental Science **964** (2022) 012023 doi:10.1088/1755-1315/964/1/012023

and CO…[28],[29],[30],[31]. Almost all these models are based on the theory of GAUSS and the turbulent dispersion theory of BERLIAND [32]. The AMS/EPA Regulatory Model (AERMOD), designing to support the US Environmental Protection Agency (EPA) management program and using the theory of GAUSS, includes a range of options for simulating air quality impacted by waste streams, building common options for many applications. The AERMOD, being assessed at high spatial resolution and medium domain extension in comparison with other models like CALPUFF, CALIOPE, and SMOKE [33], could result pollutants concentration maps in types of particular matter and gases [33]. The TAPM is the prognostic meteorological and air pollution models, being weakly nudged with a 24-hour e-folding time towards the synoptic-scale input values. The output of TAMP model can be the input of AERMOD. So that, the TAPM-AERMOD model system, being the meteorology-emission dispersion model systems, is an approach to identify the isolation and develop a sanitary buffer zone due to industrial emissions.

In Vietnam, there is Decision 3733:2002/BYT and QCVN 01:2019/BXD, regulating minimum isolation distance for manufacture process [13-14]. However, these standards are only suitable for separating processes and do not consider the wind direction factor. Therefore, the application of the model in calculating the isolation distance is a solution to overcome these drawbacks in the standards of Viet Nam. The application of model system is not only an effective solution to simulating the spread of air pollutions reliably, but also saves time, costs for planning policy.

The emission dispersion result (Table 4) indicated that simulating values of important ambient air pollutants such as TSP, NO_2, SO_2, and CO are exceeded limited thresholds. This also mean that the local area surrounding the IP with some sensitive objects like kindergartens, pagodas, residential, clinics, and hospitals is at health risk in general. So that, this research suggests four local-abatement strategies for controlling distribution air pollution in Ho Chi Minh region. These sector-specific policies, being based on replies by parties to general solution questions observation questionnaire, tend to improve the overall quality of life of local residential surrounding. The prior is the promoting energy efficiency, instead of using fossil fuels such as DO and coal for boilers, the enterprises should take advantages of agro material one like straw, sawdust or wood waste. The second is reducing the emission from transport sector by promotion of cycling and public transportation inside industrial park. The third one is reported on the measure of reducing emission from point source, which could be invested in efficiency exhaust gas treatment system and connected to automatic emission monitoring system of Tan Tao IP. Enlarging more green area inside each company also should be taken into account in abatement strategy of air pollution.

5. Conclusion

In this study, a simulation model system of air pollution dispersion was used to determine and propose hygiene isolation distance, aiming to protect public health. Emission calculation results from Tan Tao IP pollution sources are input data of the AERMOD model to simulate pollution spread. Point sources are the main air emission sources of Tan Tao IP. Total emissions of Tan Tao IP in 2019 estimated at 413.15 tons of TSP/year, 280.9 tons of NO_2/year, 621.99 tons of SO_2/year, and 2720.21 tons of CO/year.

Modeling results show that 1-hour maximum concentration of TSP, NO_2, SO_2, and CO in the simulation area is 581μg/m^3, 4.069μg/m^3, 5.478μg/m^3, 40.695μg/m^3, respectively, exceeding the standards. Noticeably, the pollution levels of NO_2 were 20 times higher than the limit value, similar trends for SO_2 (15 times higher). The simulation results suggest that the hygiene radius is widely about 2.910 meters in all directions, from the boundary of IP. Some sensitive objects such as residential areas, hospitals and kindergartens, now being located inside the affected zone, should be considered to relocated or other suitable solutions to protect their health. The method in this study would have driven an approach to create a reasonable hygiene buffer zone far from the concentrated industrial activities. The determination of isolation distance should be based on the hygiene protection requirements for these sensitive objects. Besides, several important elements that should be taken into account are (i) the background pollution caused by daily activities in the areas surrounding the industrial park, and (ii) high-rise constructions around the industrial park play as the obstacle of the pollution plume.

Author Contributions:

Nhung Pham contributes as coordinator for whole research, air quality modeling and develop abatement strategies. P.L.Vo was responsible for developing research methodology. T.T.G.N and Q.N.D collected data and process data. While H.N.K.V., T.T.N were responsible for calculating emission. H.N.K.V. is in charge of running TAPM model. B.Q.H is in charge of running AERMOD model. The manuscript was developed, revised, and edited by B.Q.H., P.L.V, Nhung Pham and H.N.K.V.

Conflicts of Interest: The founding sponsors had no role in the design of the study; in the collection, analyses, or interpretation of data; in the writing of the manuscript, and in the decision to publish the results.

Acknowledgements: The authors would like to sincerely thank you Vietnam General Confederation of Labor, Vietnam National Institute of Occupational Safety and Health for their grant to this research. The authors also acknowledged the support of time and facilities from Ho Chi Minh City (HCMUT), VNU-HCM for this study. This paper and research would not have been possible without the exceptional support of HEPZA and ITACO for their helpful and willingness to provide necessary information and relevant data. The authors also gratefully appreciate the reviewers those provided knowledgeable comments and suggestions which have largely contributed to the improvement of our works' quality. Any errors or shortcomings remain the responsibility of the authors.

References

[1] Munsif, R., Zubair, M., Aziz, A. and Nadeem Zafar, M. 2021 Industrial Air Emission Pollution: Potential Sources and Sustainable Mitigation in *Environmental Emissions*, R. Viskup (Ed.) IntechOpen doi: 10.5772/intechopen.93104.

[2] Hauptmanns, U. 2015 *Process and Plant Safety*. Berlin, Heidelberg: Springer Berlin Heidelberg doi: 10.1007/978-3-642-40954-7.

[3] B. Q. Ho *et al.* 2020 Study loading capacities of air pollutant emissions for developing countries: a case of Ho Chi Minh City, Vietnam *Sci. Rep.* **10** (1) p. 5827 doi: 10.1038/s41598-020-62053-4.

[4] Ministry of Natural Resources and Environment Malaysia 2012 *Guidelines for siting and zoning of industry and residential areas*, Second Revised Edition, Department of Environment, Malaysia.

[5] Hauptmanns, U. 2020 *Process and Plant Safety*. Berlin, Heidelberg: Springer Berlin Heidelberg doi: 10.1007/978-3-662-61484-6.

[6] Australian Capital Territory 2018 *Separation Distance Guidelines for Air Emissions,* The ACT Government, Autralia.

[7] Michalis Christou, Zsuzsanna Gyenes, Struckl, M. 2011 Risk assessment in support to land-use planning in Europe: Towards more consistent decisions? *Journal of Loss Prevention in the Process Industries* **24**(3) 219-226 doi: 10.1016/j.jlp.2010.10.001

[8] GDARD EPIA 2017 Gauteng Pollution Buffer Zone Guideline. Current review facilitated by drectorate enviromental policy planning and coordination *Gauteng Department of Agriculture and Rural Development*, Gauteng Province, Republic of South Africa.

[9] Building & Construction Authority (BCA), Economic Development Board (EDB), Energy Market Authority (EMA), JTC Corporation, National Environment Agency (NEA), Public Utilities Board (PUB), Singapore Civil Defence Force (SCDF), Urban Redevelopment Authority (URA) , "Guide on Construction of Industrial Developments in Singapore. Building and Construction Authority, Singaporedf."

[10] Che Bon Ahmad, Jamalunlaili Abdullah, Jasmee Jaafar 2018 Buffer Zone Concept and its Potential Implementation in Tasek Bera *Asian Journal of Environment-Behaviour Studies* **3**(7) 11-18 doi: 10.21834/aje-bs.v3i7.255.

[11] Đỗ Trần Hải, Phạm Quốc Quân 2020 Yêu cầu bảo vệ bầu khí quyển khi thiết kế hệ thống thông gió công nghiệp và vùng đệm vệ sinh của các cơ sở công nghiệp, khu công nghiệp *Tạp chí An toàn sức khỏe và Môi trường lao động*, **1-2-3** (in Vietnamese).

[12] Warrnambool City Council 2014 Review of land uses within industrial buffers Warrnambool City Council, Australia

[13] Ministry of Construction of the Socialist Republic of VietNam 2021 National technical regulation on construction planning: QCVN 01:2019/BXD. http://vbpl.vn/boxaydung/Pages/vbpq-toanvan.aspx?ItemID=141828&dvid=324 (accessed Aug. 27, 2021).

[14] Ministry of Health of the Socialist Republic of Viet Nam 2021 Decision No. 3733/2002/QD-BYT promulgating 21 labor hygiene standards https://thuvienphapluat.vn/van-ban/The-thao-Y-te/Decision-No-3733-2002-QD-BYT-promulgating-21-labor-hygiene-standards-142331.aspx (accessed Aug. 27, 2021).

[15] Khu công nghiệp Tân Tạo - Ban quản lý các khu chế xuất và công nghiệp thành phố Hồ Chí Minh http://www.hepza.hochiminhcity.gov.vn/web/guest/khu-cong-nghiep-tan-tao (accessed Aug. 27, 2021).

[16] Brancher, M., Piringer, M. Franco, D., Belli Filho, P., De Melo Lisboa, H. and Schauberger, G. 2019 Assessing the inter-annual variability of separation distances around odour sources to protect the residents from odour annoyance *J. Environ. Sci.* **79** 11–24 doi: 10.1016/j.jes.2018.09.018.

[17] Thoai Tam Nguyen, Hoang Ngoc Khue Vu, Thi Thu Thuy Nguyen, Thi Thuy Hang Nguyen, Quoc Bang Ho 2020 Study atmospheric pollution to develop a process for calculating the isolation safe distance for livestock activities: Applied in Ho Chi Minh City *European Journal of Geosciences* **2**(2) 1-14.

[18] Q. B. Ho, H. N. K. Vu, T. T. Nguyen, T. T. H. Nguyen, and T. T. T. Nguyen 2019 A combination of bottom-up and top-down approaches for calculating of air emission for developing countries: a case of Ho Chi Minh City, Vietnam *Air Qual. Atmosphere Health* **12**(9) 1059–1072 doi: 10.1007/s11869-019-00722-8.

[19] Bang Q. Ho 2019 *Urban Air Pollution: from theory to practice*. VNU.

[20] B. Q. Ho and D. M. Ho, *Training document: EMISEN Model, FVM Model, TAPOM Model and supporting program.* VNU.

[21] B. Q. Ho, A. Clappier, and N. Blond 2014 Fast and Optimized Methodology to Generate Road Traffic Emission Inventories and Their Uncertainties *CLEAN – Soil Air Water* **42**(10) 1344–1350 doi: 10.1002/clen.201300261.

[22] P. J. Hurley, W. L. Physick, and A. K. Luhar 2005 TAPM: a practical approach to prognostic meteorological and air pollution modelling *Environ. Model. Softw.* **20**(6) 737–752 doi: 10.1016/j.envsoft.2004.04.006.

[23] Lakesenvironmental 2021 AERMOD View *Lakes Environmental Software.* https://www.weblakes.com/software/air-dispersion/aermod-view/ (accessed Aug. 30, 2021).

[24] D. M. Ho, B. Q. Ho, and T. V. Le 208 Evaluate of air pollution dispersion and propose planing scenerios to reduce air pollution for livestock activities in Tan Thanh district, Ba Ria – Vung Tau province *Sci. Technol. Dev. J. - Sci. Earth Environ.* **2**(2) doi: 10.32508/stdjsee.v2i2.491.

[25] H. M. D. Bang Quoc Ho and Vu Hoang Ngoc Khue Nguyen Thoai Tam 2016 Air quality studies in Hoang Dieu Port, Can Tho city *ICENR 2016 and 2016 ILTER-EAP Ecological insights and environmental protection for sustainable development under climate changes in East-Asia and Pacific regions.*

[26] V. H. Borja-Aburto, J. A. Rosales-Castillo, V. Manuel, G. Corey, and G. Olaíz-Fernández, EVALUATION OF HEALTH EFFECTS OF POLLUTION p. 67.

[27] A. Al-Wahaibi and A. Zeka 2015 Health impacts from living near a major industrial park in Oman," *BMC Public Health* **15**(1) 524 doi: 10.1186/s12889-015-1866-3.

[28] Ministry of Economy, Trade and Industry 2005 Low Rise Industrial Source Dispersion Model METI-LIS Model Ver. 2.0: Technical Manual *Research Center for Chemical Risk Management, National Institute of Advanced Industrial Science and Technology* Japan.

[29] Ministry of Economy, Trade and Industry 2005 Low Rise Industrial Source Dispersion Model METI-LIS Model Ver. 2.0: Operation Manual *Research Center for Chemical Risk Management, National Institute of Advanced Industrial Science and Technology* Japan.

[30] World Health Organization (WHO-EURO) 2020 Health impact assessment of air pollution: introductory manual to AirQ+ *WHO Regional Office for Europe.*

[31] Ebel, A., Davitashvili, T., Elbern, H., Gordeziani, D., Jakobs, H. J., Memmesheimer, M. and Tavkhelidze, I. 2004 Numerical modelling of air pollution on regional and local scales *AMIM* **9**(2) 14.

[32] Leelőssy, Á., Molnár, F., Izsák, F., Havasi, Á., Lagzi, I. and Mészáros, R. 2014 Dispersion modeling of air pollutants in the atmosphere: a review *Open Geosci.*, **6**(3), doi: 10.2478/s13533-012-0188-6.

[33] E. Antuña Yudego, JL. Carús Candás, E. Álvarez Álvarez, MJ. Suárez López, L. García, VM. Fernández-Pacheco 2018 Computational Tools for Analysing Air Pollutants Dispersion: A Comparative Review *International Research Conference on Sustainable Energy, Engineering, Materials and Environment (IRCSEEME)*, Mieres, Spain.

[34] IManisalidis, I., Stavropoulou, E., Stavropoulos, A. and Bezirtzoglou, E. 2020 Environmental and Health Impacts of Air Pollution: A Review *Front. Public Health*.

Application TAPM-AERMOD system model to study impacts of thermal power plants in SouthEast and SouthWest areas to the air quality of HCMC: current status and according to Vietnam power planning VII toward 2030

Khue Hoang Ngoc Vu [1,2], **Hang Thi Thuy Nguyen** [1,2*], **Tam Thoai Nguyen** [1,2], **Bang Quoc Ho**[1,2*]

[1] Institute for Environment and Resources (IER), Add: 142 To Hien Thanh, Ward 14, District 10, Ho Chi Minh City 700000, Vietnam
[2]Department of Science and Technology, Vietnam National University-Ho Chi Minh City (VNU-HCM), Ho Chi Minh City 700000, Vietnam

*Correspondence authors: bangquoc@yahoo.com; hangnguyen6769@gmail.com.

Abstract. Vietnam's urban areas have faced serious environmental pollution issues, including water pollution, municipal waste, and air pollution. Vietnam's real gross domestic product (GDP) has been experiencing positive growth for the past five years since 2016. And in 2019, *Vietnam's* real *GDP* increased by 7.02% compared to the previous year. To maintain the growth rate, there is a huge amount of electricity required, not accounting for the other sectors. Thermal power plants generate more than 50% of total electricity in Vietnam, therefore, it is said that coal-fired power plants have been the major sources of air emissions and caused a serious impact on the environment. Recently air pollution is a hot issue in Ho Chi Minh City (HCMC), the air quality is being polluted by $PM_{2.5}$, O_3, CO, NO_2, and TSP. Despite that, the neighboring areas of the city will install more coal-fired power plants, threatening to degrade the quality of the environment. Therefore, the objectives of this study are (i) Modeling the impacts of thermal power plants in SouthEast and SouthWest areas on the air quality of HCMC for two scenarios (current status in 2019 and future according to Power planning VII (adjusted) toward 2030); And (ii) Develop interprovincial air quality protection solutions. The research applied the TAPM model for meteorological modeling and AERMOD model for air pollution dispersion. The annual average $PM_{2.5}$ concentration in the study area was approximately 0.121 $\mu g/m^3$ and the highest concentration at a location close to Vinh Tan thermal power center with 8.61 $\mu g/m^3$. NO_2 the annual average concentration from power plants in 2020 and 2030 blows to HCMC and contributes to HCMC' air quality only 0.01 and 0.03 $\mu g/m^3$, respectively. The 24 hours average concentration of SO_2 from power plants in 2030 blows to HCMC and contributes to HCMC' air quality of 10 $\mu g/m^3$. The 24 hours average SO_2 levels of HCMC in 2030 is 39.2 $\mu g/m^3$, higher than WHO's guideline (20 $\mu g/m^3$). Currently, air pollution in HCMC is polluted by $PM_{2.5}$, SO_2, and NO_2 and cause bad effect to public health. However, in the future with the contribution of 33 thermal power plants under intercity/provinces air pollution dispersion, air pollution HCMC will be worse and affect public health. Air pollution HCMC will be a huge impact on HCMC's public health in the future due to the contribution of 33 thermal power plants under intercity/provinces air pollution dispersion. The paper developed 7 main mitigation measures to reduce the impacts of air pollution from the power plan and reduce the impacts of air pollution on HCM's public health. The measures are focused on using clean fuel, advanced technology, and controlling trans-provincial air pollution.

Keywords: TAPM–AERMOD; Thermal power plants; Air quality modeling; SouthEast and SouthWest; Ho Chi Minh city; Vietnam

1. Introduction

Ho Chi Minh City (HCMC) is located in the Southeastern region of Vietnam, a contiguous area of two regions SouthEast and SouthWest. There are 20 districts and 1 city in HCMC, with a population of 8.6 million people, and about nine million private vehicles. The city's economic growth rate has been skyrocketing in recent years. Currently, there are 19 manufacturing and industrial zones, 30 industrial clusters on an area of 1,900 ha, and numerous factories and enterprises located separately in HCMC [1]. The huge volume of household cooking fuels, private traffic in the area, and industrial units in and around the city release significant amounts of pollutants into the atmosphere. The results show that the highest concentrations of TSP, $PM_{2.5}$, CO, NO_2 and O_3 in 2017 exceeded the National technical regulation in ambient air quality (QCVN 05:2013/BTNMT) 1.5, 1.5, and 1.1 times [2]. There is 93.8% of PM data recorded from the 19 traffic locations have exceeded The National Technical Regulation on Ambient Air Quality (QCVN 05: 2013/BTNMT). These high pollutant concentrations were associated with an increase in the risk of human health in HCMC [3-5]. The research of our research team in the year 2017 about the impact of air pollution on human health, showed that air pollution in HCMC was seriously affecting public health by the previous study of us [6]. The air pollution levels ($PM_{2.5}$, SO_2, and NO_2) were responsible for 1,396 deaths related to three diseases (lung cancer, cardio-pulmonary, and Ischemic Heart Disease (IHD)). In which, levels of $PM_{2.5}$ were associated with 1.136 deaths, followed by NO_2 with 172 cases and 89 cases by SO_2. Moreover, the air pollution level in HCMC is forecasted to be increased in the future under the pressure of economic development due to the National Power Development Plan (Power Master Plan VII revised) for the 2011 – 2020 period with the vision to 2030.

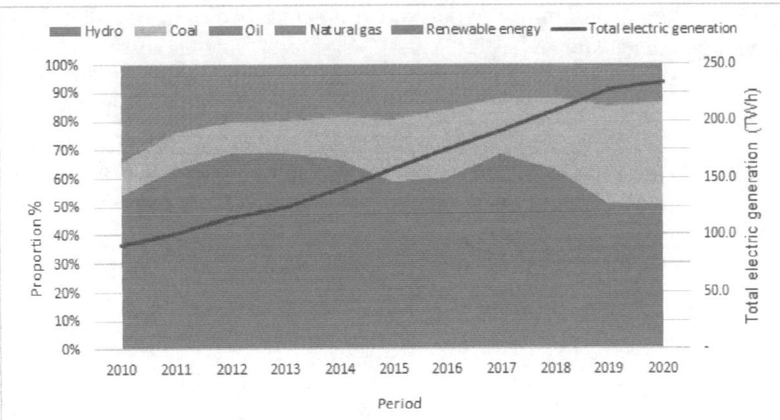

Figure 1. Proportion of fuel type to generate electricity in Vietnam from 2010 to 2020

From 1985 to 2020, the amount of electricity in Vietnam has rapidly increased 50 times from 5 TWh to 234 TWh. The average increase was 11% per year [7]. In Figure 1 shown the proportion of fuel type to generate electricity in Vietnam from 2010 to 2020, In the beginning hydroelectric was the main source (accounting for 53%) and strongly increased, fluctuating around 60% and 70% from 2012 to 2017 and then gradually decreased from 2018, then stopped at 50% in 2020. Witness the same trend, electricity from natural gas was dropped from 31% in 2010 to 10% in 2020. On the other hand, coal power plants increased quickly from 12% to 37% in 10 years, replacing the proportion of hydroelectricity. Electricity from renewable energy and Diesel Oil (DO) was negligible. Only in recent years, renewable energy has been concerned and developed, thanks to the new policy and support from the government. Resulting in the growth from 2% in 2019 to 4% in 2020.

According to the Power Master Plan VII revised, the total electricity generation will reach: 235 - 245 billion kWh in 2020, 352 - 379 billion kWh in 2025, and 506 - 559 billion kWh in 2030 [7], and the two major of energy use is hydropower and coal power. The proportion of each fuel used to generate energy is 16.9% from big and medium hydroelectric, 42.6% from coal, and 14.7% from natural gas (including Liquefied Natural Gas (LNG)), 21% from renewable energy, and 3,6% from nuclear, 1,2% from import.

Coal is the popular fuel for electricity generation in countries like India and China. The cheap and abundant supply of coal locally, together with the high prices of imported natural gas and oil, make the coal-fired generation of electricity more attractive economically [8]. Therefore, many studies on coal-fired power have been carried out in India and China. Thermal power emissions are usually calculated using the emission factor and material balance method [8-10]. The meteorological and air quality modeling system is applied to simulate the spread of air pollutants from the thermal power plants.

Vietnam has a lot of potential for renewable energy development and clean energy such as solar energy, wind energy, biomass energy, and hydroelectricity. However, there has not been any research to evaluate the overall assessment of all plants in this master plan on air pollution and the public health of HCMC. Therefore, the management agencies have not seen the impact of serious thermal power plants on air pollution and public health, therefore it is necessary to carry out this study to provide scientific data to convince the management agencies to adjustment the master plan and for inter-provincial coordination to develop an energy plan for the whole region without affecting neighboring provinces. Because the problem of air pollution is an inter-provincial issue. Therefore, the objectives of this study are (i) Modeling the impacts of thermal power plants in SouthEast and SouthWest areas on the air quality of HCMC for two scenarios (current status in 2019 and in future according to Power planning VII (adjusted) toward 2030); And (ii) Develop interprovincial air quality protection solutions. The research applied the TAPM (The Air Pollution Model) model for meteorological modeling and the AERMOD model for air pollution dispersion.

In Vietnam, there have also been studies simulating the air pollution of thermal power plants. Research by Ho Quoc Bang et al, 2019 with research on impact assessment of Duyen Hai thermal power plants. The study applied the meteorological and air quality modeling system (TAPM-CALPUFF). The study of Duong Ngoc Bach in 2017 about the simulation of pollution from Na Duong I & II thermal power plants to ambient air quality in Lang Son province. Research and apply AERMOD VIEW 8.2 software to simulate the dispersion of TSP, SO_2, and NO_2 pollutants from the chimneys of Na Duong I & II thermal power plants.

2. Materials and Methods

This research contains three main stages: air pollution emission inventory; meteorology simulation, air quality modeling, and proposed mitigation measures based on the results. Step by step process of the research is described in Figure 2.
The first step is to make a comprehensive list of power plants in the research area with detailed information such as name, fuel used, location, time in operation. Then collect all the information related to these plants including chimneys height, gas temperature, amount of fuel used, type of fuel, abatement technology, combustion system… from EIA reports, Decision of Power Master plan VII revised, all the Decision and report relating to the implementing progress the Power Master plan, etc. After that,

emission factors for each type of power plant need to gathering from AP-42, WHO, and EMEP/EEA guidelines and put into calculations, taking into consideration the effectiveness of the treatment system. The second step, meteorology, and air quality dispersion model should be set up for the research area with calibration process, detailed will be discussed after. Then, the meteorology in the year 2019 was prepared with hourly data for 365 days, presented in the surface and upper files.
Finally, input emission data and meteorology data into AERMOD model, air pollution map will be generated after finishing the calibration and validation steps. Based on these maps, inter-provinces pollution levels will be defined, and mitigation measures will be defined.

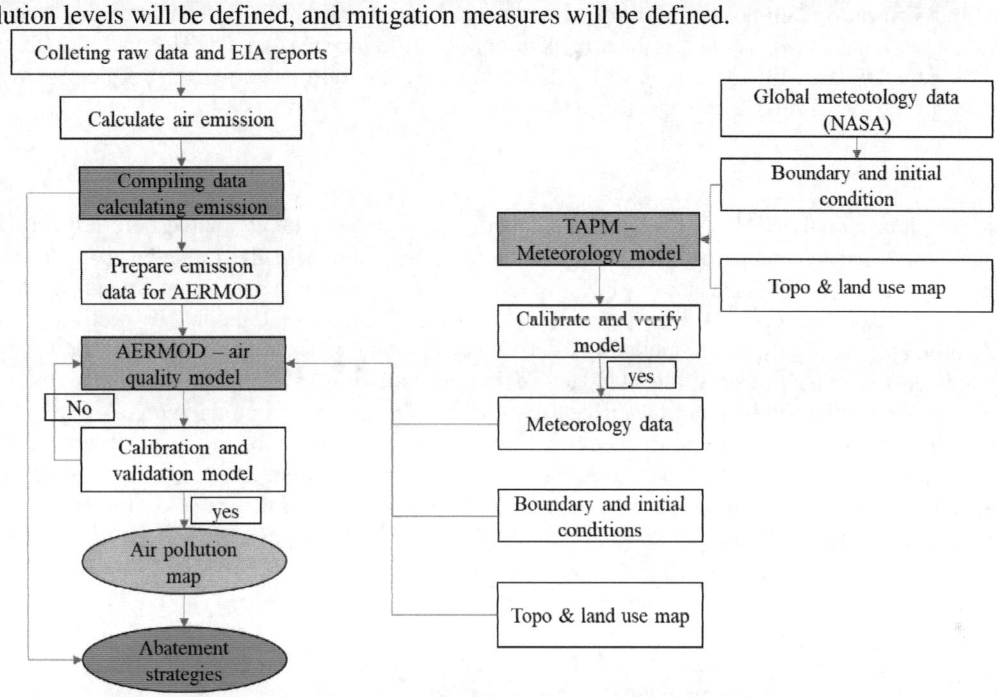

Figure 2. Research diagram of this study

2.1 Meteorological model

The Air Pollution Model (TAPM) was developed by the Commonwealth Scientific and Industrial Research Organization - Commonwealth Scientific and Industrial Research Organization (CSIRO) of Australia. TAPM is a three-dimensional Eulerian meteorological model for simulating meteorology. Detailed descriptions of the model were described by [2]. The model is non-hydrostatic and anelastic using terrain following grid with finite volume discretization [2]. The TAPM model was run for 3 domains, covering the whole study area, the smallest domain is covering the central region that focused on HCMC and surrounding provinces. Since the study aims to define the levels of pollution of thermal power plants from the outside travel to HCMC, a meteorology point in Dong Nai province (next to Formosa power plant) was adopted as the input for the AERMOD model. The model manages meteorological data on the surface and on different levels of height, allowing the calculation of atmospheric characteristics. The meteorology file consists of the following two types: surface met data file (*.sfc) contains the 1-hour observed data including wind direction, wind speed, humidity, atmospheric pressure, precipitation, cloud cover, solar radiation; and the upper air data file (*.pfl) including similar parameter but in many levels of height.

2.2 Air quality model

The AERMOD Model - The AMS/EPA Regulatory Model (AERMOD) is designed to support the US Environmental Protection Agency (EPA) management program. The model consists of three components: AERMOD (AERMIC Scattering Model), AERMOD (AERMOD Terrain Tool), and AERMOD (AERMOD Meteorological Tool). Since 1991, the AERMOD model has been developed by the Meteorological Agency and the US Environmental Protection Agency (US EPA). A team of scientists (AERMIC) has teamed up to build the AERMOD model. AERMOD was officially used on December 9, 2005, after 14 years of research and development [11]. AERMOD will produce simulation results in the form of 2-dimensional, three-dimensional images and and export them through Google Earth, making it easy for users to see the effects of emissions to the survey area. The grids domain was set 459 km by 459 km with 153 cells x 153 cells and the grid resolution was 3.0 km x 3.0 km, covering the study areas from Binh Thuan province to Ca Mau province.

2.3 Model Evaluation

Similar to other models, TAPM and AERMOD models also need to be evaluated before using. This approach was used by comparing the observations from the field with the modeled data. In which meteorological data at Bien Hoa station and air quality data monitored at 4 stations in Duyen Hai district of Tra Vinh Province were used to assess the TAPM and AERMOD model, respectively. More specifically, statistical parameters including Pearson correlation coefficient (R) between observed (O) and predicted (P) values, mean value, standard deviation, minimum value (min), and maximum value (max) were used in this study. It is imperative to evaluate the air dispersion model with observation data before using the simulated results, to check the input data and options for the research area and the uncertainty of the model. The following criteria are used to evaluate model performance. The equation below is the requirement of these indices to consider the performance of the model (table 1) [9]. Mean Normalized Bias Error (MNBE):

$$MNBE = \frac{1}{N}\sum_{i=1}^{N} \frac{C_{mod}(x_{i,t}) - C_{obs}(x_{i,t})}{C_{obs}(x_{i,t})} \times 100 \tag{1}$$

Where

$C_{mod}(x_{i,t})$: modeling value at site i, time t

$C_{obs}(x_{i,t})$: observed value at site i, time t

Table 1. Standard of evaluation indicator

Indicator	Standard
Mean Normalized Bias Error (MNBE)	± 15%

3. Results

3.1. Air emission inventory

Table 2. Emissions from thermal power plants in research area in 2020 *Sources: [13 - 16]*

	NO_X (ton/year)	CO (ton/year)	SOx (ton/year)	PM_{10} (ton/year)	$PM_{2.5}$ (ton/year)	TSP (ton/year)
	In 2020					
Coal	15.948,57	4.425,92	41.715,57	609,36	423,69	1.324,70
Natural Gas	4.882,03	14.262,12	10,28	0,65	1,30	0,33
DO	1,05	0,74	2,43	0,00	0,00	0,00
Total	**20.831,65**	**18.688,78**	**41.728,28**	**610,02**	**424,99**	**1.325,03**

	In 2030					
Coal	36.540,45	10.140,41	95.576,30	575,72	299,28	1.251,56
Natural Gas	10.410,53	30.412,79	21,91	1,39	2,78	0,69
DO	762,87	540,82	1.772,87	1,81	2,76	1,27
Total	**47.713,86**	**41.094,02**	**97.371,09**	**578,91**	**304,82**	**1.253,52**

The air emission inventory is the main data input for AERMOD air quality model. The air emission inventory is described in Table 2 for 2020 and 2030. This result includes whole thermal power plants in SouthEast and SouthWest of Vietnam with 5 coal power plants in Vinh Tan and two gases power plants in Son My of Binh Thuan province. These seven thermal power plants in Binh Thuan province are located about 218 km far from HCMC, having stacks in 210m (height) and 6m (diameter), air pollution from these stacks can disperse to HCMC (Figure 3). Therefore, this paper simulates for whole thermal power plants in SouthEast and SouthWest of Vietnam and seven power plants in Binh Thuan province. Emission calculation results is referred from other study [13] and additional calculations for seven power plants in Binh Thuan province.

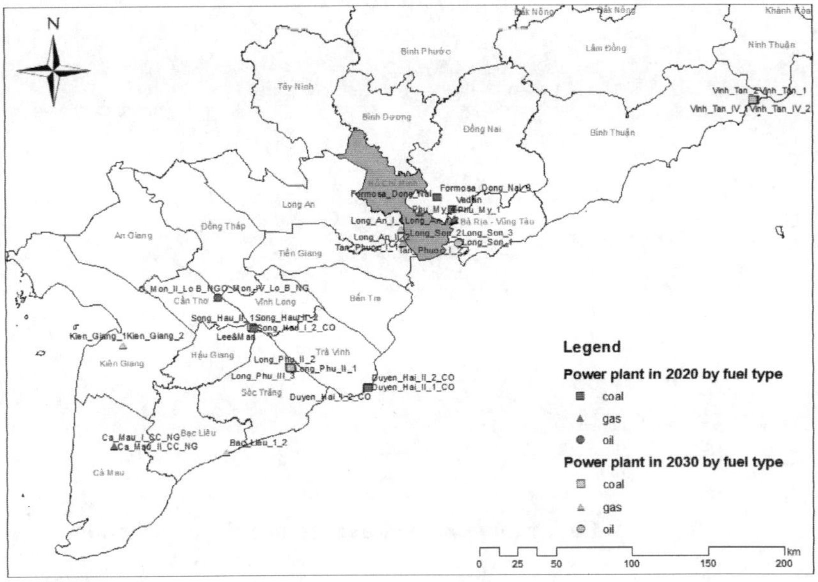

Figure 3. Distribution of Thermal power plants in study area (South-East, South-West and Vinh Tan power center)

The results of air emission inventory in Table 2 shows that air emission load in 2030 is higher than emission in 2020 more than 2 times for NOx, CO and SOx. This is because the electric production in 2030 is higher than 2020.

However, air emission inventory in Table 2 shows that air emission load in 2030 is lower than emission in 2020 for dust (TSP, PM_{10} and $PM_{2.5}$) NOx, CO and SOx. This is because the electric production in 2030 is higher than in 2020. This can be explained because in the future all plants using domestic coal will switch to using imported coal, the amount of ash in imported coal is much lower than domestic coal. Domestic coal of Quang Ninh 6A anthracite coal has 33.12% of ash, while imported coal such as from Indonesia has 5.88% of ash. Therefore, the emission load of PM_{10}, $PM_{2.5}$, and TSP for coal imported from Indonesia is lower than Quang Ninh 6A anthracite coal about 5 times.

3.2. Modeling results

3.2.1. Calibration and validation model

The TAPM meteorological model was calibrated from January 31, 2015, to February 4, 2015, for meteorology modeling at Tan Son Hoa station, and found the amplification coefficient was equal to 0.8 [2]. For validation of the TAPM model, two important parameters are surface temperature and wind speed . The model was validated from 01/01/2019 to 31/1/2019 for meteorology modeling at Bien Hoa station (Dong Nai Province). The model gives the results closest to reality with the observed and simulation temperature having R^2 coefficient = 0.86 (Figure 4). For validation of the TAPM model, two important parameters are surface temperature and wind speed. Results of the validation of the TAPM model at Bien Hoa station show that the simulation model generated reliable data with the correlation coefficient (R^2) not less than 0.7. The average temperature difference between the simulated temperature and the observed temperature in Jan 2019 is 1.4^{0}C, in December 2019 is 0.75^{0}C.

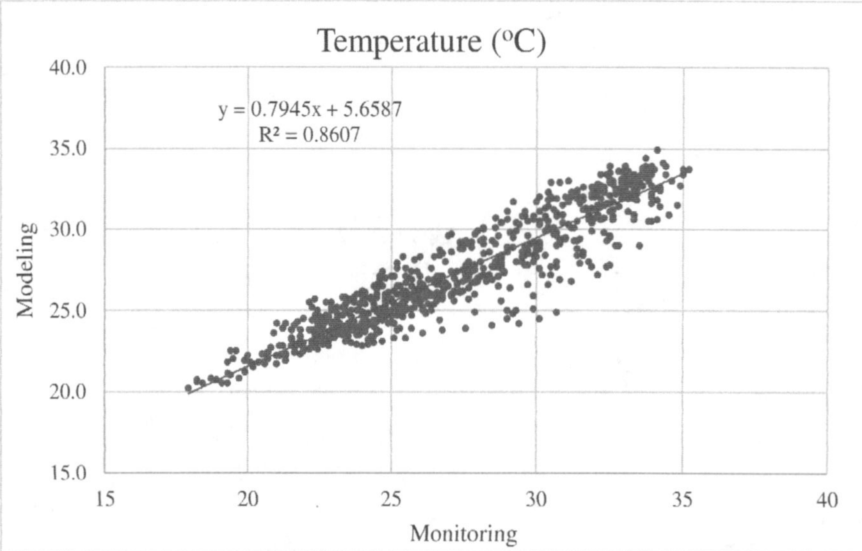

Figure 4. Calibration and validation model TAPM model

Air quality simulation results from AERMOD model have been calibrated and validated with observed value at 4 sites in Duyen Hai thermal power plant center, Tra Vinh province (Figure 5). The monitoring parameters are $PM_{2.5}$, TSP, CO, SO_2, and NO_2 hourly data from 7 am to 18 pm on 10 Jan 2018 (Table 3). The results of the MNBE error calculation between the simulated value and the average air quality monitoring value were 9.98% (ranging from 8.33% to 13.3%). This error is in the range of ± 15%. Therefore, the AERMOD model is capable of simulating the spread of air pollution by the region. The next step is to use this model to simulate air pollution for different scenarios.

Table 3. The observation sites of meteorological

Site	Coordinate	
	Lat	Long
1	9.5860960	106.5145031
2	9.575617	106.520684
3	9.591916	106.534223
4	9.5611659	106.513961

IOP Conf. Series: Earth and Environmental Science **964** (2022) 012024 doi:10.1088/1755-1315/964/1/012024

Figure 5. Sets up temporary Air quality monitoring Stations

3.2.2. Modeling of air pollution for current situation

After calibration and validation, simulation of the air quality for each hour in 2019 (8,760 hours) was conducted to determine the impacts of thermal power plants on air quality in HCMC. The simulation results included all of one-hour, average of 24-hour and annual average concentration.

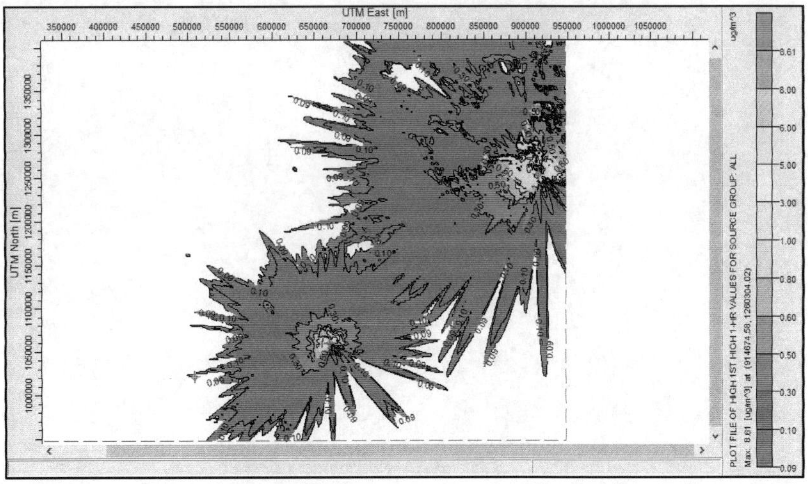

Figure 6. Average maximum 1 hours $PM_{2.5}$ concentration due to thermal power plants for the case of 2020. Red points are location of thermal power plants

The simulation results show that the $PM_{2.5}$ concentration pollution in the study area has the highest 1 hour with a concentration of 8.61 μg/m^3 close to Vinh Tan thermal power plant center (Northeast of the Figure 6).

IOP Conf. Series: Earth and Environmental Science **964** (2022) 012024 doi:10.1088/1755-1315/964/1/012024

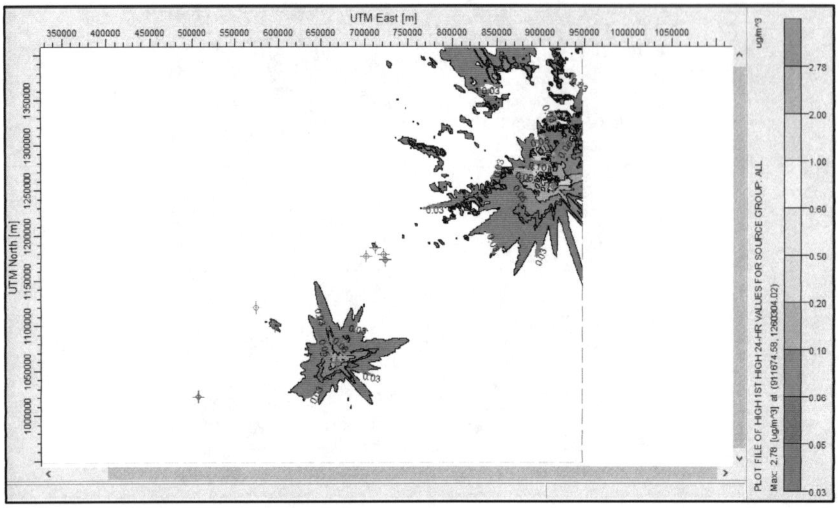

Figure 7. Average maximum 24 hours $PM_{2.5}$ concentration in 2020. Red points are location of thermal power plants

The simulation results show that the $PM_{2.5}$ concentration pollution in the study area has the highest average 24 hours with a concentration of 2.78 $\mu g/m^3$ (Figure 7). The highest 24-hour average concentration of $PM_{2.5}$ is lower than QCVN 05: 2013/BTNMT (average 24 hours for $PM_{2.5}$ is 50 $\mu g/m^3$) or WHO's guide (average 24 hours for $PM_{2.5}$ it is 25 $\mu g/m^3$).

Figure 8. Annual average $PM_{2.5}$ concentration due to thermal power plants for the case of 2020

The annual average $PM_{2.5}$ levels map was generated in Google Earth (Figure 8), after simulating for 365 days in the year 2020. The annual average concentration in this area is about 0.121 $\mu g/m^3$. However, the annual average concentration of $PM_{2.5}$ from power plants in 2020 flows to HCMC and contributes to HCMC' air quality only 0.005 $\mu g/m^3$. The average annual $PM_{2.5}$ concentration from power plants does not exceed QCVN 05:2013/BTNMT and WHO's guideline.

However, in the previous study only for HCMC by Khue [6] in Figure 9 annual average $PM_{2.5}$ concentration is 23 $\mu g/m^3$ which was higher than WHO's guideline (annual average $PM_{2.5}$ concentration is 10 $\mu g/m^3$). The $PM_{2.5}$ in HCMC is from two sources (local emission of HCMC (23 $\mu g/m^3$) and from

IOP Conf. Series: Earth and Environmental Science **964** (2022) 012024 doi:10.1088/1755-1315/964/1/012024

power plants (0.005 μg/m^3). Therefore, $PM_{2.5}$ impacts on public health of HCMC. To control and manage the air's quality and minimize the impacts from power plants to $PM_{2.5}$ in HCMC, the paper suggested many possible solutions in the next section.

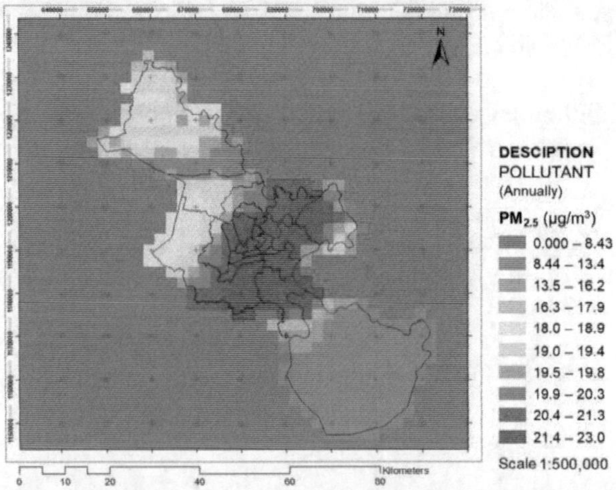

Figure 9. Annual average $PM_{2.5}$ concentration of HCMC [6]

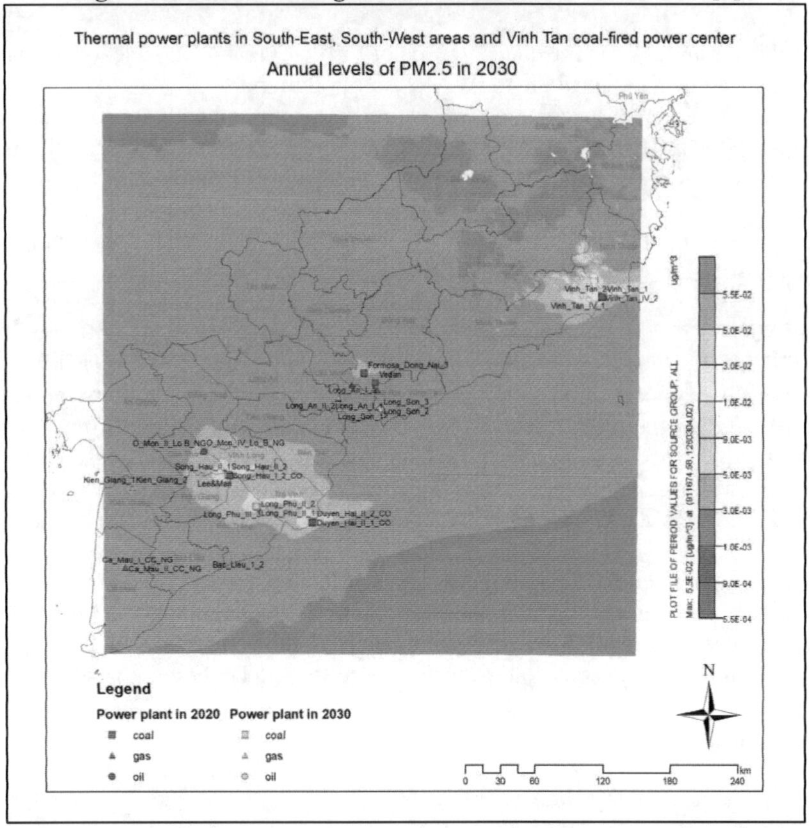

Figure 10. Annual average $PM_{2.5}$ concentration due to thermal power plants for the case of 2030

The annual average concentration of $PM_{2.5}$ due to thermal power plants for the case of 2030 in this area is about 0.055 μg/m^3 (Figure 10) However, the annual average concentration of PM2.5 from power plants in 2030 flows to HCMC and contributes to HCMC' air quality only 0.003 μg/m^3 which is less

IOP Conf. Series: Earth and Environmental Science **964** (2022) 012024 doi:10.1088/1755-1315/964/1/012024

than the case of 2020. This can be explained by two reasons: (i) the $PM_{2.5}$ emissions in 2030 are lower than 2020 about 59.12% due to the change of domestic coal into imported coal (Table 2), (ii) the second is that the number of thermal power plants in 2030 increases much compared to 2020, these plants located spread out the study area - quite close to HCMC. However, other pollutants such as NO_2 and SO_2 are different from $PM_{2.5}$ which will be discussed in detail in Table 4.

Table 4. Summary current situation 2020 and plan 2030 of pollutants

Sources	NO_2	SO_2	$PM_{2.5}$
Emission in 2020 (ton/year)	20,831.65	41,728.28	424.99
Emission in 2030 (ton/year)	47,713.86	97,371.09	304.82
Background of HCMC 2020 (μg/m^3) annual average*	67.1	19.2	23.0
Power plants contribute to HCMC 2020 (μg/m^3) annual average	0.01	0.17	0.005
Power plants contribute to HCMC 2030 (μg/m^3) annual average	0.03	0.40	0.003
QCVN 05:2013 (annual average) (μg/m^3) [17]	40	50	25
WHO guideline (μg/m^3) [18]	40	20**	10

Note: (*) is referred from Khue's study [6]; (**) is 24 hours average concentration

Table 3 shows that the annual average concentration of $PM_{2.5}$, SO_2 and NO_2 due to thermal power plants for the case of 2020 and 2030 contribute to HCMC's air pollution.
For the case of NO_2 the annual average concentration from power plants in 2020 and 2030 flows to HCMC and contributes to HCMC' air quality only 0.01 and 0.03 μg/m^3, respectively; in the previous study only for HCMC by Khue [6] in the annual average NO_2 concentration is 67.1 μg/m^3 which was higher than WHO's guideline (annual average NO_2 concentration is 40 μg/m^3). The NO_2 in HCMC is from two sources (local emission of HCMC (67.1 μg/m^3) and from power plants (0.03 μg/m^3). Therefore, NO_2 impacts on public health of HCMC.

For the case of SO_2 the 24 hours average concentration from power plants in 2030 flows to HCMC and contributes to HCMC' air quality only 10 μg/m^3; in the previous study only for HCMC by Khue [6] in the annual average SO_2 concentration is 19.2 μg/m^3 (WHO's guideline doesn't have annual average SO_2 guideline). If HCMC has the annual average SO_2 concentration is 19.2 μg/m^3, then 24 hours average concentration SO_2 of HCMC should be higher than 19.2 μg/m^3. That's the reason why the 24 hours average SO_2 of HCMC in 2030 is higher than 39.2 μg/m^3 which is higher than WHO's guideline (24 hours average SO_2 concentration is 20 μg/m^3). Therefore, SO_2 impacts on public health of HCMC.

With the above discussion, currently air pollution in HCMC is already polluted by $PM_{2.5}$, SO_2 and NO_2 and impacts on public health. However, in the future with the contribution of 33 thermal power plants under intercity/provinces air pollution dispersion, air pollution HCMC will be a huge impact on HCMC's public health. It's necessary to develop measures to reduce these impacts.

3.3. Abatement strategies

Thermal power plants with high chimneys up to 150 – 210 meters in height (4 – 6 meters in diameter) will be able to spread the pollution plum to inter-provincial areas, especially the neighbor. Therefore, in the stage of appraisal and investment of new projects, it is necessary to consider the background contaminated concentration of surrounding thermal power plants in the Environment Impact Assessment (EIA).

IOP Conf. Series: Earth and Environmental Science **964** (2022) 012024 doi:10.1088/1755-1315/964/1/012024

In recent years, several cities/areas in the Southeast and Southwest regions have had the air polluted by TSP, PM_{10}, and $PM_{2.5}$ such as HCMC, Dong Nai, Binh Duong, Ba Ria Vung Tau, and Can Tho. Therefore, constructing more thermal power plants using dirty fuels will worsen the air quality. Meanwhile, other clean and renewable energy such as wind, solar, gas, biomass, and garbage should be considered to develop in the local area, to address energy needs and co-benefits solutions for waste treatment. Thus, more policies and programs to support the development of clean energy and renewable energy (to replace the role of the coal power plant) should be issued by the government, in similar circumstances to achieve the rapid growth of solar power nationwide.

The air quality management plan in the 5-year period of each province must consult with neighboring provinces to mitigate the impact of trans-provincial air pollution.

The competent person who decides on investment projects, selects power generation plans, technology, investment locations ... , it is necessary to conduct a strategic environmental assessment of energy at the regional level. The strategic environmental assessment should be conducted before approving the project investment policy. It is necessary to assess the cumulative impact of all projects at the same time to make strategic decisions to minimize the wide-ranging negative impact and severity.

In the overview, the current status and in the future plan, coal power plants account for a very large proportion of 42.6% of total energy. Meanwhile, particulate matter such as TSP, PM_{10}, and $PM_{2.5}$ was estimated using the emission factor based on fuel type consumption and technology of the abatement system. It shows that emissions strongly depend on the quality of input fuels and the effectiveness of the treatment system. For example, Duyen Hai I has unit 1 using Anthracite 6A Quang Ninh coal-bran fuel, so the ash level A = 33.12%. Meanwhile, Duyen Hai III Expanded at the same capacity as Duyen Hai I, mainly using Bituminous coal and Sub-Bituminous coal imported from Indonesia with ash level A = 5.88%. In the results, the emissions of TSP, PM_{10}, and $PM_{2.5}$ of Duyen Hai I are greater than Duyen Hai III Expanded up to 2.98, 2,0, and 2.0 times, respectively. Therefore, one of the main solutions to reduce dust pollution TSP, PM_{10}, and $PM_{2.5}$ is to use clean coal fuel with a low ash ratio.

According to the calculation results, based on the amount of emissions per unit of electricity (Ex. ton PM2.5/MW.year), The suggested order of precedence is as follows: gas power, oil, biomass, imported coal, domestic coal.

In spatial planning of power plants, to ensure environmental safety, sustainable ecosystem, and protecting public health, the application of advanced technologies (SC, USC) is very essential. Especially for coal-fired power, as an important step forward, this technology will increase power generation efficiency, reduce fuel use, thus, reduce the emissions and slag ash with the same amount of electricity generated. But, the ratio of emissions per ton of fuel does not change, so does the amount of slag ash and the amount of cooling heat. As a result, many factories are concentrated in one geographic area, and total emissions can still exceed public health safety.

4. Conclusions

The paper studied the impacts of thermal power plants in SouthEast, SouthWest, and 7 power plants in Binh Thuan province on the air quality of HCMC for two scenarios (current status of 2020 and according to Power planning VII (adjusted) toward 2030) by using the TAPM-AERMOD system model.

The results show that the simulation results of $PM_{2.5}$ concentration pollution in the study area has the highest 1 hour with a concentration of 8.61 $\mu g/m^3$ close to Vinh Tan thermal power plant center. The $PM_{2.5}$ concentration pollution in the study area has the 24-hour average with a concentration of 2.78 $\mu g/m^{3.}$ A map of simulating annual average $PM_{2.5}$ was generated by extracting the results from these

8,760 hours of the year with the average concentration in this area is about 0.121 μg/m^3. NO_2 the annual average concentration from power plants in 2020 and 2030 flows to HCMC and contributes to HCMC' air quality only 0.01 and 0.03 μg/m^3, respectively. The 24 hours average concentration of SO_2 from power plants in 2030 flows to HCMC and contributes to HCMC' air quality of 10 μg/m^3. The 24 hours average SO_2 of HCMC in 2030 is higher than 39.2 μg/m^3 which is higher than WHO's guideline (24 hours average SO_2 concentration is 20 μg/m^3). Therefore, SO_2 impacts on public health of HCMC. Currently, air pollution in HCMC is already polluted by $PM_{2.5}$, SO_2 and NO_2 and impacts public health. However, in the future with the contribution of 33 thermal power plants under intercity/provinces air pollution dispersion, air pollution HCMC will be worse and have impacts on HCMC's public health. The paper developed 7 main mitigation measures to reduce impacts of air pollution from the master plan of Power planning VII (adjusted) toward 2030 and reduce the impact of air pollution on HCM's public health. The measures are focused on using clean fuel, advanced technology and controlling trans-provincial air pollution.

Author Contributions:
Khue Vu contributes as coordinator for whole research, air quality modeling and develop abatement strategies. While H.T.T.N. were responsible for calculating emissions. Bang Q. HO is in charge of running the TAPM-AERMOD model. T.T.N. contributed to the overview and discussion tasks. The manuscript was developed, revised, and edited by Bang Q. HO and Khue Vu.

Funding: This research is funded by Vietnam National University Ho Chi Minh City (VNU-HCM) under grant number C2021-24-03.

Conflicts of Interest: The founding sponsors had no role in the design of the study; in the collection, analyses, or interpretation of data; in the writing of the manuscript, and in the decision to publish the results.

References

[1] Bang, H.Q., Khue, V.H.N., Tam, N.T., Thuy, N.T.T., Hang, N.T.T. (2019). A combination of bottom-up and top-down approaches for calculating of air emission for developing countries: A **case** of Ho Chi Minh city, Vietnam. *Air Quality, Atmosphere & Health*. DOI: 10.1007/s11869-019-00722-8.

[2] Bang, H.Q., Hiep, N.D., Khue, V., Hien, T.T. 2018. Photochemical smog modelling using the air pollution chemical transport model (TAPM-CTM) in ho chi minh city, vietnam. Environmental Modeling & Assessment Springer International Publishing: 1-16.

[3] HEI. Effects of Short-Term Exposure to Air Pollution on Hospital Admissions of Young Children for Acute Lower Respiratory Infections in Ho Chi Minh City, Vietnam; Health Effects Institute: Boston, MA, USA, 2012.

[4] Mehta, S.; Long, H.N.; Dzung, D.V.; Cohen, A.; Thach, T.Q.; Dan, V.X.; Tuan, N.D.; Giang, L.T. Air pollution and admissions for acute lower respiratory infections in young children of Ho Chi Minh city. *Air Qual. Atmos. Health* **2013**, *6*, 167–179.

[5] Bang, H.Q. Modeling PM_{10} in Ho Chi Minh City, Vietnam and evaluation of its impact on human health. *Sustain. Environ. Res.* **2017**, *27*, 95–102.

[6] Vu Hoang Ngoc Khue, Quang Phuc Ha, Nguyen Duc Hiep, Nguyen Thi Thu Thuy, Nguyen Thoai Tam, Nguyen Thi Thuy Hang, Nguyen Thi Thanh Nguyet, Tran Ngoc Dang, Bang Q. Ho. 2020. Poor air quality and its association with mortality in Ho Chi Minh city: Case Study. Atmosphere 2020, 11, 750.

[7] Dicision 428/QD-TTg dated 18/3/2016 of the Prime Minister on approving the adjustment of the National electricity development planning for the period 2011 - 2020 with a vision to 2030.

[8] Moti L. Mittal, Sharma C., Singh R. 2012. Estimates of Emissions from Coal Fired Thermal Power Plants in India. ResearchGate.
https://www.researchgate.net/publication/267687877.

[9] Fang, C., Wang, L., Gao, H., & Wang, J. 2020. Analysis of the PM2.5 emission inventory and source apportionment. Environmental Science and Pollution Research in Jilin City, Northeast of China. doi: https://doi.org/10.1007/s11356-020-07605-x.

[10] Zheng, J., Zhang, L., Che, W., Zheng, Z., Yin, S. 2009. A highly resolved temporal and spatial air pollutant emission inventory for the Pearl River Delta region, China and its uncertainty assessment. Atmospheric Environment 43 (2009) 5112–5122.

[11] Lakesenvironmental (2013). AERMOD model information. http://www.weblakes.com

[12] Bang, H.Q. 2016. Book "Urban Air Pollution: from theory to practice". 420 pages, VNU.

[13] Vu Hoang Ngoc Khue and Nguyen Thi Thuy Hang. 2021. Air pollution emission inventory of thermal power plants in South-east and South-west areas, and proposed solutions to reduce emissions. VNU Journal of Science: Earth and Environmental Sciences (Accepted)

[14] EIA (Environmental impact Assessment). Vinh Tan Thermal Power Project Management Board. 2009. Environmental Impact Assessment Report of Vinh Tan 2 Thermal Power Plant

[15] EIA (Environmental impact Assessment). Vinh Tan Thermal Power Project Management Board. 2013. Environmental Impact Assessment Report of Vinh Tan 4 Thermal Power Plant

[16] EVN. 2015 Vietnam Electricity Corporation. 2015. Report on environmental impact assessment of Vinh Tan 4 thermal power plant expanded

[17] QCVN 05:2013. 2013. National Technical Regulation on Ambient Air Quality. Ministry of Natural resources and Environment.

[18] WHO. 2018. Ambient (outdoor) air pollution. https://www.who.int/news-room/fact-sheets/detail/ambient-(outdoor)-air-quality-and-health

IOP Conf. Series: Earth and Environmental Science **964** (2022) 012025 doi:10.1088/1755-1315/964/1/012025

Zoning groundwater potential recharge using remote sensing and GIS technique in the Red river delta plain

Pham Quy Nhan[1*], Le Viet Hung[1,4], Tran Thanh Le[1], Thi Van Le Khoa[1], Dang Duc Nhan[2], Tran Quoc Cuong[3]

[1] Hanoi University of Natural Resources and Environment (HUNRE)
[2] Institute of Water Resources and Environment (IWRE)
[3] Institute of Geological Sciences, Vietnam Academy of Science and Technology (IGS)
[4] Graduate University of Sciences and Technology (GUST)

Corresponding author's e-mail address: *pqnhan@hunre.edu.vn*

Abstract. The Red River delta plain is the second largest delta in Vietnam and is located in the North of the country with an area of 14,860 km² and residing more than 22.5 million inhabitants. Groundwater is mainly exploited in Quaternary sedimentary aquifers with a total discharge of about 3 million m^3/day. Some localities have shown signs of over-exploitation such as in Hanoi and in Nam Dinh, which may lead to related problems such as depletion, subsidence, saltwater intrusion, and water pollution. In order to be able to sustainably exploit groundwater, the groundwater potential recharge needs to be estimated. There have been many studies using different methods to estimate the groundwater recharge and to zone potential recharge. In the study area, there are several studies for groundwater recharge, but some are still uncertain because of using indirect methods, some are locally estimated in specific areas. Therefore, the objective of this study is to apply remote sensing and GIS to zone the groundwater potential recharge and its verification by using radioactive isotope ^3H analysis in the Red River delta plain. Various types of satellite images have been used and interpreted to detect the different thematic layers which concern the groundwater potential recharge. GIS has been applied as a platform for analysis and integration of thematic layers for zonation, finally. Field trip and water sampling for chemical and radioactive ^3H analysis were also conducted. Zones with low, moderate, and high groundwater potential recharge have been delineated with good agreement from the direct estimation of groundwater recharge by radioactive isotopes ^3H.

1. Introduction

The groundwater recharge of a certain area is a fundamental component of the balance of a certain catchment, which contributes to the sustainable exploitation of water resources in general and groundwater in particular. Lerner et al. [1] and Scanlon et al. [2] reviewed numerous methods have been used to estimate groundwater recharge. In recent years, many researchers have applied remote sensing methods and GIS techniques to assess potential groundwater [3, 4, 5, 6], groundwater recharge [7, 8, 9], potential areas for artificial recharge [7], and potential sites for groundwater exploration [10,11]. This approach is based on the interpretation and integration of factors related to the influence of groundwater potential, groundwater recharge, conditions for artificial recharge, and the possibility of groundwater exploration. The limitation of this approach is that the methods are indirect with some uncertainties. To improve the reliability of these methods, it is very necessary to determine the weight of the relative importance of the relevant indicators objectively by analytical hierarchical process (AHP). Furthermore, verifying the evaluation results by the direct method is also very necessary. However, due to large uncertainties involved in the measurement of individual parameters of each method, a common recommendation is that recharge should be estimated by the use of multiple methods and the results compared [2, 12].

The Red River Delta Plain (RRDP) is the second largest delta in Vietnam with an area of over 14,860 km2 accounting for about 4.5% of the total area of the country, residing 22.5 million people and the place with the largest population density in the country [13]. Groundwater is primary water supply source for the RRDP with a total exploitation discharge about 3 million m^3/day [14]. However, long-term excesses groundwater exploitation in the delta has resulted in many problems such as ground subsidence, pollution, depletion, and saltwater intrusion related to groundwater have also occurred in this area [15].

There have been a number of studies to assess the amount of groundwater recharge in this area. For examples Pham [16] estimated the groundwater recharge in RRDP by using a modeling method. Tran [17] indicated that the groundwater recharge and the interaction between aquifers in Quaternary sediments in Thach That - Dan Phuong area, Hanoi were based on determining the average travel time by using stable isotope 18O/2H and radioactive isotope 3H. Pham et al [15] conducted isotope sampling to evaluate the interaction between river and groundwater. The authors showed that the water ratio from Red River recharged to well fields Ha Dinh, Mai Dich, Phap Van extracting from Pleistocene were 50%, 52% and 57%, respectively. Postma et al. [18] determined the recharge from the Red River to aquifers by Tritium/Helium dating in Nam Du area. Larsen et al. [19] using isotopes and modeling to determine the recharge from Red River and rainwater to Quaternary aquifers in Dan Phuong area. Although there have been several studies to evaluate groundwater recharge as mentioned above, either the reliability of the research results is limited, or the research results are just focused on specific areas in RRDP. The objective of this study is to apply remote sensing and GIS technique to zone groundwater recharge potential in the RRDP. In this study, we only focus on diffuse (direct) recharge derived from precipitation or irrigation, etc. that occurs fairly uniformly over large areas.

2. Study area

The RRDP is located between latitude 21°34′ to 19°5′N and longitude 105°17′ to 107°7′10’E as its extremities. The overall area of the plain is approximately 14,860 km² (Figure 1). Contiguous to the North and Northeast is the Northeast region, to the West and Southwest is the Northwest region, the East is the Gulf of Tonkin and the South is the North central region. The delta gradually lowers from the Northwest to the Southeast, from the ancient alluvial shelves with an elevation of 10-15 m down to the alluvial flats of 2-4 m in the center where the tidal flats are still flooded every day. In the center of the delta terrain elevation varies from 8-10 m, flat terrain. In the North, the delta is limited by the Tam Dao - Yen Tu Mountain range, the South is limited by the Ba Vi-Vien Nam Mountain range, in the East is limited by the coastline. In the middle of the plain, there are round top hills with gentle slopes and elevations from 25 to 100 m. RRDP belongs to a humid tropical monsoon climate with two distinct seasons, the dry season and the rainy season. The rainy season is from May to October and the dry season is from November to April next year. The characteristics of rainfall, evaporation, and sunshine hours have markedly changed according to the seasons of the year and for different areas in the whole region. The average annual rainfall for the whole delta is from 1033 to 2338.7 mm and the amount of evaporation is from 828.2mm to 1057.1mm. The highest air temperature is 39°C and average is 23.5°C. The daily temperature difference is relatively large 8°C-15°C in summer and 4°C-13°C in winter season. RRDP has a dense network of rivers with an average density of 0.7 to 1 km/km^2. The whole region has two main river systems: the Red river system and the Thai Binh river system. Research documents as well as satellite maps show that the Red River bed has moved and as a result formed a system of oxbow lakes as well as affecting the distribution of sedimentary facies in the RRDP. The coast is located in the

East and South East of the RRDP running from Quang Ninh to Thanh Hoa with a total length of about 200 km. Due to the impact of waves, tides and river systems, the surface water has been significantly saline intrusion in the estuaries. This saline intrusion in the river system not only affects the coastal ecosystem, irrigation water supply, but also affects the shallow aquifers in the vicinity [18, 20].

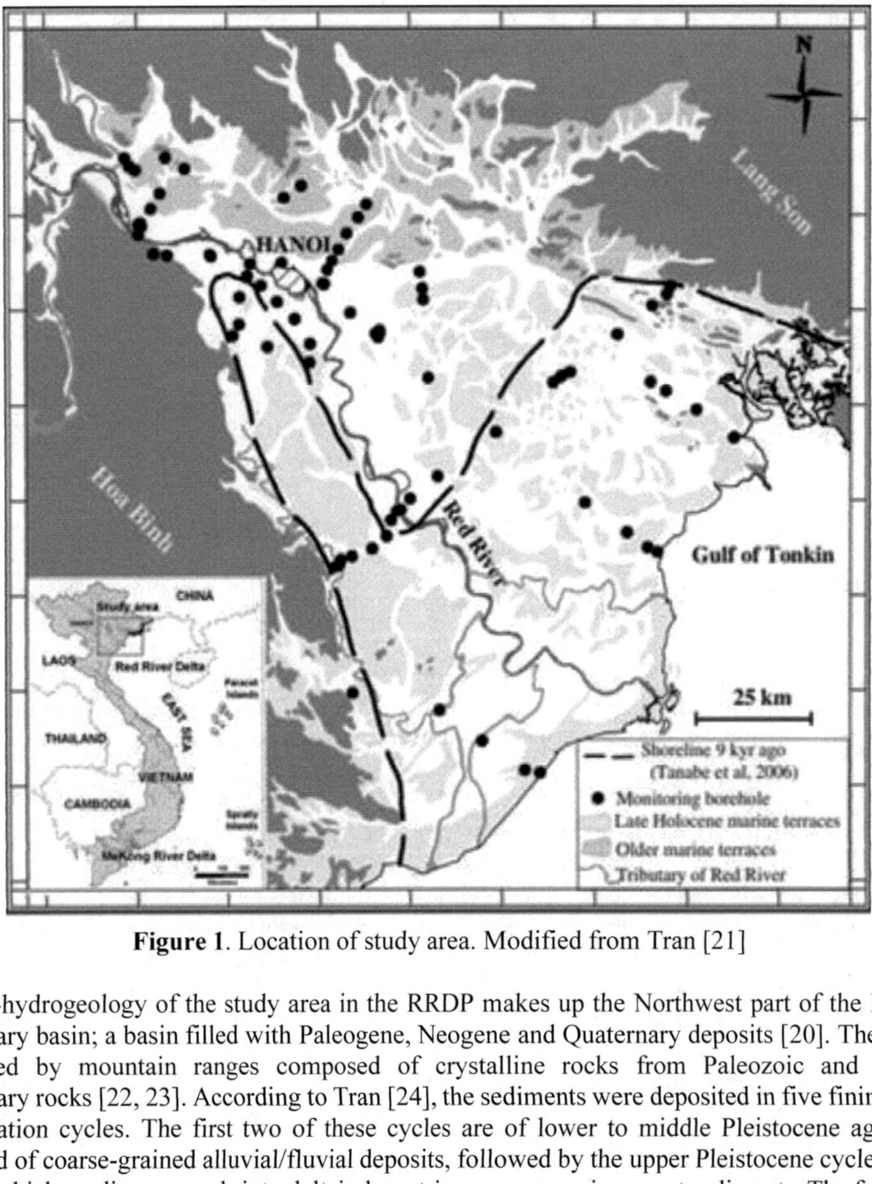

Figure 1. Location of study area. Modified from Tran [21]

The Geo-hydrogeology of the study area in the RRDP makes up the Northwest part of the Red River sedimentary basin; a basin filled with Paleogene, Neogene and Quaternary deposits [20]. The RRDP is surrounded by mountain ranges composed of crystalline rocks from Paleozoic and Mesozoic sedimentary rocks [22, 23]. According to Tran [24], the sediments were deposited in five fining-upward sedimentation cycles. The first two of these cycles are of lower to middle Pleistocene age and are composed of coarse-grained alluvial/fluvial deposits, followed by the upper Pleistocene cycle of fluvial deposits which grading upwards into deltaic-lacustrine swamp environment sediments. The fourth cycle has a lower to middle Holocene age and is composed of fine-grained sand and clay that formed in a deltaic environment; The top fifth cycle from the upper Holocene is dominated by coarse-grained sediments that form in delta plain and delta front environments. The deeper Pleistocene aquifers are recharged by the surrounding mountain range and the average annual recharge to the aquifers is in the range of 100-400 mm. Hydraulic gradients in the range of 0.05-0.15 % are typical and groundwater flow velocities in the Holocene aquifers are a few tens of meters per year [19].

3. Methods and Materials

3.1. Data collection

The data on the groundwater monitoring network in the RRDP were collected, including data on the structure of the monitoring boreholes, the arrangement of screen at the aquifer. Monitoring data on rainfall and evaporation at 6 meteorological stations, 7 rivers, water levels at hydrological stations and groundwater level data at 128 groundwater monitoring boreholes were also collected. These data were included in the computation, assessment of groundwater recharge on the basis of data of stable isotopes and radioactive isotopes.

3.2. Delineation of groundwater recharge potential zones

Delineation of groundwater recharge potential zones was conducted by integrating different thematic maps obtained from satellite interpretation and field data analysis on GIS platform. The satellite images selected for interpretation depend on their application and are shown in the table below:

Table 1 Type of satllite images, their parameters and application

Satellite Images	Resolution	Number of scenes	Time	Applications
Multi-temporal PALSAR-/ScanSAR	50 m	450 (1° x 1°)	2016	Land use and land cover, topography, geology, climat and hydrology
ALOS-2 /PALSAR-2	25m	60 (1° x 1°)	2016	Land coverage observation, disaster monitoring and resource surveying and elevation data
AW3D30	30m	60 (1°x 1°)	2006-2011	Digital surface models
Multi-temporal Sentinel-1 SAR GRD C-band	10m	480 (1° x 1°)	2016	Not hindered by atmospheric effects and are capable of imaging through tropical clouds and rain showers to maritime monitoring, land monitoring
Multi-temporal Landsat 8 OLI Surface Reflectance Tier 1	30m	480 (1° x 1°)	2016	To create higher-level science data such as surface temperature, surface water, land change and use analysis.
Multi-temporal Sentinel-2 MSI, MultiSpectral Instrument, Level-1C	10m	480 (1° x 1°)	2016	Urban land cover and urban extent mapping

The thematic maps as the factors of groundwater recharge influence were established on the same scale as i) Lithology map ii) Geomorphological map iii) Topographic slope map iv) Drainage system map v) Soil classification map vi) Land use map vii) Rainfall map and viii) Groundwater level map. The above maps were built on the same coordinate system and projection zone so that they can be integrated by the ArcGIS 10.5. Total value was applied to develop the groundwater recharge potential index of the area using the equation proposed by Malczewski, 1999 [25]

$$\mathrm{GRI} = \sum_{w=1}^{m}\sum_{j=1}^{n} W_j X_i \quad (1)$$

Where GRI is groundwater recharge index, W_j is normanlized weight of the jth thematic factor, X_i is the normalized weight of the ith feature of the thematic factor, m is the total number of thematic factors, and n is the total number of features of a given factor.

The analytic hierarchy process (AHP), developed by Saaty [26] is one of the most popular and widely employed multi-criteria methods. In this technique, the processes of rating alternatives and aggregating to find the most relevant alternatives are integrated. The technique is employed for ranking a set of alternatives or for the selection of the best in a set of alternatives. The ranking/selection is done with respect to an overall goal, which is broken down into a set of criteria. The application of the methodology consists of establishing the importance weights to be associated with the criteria in defining the overall goal. This is done by comparing the criteria pairwise. Let us consider two criteria, C_j and C_k. The comparative judgment is captured on a semantic scale (equally important/moderately more important/strongly important and so on) and is converted into a numerical integer value ajk. The relative importance of C_k over C_j is defined as its reciprocal, i.e., $a_{kj}=1/a_{jk}$. A reciprocal pairwise comparison matrix A is then formed using ajk, for all j and k. Note that $a_{jj}=1$. It has been generally agreed that the weights of criteria can be estimated by finding the principal eigenvector w of matrix A. When the vector w is normalized, it becomes the vector of priorities of the criteria with respect to the goal; λ_{max} is the largest eigenvalue of the matrix A and the corresponding eigenvector w contains only positive entries. The methodology also incorporates established procedures for checking the consistency of the judgments provided by the decision-maker. Using similar procedures, the weights of alternatives with respect to each criterion are computed. Then, the overall weights of alternatives are computed using the weighted summation. A method developed by structuring a decision-making problem as a hierarchy in the form of an upside-down tree where the main goal is placed on top. Partial objectives that meet the main objective are placed at the second level. Each partial objective at the second level can be decomposed into third-level objectives, and each set at each level meets the objective of the level to which they are subordinate. These partial objectives are treated as criteria in this text. At a lower level, the alternatives are listed and then compared pairwise according to their contribution to reaching each objective, or criterion, from the lower level [26]

In this case, the weights which got from expert opinions and from Tesfa and Girum [6], Senanayake et al.[9], Jaiswal et al. [11], Magesh et al. [[8] and Chuma et al.[3]) were assigned to each factor. Pairwise comparisons between this each thematic factor obtained based on methods used by Machiwal et al. [4] and Sahoo et al. [10], Chowdhury et al. [5], and Chowdhury et al. [7] were presented in Table 2. The normalized weights were determined by dividing each geometric mean by the its total column. This ratio reflects a reasonable level of consistency in the pair-wise comparison phase. Similarity, each of the thematic factor has been divided into different features based on indicators which got references from Tesfa and Girum [6] and Senanayake et al. [9]. Nomanlized weights were also calculated by the AHP method as described above.

3.3. Water sampling and analysis

Groundwater was taken using a a Grundfos MP1 submersible pump for main ion and tritium (^3H) samples. Before taking samples, static water level in each sampling borehole was measured. The samples were taken once pH and electrical conductivity (EC) in water became unchanged. For ^3H analysis, the water samples were first subjected to distillation to remove the minerals dissolved until the electric conductivity was less than 10 mS cm-1. Around 500 ml of the distilled water samples were then subjected to the electrolytic enrichment for tritium at 4°C till around 10 ml was attained [27]. These samples were analysed at Institute of Nuclear Science and Technology, Hanoi (INST). Sixteen of ^3H samples and major ions samples were collected from national monitoring network of groundwater with different screen depth in Holocene aquifer. Based on the analysed results groundwater recharge rate can be estimated and also validated for zoning groundwater potential recharge.

3.4. Estimation of groundwater recharge using isotopes technique

According to David [28] the groundwater recharge calculated on the basis of radioactive isotope 3H is determined by the following formula:

$$W \geq \frac{CD-(EL-WT)}{A} n \tag{2}$$

Where:

- -W: Groundwater recharge rate (mm/year)
- -CD: Depth of borehole's screen, isotope sampling position (m)
- -EL: Elevation of borehole top (m)
- -WT: Elevation of water table (m)
- -A: The age of the water in the borehole is determined by measuring Tritium ($T/^3He$).
- -n: Formation porosity (%).

4. Results and discussion

4.1. Building the related factors to groundwater potential recharge

i. Rainfall

Rainfall distribution in the delta also directly affects the groundwater recharge. Heavy rainfall increases the groundwater recharge rate and vice versa. Using CHIRPS: Rainfall Estimates from Rain Guage and satellite Observation data with 08 meteological stations' data in the RRDP and associated with Multi-temporal Sentinel-1 SAR GRD C-band interpretated data supported to devide the RRDP into 5 areas with different rating presented at Table 4 and Figure 2 a. The heavy rainfall were located in the edges and near coastal areas.

ii. Lithology

The aquifer properties vary from one rock type to another rock type. The lithology influences both the permeability of aquifers while the distribution of weathered regolith is also known to be important for groundwater yields in crystalline rocks as decscribed by Wright [29]; McFarlane et al. [30]. The weathered regolith may extend to greater depths in a highly fractured terrain like in RRDP margin. By using Multi-temporal PALSAR-/ScanSAR images interpretation associated with collected data from previous studies [31-34] the lithology was divided into 06 areas and presented in Table 4 and Figure 2b. The RRDP consists of metamorphic rocks of khondalitic group (Fig. 2) which include garnetiferous biotite gneiss (48.03%), garnet biotite gneiss of migmatite complex (41.34%)

iii. Geomorphology

The identification and characteristics of various landforms and structural features in the study area was very important from geomorphological study point of view. Many of these features are favourable for occurrence and recharge of groundwater and are classified in terms of groundwater recharge potentiality. The geomorphic imprints can be considered as surface indicators for identification of subsurface water conditions. This information provides a reliable base for effective planning, development and management of groundwater resources of an area. By using Digital Elevation, Shuttle Radar Topography Mission (SRTM) 1 Arc- Second Global and Multi-temporal PALSAR-/ScanSAR data the geomorphology were founded and were divided into 04 areas and presented in Table 4 and Figure 2c. The study area are covered with plain terrain and high gradient hill in edge North-Western and South-Western of the deltas.

iv. Slope

The precipitous terrain causes rapid runoff and does not store water easily. Slope of any terrain is one of the factors allowing the infiltration of groundwater into subsurface or in other words groundwater recharge. In the gentle slope area the surface runoff is slow allowing more time for rainwater to percolate whereas steep slope area facilitates high runoff allowing less residence time for rainwater to percolate and hence comparatively less infiltration. Slope is the rate of change of elevation and considered as the principal factor of the superficial water flow since it determines the gravity effect on the water movement. The slope is directly proportional to runoff and groundwater recharge will be lesser in the areas with steep slope. The water flow over the gently undulating plains is slow and adequate time is available to enhance the infiltration rate of water to the underlying fractured aquifer. With the data from

Digital Elevation, Shuttle Radar Topography Mission (SRTM) 1 Arc- Second Global, the slope percentage map of zone research was created with 05 ranks from 0 to plus of 26% and presented in Table 4 and Figure 2d. The major deltas were found in class 1 (slope from 0 to 5 %) which covers 70% of the study area while the class 5 (slope plus 26%) covers only 5%.

v. Groundwater level

Groundwater level is related to the unsaturated zone thickness distribution in unconfined aquifers. Obviously, the shallower the groundwater level the greater the ability to groundwater recharge and vice versa. Groundwater levels at 128 monitoring boreholes were measured during sampling and divided in 05 areas and presented Table 4 and Figure 2e. Near the coastal areas, the grounwater level found from 1 to 3m from the ground covered 32% of areas. The zone deepest values were found in Ha Noi (> 10m), covers 1.5% zone research.

vi. Drainage

The drainage is dominated by the deeply incised, sand/clay filled seasonal streams. The development of stream segments is affected by slope and local relief and these may produce differences in drainage density from place to place. The spatial variation of drainage density was estimated by following the methodology adopted for calculating lineament density. Drainage data in the RRDP is available in annual statistic report for every province. The drainage density of Red River Delta Plain was prepared from Digital Elevation, Shuttle Radar Topography Mission (SRTM) 1 Arc- Second Global, the digital elevation AW3D30 (30 m x30 m resolution) in ArcGIS 10.3.1 platform and presented in Table 4 and Figure 2e. About 56.6% and 20.2% of the area were found in very low (0-0.25 km/km2) and low (0.25-0.5 km/km2) drainage density classes, respectively.

vii. Landuse/Land cover

Land use/land cover features control the occurrence of groundwater and also causes for infiltration for recharge with variety of classes among itself. Standard visual interpretation methods were applied to identify and interpret the land use pattern of the area and various land use classes delineated includes agriculture, forest, waste land, built-up land and water body. With the data from Japan Aerospace Exploration Agnecy - Earth Observation Research Center and using Multi-temporal Sentinel-2, Landsat 8 imagery, we were interpeted land use, land cover map (see Table 4 and Figure 2e). The study area consists of seven types of land cover: rice, crop, forest, wetland, bare land, urbanization and water. The major parts of the deltas were dominated by agriculture covering with 32% of the study area.

viii. Soil

Soil characteristics have a considerable role on the infiltration of water. The rate of infiltration largely depends on the grain size of soils. Wetland hydromorphic soils are poorly drained with a depth and they are developed on fluvial sediments. The low land plains are differentiated by the presence of lateritic soils and riverine alluvium and moderately well drained nature with depth and texture varies from clay loam to clay texture. Soil maps are available in DONRE and FAO of every province. This soil types was divided into 08 areas and presented in Table 4 and Figure 2f. All of the most of deltas is the alluvial soil (45% of areas). The feralit soil located on the North-Western and South-Western edges of the plain (7%) and the acidic, salty soil distributed in a coastal strip from Hai Phong to Ninh Binh (28%). These soils are characterized by poor to moderate infiltration property.

IOP Conf. Series: Earth and Environmental Science **964** (2022) 012025 doi:10.1088/1755-1315/964/1/012025

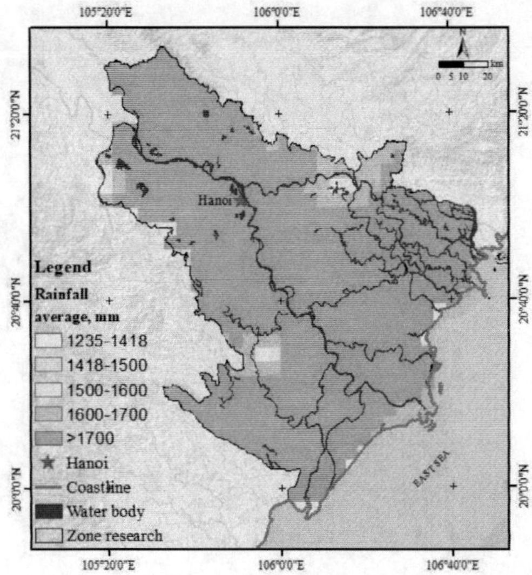

(a)

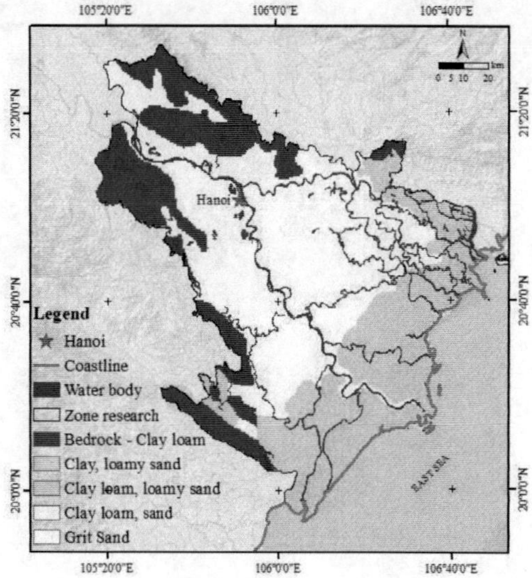

(b)

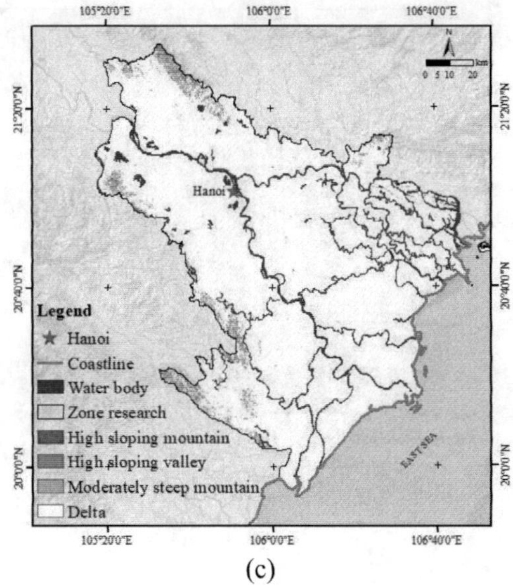

(c)

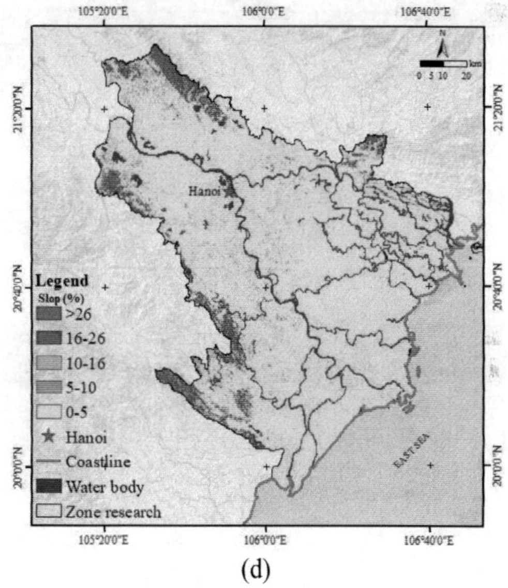

(d)

IOP Conf. Series: Earth and Environmental Science **964** (2022) 012025 doi:10.1088/1755-1315/964/1/012025

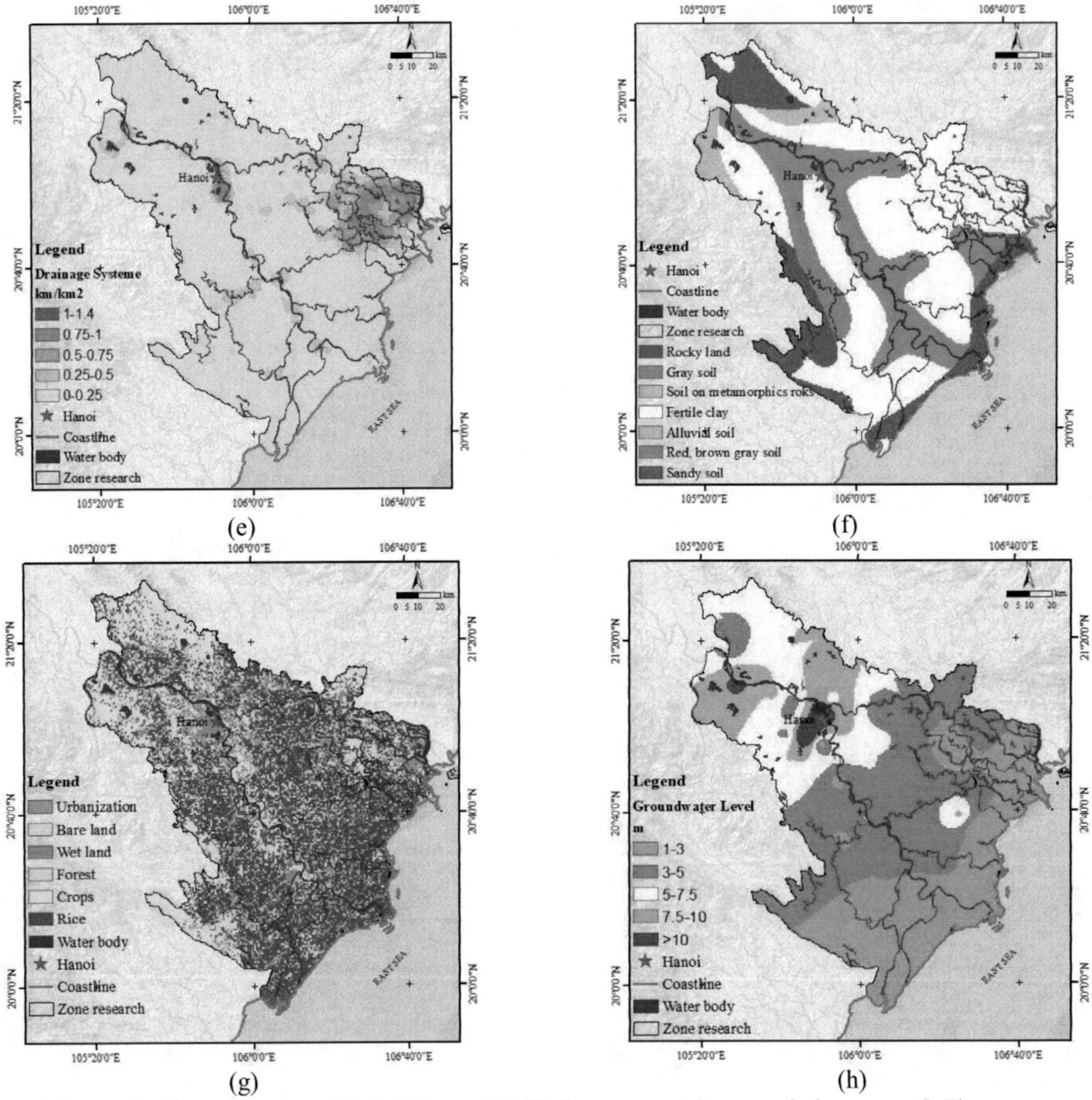

Figure 2. Thematic maps a) Rainfall map b) Lithology map c) Geomorphology map d) Slop map e) Drainage system map f) Soil map g) Land use/land cover map h) Groundwater level map

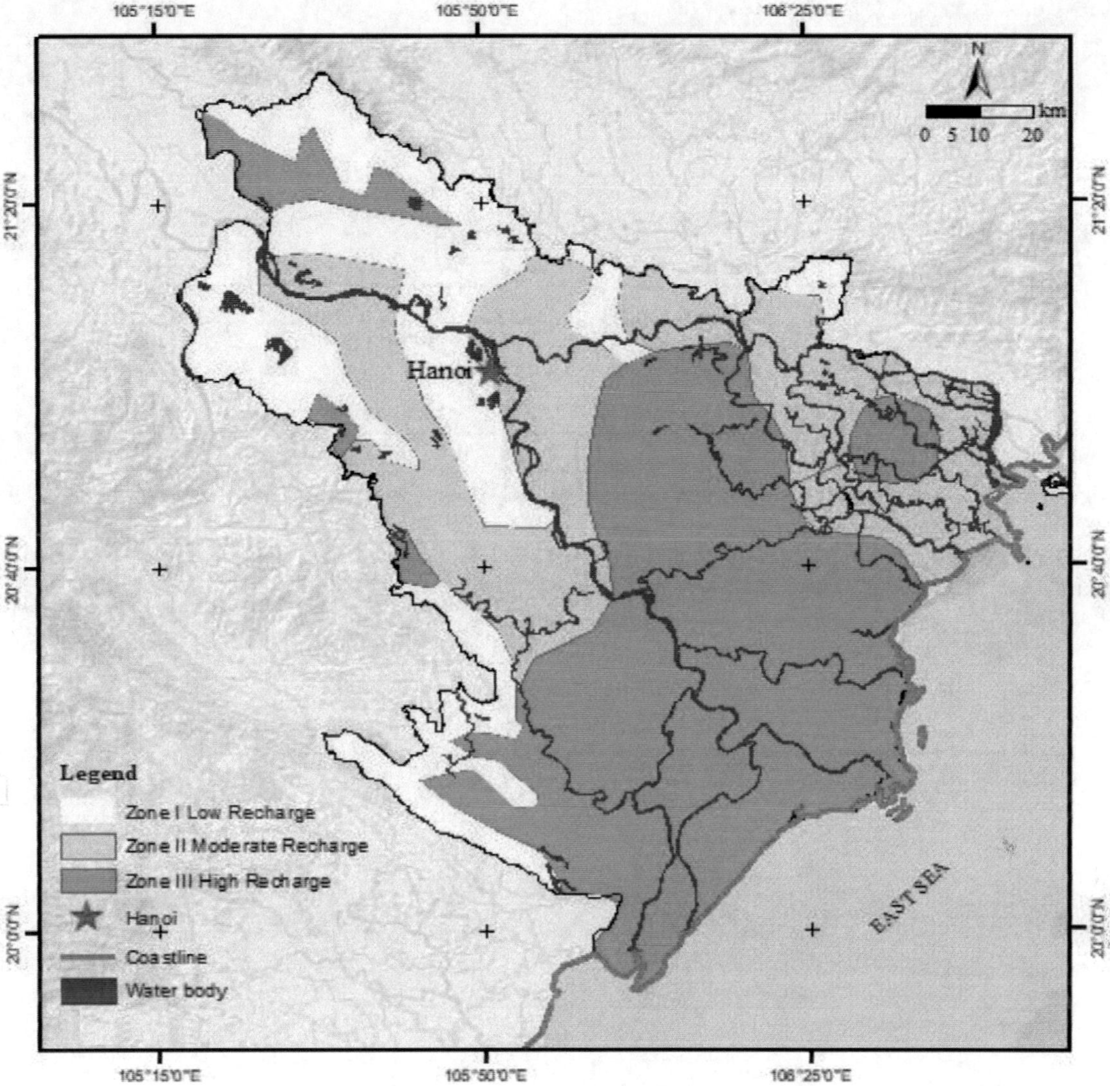

Figure 3. Map of groundwater recharge potential zones

4.2. Building the related factors to groundwater potential recharge

The groundwater recharge potential map is developed by overlaying thematic factors. Groundwater recharge zones were evaluated based on the groundwater potential index (GRI) calculated from the integration of all factors affecting groundwater recharge by thematic indicator computed according to Formula 1. Results of pairwise comparisons between thematic factors and its normalized weights are presented in Table 2. Similarly, weights are assigned respectively from expert opinion and references to the indicator variables and the corresponding normalized weights are presented in Table 3.

Table 2 Pairwise comparison between thematic factors

Factor *	RF	LG	GG	SP	GL	DS	LC	SL	Assigned weight	Geom-etric mean	Normal i-zed weight
RF	8/8	8/7	8/4.5	8/5	8/5	8/3	8/3.5	8/2.5	8	1.66	0.21
LG	7/8	7/7	7/4.5	7/5	7/5	7/3	7/3.5	7/2.5	7	1.45	0.18
GG	4.5/8	4.5/7	4.5/4.5	4.5/5	4.5/5	4.5/3	4.5/3.5	4.5/2.5	4.5	0.94	0.12
SP	5/8	5/7	5/4.5	5/5	5/5	5/3	5/3.5	5/2.5	5	1.04	0.13
GL	5/8	5/7	5/4.5	5/5	5/5	5/3	5/3.5	5/2.5	5	1.04	0.13
DS	3/8	3/7	3/4.5	3/5	3/5	3/3	3/3.5	3/2.5	3	0.62	0.08
LC	3.5/8	3.5/7	3.5/4.5	3.5/5	3.5/5	3.5/3	3.5/3.5	3.5/2.5	3.5	0.73	0.09
SL	2.5/8	2.5/7	2.5/4.5	2.5/5	2.5/5	2.5/3	2.5/3.5	2.5/2.5	2.5	0.52	0.06
Total										8.00	1.00

*RF: rainfall, LG: Lithology, GG: Geology, SP: Slope, GL: Groundwater level, DS: Drainage system, LC: Land cover, SL: Soil

Table 3 Normalized weights corresponding to indicator variables of each factor

Factors	Indicator variables	Assigned weight	Geometric mean	Normalized weight
Rainfall (RF) (mm/years)	1,700.01 - 2.026.00	9	1.73	0.35
	1,600.01 - 1.700.00	7	1.35	0.27
	1,500.01 - 1.600.00	5	0.96	0.19
	1,418.01 - 1,500.00	3	0.58	0.12
	1,235.00 - 1,418.00	2	0.38	0.08
Lithology (LG)	Grit sand	6	2.32	0.39
	Clay loam, sand	3	1.16	0.19
	Clay loam, loamy sand	2.5	0.97	0.16
	Clay, loamy sand	2	0.77	0.13
	Clay loam	1	0.39	0.06
	Bedrock	1	0.39	0.06
Geomorphology (GG)	Delta	8	2.13	0.53
	Moderately steep mountains	4	1.07	0.27
	High sloping valley	2	0.53	0.13
	High sloping mountain	1	0.27	0.07
Slope (SP), %	0.00 - 5.00	7	1.84	0.37
	5.01 - 10.00	5	1.32	0.26
	10.01 - 16.00	4	1.05	0.21
	16.01 - 26.00	2	0.53	0.11
	> 26.00	1	0.26	0.05
Groundwater level (GL) m bgs	>10	8	1.74	0.35
	7.51 - 10	6	1.30	0.26
	5.01 - 7.5	4	0.87	0.17
	3.01 - 5	3	0.65	0.13

Factors	Indicator variables	Assigned weight	Geometric mean	Normalized weight
	1 - 3	2	0.43	0.09
Drainage system (DS), km/km^2	0.00 - 0.25	8	2.1	0.41
	0.26 - 0.50	5	1.3	0.26
	0.51 - 0.75	3.5	0.9	0.18
	0.76 - 1.00	2	0.5	0.10
	1.01 - 1.40	1	0.3	0.05
Landcover (LC)	Water bodies	8	1.72	0.25
	Rice	7	1.51	0.22
	Crop	6	1.29	0.18
	Forest	4	0.86	0.12
	Wetland	3.5	0.75	0.11
	Bare land	3	0.65	0.09
	Urbanization	1	0.22	0.03
Soil (SL)	Sandy soil	7	2.04	0.25
	Red brown silver gray soil	5	1.45	0.18
	Fertile alluvial soil	4	1.16	0.15
	Salty brackish alluvial soil	4	1.16	0.15
	Rich in nutrients and fertile clay	3	0.87	0.11
	Discolored gray soil on sedimentary and metamorphic rocks	2	0.58	0.07
	Gray soil. silver color shale	1.5	0.44	0.05
	Rocky land	1	0.29	0.04

All the thematic maps were converted into raster format and superimposed by weight overlay method which consists of GRI and integration of them through GIS. Integration of thematic maps for carrying out multi-criteria or overlay analysis in the GIS environment was done using ArcGIS software. The GRI value of groundwater recharge potential was calculated as Formula 1 and got in range from 0.14 to 0.34. It has been reclassified into 3 zones of groundwater recharge potential, namely low (zone I), moderate (zone II), and high (zone III). GRI value less than 0.22 was assigned for a low potential zone whereas it from 0.22 to 0.27 and higher than 0.27 were assigned for moderate and high recharges, respectively.

4.3. Estimation of groundwater potential recharge using radioactive isotope ^3H

Table 4 showed results of ^{3}H-age of the groundwater samples under this study and these results were separated into 3 groups which attributed to 3 zones.

Table 4 ^{3}H-age of groundwater samples taken from boreholes with different water table elevation H (mbs: meter below the ground surface)

Zone I: Low recharge			**Zone II: Moderate recharge**			**Zone III: High recharge**		
Borehole ID	H, mbs	^{3}H-Age, y	Borehole D	H, mbs	^{3}H-Age, y	Borehole ID	H, mbs	^{3}H-Age, y
Q68a	-8	2.8	Q83	-7.7	0.0	Q108	-12	11.0
Q1	-9	6.4	Q89	-8.4	1.4	Q109	-9	5.5

Q67	-7	0.4	Q115	-14	22.0	Q110	-9	3.5
Q33	-8.94	20.0				Q159a	-7.5	3.0
Q66	-12	27.0				Q158	-7.4	1.7
						Q164	-9	3.5
						Q145	-9	3.5
						Q147	-9	2.5

With results in Table 4 potential recharge rates for the three expected zones in the RRDP were estimated based on the relationship between the ^{3}H-age of groundwater samples and the elevation of the water table in boreholes from where the samples were taken in this study. Figure 4 presents this relationship.

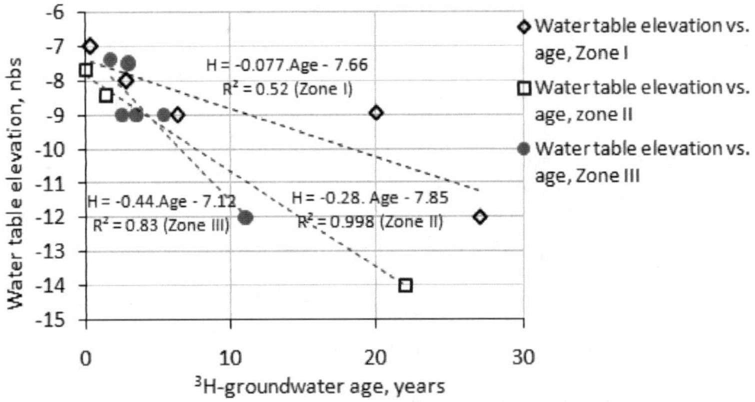

Figure 4 A graph showing a relationship between groundwater table elevation and ^{3}H age of groundwater samples taken from 3 zones with different potential recharge rates; mbs: meter below the surface; ^{3}H-age: groundwater age estimated by the ^{3}H method

As seen from Fig 4 in zone I, II, and III the relationships between groundwater table elevation and 3H-age followed three models as:

Zone I: H = -0.077.Age - 7.66 (m) (R^2=0.52) (3)

Zone II: H = -0.28.Age - 7.85 (m) (R^2= 0,998) (4)

Zone III: H = -0.44.Age - 7.12 (m) (R^2 = 0.83) (5)

This means that in zone I, II, and III the potential recharge rates could be as high as 77 mm/year, 280 mm/year and 440 mm/year, respectively.

4.4. Verification

The estimation of groundwater recharge rate by radioactive isotope ^{3}H shows that the zonation of groundwater potential recharge based on remote sensing combined with GIS techniques and determining the weights for overlay analysis by mean of AHP is completely reliable and acceptable. However, due to the limited number of samples in the regions and the uneven distribution of screen depth of the boreholes, the correlation coefficient is not good enough, especially for zone I. The reason for the small correlation coefficient is large differences of the 3H isotope compositions of the groundwater samples at boreholes Q67 and Q33 where the screen depth of boreholes are almost located at the same depth in different locations. It may be the lithology of the unsaturated layer at borehole Q33, which is mainly clay so that the travel time of rainwater is also longer than borehole Q67.

Some previous studies of groundwater recharge in RRDP also verified the above mentioned zonation. Postma et al. [18] estimated the recharge from the Red River to aquifers by Tritium/Helium dating in Dan Phuong where it is located in Noth Hanoi and in between zone I and zone II. Water samples for Tritium/Helium dating of the groundwater were taken from screens placed at different depths in the distance range from 64 to 75 m. The results suggest the ground waters to be less than 40 years old and a downward groundwater velocity of 0.5 m/yr. If formation porosity is 30% groundwater recharge rate could be about 150mm/year. Tran [11] used water balance site with 05 boreholes and groundwater level measurements in period 2008-2011 in Dan Phuong - Thach That area where is margin of the delta. Groundwater recharge rate were estimated as 175 mm/year. These results also proved that delineation is fair enough for groundwater potential recharge in the RRDP.

5. Conclusions

The Quaternary aquifer system in the RRDP is recharged mainly from rainwater, river water, and surrounding bedrocks. The topmost Holocene aquifer is an unconfined and semi-unconfined aquifer and recharged from mainly rainwater and surface water bodies. In this case of a large area with a lack of related information for delineation of groundwater potential recharge, a holistic approach based on the factors by using remote sensing interpretation on conditions of groundwater recharge and overlay analysis by GIS technique. To improve the reliability of this technique, it is very necessary to determine the weight of the relative importance of the relevant factors objectively by AHP. Furthermore, verifying the evaluation results by the direct method of radioactive isotope ^3H is also very effective. By this approach, groundwater potential recharge in the RRDP has been classified into 03 zones: low potential, moderate potential and high potential recharge. Using the results of 3H radioisotope analysis the groundwater recharge rate from high, moderate, and low potential zone were roughly estimated 77; 280; and 440 mm/year, respectively. However, 3H analysis samples are quite few and have only been conducted in the dry season, so they need to continue to be followed up in the near future. Weight value for thematic factors need to be improved if necessary when the data to estimate the groundwater recharge rate is large enough to have a more suitable.

6. Acknowledgments

This research is supported by the Vietnam Ministry of Natural Resources and Environment entitled "Applying Artificial Intelligence (AI) for flow routing to support water resources allocation in the river basin, testing on the Red river basin", Code number: TNMT.2021.04.05 and the collaboration project with Delft University of Technology (Netherland) under Project code OP-VNM-10005 for sampling and analyses. The authors would like to express their sincere thanks to anonymous reviewers for their helpful comments and review of the manuscript.

References

[1] Lerner, D.N., Issar, A.S., Simmers, I., 1990. Groundwater recharge. A guide to understanding and estimating natural recharge. International contributions to hydrogeology. Verlag Heinz Heise.8

[2] Scanlon, B.R., Richard, W., Healy, P. G., Cook., 2002. Choosing Appropriate Techniques for Quantifying Groundwater Recharge. Hydrogeology Journal, 10,18-39. DOI 10.1007/s10040-0010176-2

[3] Chuma, C., Orimoogunje, O.O., Hlatywayo, D.J. and Akinyede, J.O., 2013. Application of Remote Sensing and Geographical Information Systems in Determining the Groundwater Potential in the Crystalline Basement of Bulawayo Metropolitan Area, Zimbabwe. Advances in Remote Sensing, 2,149 - 169. DOI:10.4236/ars.2013.22019

[4] Machiwal, D., Jha, M.K., Mal, B.C., 2011. Assessment of groundwater potential in a semi-arid

region of India using remote sensing, GIS and MCDM technique. Water Resource, Manag 25, 1359-1386.

[5] Chowdhury, A. Jha, M.K Chowdary, V.M Mal, B.C, 2009. Integrated remote sensing and GIS based approach for assessing groundwater potential in West Medinipur district, West Bengal, India. Int.J.Remote Sens., 30, 231-250.

[6] Tesfa G.A, Girum G.D, 2019. Groundwater potential assessment using GIS and remote sensing: A case study of Guna tana landscape, upper blue Nile Basin, Ethiopia, Journal of Hydrology: Regional Studies, 24, 100610.

[7] Chowdhury, A., Jha, M.K., Chowdary, 2010. Delineation of groundwater recharge zones and identification of artificial recharge site in West Medinipur district, West Bengal using RS, GIS and MCDM techniques. Environ. Earth Sci. 59, 1209-1222.

[8] Kaliraj, S., Chandrasekar, N. and Magesh, N.S., 2014. Identification of potential groundwater recharge zones in Vaigai upper basin, Tamil Nadu, using GIS-based analytical hierarchical process (AHP) technique. Arabian Journal of Geosciences, 7(4), pp.1385-1401.

[9] Senanayake I.P., Dissanayake, D.M.D.O.K., Mayadunna, B.B., Weerasekera, W.L., 2016. An approach to delineate groundwater recharge potential sites in Ambalantota, Sri Lanka using GIS techniques. Geoscience Frontiers 7, 115-124.

[10] Sahoo, S. Jha, M.K., Kumar, N., Chowdary, V.M., 2015. Evaluation of GIS based muicriteria decision analysis and probabilitistic modeling for exploring groundwater prospects. Environ. Earth Sci. 74, 2223-2246.

[11] Jaiswal, R.K., Mukherjee, S., Krishnamurthy, J., Saxena, R., 2003. Role of remote sensing and GIS techniques for generation of groundwater prospect zones towards rural development approach. Int J Remote Sens, 24, 993-1008.

[12] Healy, R.W. and Cook, P.G., 2002. Using groundwater levels to estimate recharge. Hydrogeology journal, 10, 91-109.

[13] MPI, 2020. Annual Statistical Report (in Vietnamese)

[14] MONRE, 2015. National Report on Water Resources (in Vietnamese).

[15] Pham, Q.N., Dang, T.T, Tran, T.L, 2019. Susstainable groundwater development in Hanoi city. Science and Technics Publishing House. ISBN:978-604-67-1284-8

[16] Pham, Q.N., 2000. Groundwater reserves in Red River delta plain and its sustainable development. PhD Thesis. Hanoi University of Mining and Geology. Vietnam. p 144

[17] Tran., T.L., 2011. Estimation of groundwater recharge and hydraulic interaction between Quaternary aquifer in Thach That - Đan Phuong, Hanoi by using isotopes. MSc Thesis. Hanoi University of Mining and Geology.

[18] Postma, D., Larsen, F., Nguyen T. M. H., Mai, T. D., Pham, H. V., Pham, Q. N., Jessen, S., 2017. Arsenic in groundwater of the Red River floodplain, Vietnam: Controlling geochemical processes and reactive transport modeling. Geochimica et Cosmochimica Acta 71, 5054-5071.

[19] Larsen, F., Pham, N.Q., Dang, N.D., Postma, D., Jessen, S., Pham, V.H., Nguyen, T.B., Trieu, H.D., Tran, L.T., Nguyen, H., Chambon, J., Nguyen, H.V., Ha, D.H., Hue, N.T., Duc, M.T., Refsgaard, J.C., 2008. Controlling geological and hydrogeological processes in an arsenic contaminated aquifer on the Red River flood plain. Vietnam Appl Geochem 23, 3099-3115.

[20] Nielsen, L.H., Mathiesen, A., Bidstrup, T., Vejbæk, O.V., Dien, P.T. and Tiem, P.V., 1999. Modelling of hydrocarbon generation in the Cenozoic Song Hong Basin, Vietnam: a highly prospective basin. Journal of Asian Earth Sciences, 17, 269-294.

[21] Tran, L.T., Larsen, F., Pham, N.Q., Christiansen, A.V., Tran, N., Vu, H.V., Tran, L.V., Hoang, H.V. and Hinsby, K., 2012. Origin and extent of fresh groundwater, salty paleowaters and recent saltwater intrusions in Red River flood plain aquifers, Vietnam. Hydrogeology journal, 20, 1295-1313.

[22] Mathers, S.J., Davis, J., McDonald, A., Zalasiewitz, J.A., Marsh, S., 1996. The Red River Delta of Vietnam. Technical report WC/96/02, British Geological Survey, Keyworth, UK, p 41

[23] Mathers, S. and Zalasiewicz, J., 1999. Holocene sedimentary architecture of the Red River delta,

IOP Conf. Series: Earth and Environmental Science **964** (2022) 012025 doi:10.1088/1755-1315/964/1/012025

Vietnam. Journal of Coastal Research, pp.314-325.

[24] Nghi, T., Toan, N.Q., Minh, N.D. and Van Vuong, N., 1991. Quaternary sedimentation of the principal deltas of Vietnam. Journal of Southeast Asian Earth Sciences, 6(2), pp.103-110.

[25] Malczewski, J., 1999. GIS and multicriteria decision analysis. John Wiley & Sons.

[26] Saaty, T.L., 1980. The Analytic Hierarchy Process. McGraw-Hill, New York

[27] IAEA, 2002. International Atomic Energy Agency. Water and Environment Newsletter of the Isotope Hydrology Section; IAEA: Vienna, Austria, 2002; 8p

[28] David J.T., 1995. Groundwater recharge rates calculatated from the isotopic content of ground water a Pilot study, St Johns river water management district Palatka, Florida;

[29] Wright, E. P., 1992. The hydrogeology of crystalline basement aquifers in Africa, Geol. Soc. London Spec. Publ., 66, 1 –27

[30] McFarlane, M.J., Chilton, P.J., Lewis MA., 1992. Geomorphological controls on borehole yields; a statistical study in an area of basement rocks in central Malawi. In: Wright, E.P., Burgess, W.G. (eds). Hydrogeology of crystalline basement aquifers in Africa. Geological Society, London, SpecPubl 66, p131-154

[31] Ky, N.H., 2010. Vietnam Quaternary Geology and Environment, Science and Technology Publishing House

[32] Lam, D.D., Evolutionary history of Holocene sediments of the Red River structure, PhD thesis in Geology, National Library, Hanoi

[33] Lam, D.D., Boyd, W.E., 2003. Holocene costal stratigraphy and the sedimentary development of the Hai Phong area of the Bac Bo plain (Red River delta). Vietnam Aust Geogr, 34,177-194.

[34] Nghi, T., Toan.N.Q., 1991. Characteristics of sedimentary cycles and Quaternary geological evolutionary history of the Red River Delta, Journal of Geology, 206-207.

Catalytic oxidation of formaldehyde over silver supported on ZSM-5: The role of Ag and mesopores

Nguyen Tan Luon,[1,2] Le Nguyen Quang Tu,[1,2] and Nguyen Quang Long[1,2,*]

[1] Faculty of Chemical Engineering, Ho Chi Minh City University of Technology (HCMUT), Ho Chi Minh City, Vietnam
[2] Vietnam National University, Ho Chi Minh City, Vietnam

*Email: nqlong@hcmut.edu.vn

Abstract. Silver nanoparticles (AgNPs) are increasingly drawing a great deal of attention because of their exclusive properties and a huge variety of applications. In recent years, using AgNPs supported on various carriers as heterogeneous catalysts has become promising for treating some toxic gases in the environment, such as HCHO. This study has successfully synthesized AgNPs onto ZSM-5 microporous zeolite and ZSM-5 mesopore-modified zeolite (Meso-ZSM-5) by ion-exchange method using sodium borohydride as a reducing agent. The resulting catalysts were then characterized by N_2 adsorption-desorption method. In order to evaluate HCHO adsorption, desorption, and the surface reaction of these catalysts, temperature-programmed desorption (TPD) and temperature-programmed surface reaction (TPSR) were employed. The TPD and TPSR experiments were conducted with different relative humidity. The results showed that Ag/Meso-ZSM-5 exhibited higher catalyst activity in HCHO complete oxidation than Ag/ZSM-5 at high temperatures because of a new larger pore system within the zeolite. Furthermore, TPD and TPSR experiments provided an explanation for the poor performance of the catalysts at low temperatures, which was associated with the high adsorption capacity of the zeolite.

Keywords: Ag nanoparticles, mesoporous zeolite, HCHO catalytic oxidation, TPD experiment, TPSR experiment.

1. Introduction

Formaldehyde is one of the most common indoor gaseous pollutants, which can affect human health and comfort. The main sources of HCHO emission are industrial processes, product use, agriculture and transport. Long-time exposure to HCHO can cause headache, respiratory tract, pneumonia, eye irritation and even cancer. According to World Health Organization, the indoor HCHO concentration must be below 0,1 mg/m^3 [1-3]. Several approaches have been used to reduce the amount of HCHO in the air, such as adsorption, bio-filtration, thermal oxidation…Adsorption is the simplest method to remove formaldehyde from the air at a rapid rate and low cost, but limited adsorption capacity is their major problem. Destruction methods are preferable to separation methods as they allow to eliminate formaldehyde completely. Thermal oxidation is one of them, which permits to oxidize formaldehyde to harmless and less harmful products (H_2O and CO_2) by rising the temperature above the ignition point in O_2 flow. However, they require very high operation temperature and choosing the feed composition to avoid explosion [4-6]. Catalytic oxidation is a promising technique to completely convert HCHO into H_2O and CO_2 at lower operating temperature and lower capital cost. Two types of catalysts are single/mixed oxides (Cu, Mn, Cr, Fe, Ni) and supported noble metal catalysts. While the main pros of metal oxides are availability and cost, higher activity at low temperature, resistance to deactivation and ability to be regenerated can be achieved with transition metal catalysts. The current challenges lay in developing a new catalyst that is active at low temperature with low cost and good usability [7-11]. Many studies have been taken in recent years and achieved great results, such as Al-rich Beta zeolite supporting Pt (1%) and Pt (0.2%) supported onto pure silica Beta zeolite, which could completely oxidize HCHO at shallow temperature [12, 13]. However, the extortionate cost of noble metals for wide applications is still the main problem. More recently, Ag-based catalysts have been taken into consideration because silver is significantly cheaper than other metals (Pt, Pd and Au) and shows high catalyst activities for HCHO oxidation at low temperatures [14, 15]. For example, 10%Ag/Beta-Si can oxidize HCHO (80 ppm) completely to CO_2 and H_2O at around 70°C, or 8%Ag/TiO_2 can reach 100% conversion of HCHO (110 ppm) at about 95°C [16, 17]. Since the amount of noble metal used as a catalyst is minimal, they need a support to maintain their good performance. Moreover, supports play a vital role in forming, distributing, and accumulating small particles. Zeolite with high surface area, rigid frameworks and shape selectivity have been considered as an ideal support. However, micropores in conventional zeolites have high mass transfer resistance for reactants, intermediates and products to go through their channels and cages. This problem could be addressed by expanding the micropores within zeolites [5, 18-20]. Furthermore, introducing a more extensive pore system into the zeolites can also help reduce coke formation since reactants, intermediates and products can rapidly diffuse through their channels [21, 22]. Several methods have been used to introduce larger pores into zeolites, such as hydrothermal dealumination and other chemical treatments. Among these, desilication is widely applied since they allow to synthesize zeolites with uniform mesopores and little change in the number of acid sites. The Si/Al ratio has a big influence on the structural properties of zeolites modified by alkaline treatment [23-25].

In this paper, we report the catalytic performance of HCHO complete oxidation over Ag/Meso-zeolite compared to that over Ag/zeolite at different temperatures. We also investigated the HCHO adsorption capacity of these materials at various relative humidity. The temperature-programmed desorption (TPD) and temperature-programmed surface reaction (TPSR) were employed in further studies so as to understand the desorption behaviour and surface reaction for formaldehyde oxidation on these catalysts as well as the effect of the support on catalyst activity. The prepared catalysts were characterized by N_2 adsorption/desorption, and experiments related to adsorption and surface reaction of catalysts were carried out in a continuous flow fixed bed reactor.

2. Experimental

2.1. Sample preparation

For synthesis of Ag/Meso-ZSM-5, 100 mL of 0.2M NaOH solution was heated at 65°C. 3.3g of zeolite ZSM-5 (Si/Al = 200) was then added into the heated solution under vigorous stirring. After one hour, the slurry was cooled down immediately using an ice bath. It was then filtered and washed until neutral pH and dried at 80°C overnight. The solid was grinded and labelled Meso-ZSM-5. 2g of Meso-ZSM-5 was suspended in 231.5 mL of 1×10^{-3}M $AgNO_3$ solution and stirred for 24 hours at room temperature for complete cation exchange. Then, 23.1 mL of 4×10^{-2}M $NaBH_4$ solution was added into the mixture under continuous stirring and left for one hour at room temperature. The obtained suspensions of Ag/ZSM-5 were dried under vacuum at 80°C until the water was removed. The solid was rinsed and filtered with 200 mL of distilled water. After drying at 80°C overnight, the solid product was calcined at 300°C for two hours in the air.

A similar procedure was used for synthesis of Ag/ZSM-5. 2g of NH_4-ZSM-5 was calcined at 450°C for 1 hour in order to completely convert NH_4-ZSM-5 to H-ZSM-5. Then, H-ZSM-5 was suspended in 231.5 mL of 1×10^{-3}M $AgNO_3$ solution and stirred for 24 hours at room temperature for complete cation exchange. The following steps were the same as the procedure of Ag/Meso-ZSM-5 preparation.

2.2. Catalyst characterization and catalytic tests
The specific surface area, pore volume, and pore size distribution were investigated by N_2 adsorption-desorption isotherms using Nova 2200e Quantachrome.

The catalytic performance was tested in a fixed bed reactor with 0.1g of catalyst (granule's diameter of 1-2 mm). Gaseous formaldehyde was generated by flowing continuously pure N_2 (> 99.99%) over a glass bubbler containing formaldehyde solution, which was put in a water bath. The formaldehyde concentration was controlled by the flow rate of nitrogen and the water bath temperature. Relative humidity (RH) was also generated by consecutive bubbling method. The amount of water vapour was only controlled by the nitrogen flow rate. These gaseous streams were then mixed with the flows of nitrogen and oxygen to form a typical feed consisting of 90 ppm HCHO, 20 vol.% O_2, and N_2 for balance with a total flow rate of 50 mL/min going into the reactor. Before conducting the experiments, the catalysts were pretreated by N_2 stream at 300°C for 1 hour. The HCHO conversion was analyzed by gas chromatography equipped with a flame ionization detector (FID).

2.3. Adsorption/desorption and surface reaction experiments
These experiments were carried out in a fixed bed reactor system. Before the experiments, the catalyst (0.1g, diameter of 1-2 mm) was activated at 300°C for 1 hour in a flow of N_2 gas. The adsorption of HCHO was carried out under a continuous flow of HCHO/N_2 with 338 ppm of HCHO concentration. Until the adsorption completion indicated by stable peaks in the gas chromatography with the FID, the HCHO-TPD and HCHO-TPSR experiments were carried out. The HCHO-TPD was conducted in a continuous flow of N_2 (F = 50 mL/min), and the HCHO-TPSR was also carried out similarly in a continuous flow of O_2/N_2 (20 vol.% O_2, F = 50 mL/min) with the temperature ramped at 10°C/5 min from room temperature to 400°C. The desorption peaks were recorded at each 10°C.

3. Results and Discussion
Figure 1 presents the N_2 adsorption-desorption isotherms of the HZSM-5 and Meso-ZSM-5 pre and post-doping of Ag nanoparticles. All samples exhibited the type I isotherms which is typical for microporous structures. In addition, the type H4 hysteresis loops in the relative pressure ranging from 0.4 to 1 implied the presence of mesopores in the materials. The hysteresis loop of Meso-ZSM-5 was more extensive than that of ZSM-5, indicating significant mesoporosity formed after zeolite modification. The N_2 adsorption-desorption isotherm of the ZSM-5 was similar to that of Ag/ZSM-5, implying the structure of zeolite was maintained after introducing Ag nanoparticles onto zeolite. This was also true for Meso-ZSM-5 and Ag/Meso-ZSM-5.

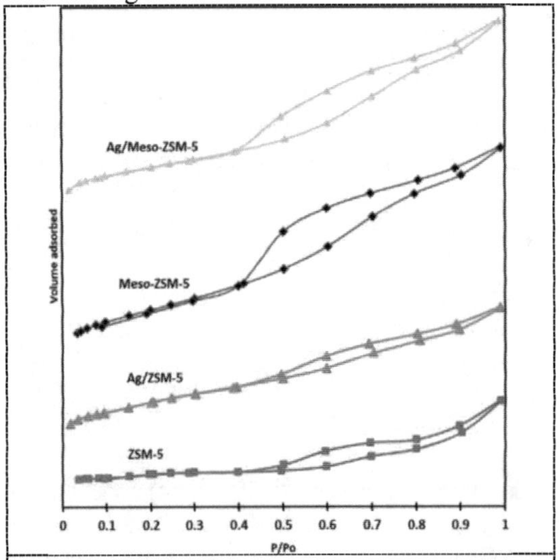

Figure 1. N_2 adsorption-desorption isotherms.

Regarding the pore size distribution of initial ZSM-5 zeolite and mesopore-modified ZSM-5, it could be seen that a huge number of mesopores with the size from 3 nm to 15 nm have been introduced into the zeolite after alkaline treatment. However, no single peak could be seen in the mesopores' size distribution after desilication, which means that the removal of Si from the zeolite framework is not selective. As for the pore size distribution of Meso-ZSM-5 pre and post doping, the material didn't show a significant change after dispersing Ag. Nevertheless, there was a decrease in the volume of pores ranging from 4 -7 nm, indicating that Ag nanoparticles were deposited in these mesopore cavities.

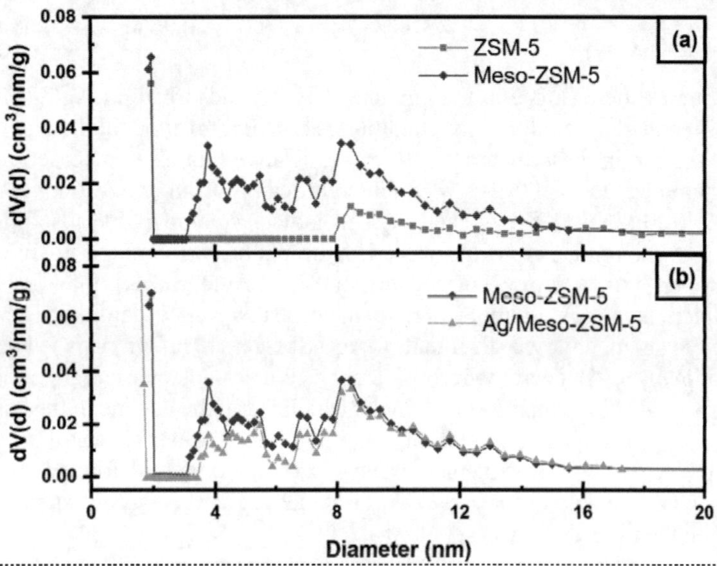

Figure 2. Pore size distribution of ZSM-5 zeolite before and after alkaline treatment (a) and pore size distribution of Meso-ZSM-5 pre and post doping Ag (b).

Figure 3 illustrates the pore formation of the zeolite by alkaline treatment. It is evident that the high ratio of silica to aluminum will lead to excessive Si dissolution and thus lead to the broad pore size distribution. Uniform mesopore-modified ZSM-5 is more desirable because it could offer better catalytic properties (selectivity and stability). So it is advised to choose another zeolite with lower Si/Al ratio in desilication treatment to achieve a better support for the catalyst.

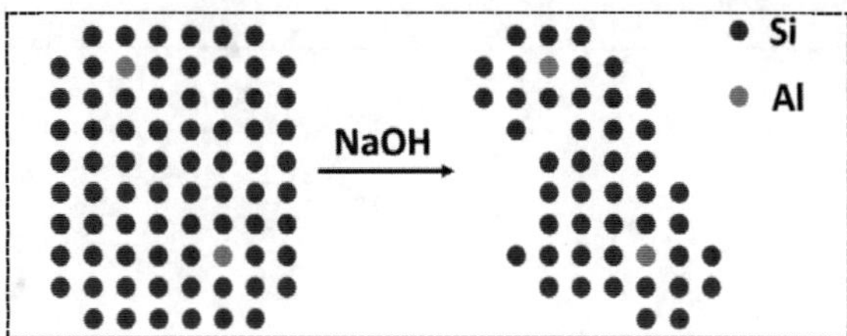

Figure 3. Schematic representation of creating mesopores in the zeolite by desilication [23, 24].

In table 1, it is notable that, after surface modification, the surface area of zeolite was significantly higher than the initial zeolite, which will facilitate the dispersion of Ag nanoparticles onto the zeolite. However, after introducing Ag onto zeolite, the surface area substantially decreased, especially in mesopores. Meanwhile, there was almost no change in the surface area of micropores. This proves that the Ag nanoparticles' size was as large as the dimension of the mesopores, so they have blocked a part of these mesopores entirely. This could also be observed in the volume of mesopores and micropores pre and post-doping Ag.

Table 1. Textual properties of HZSM-5 zeolite, Meso-ZSM-5 and Ag supported catalysts.

Sample	S_{BET} (m^2/g)	S_{micro} (m^2/g)	S_{meso} (m^2/g)	V_{total} (cm^3/g)	V_{micro} (cm^3/g)	D_{pore} (nm)
ZSM-5	288.3	259.4	28.9	0.1071	0.1031	1.927
Meso-ZSM-5	362.3	144.7	217.6	0.3450	0.0595	3.810
Ag/Meso-ZSM-5	287.1	145.1	142.0	0.2934	0.0594	4.086

Figure 4a illustrates the changes in the amount of HCHO adsorbed on Ag/ZSM-5 and Ag/Meso-ZSM-5 at two different relative humidity conditions according to time, and figure 4b calculates the amount of HCHO stored in these materials. When the relative humidity increased from 18% to 43%, the HCHO adsorption capacity of Ag/ZSM-5 fell significantly from 555.4 μmol-HCHO/g-catalyst to only approximately 140 μmol-HCHO/g-catalyst. This indicated comparative adsorption between water and formaldehyde. Although the ZSM-5 surface is hydrophobic with around 5Å in capillary size, this was easier for water with the molecular diameter of 2.8Å to diffuse into pores as opposed to formaldehyde which is 3.73Å in molecular diameter. However, regarding Ag/Meso-ZSM-5, the adsorption capacity was much lower than that of Ag/ZSM-5 at 18% RH. This is because after NaOH treatment, the ratio of Si/Al decreased considerably. Water molecules preferentially adsorbed on hydrophilic centers (Al^{3+}) more than formaldehyde, which led to the decline in the adsorption capacity. On the other hand, when the concentration of water rose, more water appeared on the surface of the material. These molecules bonded together via hydrogen bonding and formed small water clusters [26]. The clusters could be able to interact with formaldehyde in the gaseous stream, so slightly more HCHO adsorbed on the material.

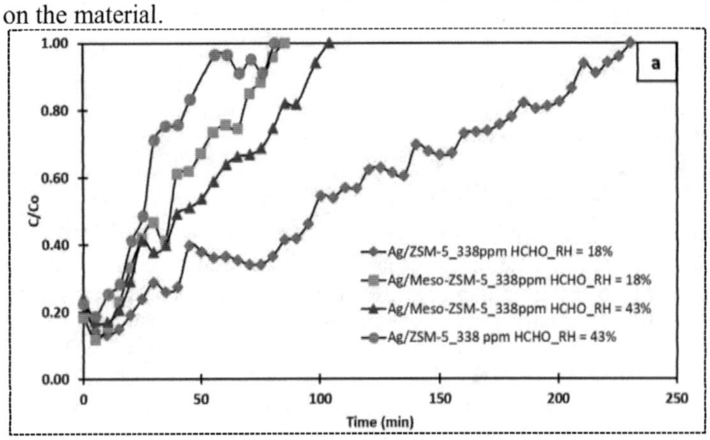

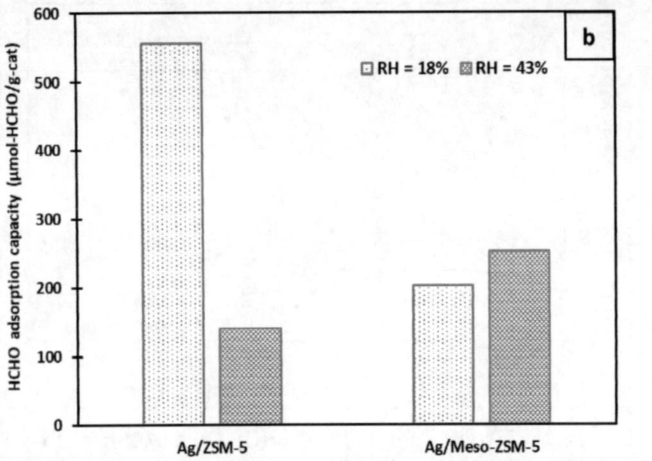

Figure 4. Graph showing the adsorption capacity of materials over time with different relative humidity (C_{HCHO} = 338 ppm_v, F = 40 mL/min, T = 32°C) (a) and HCHO adsorption capacity of materials at different relative humidity conditions.

Figure 5 and figure 6 show desorption characteristics for HCHO during TPD and TPSR processes on Ag/Meso-ZSM-5 at 43% RH and 18% RH, respectively. It is clear that part of HCHO adsorbed reversibly on the catalyst at room temperature. Despite the fact that the desorption significantly took place from room temperature to 160°C in both cases, there was a difference in the desorption trend in the various presences of moisture. The peak of desorption at RH of 43% was around 50°C, while the desorption at 18% RH happened in a wide range of temperatures, from 40°C to 240°C. An answer could be the desorption of water vapor at high temperatures, which means that the interaction between water and Ag/Meso-ZSM-5 is stronger than that between the material and formaldehyde. It is remarkable that there was no change between the TPD curve and TPSR curve in both situations. This means that the catalyst showed almost no activity at low temperatures.

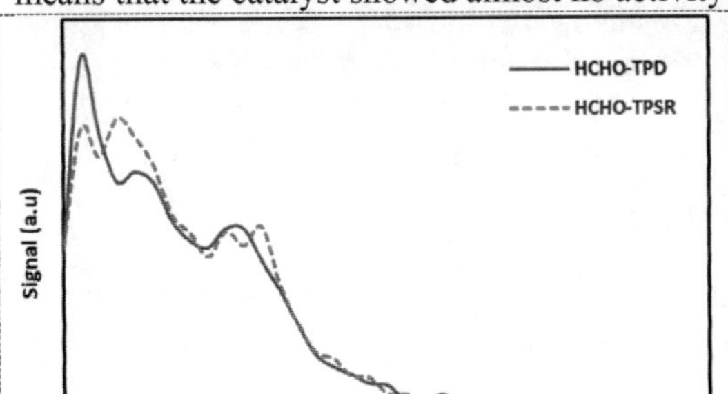

Figure 5. HCHO-TPD and HCHO-TPSR profiles for Ag/Meso-ZSM-5 with 43% RH.

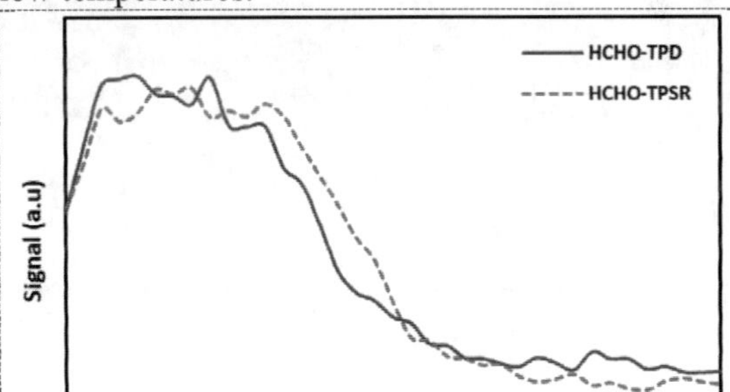

Figure 6. HCHO-TPD and HCHO-TPSR profiles for Ag/Meso-ZSM-5 with 18% RH.

Figure 7 shows data about the amount of HCHO desorbed from the zeolite surface. The adsorption and desorption amount has a great impact on catalyst activity. The catalyst will have better performance if the desorption is considerable at low temperatures. It will recover new sites for HCHO to be adsorbed on the catalyst [19]. However, if we look at Ag/Meso-ZSM-5, the desorption was still intense until 160°C, so there are a lot active sites covered by HCHO at high temperatures.

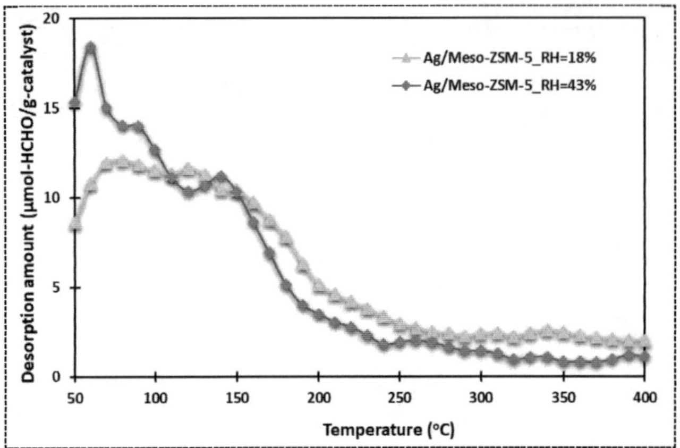

Figure 7. The amount of HCHO desorbing from Ag/Meso-ZSM-5 with different relative humidity.

A lot of researches showed a strong association between catalyst activity and the formation of DOM (intermediate species), which we can see in figure 8 [5, 7, 27]. More specifically, the more and faster the number of DOM species is generated, the better the catalystic performance. Most formaldehyde molecules adsorb onto the zeolite since the amount of Ag doping on the catalyst is relatively small. No difference in the TPD and TPSR curves points out that the support is inner in the formaldehyde oxidation. Ag nanoparticles are the only active sites for the reaction, which formaldehyde molecules adsorb on and converse into CO_2 and H_2O in the presence of O_2. In addition, high adsorption capacity of the carrier could influence catalytic performance because a majority of formaldehyde species adsorbed physically on the catalyst, which can prevent formaldehyde from approaching active centers.

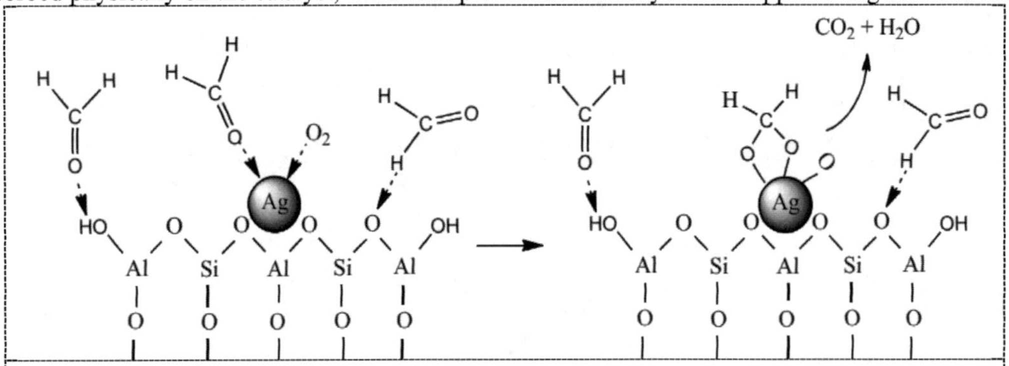

Figure 8. Proposed mechanism for formaldehyde oxidation [7, 19].

Figure 9 shows HCHO complete conversion as a function of temperature over Ag/ZSM-5 and Ag/Meso-ZSM-5 at 65% relative humidity. There was no catalyst activity until the temperature above 120°C in both Ag/Meso-ZSM-5 and Ag/ZSM-5, which suited the discussion in desorption character. Furthermore, Ag/Meso-ZSM-5 showed a better catalyst than Ag/ZSM-5 at the same condition. Especially at 200°C, the HCHO conversion to CO_2 over Ag/Meso-ZSM-5 achieved 70% in comparison with 20% of conversion Ag/ZSM-5 exhibited. It could be assumed that the diffusion has been improved after surface modification, which permits organic reactants and products to diffuse easily throughout the nanochannels of the zeolite. The overall rate of reaction includes both the reaction rate on the exterior and interior surface. The external surface reaction rate simply depends on the reaction rate constant and the concentration of reactants. Meanwhile, the reaction rate at the interior surface relies on not only these factors but also pore diffusion. At high reaction temperature, the reaction process is relatively fast to the diffusion processes. In this case, the rate of reaction is dominated by diffusion [28]. Introducing extra mesopores into conventional zeolite helped to reduce diffusion limitation and thus enhanced catalyst efficiency at high temperatures.

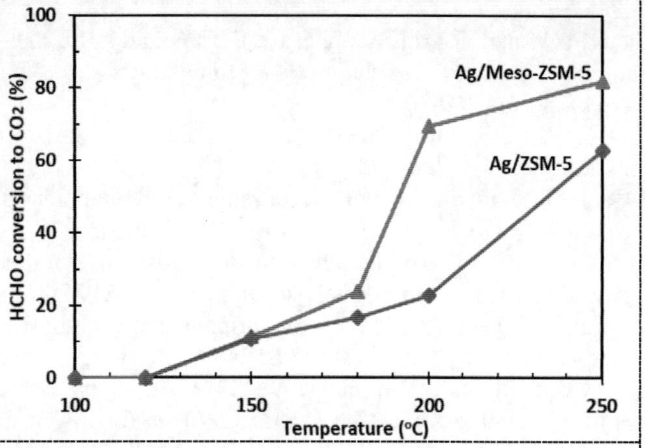

Figure 9. HCHO complete oxidation over Ag/Meso-ZSM-5 and Ag/ZSM-5 (RH = 65%).

4. Conclusion

In short, we have already prepared silver nanoparticles supported onto ZSM-5 and Meso-ZSM-5 and investigated their performance in HCHO complete oxidation. Both Ag/ZSM-5 and Ag/Meso-ZSM-5 have not exhibited high catalytic performances at low temperatures. However, at high temperatures, Ag/Meso-ZSM-5 showed better HCHO oxidation than Ag/ZSM-5. In addition, this work has studied HCHO adsorption-desorption properties and surface reaction of Ag/Meso-ZSM-5. There is clear that Ag nanoparticles are the only active sites of the catalyst, and the high adsorption capacity of the zeolite caused the poor catalytic performance. Introduction of an additional mesopores system enhanced catalyst activity due to reducing diffusion resistance.

Acknowledgements

We acknowledge the support of time and facilities from Ho Chi Minh university of Technology (HCMUT), VNU-HCM for this study.

References

[1] Z L Que, F B Wang, L F Ma and T Furuno 2013 Effect of loading ratio, conditioning time and air exchange rate on the formaldehyde emission from wood-based board using large chamber and desiccator method. *Composites, Part B* **47** 278-82

[2] T Salthammer 2019 Formaldehyde sources, formaldehyde concentrations and air exchange rates in European housings. *Build.Sci.* **150** 219-32

[3] Z Yan, Z Xu, J Yu and M Jaroniec 2015 Highly active mesoporous ferrihydrite supported Pt catalyst for formaldehyde removal at room temperature. *Environ.Sci.Technol.* **49**(11) 6637-44

[4] Santos V P D C O 2010 *PhD Dissertation* Catalytic oxidation of volatile organic compounds. Universidade do Porto (Portugal)

[5] J Ye, Y Yu, J Fan, B Cheng, J Yu, W Ho 2020 Room-temperature formaldehyde catalytic decomposition. *Environ.Sci.:Nano* **7**(12) 3655-709

[6] J Zhang, Y Li, L Wang, C Zhang, H He 2015 Catalytic oxidation of formaldehyde over manganese oxides with different crystal structures. *Catal.Sci.Technol.* **5**(4) 2305-13

[7] J Pei and J S Zhang 2011 Critical review of catalytic oxidization and chemisorption methods for indoor formaldehyde removal. *Hvac&R Res* **17**(4) 476-503

[8] L Torrente-Murciano, B Solsona, S Agouram, R Sanchis, J M López, T García and R Zanella 2017 Low temperature total oxidation of toluene by bimetallic Au–Ir catalysts. *Catal. Sci.Technol.* **7**(13) 2886-96

[9] H Huang, C Zhang, L Wang, G Li, L Song, G Li, S Tang and X Li 2016 Promotional effect of HZSM-5 on the catalytic oxidation of toluene over MnO_x/HZSM-5 catalysts. *Catal.Sci.Technol.* **6**(12) 4260-70

[10] S Valange and J C Védrine, 2018 General and prospective views on oxidation reactions in heterogeneous catalysis. *Catalysts* **8**(10) 483

[11] L Liotta 2010 Catalytic oxidation of volatile organic compounds on supported noble metals. *Appl.Catal.,B* **100**(3-4) 403-12

[12] L Zhang and et al 2017 Complete oxidation of formaldehyde at room temperature over an Al-rich Beta zeolite supported platinum catalyst. *Appl.Catal.,B* **219** 200-8

[13] L Zhang and et al 2020 Exceptional activity for formaldehyde combustion using siliceous Beta zeolite as a catalyst support. *Catal.Today* **339** 174-80

[14] L Yu, R Peng, L Chen, M Fu, J Wu and D Ye 2018 Ag supported on CeO_2 with different morphologies for the catalytic oxidation of HCHO. *Chem.Eng.J.* **334** 2480-87

[15] Z Qu, S Shen, D Chen and Y Wang 2012 Highly active Ag/SBA-15 catalyst using post-grafting method for formaldehyde oxidation. *J. Mol.Catal.A: Chem.* **356** 171-7

[16] L Zhang, Y Xie, Y Jiang, Y Li, C Wang, S Han, H Luan, X Meng and F-S Xiao 2020 Mn-promoted Ag supported on pure siliceous Beta zeolite (Ag/Beta-Si) for catalytic combustion of formaldehyde. *Appl.Catal.,B* **268** 118461

[17] J Zhang, Y Li, Yan Z, M Chen, L Wang, C Zhang and H He 2015 Effect of support on the activity of Ag-based catalysts for formaldehyde oxidation. *Sci.Rep* **5**(1) 1-10

[18] D Xu, H Lv and B Liu 2018 Encapsulation of metal nanoparticle catalysts within mesoporous zeolites and their enhanced catalytic performances: a review. *Front.Chem.* **6** 550

[19] D Chen, Z Qu, S Shen, X Li, Y Shi, Y Wang, Q Fu and J Wu 2011 Comparative studies of silver based catalysts supported on different supports for the oxidation of formaldehyde. *Catal.Today* **175**(1) 338-45

[20] D Farrusseng and A Tuel 2016 Perspectives on zeolite-encapsulated metal nanoparticles and their applications in catalysis. *NewJ.Chem* **40**(5) 3933-49

[21] T Weissenberger, A G F Machoke, J Bauer, R Dotzel, J L Casci, M Hartmann and W Schwieger 2020 Hierarchical ZSM-5 catalysts: The effect of different intracrystalline pore dimensions on catalyst deactivation behaviour in the MTO reaction. *ChemCatChem*, **12**(9) 2461-68

[22] P Peng, X H Gao, Z F Yan, S Mintova 2020 Diffusion and catalyst efficiency in hierarchical zeolite catalysts. *Natl.Sci.Rev* **7**(11) 1726-42

[23] Y Tao, H Kanoh, L Abrams, K Kaneko 2006 Mesopore-modified zeolites: preparation, characterization, and applications. *Chem.Rev* **106**(3) 896-910

[24] L Zhao, J Gao, C Xu, B Shen 2011 Alkali-treatment of ZSM-5 zeolites with different SiO_2/Al_2O_3 ratios and light olefin production by heavy oil cracking. *Fuel Process.Tech.* **92**(3) 414-20

[25] M Ogura, S Shinomiya, J Tateno, Y Nara, M Nomura, E Kikuchi, M Matsukata 2001 Alkali-treatment technique-new method for modification of structural and acid-catalytic properties of ZSM-5 zeolites. *Appl.Catal.,A* **219**(1-2) 33-43

[26] S Xian, Y Yu, J Xiao, Z Zhang, Q Xia, H Wang and Z Li 2015 Competitive adsorption of water vapor with VOCs dichloroethane, ethyl acetate and benzene on MIL-101 (Cr) in humid atmosphere. *RSC Adv.* **5**(3) 1827-34

[27] H Yunbing and J Hongbing, 2010 In-situ DRIFTS study on catalytic oxidation of formaldehyde over Pt/TiO_2 under mild conditions. *Chin.J.Catal.* **31**(2) 171-5

[28] C H Bartholomew and R J Farrauto 2011 *Fundamentals of industrial catalytic processes* (John Wiley & Sons)

Au nanoparticles loaded Hydroxyapatite catalyst prepared from waste eggshell: synthesis, characterization and application in VOC removal

Luu Trung Thien[1,2], Le Nguyen Quang Tu[1,2], Bien Cong Trung[1,2], Nguyen Quang Long[1,2,*]

[1] Faculty of Chemical Engineering, Ho Chi Minh City University of Technology (HCMUT), Ho Chi Minh City, Vietnam
[2] Vietnam National University, Ho Chi Minh City, Vietnam

*Email: nqlong@hcmut.edu.vn

Abstract. Eggshell, which is made almost entirely of calcium carbonate, is among the most abundant waste materials from poultry production. In 2018, the worldwide egg production exceeded 78 million metric tons which contributed to over 8 million metric tons of eggshell waste. However, this waste can be converted into useful materials for several industrial applications. Herein, hydroxyapatite (HAp) of fine particles was prepared using a green synthesis procedure in water medium from eggshell waste. Then, HAp was utilized as support material for loading different contents of Au nanoparticles (0.2 and 0.5 wt%). The materials were characterized by Brunauer-Emmett-Teller (BET) surface area analysis, Scanning Electron Microscopy with Energy Dispersive Spectroscopy (SEM-EDS), X-ray diffraction (XRD). The prepared catalysts were used for VOCs oxidation of formaldehyde in continuous flow and humid condition. In this study, Au/HAp (0.2 wt%) exhibited superior catalytic activity with good low-temperature reproducibility and high dispersion of particles on the surface of hydroxyapatite. Hence, hydroxyapatite prepared from eggshell waste is considered as a promising support material for noble metal catalysts for VOCs oxidation.

1. Introduction

Volatile organic compounds (VOCs) are organic chemical compounds whose composition makes them volatile under normal, indoor atmospheric conditions of temperature and pressure [1]. VOCs are usually industrial solvents such as chloroform, dichloromethane, benzene, toluene, xylenes, etc. In recent years, the industries of chemicals, pharmaceuticals, automobiles, oil and gas, etc. are growing with the number of factories constantly increasing. The manufacturing processes of these plants produce large amounts of emissions containing VOCs which cause environmental problems. Indoor emission of VOCs are mainly released into the air through the use of products and materials that contain organic solvents. Exposure to VOCs can cause immediate or long-term harm to human health [2]. Today, human are more and more concerned about the quality of air indoor. Indoor air pollutants mainly contain nitrogen oxides (NO_x), carbon oxides (CO and CO_2), volatile organic compounds and particulates. The overall rate of elimination of a VOC species from the atmosphere can be used to calculate its atmospheric lifespan. This rate can be calculated by adding the rates of radical species reactions, photolysis rates, and wet and dry deposition rates. Therefore, VOCs need to be analyzed, evaluated, treated and controlled from sources that emit VOCs into the environment. Due to the negative effects on human health, it is necessary to take measures to treat and remove VOCs from the air. Methods for removing VOCs from the air include physical, chemical or biological methods and can be divided into two main groups: recovery or decomposition [3]. It can be seen that with the recovery method, there will be many ways to remove VOCs, typically absorption, adsorption, condensation and finally membrane filtration. Another side, with the decomposition method, thermal oxidation processes are preferred in most cases because of their reliability, the fact that no further treatment is usually required, and because of their ability to achieve high efficiency destruction of VOCs. The prevalence of combustion systems such as boilers, furnaces, furnaces and incinerators in a multitude of manufacturing and fabrication facilities has enabled operators to accept thermal oxidants. Most other treatment technologies require additional treatment of VOCs after they have been removed from the gas stream. In this work, decomposition method is chosen to formandehyde removal. The process of catalytic oxidation is comparable to the method of thermal oxidation. Despite its significant cost savings, catalytic oxidation has a number of drawbacks, including high catalyst costs, additional byproducts, the possibility of poisoning the catalyst after use if not properly managed, and sensitivity to non-VOCs (water, halogen and sulfur compounds). As a result, thermal oxidation methods are employed more frequently than catalytic oxidation methods.

Catalytic oxidation of VOCs is the most promising approach among them, as it can be done at lower temperatures, at a cheaper cost, and with a higher oxidation efficiency. Gold has long won the hearts of mankind. Gold, abbreviated Au, is the 79th element in the periodic table, belonging to group IB and having the structure $[Xe]5d^9 6s^2$. Because the energy of the 5d and 6s subclasses of Au is nearly identical, there is competition between them. As a result, electrons are extremely malleable, able to migrate between both states, resulting in a unique plasticity in Au and a diverse optical spectrum. Gold is found in ores created from rock with very small or very small fractions of gold. Au has metal properties such as being soft, ductile, and easy to laminate. When carefully cut, the golden tint of this metal can turn black, ruby, or purple. Gold is inert to most chemicals, however it dissolves in Aqua Regia and forms cloroauric acid, as well as reacting with alkali metal cyanide solutions. For millennia, it has been used in art, cosmetics, metallurgy and architecture. This element is considered to be the ultimate statement of prosperity and beauty. Initial experimental work involving surface characteristics and related calculations of density functions showed that, for Au, dissociative adsorption of H_2 and O_2 does not take place at temperatures lower than 473K , and thus it will not exhibit catalytic activity for hydrogenation and oxidation reactions [4]. This is why Au has not been of catalytic interest up to now, but this property disappears when gold is synthesized in the nanoscale, in 1987 when Haruta et al. found catalytic activity particularly unexpected effect of nanometer-sized gold nanoparticles on CO oxidation even at lower room temperature [5]. With the advancement of science and technology, numerous important properties of gold at the nanoscale, as well as their practical applications, have been found. In 1991, Hutching and his colleagues used gold nanoparticles (AuNPs) as catalysts in the hydrochlorination of acetylene [6]. When compared to its bulk state, golden has a number of unique features at the nanoscale. The power to change hue, from golden to red or light purple, is one of them. This is due to the fact that gold nanoparticles absorb and scatter light in a different wavelength spectral region than solid gold. Furthermore, even at high sulfur conditions, gold catalyst is difficult to poison. Recently, gold nanoparticles (Au NPs) gain valuable interest in various fields such as impact, sensing, optics and electrochemistry as active features high impact, high electrical conductivity, modification, stability, adjustable spoilage and good biocompatibility [7]. Furthermore, it is known that Au can exhibit the lowest temperature response compared to other metals, for which it is listed as the most "noble" [8]. Similar to most nanoparticle catalysts, Au NPs still exhibited limitation in uniformed distribution, difficulty to recover at nanoscale [9]. Thus, immobilizing gold nanoparticles on a suitable substance is important.

Despite the various potential uses of supported gold catalysts, sintering of gold nanoparticles at high temperatures is one of the most significant restrictions in their utilization. Certain applications would necessitate the catalyst's ability to withstand high temperatures and remain functional for extended periods of time. It is well known that the choice of supports and the precise interaction between gold and the support have a significant impact on the catalytic activity and stability of supported gold. Hydroxyapatite (HAp) $[Ca_{10}(PO_4)_6(OH_2)]$, which has been extensively used for ion exchange, adsorption, dissolution/precipitation and formation of surface complexes have been recognized as removal mechanisms. The application of HAp in the fields of heterogeneous has received a lot of attention in recent years. The intrinsic acid-basic characteristics of HAp, as well as the possibility of a large specific surface area and thermal stability, make it a good catalytic material. [11-15]. The ability of HAp to remove hazardous metals from aqueous solutions has also been established at a laboratory scale. HAp has strong affinity for the fixation of different metals including lead, cadmium, copper, zinc, strotinum, mecury, cobalt, etc. [16-21]. Furthermore, because of its outstanding chemical stability, bioactivity, and biocompatibility, the HAP nanomaterial has been extensively studied in biomedical and catalyst-promoter applications [22-24]. However, to the best of our knowledge, there are just a few industrial applications of HAp for the usage of support material. Furthermore, functional composite catalysts with different surface acid-base characteristics, which may display unexpected HCHO capture and auxiliary oxidation performance, have yet to be thoroughly explored for HCHO elimination. HAP has been shown to have superior CO oxidation activity and to improve the stability of gold nanoparticles against sintering at temperatures as high as 600°C when used as a support for nano-gold [25,26]. Although the cause of this stabilization is unknown, it has been attributed to the existence of the PO_4^{3-} and OH^- groups [27]. This catalyst's exceptional sintering resistance opens up new possibilities for the production of stable nano-gold catalysts. These factors lead to HAp is decided to use as support material for Au. In fact, for a process to be viable, costs must be kept to a minimum, especially for environmental reasons. There are various way to synthesis HAp, as synthesized by precipitation method, sol-gel method, chemical ultrasound, spray drying method, electrochemical method, hydrothermal method, composite method or solid phase reaction method [28]. Precipitation method is commonly used to prepare HAp materials commercial because it is a simple method, has high economic efficiency and does not require complicated equipment. The principle of the precipitation method whereby starting materials containing Ca^{2+} and PO_4^{3-} ions are mechanically mixed together, using suitable agents to adjust the pH of the solution to obtain HAp. Thus, this method is beneficial for preparing HAp from rich sources of Ca^{2+} such as eggshells which are made of 95% of $CaCO_3$. In this process precipitation method is used for samples preparation, which is known as the possibility of creating pure and homogenous. The advantage of this method is that it is possible to control the reaction parameters by interfering with the crystallization process to obtain nanometer-sized HAp products. The essence of this method is to go from ions in solution to form precipitated products with sizes that gradually increase from the original size. Therefore, the shape growth and crystallinity of the products can be controlled by changing the reaction conditions such as temperature, time, pH, solvent. Moreover, the reaction may not require frequent pH adjustment. In addition, this method uses easy-to-find chemicals, simple equipment and thus it is environmentally friendly and suitable for industrial production. It has been also demonstrated that materials prepared by precipitation method have many attractive properties for high technology applications, including high crystallinity, small particle size, high specific surface area, and high average pore diameter [29]. Over 78 million tons of eggs was produced in 2018 and created over 8 million eggshells as solid waste [30]. The majority of eggshells is usually disposed at landfill and considers as a toxic waste for the environment by the EU [31]. However, eggshells can be a valuable source of calcium for the production of hydroxyapatite because they are composed of primarily of calcium carbonate (>94%). Eggshell-derived hydroxyapatite can be very beneficial as it reduces the solid waste discharged into the environment and the prepared material has excellent properties for environmental remediation (e.g. heavy metallic ion adsorption [32]).

In this paper, hydroxyapatite was prepared from eggshells waste collected from a local restaurant in Ho Chi Minh City Vietnam. The prepared hydroxyapatite was used as support material for the loading of Au nanoparticles for the catalytic oxidation of HCHO in gaseous phase. Formaldehyde (HCHO), which can be emitted for a long time in an airtight area from adhesives, paints, and ornamental materials, is well known for its harmful effects on human health [33-37]. The effect of Au loading content, temperature and relative humidity on the catalytic activity was investigated. XRD, SEM-EDS and N_2 adsorption-desorption isotherms were used to characterize the prepared materials.

2. Experimental

2.1. Samples preparation

Hydroxyapatite (HAp). Eggshells were collected, soaked and washed several times with deionized water, before being ground and calcined at 800°C to form CaO. Then, the CaO was dispersed 100 mL of H_2O, while stirring continuously at 80°C. Next, a solution of H_3PO_4 (1M) was added dropwise into the above mixture. The reaction was carried out for 2 h and was settled overnight. The mixture was then filtered and dried overnight at 80°C. The solid was then grinded and calcined at 350°C to stabilize and increase crystallinity. The final white powder was labeled HAp. The HAp have ratio Ca/P = 1.67 in theory.

Au loaded hydroxyapatite (Au/HAp). 1.0g of HAp powder was dispersed in 100mL of H_2O. The temperature of the mixture was raised to 60°C and was kept throughout the procedure. An appropriated amount of $HAuCl_4$ (0.0313M) solution was added corresponding to the desired final mass ratio (0.2 and 0.5 wt%) following by the addition of citric acid (1M) solution. The reaction was carried out for 2 h and the mixture was settled overnight. The pinkish slurry was filtered and dried overnight. The solid was ground and calcined at 350°C for 2 h and the final purple powder was labeled XAu/HAp with X=0.2 or 0.5.

2.2. Catalyst characterization

The crystalline structure of the prepared hydroxyapatite was determined using Bruker D2 diffractometer, with Cu Kα radiation (λ = 1.54184Å) operated at 30 kV and 10 mA, the range 2θ of 20° - 70°, the scanning step is 0.03° for a period of 0.7s per step in the temperature of 25°C. SEM-EDS images were obtained using Hitachi FE SEM S-4800 Scanning Electron Microscope with Energy Dispersive Spectroscopy. The specific surface area was determined using the NOVA 2200e Surface area and Pore size analyzer (Quantachrome Corp.).

2.3 Catalytic oxidation experiments

The initial nitrogen and oxygen sources of the system have the purity of >99.9% and each stream was controlled using a set of needle valve and flow meter. Three streams were divided from the original nitrogen source which included a HCHO feed stream, a water vapor feed stream and a dilute stream. The HCHO tank and water tank were placed in liquid baths with thermometers. Thus, the relative humidity (RH=5% or 65%)

could be controlled by changing the bath's temperature and ratio of the streams. The three nitrogen streams and an oxygen stream were uniformly mixed in a chamber to form a model gas with C_{O2}=20 % v/v. The outlet of the mixing chamber was divided into two streams, a bypass stream and a stream to the reactor using a 3-way valve. The catalyst samples were pressed, which pelleting equipment that are mainly used to produce pellets. The ring die pellet mill has a slot all over the die. It works by distributing the powder once it is fed into the die. It has one roller that are responsible for compressing the powder through the hole. Afther that crushing, and sieving, 0.1 g of HAp or Au/HAp was fixed in the reaction tube with 2 layers of glass wool (1cm thick of catalyst). Catalyst was activated at 300°C for 1h in a flow of N_2 gas before reaction.

The experiment procedure consisted of three phases. Firstly, the catalyst was activated for 2h at 300°C and cooled down in a nitrogen stream. Next, the model gas was fed to the reactor by adjusting the 3-way valve for adsorption of HCHO. After reaching saturation, the temperature of the reactor was increased from 50 to 260°C to perform the catalytic oxidation reaction. At each temperature, the reaction is maintained until equilibrium reaching. The outlet of the reactor was connected with gas chromatography (Hewlett Packard 5890plus) equipped with a Flame Ionization Detector (FID) via a 6-way valve for online injection and monitor of HCHO concentration.

The conversion efficiency of formaldehyde was calculated using the following equation:

$$\chi = \left(1 - \frac{C_{out}}{C_{in}}\right) \times 100$$

where χ is the conversion efficiency, $C_{out/in}$ is the outlet/inlet concentration of formaldehyde.

3. Results and discussion

Fig. 1 illustrated the XRD pattern of the prepared HAp sample. The prepared-HAp sample exhibited characteristic peaks at 2θ = 26.2°; 32.3°; 39.5°; 47.7°; 48.7° and 54.6° which was corresponding to hydroxyapatite (JCPDS no.09-0432). Furthermore, there is no record of secondary phase on the pattern indicating the prepared hydroxyapatite was pure. Washed eggshells had a specific surface of 4 m^2g^{-1}, which were investigated in the laboratory, while that of prepared HAp is 85 m^2g^{-1}, which were also analysed in the laboratory. Thus, it can be seen that the specific surface area had increase significantly by about 21 times. Thus, the prepared HAp sample was more suitable to be used as catalyst support then eggshell waste.

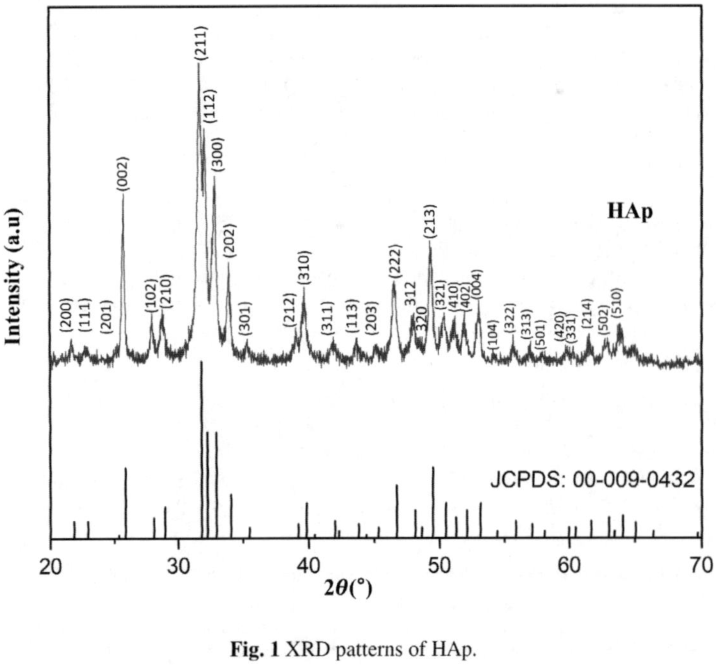

Fig. 1 XRD patterns of HAp.

The prepared HAp nanoparticles were relatively uniform in shape and size as shown in Fig. 2a and Fig. 2b. The nanoparticles in Fig. 2b had the shape of rice grains with an average length of 120±5 nm and an average width of 40±5 nm. Au NPs were not detected in Fig. 2 suggesting that they were small in size and had good distribution on the surface of prepared HAp NPs. EDS analysis also confirmed the presence of Au element on the final material which was associated with the successful doping of Au element on the prepared HAp nanoparticles.

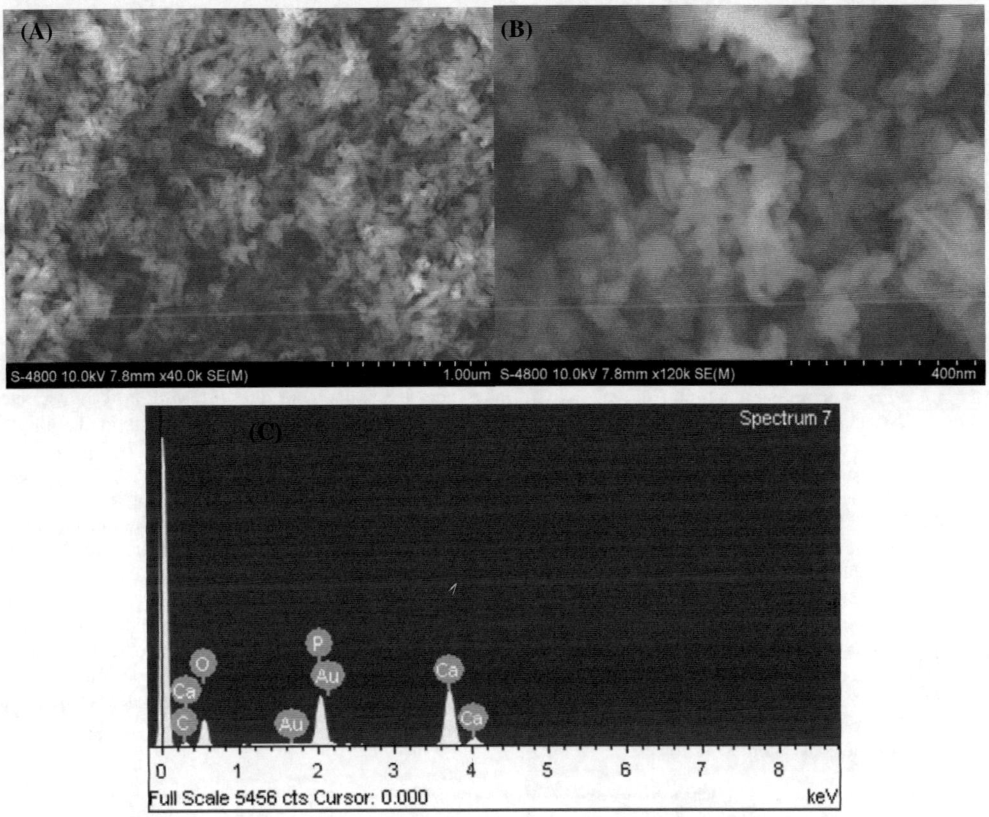

Fig. 2 SEM image of Au/HAp and the corresponding EDS elemental image

Fig. 3 illustrated the HCHO adsorption capacity of the prepared materials at C_{HCHO} = 90 ppm, RH = 65% T = 30°C and F = 3 L h^{-1}. HAp, 0.2Au/HAp and 0.5Au/HAp all exhibited similar adsorption behavior with the highest adsorption percentage at the first 5 minutes and they gradually decreased overtime. This behavior was to be expected as the experiment was carried in a continuous flow as the adsorption sites would be occupied eventually with a continuous feed of HCHO. HAp sample had the highest adsorption performance after 5 minutes at nearly 45%. The presence of Au nanoparticles had reduced the adsorption capacity of the material. Because Au obstructs the location where HCHO can connect, the adsorption capacity is lowered. When the temperature is elevated, the Au's activity is clearly visible (as shown in figure 4). The role of Au nanoparticles in this work can clearly be observed as a catalyst for the production of oxidizing agents. The highest HCHO removal percentage by adsorption of 0.2Au/HAp and 0.5Au/HAp was 36% and 29%, respectively.

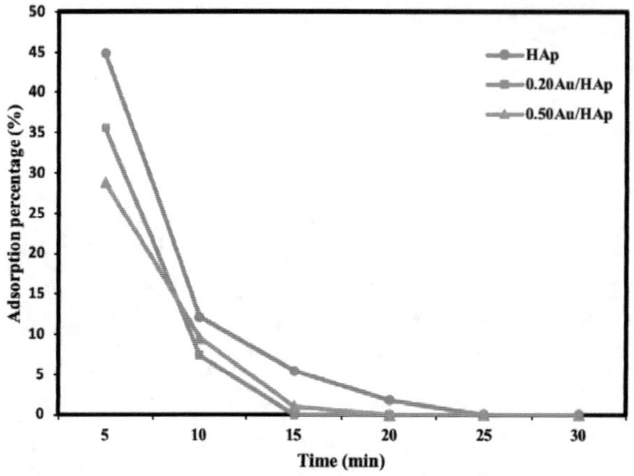

Fig. 3 Formandehyde removal by adsorption at C_{HCHO}= 90 ppm_v, RH=65%, F=3Lh^{-1}.

The experiments to investigate the effect of temperature on the catalytic performance were carried out at the temperature 50-260°C with C_{HCHO} = 90ppm and RH = 65% and the results were shown in Fig.4. As shown in Fig. 4, HAp exhibited no catalytic activity throughout the experiment due to having no catalytic sites to generate oxidizing agents. Thus, HAp only plays the role of support material for distribution of Au NPs and adsorption sites of gas molecules (H_2O, O_2 and HCHO) in the feed flow. This would also highlight importance of the presence of Au NPs in improving the thermal catalytic oxidation of HCHO using the prepared composite. The HCHO catalytic oxidation over Au/HAp can be divided into two regimes: (I) 100-150°C, (II) 150-200°C and (III) 220-260°C. In the first (I) regime, the catalytic activity of both samples was insignificant. They both converted about 10% of HCHO at 120°C but reduced to about 10% of HCHO at 120°C. The second (II) regime exhibited a jump in the efficiency of the catalytic activity and it increased with the rise of temperature. Specifically, the HCHO conversion of the 0.2Au/HAp at 150, 180, 200°C was 44%, 53% and 65%, respectively. 0.5Au/HAp also achieved a small difference (2-5%) in the catalytic efficiency. However, in the third (III) regime, there was a difference in the performance of two samples. They both continued the positive trend of converting more HCHO but 0.2Au/HAp was better. The difference in performance ranged from 10-20%. At the end of the experiment, at 260 °C, 0.2Au/HAp exhibited over 85% in efficiency while that of 0.5Au/HAp was about 75%. It can be seen that increasing the Au content on HAp resulted in a negative effect on the conversion of HCHO.

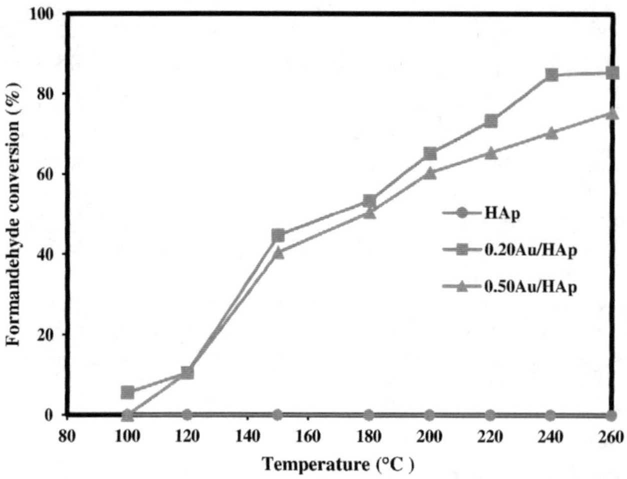

Fig. 4 Formandehyde conversion as a function of reaction temperature

IOP Conf. Series: Earth and Environmental Science **964** (2022) 012027 doi:10.1088/1755-1315/964/1/012027

From Fig. 5 illustrated the effect of temperature on formaldehyde conversion. Experiments were carried at temperatures from 50-260°C, C_{HCHO} = 90ppm, RH = 65% and 5%. HAp have OH- anion, which may affect the formandehyde removal. Because on the surface of HAp have hydroxyl ion (OH-). Which may adsorbed oxygen moleculars, it leads to affect the formandehyde conversion. So when hunidity is changed, this will influence the catalyst performance. In this work the humidity is decided to investigated is RH = 65% and 5%. RH = 5% is chosen, it means at our laboratory did not supply the vapor water through the reaction process. Which have a huge different condition to compare the affect of humidity between RH = 65% and 5% on the catalytic oxidation performance. It can also be divided into three temperature regimes: (I) 50-120°C, (II) 150-200°C and (III) 220-260°C. Remarkably, the prepared HAp exhibited no catalytic activity throughout the experiment on both different hunidity. In the first (I) regime, the reduction in relative humidity seemed to hinder the activity of Au/HAp samples by 2-5%. In the second (II) regime, the effect of RH was also negligible for the performance of both samples with the HCHO conversion at 150, 180 and 200°C were about 40, 50 and 60%, respectively. However, in the third (III) regime, there were opposite effects on the performance of 0.2Au/HAp and 0.5Au/HAp. The performance of 0.5Au/HAp catalyst has an improvement in activity. At RH=5%, the catalytic activity of 0.5Au/HAp at 240 and 260°C had increased by a margin of 7-9% compared to those at RH=65% and ended at over 82%. The changes in the HCHO conversion at RH=5% over 0.2Au/HAp in this regime was insignificant and also ended at over 82% at 260°C.

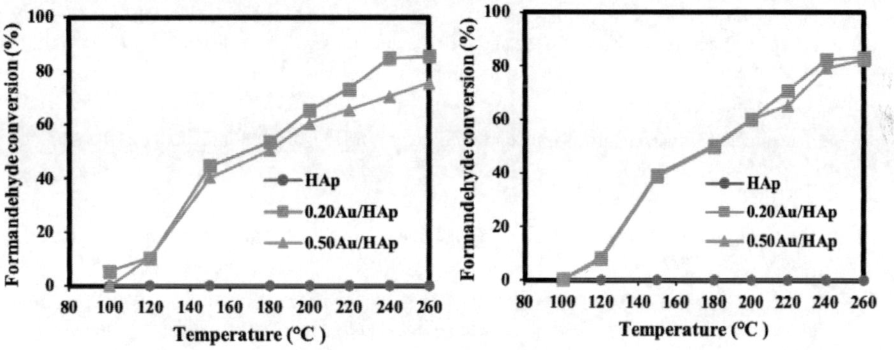

Fig. 5 Effect of relative humidity on the catalytic oxidation performance

RH=65% (left) and RH=5% (right)

Table 1. Summarization of related researches

Year	Catalyst	Application	Removal efficiency	Reference
2018	Ag/egshell	Benzene oxidation	Above 90% at 270°C	[38]
2016	Pt/HAp	Formandehyde oxidation	54% after 1h reaction	[39]
2015	Au/HAp	Folamdehyde and benzene oxidation.	Above 90% at 270°C	[40]
2015	Rh/HAp	N_2O decomposition	Above 90% at 270°C with pH = 10.5	[41]

Comparision between Au/HAp with Pt/HAp, which is pulished by Zhihuaxu and patners [39]. At 0.2Pt/HAp the conversion of formandehyde reached 54% after 1h reaction. Furthermore, reaching with Ag/egshell of Yunlong Guo and partners show that the benzene conversion when increasing the temperature to 260°C reach over 90%. Beside that when decreasing the humidity the benzen elimination dropped [38]. Although HAP has previously been demonstrated to be an effective support for nano-gold (1 wt%) for HCHO removal[40], it was also discovered to help gold NPs resist sintering. Nonetheless, due of its depletable resource and high price, the catalyst's high proportion of noble metal (1 wt%) as one of the principal faults hampers its practical implementation. However, achieving functional composite catalysts with modest noble metal loading while maintaining catalytic performance remains critical. The HAP-supported lowcontent Au (0.2Au/HAp and 0.5Au/HAp) catalyst was prepared and investigated for HCHO removal, taking into account its environmental friendliness, rich resource, and unique surface acid-base character; and the effect of the modifier introduced during the preparation process on the catalyst surface property and activity toward HCHO oxidation was investigated. Futhermore, with RH/HAp was investigated to N_2O removal [41], which show above 90% N_2O decomposition. This leads to a promising catalyst with HAp suppoted nobel metal for removal VOCs.

4. Conclusion

This study had successfully synthesis HAp nanomaterial from waste eggshells which was confirmed in XRD pattern and prepared Au loading HAp nanomaterial for the catalytic oxidation of HCHO. The BET specific area of HAp was 85 m^2g^{-1} which was a significant improvement when compared with the eggshells. The oxidation activity of HCHO over Au/HAp catalyst was investigated based on the change of factors: reaction temperature (50-260°C), Au loading content (0.2 and 0.5 wt%) and relative humidity (RH = 5% and 65%). They both exhibited good catalytic activity throughout the experiments with at least 40% of HCHO was converted at 150°C. Furthermore, the 0.2Au/HAp sample had the most outstanding activity and the highest efficiency of 85% at 260°C and RH = 65%. In the future wok for this research, different mass of Au loading on HAp, the stability of the catalyst are and different humidity will be investigated to find out the best condition of the catalyst.

Acknowledgments
We acknowledge the support of time and facilities from Ho Chi Minh university of Technology (HCMUT), VNU-HCM for this study.

References

[1] Holgate, S. T. (2017). ‘Every breath we take: the lifelong impact of air pollution’–a call for action. Clinical Medicine, 17(1), 8.

[2] Rajabi, H., Mosleh, M. H., Mandal, P., Lea-Langton, A., & Sedighi, M. (2020). Emissions of volatile organic compounds from crude oil processing–Global emission inventory and environmental release. Science of The Total Environment, 727, 138654.

[3] Kamal, M. S., Razzak, S. A., & Hossain, M. M. (2016). Catalytic oxidation of volatile organic compounds (VOCs)–A review. Atmospheric Environment, 140, 117-134.

[4] Nouailhat, A. (2010). An introduction to nanoscience and nanotechnology (Vol. 10). John Wiley & Sons.

[5] Hebié S, Kokoh KB, Servat K, Napporn T. Shape-dependent electrocatalytic activity of free gold nanoparticles toward glucose oxidation. Gold Bulletin. 2013;46:311–318.

[6] J. Wang and H. Gu, "Novel metal nanomaterials and their catalytic applications," Molecules, vol. 20, no. 9, pp. 17070-17092, 2015.

[7] Hsiao, M. W., Adzic, R. R., & Yeager, E. B. (1992). The effects of adsorbed anions on the oxidation of D-glucose on gold single crystal electrodes. Electrochimica acta, 37(2), 357-363.

[8] Hammer, B., & Norskov, J. K. (1995). Why gold is the noblest of all the metals. Nature, 376(6537), 238-240

[9] Schubert, M. M., Hackenberg, S., Van Veen, A. C., Muhler, M., Plzak, V., & Behm, R. J. (2001). CO oxidation over supported gold catalysts—“Inert” and “active” support materials and their role for the oxygen supply during reaction. Journal of Catalysis, 197(1), 113-122.

[10] Matsumura, Y., & Moffat, J. B. (1993). Catalytic oxidative coupling of methane over hydroxyapatite modified with lead. Catalysis letters, 17(3), 197-204.

[11] Tsuchida, T., Kubo, J., Yoshioka, T., Sakuma, S., Takeguchi, T., & Ueda, W. (2008). Reaction of ethanol over hydroxyapatite affected by Ca/P ratio of catalyst. Journal of Catalysis, 259(2), 183-189.

[12] Verwilghen, C., Rio, S., Nzihou, A., Gauthier, D., Flamant, G., & Sharrock, P. J. (2007). Preparation of high specific surface area hydroxyapatite for environmental applications. Journal of materials science, 42(15), 6062-6066.

[13] Venugopal, A., & Scurrell, M. S. (2003). Hydroxyapatite as a novel support for gold and ruthenium catalysts: Behaviour in the water gas shift reaction. Applied Catalysis A: General, 245(1), 137-147.

[14] Kanai, H., Nakao, M., & Imamura, S. (2003). Selective photoepoxidation of propylene over hydroxyapatite–silica composites. Catalysis Communications, 4(8), 405-409.

[15] Sandrine, B., Ange, N., Didier, B. A., Eric, C., & Patrick, S. (2007). Removal of aqueous lead ions by hydroxyapatites: equilibria and kinetic processes. Journal of Hazardous Materials, 139(3), 443-446.

[16] Fernane, F., Mecherri, M. O., Sharrock, P., Hadioui, M., Lounici, H., & Fedoroff, M. (2008). Sorption of cadmium and copper ions on natural and synthetic hydroxylapatite particles. Materials Characterization, 59(5), 554-559.

[17] Nzihou, A., & Sharrock, P. (2010). Role of phosphate in the remediation and reuse of heavy metal polluted wastes and sites. Waste and Biomass Valorization, 1(1), 163-174.

[18] Fernane, F., Mecherri, M. O., Sharrock, P., Fiallo, M., & Sipos, R. (2010). Hydroxyapatite interactions with copper complexes. Materials Science and Engineering: C, 30(7), 1060-1064.

[19] Minh, D. P., Rio, S., Sharrock, P., Sebei, H., Lyczko, N., Tran, N. D., ... & Nzihou, A. (2014). Hydroxyapatite starting from calcium carbonate and orthophosphoric acid: synthesis, characterization, and applications. Journal of Materials Science, 49(12), 4261-4269.

[20] Sharrock, P., Brumas, V., & Fiallo, M. M. L. (2013). Wastewater sorption on HA: old recipes for new tastes. Procedia Earth and Planetary Science, 7, 256-259.

[21] Ferraz, M. P., Monteiro, F. J., & Manuel, C. M. (2004). Hydroxyapatite nanoparticles: a review of preparation methodologies. Journal of Applied Biomaterials and Biomechanics, 2(2), 74-80.

[22] Deshmukh, K., Shaik, M. M., Ramanan, S. R., & Kowshik, M. (2016). Self-activated fluorescent hydroxyapatite nanoparticles: a promising agent for bioimaging and biolabeling. ACS Biomaterials Science & Engineering, 2(8), 1257-1264.

[23] Liu, Y., Tsunoyama, H., Akita, T., Xie, S., & Tsukuda, T. (2011). Aerobic oxidation of cyclohexane catalyzed by size-controlled Au clusters on hydroxyapatite: size effect in the sub-2 nm regime. Acs Catalysis, 1(1), 2-6.

[24] Chong, R., Fan, Y., Du, Y., Liu, L., Chang, Z., & Li, D. (2018). Hydroxyapatite decorated TiO2 as efficient photocatalyst for selective reduction of CO2 with H2O into CH4. International Journal of Hydrogen Energy, 43(49), 22329-22339.

[25] Zhao, K., Qiao, B., Wang, J., Zhang, Y., & Zhang, T. (2011). A highly active and sintering-resistant Au/FeO x–hydroxyapatite catalyst for CO oxidation. Chemical Communications, 47(6), 1779-1781.

[26] J.-D. Wang, J.-K. Liu, Y. Lu, D.-J. Hong, X.-H. Yang, Mater. Res. Bull. 55 (2014) 190–197.

[27] Yan, W., Brown, S., Pan, Z., Mahurin, S. M., Overbury, S. H., & Dai, S. (2006). Ultrastable Gold Nanocatalyst Supported by Nanosized Non-Oxide Substrate. Angewandte Chemie, 118(22), 3696-3700.

[28] Crepaldi, E. L., Pavan, P. C., & Valim, J. B. (2000). Anion exchange in layered double hydroxides by surfactant salt formation. Journal of materials chemistry, 10(6), 1337-1343.

[29] M.Shahbandeh, Egg production worldwide 1990-2019, Mar 3, 2021, access 02 May 2021

[30] Food and Agriculture Organization of the United Nations: Statistics Division (FAOSTAT) Production: Livestock Primary: Eggs Primary, 2014,

[31] Ibrahim, M., Labaki, M., Giraudon, J. M., & Lamonier, J. F. (2020). Hydroxyapatite, a multifunctional material for air, water and soil pollution control: A review. Journal of hazardous materials, 383, 121139.

[32] Minh, D. P., Rio, S., Sharrock, P., Sebei, H., Lyczko, N., Tran, N. D., ... & Nzihou, A. (2014). Hydroxyapatite starting from calcium carbonate and orthophosphoric acid: synthesis, characterization, and applications. Journal of Materials Science, 49(12), 4261-4269

[33] Jiang, C., Li, D., Zhang, P., Li, J., Wang, J., & Yu, J. (2017). Formaldehyde and volatile organic compound (VOC) emissions from particleboard: identification of odorous compounds and effects of heat treatment. Building and Environment, 117, 118-126.

[34] Suresh, S., & Bandosz, T. J. (2018). Removal of formaldehyde on carbon-based materials: A review of the recent approaches and findings. Carbon, 137, 207-221.

[35] Xu, Z., Yu, J., Liu, G., Cheng, B., Zhou, P., & Li, X. (2013). Microemulsion-assisted synthesis of hierarchical porous Ni (OH)2/SiO2 composites toward efficient removal of formaldehyde in air. Dalton Transactions, 42(28), 10190-10197.

[36] Wang, Y., Jiang, C., Le, Y., Cheng, B., & Yu, J. (2019). Hierarchical honeycomb-like Pt/NiFe-LDH/rGO nanocomposite with excellent formaldehyde decomposition activity. Chemical Engineering Journal, 365, 378-388.

[37] Guo, J., Lin, C., Jiang, C., & Zhang, P. (2019). Review on noble metal-based catalysts for formaldehyde oxidation at room temperature. Applied Surface Science, 475, 237-255.

[38] Guo, Y., Yang, D. P., Liu, M., Zhang, X., Chen, Y., Huang, J., ... & Luque, R. (2019). Enhanced catalytic benzene oxidation over a novel waste-derived Ag/eggshell catalyst. Journal of Materials Chemistry A, 7(15), 8832-8844.

[39] Xu, Z., Huang, G., Yan, Z., Wang, N., Yue, L., & Liu, Q. (2019). Hydroxyapatite-supported low-content Pt catalysts for efficient removal of formaldehyde at room temperature. ACS omega, 4(26), 21998-22007.

[40] Wang, Y., Chen, B. B., Crocker, M., Zhang, Y. J., Zhu, X. B., & Shi, C. (2015). Understanding on the origins of hydroxyapatite stabilized gold nanoparticles as high-efficiency catalysts for formaldehyde and benzene oxidation. Catalysis Communications, 59, 195-200.

[41] Huang, C., Ma, Z., Xie, P., Yue, Y., Hua, W., & Gao, Z. (2015). Hydroxyapatite-supported rhodium catalysts for N2O decomposition. Journal of Molecular Catalysis A: Chemical, 400, 90-94.

Combination of photocatalytic degradation and adsorption in dye removal by TiO_2-chitosan-glycerol beads under natural sunlight

Anh Thi-Kim Tran, Nhung Thi-Tuyet Hoang,* and Tien Thuy Hoang Nguyen

Faculty of Chemical and Food technology, Ho Chi Minh City University of Technology and Education, 01 Vo Van Ngan, Thu Duc city, Ho Chi Minh city, Vietnam.

*Email: nhunghtt@hcmute.edu.vn

Abstract. The synthesis of TiO_2-chitosan-glycerol (TiCsG) beads was prepared to determine the optimum condition for dye removal under natural sunlight. By using the orthogonal array design with five parameters (ratio of TiO_2/Chitosan, initial concentration of dyes, pH of solution, irradiation time, dose of TiCsG) to analyze their interaction such as well as optimizing the photocatalytic process. The significant parameters influent in the removal of Acid Blue 193, MO, NWY dyes were irradiation time ($p = 0.011$) and the ratio of TiO_2/chitosan ($p = 0.025$), while the initial concentration of dyes, pH of dye solution and dose of TiCsG were unimportant factors with $p > 0.05$. The optimum of factors based on the ratio of signal per noise to attain the highest dye removal ability of TiCsG was the percentage of TiO_2/chitosan = 2 % w/w (level 2), initial concentration of dye at level 5 (10 mg/L of AB 193, 5 mg/L for MO and 100 mg/L for NWY), pH =4.5 (level 2), irradiation time =1.5 h (level 5) and dose of TiCsG =0.2 g/mL (level 4).

Keywords: TiO_2, chitosan, glycerol, dye, Taguchi method

1. Introduction

Textile wastewater contains various dyes which are difficult to biological degrade due to their synthetic origin and complex aromatic molecular structures [1]. This poisonous waste effluent into the environment and damage human and aquatic life. Furthermore, colored wastewater can limit the penetration of sunlight and oxygen into the aquatic area [2]. Many approaches of dye treatment have been used, such as adsorption, sedimentation, chemicoagulation, biological methods, and advanced oxidation procedures. Adsorption is one of the acceptable physical methods for dye removal. Chitosan has been widely applied in textile wastewater treatment because of its high capacity on color and pollutant degradation [3, 4]. However, it is a high cost due to not reusable in general. The applications of photocatalysis have also received substantial attention in dye removal [5, 6]. Many photocatalysts such as TiO_2, ZnO, Fe_2O_3, WO_3, and Cu_2O were studied to degrade the contaminants under UV and visible radiation. Among those, TiO_2 has been the most concerned due to its relatively efficient, low price, non-toxic, chemically and biologically inert, and high photocatalytic activity under UV light irradiation [7]. The TiO_2 combinating with chitosan can reduce the drawback of chitosan in regenerating the adsorbent. The functional groups of chitosan matrix captured dye. TiO_2 could degrade these adsorbed pollutants leading to the recovery of bio-sorbent chitosan [8]. Glycerol was added to stronger the hydrogen bond in the chitosan matrix and increased the mechanical properties of chitosan [9]. Many research studied the ability of photocatalysis and adsorption, mechanism and kinetics of pollutant treatment, synergistic effects, and the TiO_2-chitosan characteristics [4, 10-12]. The dye degradation activities of TiO_2-chitosan-glycerol were significantly higher than that of TiO_2-chitosan [13].

In the present work, the synthesis of TiO_2-chitosan-glycerol beads was studied for dye removal under sunlight irradiation. Experimental factors for degradation of Methyl orange (MO), acid blue 193 (AB 193), and NearWash Yellow RCL (NWY), including dyes and photocatalyst concentrations, ratios of components in composite, pH, and irradiation time were investigated. The Taguchi method was used to investigate the optimum conditions and the correlation of factors for the dye removal ability, which was the target response variable.

2. Material and methods

2.1. Chemicals

High molecular weight chitosan was obtained from a company in Vietnam, degree of deacetylation was 91.6%. Commercial TiO_2 Degussa P25 (Merck, 80% anatase, 20% rutile) and glycerol (Merck, 99.5%) was used as a material for strengthening the mechanical properties of chitosan. Sodium triphosphate (TPP, China, > 98.0%) was used as solidifying chemical. Acid blue 193 (AB 193, Merck), methyl orange (MO, Merck), and Near Wash Yellow RCL (NWY, NearChimica) was chosen as dyes for analyzed the dye removal of TiCsG beads.

Table 1. Properties of Dyes Selected for Study

Properties	Acid blue 193 (AB193)	Methyl orange (MO)	Near Wash Yellow (NWY)
Molecular formula	$C_{40}H_{22}CrN_4Na_2O_{10}S_2^-$	$C_{14}H_{14}N_3NaO_3S$	$C_{24}H_{19}ClFN_9Na_2O_9S_2$
Molecular weight (g/mol)	880.7 g/mol	327.34 g/mol	742 g/mol
Synonym	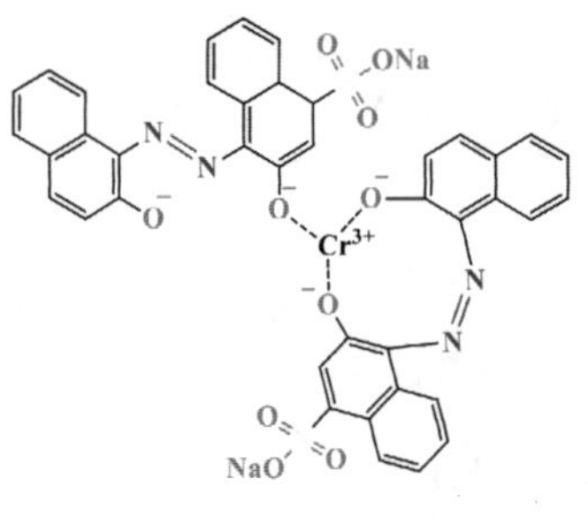	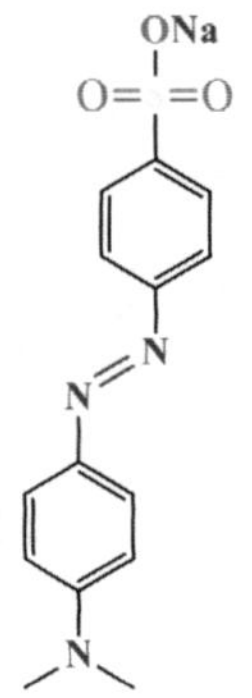	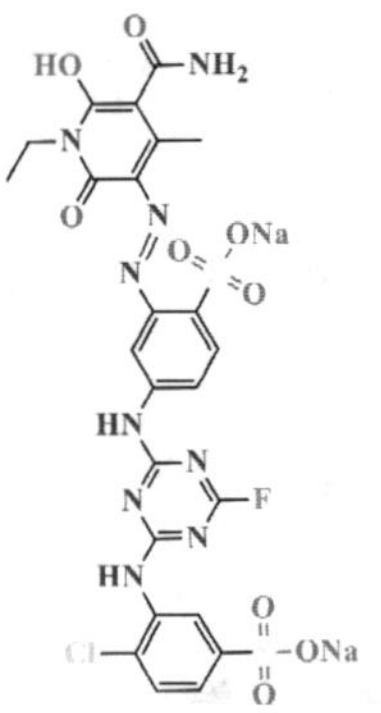
Maximum absorbance values (nm)	598	463	421

2.2. Methods

Chitosan solution was prepared by dissolving in acetic acid 5% (v/v) solution. The ratio of the TiO_2 and glycerol was controlled at 1:10 (w/w) [13]. This mixture was poured into chitosan solution with various percentages of TiO_2 and chitosan from 1 to 5 (w/w). After that, this solution was stirred for 30 min, and homogenized by an ultrasonic device (Hielscher UP 100H) in 30 minutes to form TiO_2-chitosan-glycerol gel. This gel was added dropwise into the solution of TPP 1% (w/v) to to form spherical particles (Figure

1)[14]. The formed TiO_2-chitosan-glycerol beads were washed with distilled water and dried at 70°C for 8 h.

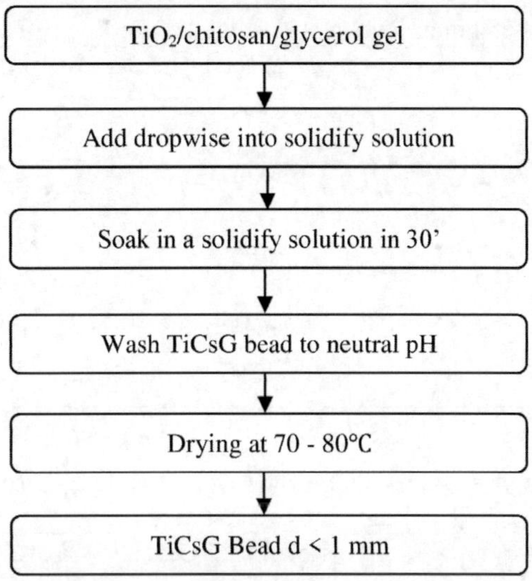

Figure 1. Procedure of TiCsG bead forming

2.3. *Characterization*

The crystal phase of the TiO_2-chitosan-glycerol beads was analyzed by X-ray diffraction (XRD) on a BRUKER D2-212917. Scanning Electron Microscope (FE SEMS-4800 Hitachi, Japan) was used to examine the materials' morphology. FT-IR spectra (Perkin-Elmer FT-IR) was used to assess the functional groups present in the TiCsG bead FT-IR

2.4. *Photocatalytic degradation experiments*

The photocatalytic degradation of TiCsG was examined by taking three different dyes into glass tubes (*d* = 2.0 cm) under natural sunlight irradiation. The fixed amount of dyes (10-mL with specific material concentrations was axially conducted by shaker at noon with the sunlight intensity > 30 W/m2).

After a fixed irradiated time, the sample was withdrawn and centrifuged to remove the minus material from the solution. The standard curve using UV visible spectrophotometer to analyze the dye concentration.

The dye removal efficiency (E%) of dyes was determined by the following equation (1):

$$E\% = \frac{(C_o - C)}{C_o} \; x\ 100 \qquad (1)$$

where Co is the initial concentration of dye solution before photocatalysis and C (mg/L) is the concentration of dye solution after a certain irradiation time.

2.5. *Experimental design methodology*

The orthogonal array was determined by the Taguchi approach to design several experiments and to analyze the quality through the signal-to-noise ratio (S/N). Five controllable factors included the percentage of TiO_2/chitosan, initial concentration of dye, pH of dye solution, irradiation time, and the dosage of TiCsG bead. The levels of initial dye concentration for AB, MO, and NWY dyes are 2 – 10 mg/L, 1 – 5 mg/L, and 20 – 100 mg/L, respectively. The levels of initial dye concentration for AB, MO, and NWY dyes are a, b, c, d, and e, which are presented in table 3. The difference in initial dye concentration for AB 193, MO, and NWY is due to the dark color when changing the concentration. The experiments were designed to maximize the dye removal efficiency of TiO_2-chitosan-glycerol beads. According to the Taguchi approach, the number of experiments to analyze the factor effect of photocatalysis with 5 factors and 5 levels is 25. The statistical results using MINITAB software V16 based on the Taguchi method was done.

Table 2. Factors and their level was designed according to Taguchi approach.

Independent factors		Level 1	Level 2	Level 3	Level 4	Level 5
TiO_2/chitosan % (w/w)	A	1	2	3	4	5
Initial concentration of dye (mg/L)	B	a	b	c	d	e
pH of dye solution	C	3	4.5	6	7.5	9
Irradiation time (hours)	D	0.5	1	1.5	2	2.5
TiCsG loading (g/mL)	E	0.05	0.1	0.15	0.2	0.25
		TiCsG1	TiCsG2	TiCsG3	TiCsG4	TiCsG5

Table 3. The levels of initial dye concentration for AB, MO and NWY dyes applied for experimental design by Taguchi method

Initial dye concentration	Level 1	Level 2	Level 3	Level 4	Level 5
	a	b	c	d	e
AB dye (mg/L)	2	4	6	8	10
MO dye (mg/L)	1	2	3	4	5
NWY (mg/L)	20	40	60	80	100

3. Results and discussion

3.1. TiCsG bead characterization

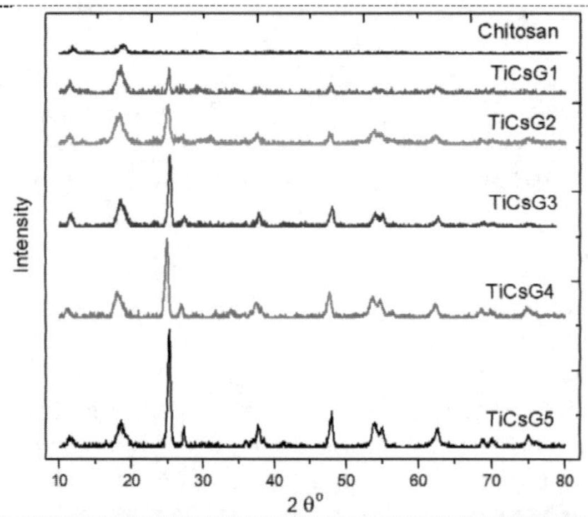

Figure 2. X-ray diffraction patterns of chitosan beads and TiCsG (TiCsG 1- TiCsG 5) with various ratios of TiO_2 and chitosan (from 1 to 5%, w/w).

X-ray diffraction patterns of powder TiCsG with percentage ratios of TiO_2 andchitosan ranging from 1 to 5% (w/w) are shown in Figure 2. With the existence of TiO_2 in the matrix of TiCsG, the crystallographic structure of chitosan changed. The peaks of the amorphous state of chitosan in TiCsG patterns changed from 18.8° to 18.5°[13]. The X-ray diffraction pattern for anatase phase TiO_2 (1 1 0) gives single diffraction lines at 25.2, 37.6, 47.9, 53.9, 54,2, 62.5, and 74.9°. The intensity of these diffraction lines increased with the TiO_2 loading because the amorphous region increases in the chitosan matrix during the synthesis of material.

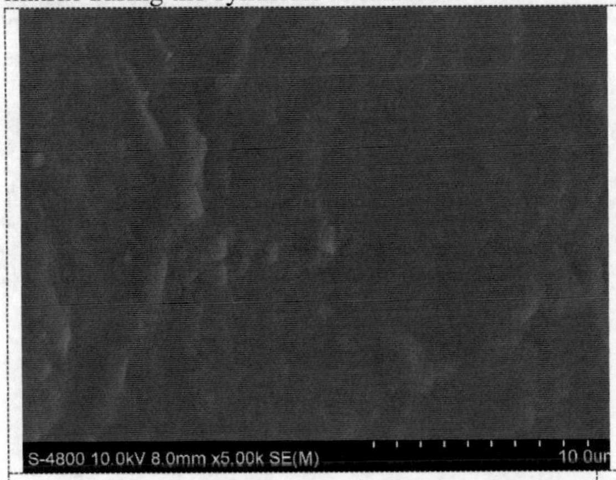

Figure 3a SEM of the surface morphologies of Chitosan

Figure 3b SEM of the surface morphologies of TiCsG (2% TiO_2-chitosan)

The shape and morphology of the TiCsG were presented in Figure 3. SEM analysis shows rough surface (Figure 3b) compared to chitosan beads (without TiO_2) (Figure 3a). The formation of minute granules is due to TiO_2 the aggregation (0.5 μm–5 μm in diameter) wrapped with chitosan [13]. The TiCsG beads are non-spherical form with the color of white-yellow and the mean diameter around 1mm.

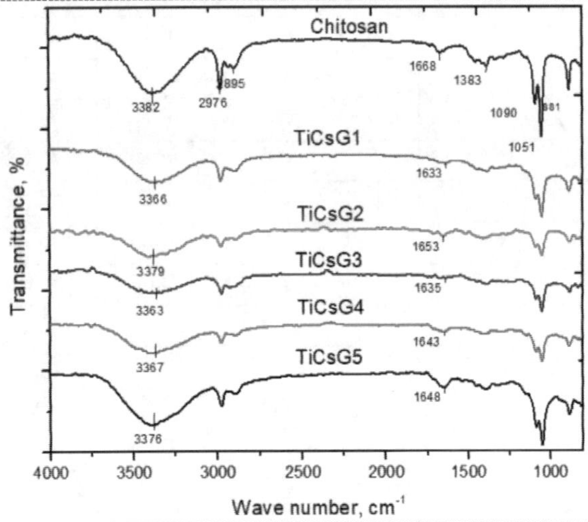

Figure 4. FTIR spectra of chitosan beads and TiCsG (TiCsG 1- TiCsG 5) with various ratios of TiO_2 and chitosan (1-5%, w/w).

The functional groups of chitosan and TiCsG (TiO_2/chitosan ratios varied from 1 to 5%, w/w) determined by FTIR spectra are presented in Figure 4. The strong band at 3382 cm^{-1} can be correspond to the O-H and N-H stretching as the intramolecular hydrogen bonds in chitosan [14]. The signals at around 2976 and 2895 cm^{-1} can be attributed to the stretching vibrations of C-H groups and asymmetric stretching, respectively [14]. The peaks at 1668 and 1383 cm^{-1} assigned to residual N-acetyl groups including the amide-I band (C=O stretching) and CH_2 group (C-H bending) in chitosan, respectively.

The bands at 1090 and 1051 cm^{-1} belong to C-O stretching. The C-H bending out of monosaccharides was confirmed by the band at 881 cm^{-1} [15].

3.2. Taguchi approach to optimize the highest color degradation

A Taguchi approach with the L_{25} orthogonal array design (25 test runs) was used to determine significant parameters and their intercorrelation for dye removal. Experiments were carried on under constant solar radiation (solar intensity > 30 W/cm^2). The results of 25 experiments for three dyes degradation are shown in Table 4. Experiments were performed in triplicate.

Table 4. L25 orthogonal array response value and predicted value.

Run no.	Factors					AB 193 removal efficiency (%)	MO removal efficiency (%)	NWY removal efficiency (%)
	TiO_2/chitosan, w/w	Initial concentration dye of (mg/L)	pH	Irradiation time (hour)	Dosage of TiCsG (g/mL)			
1	1	2	3.0	0.5	0.05	16.34	57.72	23.52
2	1	4	4.5	1.0	0.10	37.78	74.07	61.08
3	1	6	6.0	1.5	0.15	48.40	90.34	65.75
4	1	8	7.5	2.0	0.20	52.13	93.82	73.24
5	1	10	9.0	2.5	0.25	51.80	82.40	72.65
6	2	2	4.5	1.5	0.20	79.02	84.57	70.35
7	2	4	6.0	2.0	0.25	81.36	93.97	72.38
8	2	6	7.5	2.5	0.05	67.58	88.93	56.75
9	2	8	9.0	0.5	0.10	28.60	77.19	35.14
10	2	10	3.0	1.0	0.15	77.45	83.07	72.62
11	3	2	6.0	2.5	0.10	58.77	84.08	68.35
12	3	4	7.5	0.5	0.15	24.68	70.17	33.98
13	3	6	9.0	1.0	0.20	52.55	84.86	64.03
14	3	8	3.0	1.5	0.25	68.95	94.55	72.67
15	3	10	4.5	2.0	0.05	55.01	83.49	72.16
16	4	2	7.5	1.0	0.25	21.07	77.69	59.49
17	4	4	9.0	1.5	0.05	39.56	77.03	53.20
18	4	6	3.0	2.0	0.10	45.38	93.79	71.42
19	4	8	4.5	2.5	0.15	55.91	93.74	73.33
20	4	10	6.0	0.5	0.20	28.27	77.80	46.77
21	5	2	9.0	2.0	0.15	54.28	78.48	65.15
22	5	4	3.0	2.5	0.20	69.48	81.89	72.73
23	5	6	4.5	0.5	0.25	36.46	62.77	51.55
24	5	8	6.0	1.0	0.05	29.34	73.58	33.84
25	5	10	7.5	1.5	0.10	48.81	75.56	60.34

The analysis of variance (ANOVA) uses a general linear model to determine the influent of factors and their confidence levels. The p-value evaluated the significance of factors. The statistically significant test of factors was evaluated by the p-value ≤ 0.05 [16]. From the results in Table 5, the color degradation efficiencies of TiCsG were significantly affected by the irradiation time (p = 0.001 for MO, p = 0.011 for AB 193 and p = 0.005 for NWY). The TiO_2/chitosan ratio was also an important parameter for AB 193 and MO dye, while the TiCsG loading was a significant factor for MO and NWY. The initial dye concentration and pH were insignificant factors with p > 0.05.

Table 5. Percentage contribution of each parameter on dye removal.

Dyes	AB 193		MO		NWY	
Factors	**F**	**p**	**F**	**p**	**F**	**p**
TiO_2/chitosan ratio (% (w/w)	9.50	**0.025**	18.73	**0.007**	0.64	0.664
Initial concentration of dye (mg/L)	0.52	0.728	14.87	**0.011**	1.4	0.376
pH of dye solution	2.03	0.255	2.83	0.169	2.01	0.258
Irradiation time (hour)	14.86	**0.011**	57.53	**0.001**	23.16	**0.005**
TiCsG loading (g/mL)	2.87	0.166	9.87	**0.024**	7.16	**0.041**

The average removal percentages of dyes were analyzed using the MINITAB 19 software to evaluate the effect of each factor on the optimization criteria. Here, the output *S/N* ratio from the analysis of variance (ANOVA) would be estimated for every experiment to determine the distinguishing characteristic between control or signal parameters. The ratio of S/N for every results of experiments in the orthogonal array was chosen to be "the larger, the better" to get the maximum in the expected dye removal efficiencies of TiCsG. From this, control factors (significant effects on the signal to noise ratio) and signal factors (most minor effect on the S/N ratio) were determined [17].

3.2.1. Effect of irradiation time

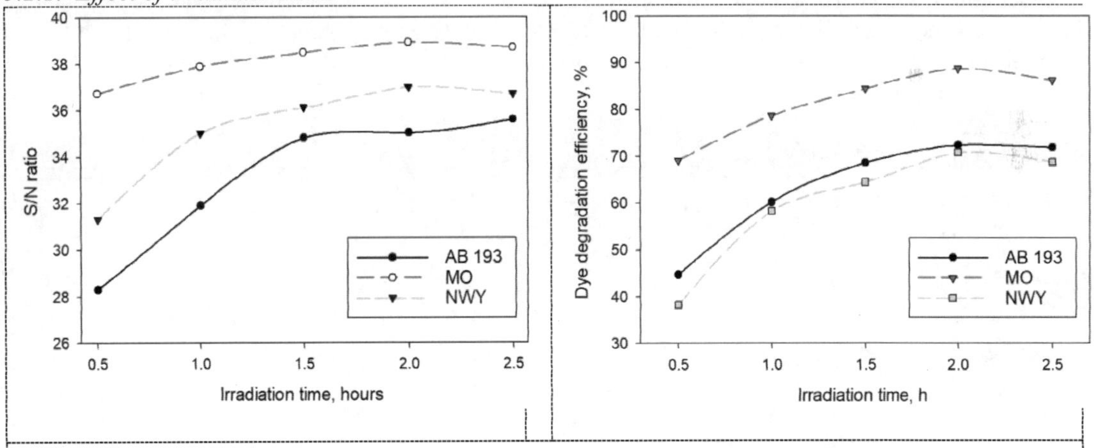

Figure 5. The effect of irradiation time on the dyes degradation

As can be seen in Figure 5, with 0.5 to 2.5 h of irradiation time, there was an extreme increase in S/N ratio from 36.73 to 38.7 for MO, 31.32 to 36.71 for NWY, and 28.3 to 35.61 for AB 193, respectively. Obviously, the higher the photocatalytic time, the more interaction between the materials and the light energy to produce OH^*, and the interaction between dyes and the hydroxyl radicals [18]. Therefore, the efficiencies of dye removal were increased with the irradiation time. On the other hand, there are no significant differences in S/N ratio more than 1.5 h illuminating because of the termination reaction that reduce the quantity of free radicals. Therefore, the optimum irradiation time for three dyes was 1.5 h. Under 1.5h irradiation, the dye removal percentage of MO reached 84%, while that of AB 193 and NWY were 68 and 63%, respectively.

3.2.2. Effect of TiO_2/chitosan

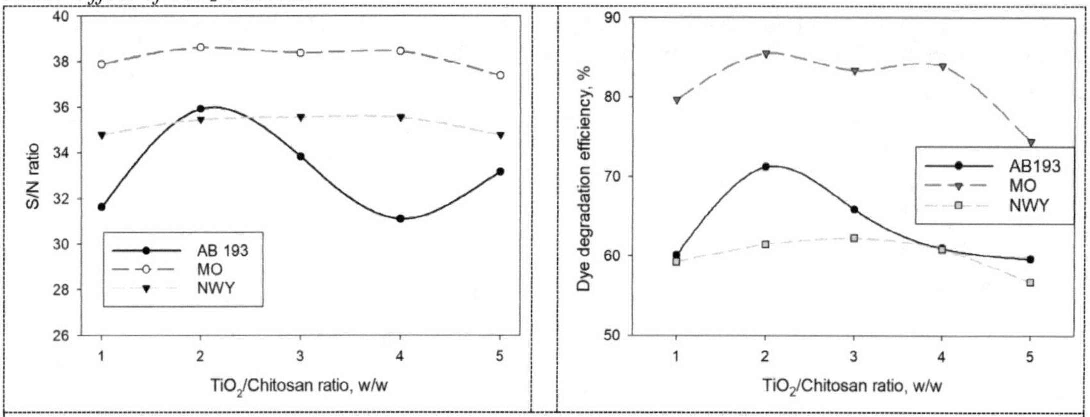

Figure 6. The effect of the percentage of TiO_2 and Chitosan on the dyes degradation.

The amount of catalyst amount could be positive and negative influent on the photodegradation rate [19]. The percentage of TiO_2/chitosan was tested with different concentrations in the range 1 – 5 % (w/w). Figure 6 show that the S/N of TiO_2/chitosan ratio for AB 193 fluctuated while those for other dyes were stable. The S/N increased when the dosage of TiO_2/chitosan increased to the optimum point of 2 % (w/w) and then reduced at higher values. Other researchers[20, 21] also found the same results. The initial efficiencies were directly proportional to catalyst concentration due to increasing TiO_2 and, consequently, the number of photons adsorbed. However, with the excessive amounts of catalysts, the aggregation of TiO_2 particles leads to a decreased number of active sites. Furthermore, the opacity of beads limits the penetration of light when the excess catalyst was mixed [19, 20]. Thus, an optimum amount of catalyst should be mixed to ensure the absorption of light photons for efficient photomineralization [22].

3.2.3. Effect of TiCsG loading

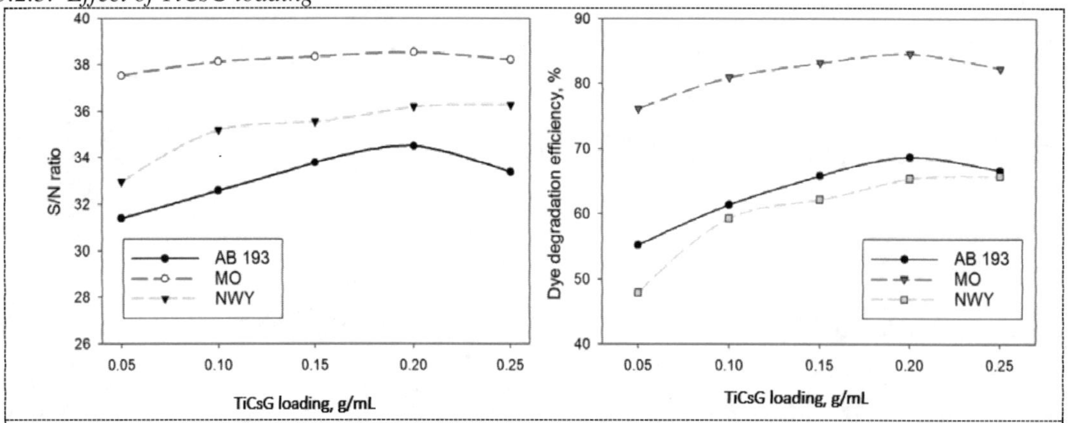

Figure 7. The effect of TiCsG loading on the dyes degradation.

The interaction of TiCsG bead on the dye removal efficiencies was determined by S/N ratio, and the results are illustrated in Figure 7. The increase in the mass of material from 0.05 to 0.2 g/mL led to an increase in the ratio of S/N and then reduced with 0.25 g TiCsG per mL. The highest S/N of TiCsG for MO, AB 193 and NWY were 34.51, 38.53 and 36.2, respectively, with 0.2 g/mL. The higher the adsorbent dose (chitosan), the higher surface area and the more functional groups of chitosan, consequently leading to higher dye removal. Nonetheless, an excessive quantity of chitosan (0.2 g/mL) reduced the dye removal of TiCsG beads. The downtrend of S/N ratio showed that the conglomeration and interaction of chitosan and TiO_2 particles resulting from a high dose of TiCsG beads [22]. Therefore, the optimum amount of TiCsG beads was fixed at 0.2 g/mL based on the largest S/N ratio and minimum material mass. The dye degradation efficiency of TiCsG at the optimum loading was 84.6% for MO, while that for AB 193 and NWY was only 68.8% and 65.4%, respectively. These results could be explained by that AB 193 and NWY formula have double (SO_3Na) group compared to MO, making MO be degraded faster.

3.2.4. Effect of pH

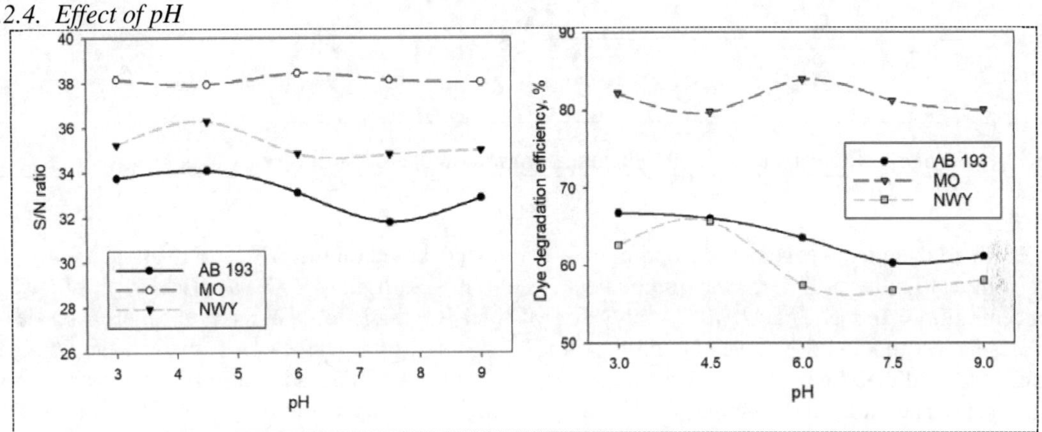

Figure 8. The effect of pH on the S/N ratio for dyes degradation.

The pH of dye solution plays a significant role in adsorption and photocatalysis. According to the S/N ratio analyzed from test runs with various pH in range 3 to 9, the strong dependence of pH on dye removal abilities of AB 193 and NWY while a bit of influence on MO dye degradation. At low pH, the presence of appropriate quantities of protons protonates the amino groups of chitosan molecules to formed NH_3^+ and changes TiCsG bead to positive charge, consequently [23]. Furthermore, it is noted that the dyes have negatively charged sulfonic groups in acidic conditions. Therefore, these groups could be adsorbed by TiCsG beads [24]. On the other hand, with the pH value less than pH_{pzc} of TiO_2 Degussa P25 (pH_{pzc} = 6.5) [25], the surface of TiO_2 particles gain a positive charge ($TiOH_2^+$) [26] and also attracted negative dye more and faster.

$$Dye - SO_3Na + H_2O \rightarrow Dye - SO_3^- + Na^+$$
$$Adsorbent - NH_2 + H^+ \text{ (acid condition)} \rightarrow Adsorbent - NH_3^+$$
$$Adsorbent - NH_3^+ + Dye - SO_3^- \rightarrow Adsorbent - NH_3^+ \text{ - } O_3S \text{ - } Dye$$

3.2.5. Effect of initial dye concentration

As shown in Figure 9, the ratio of S/N increased with increasing the initial dye concentration up to level 4 with MO dye and level 3 with AB 193 and NWY dye and then slightly declined. These results agree with other studies [27-29]. The higher the initial concentration of dye, the more the driving force of the concentration gradient. However, the limit of dye adsorption per unit of chitosan molecule would reduce at high dye concentration. Furthermore, the more initial concentration of dye, the more color intensity. Therefore, the dye molecules are adsorbed on the surface of the adsorbent, and dye molecules absorb UV irradiation rather than TiCsG beads [27]. The color intensity reduced the penetration of light passing through the chitosan medium to reach the catalyst's surface [30]. This leads to less photon's energy reaching the active sites on the catalyst, resulting in the number of hydroxyl radicals produced.

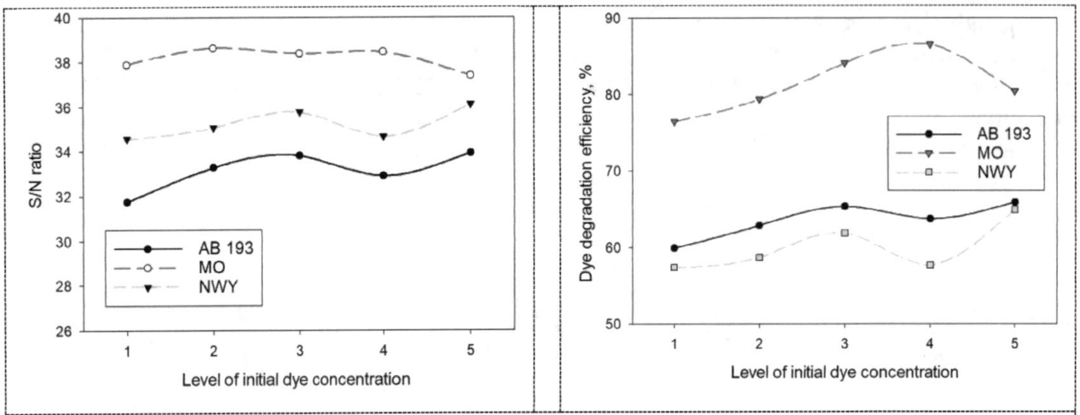

Figure 9. The effect of initial dye concentration on the S/N ratio for dyes degradation.

From all the effect of factors, the optimum of parameters based on the S/N ratio to attain the highest dye removal ability of TiCsG was the percentage of TiO_2/chitosan = 2 % w/w (level 2), initial dye concentration = 10 mg/L of AB 193, 5 mg/L for MO and 100 mg/L for NWY (level 5), pH =4.5 (level 2), irradiation time =1.5 h (level 3) and dose of TiCsG =0.2 g/mL (level 4). However, the obtained results showed that the effects of these factors on MO and NWY degradation efficiency were not so significant. The molecular formula of MO has just one SO_3^- group while that of NWY has two. This could make higher efficicency on dye removal of MO than NWY. In addition, molecular formula of AB 193 with Cr^{3+} is more complicate to degradate with the composite. Therefore, the optimum of factors depends on AB 193.

4. Conclusions

The photocatalytic efficiency of TiCsG under natural sunlight was investigated for degradation of three different dyes (methyl orange, acid blue 193 and Near Wash Yellow). A Taguchi approach using the orthogonal array experimental design was carried out to optimize the factors affecting the photocatalytic efficiency of CS/PVA/TiO_2 nanocomposite included: ratio of TiO_2/chitosan, initial concentration of dye, pH, irradiation time, and dose of TiCsG. The optimum conditions were found from the statistical results of L25 test runs: TiO_2/chitosan = 2 % w/w, initial dye concentration =10 mg/L of AB 193, 5 mg/L for MO and 100 mg/L for NWY, pH =4.5, irradiation time =1.5 h and dose of TiCsG. The time irradiation was the most significant factor for the dye removal ability of AB 193 and MO dye, followed by the percentage of TiO_2/chitosan. The initial concentration dye and pH were unimportant factors with $p > 0.05$. The combination of TiO_2 and chitosan in dye removal offers the potential to reduce the drawback of chitosan in regenerating the adsorbent. The results of TiCsG under natural sunlight show the effective ability of this combination in textile wastewater treatment.

References

1. Yaseen, D.A. and M. Scholz, *Textile dye wastewater characteristics and constituents of synthetic effluents: a critical review.* International Journal of Environmental Science and Technology, 2019. **16**(2): p. 1193-1226.
2. Berradi, M., et al., *Textile finishing dyes and their impact on aquatic environs.* Heliyon, 2019. **5**(11): p. e02711.
3. Pal, P. and A. Pal, *Dye removal using waste beads: Efficient utilization of surface-modified chitosan beads generated after lead adsorption process.* Journal of Water Process Engineering, 2019. **31**: p. 100882.
4. Rasoulifard, M., M. Dorraji, and V. Mozafari, *Visible light photocatalytic activity of*

chitosan/poly(vinyl alcohol)/TiO2 nanocomposite for dye removal: Taguchi-based optimization. Environmental Progress & Sustainable Energy, 2016. **36**.

5. Orozco, S.L., C.A. Arancibia-Bulnes, and R. Suárez-Parra, *Radiation absorption and degradation of an azo dye in a hybrid photocatalytic reactor.* Chemical Engineering Science, 2009. **64**(9): p. 2173-2185.
6. Malekshoar, G. and A.K. Ray, *In-situ grown molybdenum sulfide on TiO2 for dye-sensitized solar photocatalytic hydrogen generation.* Chemical Engineering Science, 2016. **152**: p. 35-44.
7. Nhung Thi-Tuyet Hoang , N.V.S., The-Vinh Nguyen *Antibacterial activities of gel-derived Ag-TiO2-SiO2 nanomaterials under different light irradiation.* AIMS Materials Science, 2015. **3**(2): p. 339-348.
8. Bahrudin, N.N. and M.A. Nawi, *Mechanistic of photocatalytic decolorization and mineralization of methyl orange dye by immobilized TiO2/chitosan-montmorillonite.* Journal of Water Process Engineering, 2019. **31**: p. 100843.
9. Quijada-Garrido, I., et al., *Characteristic Features of Chitosan/Glycerol Blends Dynamics.* Macromolecular Chemistry and Physics, 2006. **207**(19): p. 1742-1751.
10. Ispirli Dogac, Y., et al., *TiO(2) beads and TiO(2)-chitosan beads for urease immobilization.* Mater Sci Eng C Mater Biol Appl, 2014. **42**: p. 429-35.
11. Farzana, M.H. and S. Meenakshi, *Synergistic Effect of Chitosan and Titanium Dioxide on the Removal of Toxic Dyes by the Photodegradation Technique.* Industrial & Engineering Chemistry Research, 2014. **53**(1): p. 55-63.
12. Wiącek, A.E., A. Gozdecka, and M. Jurak, *Physicochemical Characteristics of Chitosan–TiO2 Biomaterial. 1. Stability and Swelling Properties.* Industrial & Engineering Chemistry Research, 2018. **57**(6): p. 1859-1870.
13. Hoang, N.T.-T., et al., *Synergistic effect of TiO2 /chitosan/glycerol photocatalyst on color and COD removal from a dyeing and textile secondary effluent.* Environmental Technology & Innovation, 2021. **21**: p. 101255.
14. Nitsae, M., et al., *Preparation of Chitosan Beads using Tripolyphosphate and Ethylene Glycol Diglycidyl Ether as Crosslinker for Cr(VI) Adsorption.* Chemistry & Chemical Technology, 2016. **10**: p. 105-113.
15. Queiroz, M., et al., *Does the Use of Chitosan Contribute to Oxalate Kidney Stone Formation?* Marine drugs, 2014. **13**: p. 141-158.
16. Taguchi, G., *The System of Experimental Design: Engineering Methods to Optimize Quality and Minimize Costs*, ed. 1. Vol. 1. 1987: Quality Resources.
17. Kim, K.D., S.H. Kim, and H.T. Kim, *Applying the Taguchi method to the optimization for the synthesis of TiO2 nanoparticles by hydrolysis of TEOT in micelles.* Colloids and Surfaces A: Physicochemical and Engineering Aspects, 2005. **254**(1): p. 99-105.
18. Abdullah, A.M., et al., *Carbon/nitrogen-doped TiO2: New synthesis route, characterization and application for phenol degradation.* Arabian Journal of Chemistry, 2016. **9**(2): p. 229-237.
19. Mortazavian, S., A. Saber, and D.E. James, *Optimization of Photocatalytic Degradation of Acid Blue 113 and Acid Red 88 Textile Dyes in a UV-C/TiO2 Suspension System: Application of Response Surface Methodology (RSM).* Catalysts, 2019. **9**(4).
20. Sohrabi, M.R. and M. Ghavami, *Photocatalytic degradation of Direct Red 23 dye using UV/TiO2: Effect of operational parameters.* Journal of Hazardous Materials, 2008. **153**(3): p. 1235-1239.
21. Daneshvar, N., H. Ashassi-Sorkhabi, and A. Tizpar, *Decolorization of Orange II by Electrocoagulation Method.* Separation and Purification Technology, 2003. **31**: p. 153-162.
22. Vakili, M., et al., *Effective Adsorption of Reactive Black 5 onto Hybrid Hexadecylamine Impregnated Chitosan-Powdered Activated Carbon Beads.* Water, 2020. **12**(8).
23. Rocher, V., et al., *Dye removal from aqueous solution by magnetic alginate beads crosslinked*

with epichlorohydrin. Journal of hazardous materials, 2010. **178**: p. 434-9.
24. Wu, X.-J., J.-D. Wang, and L.-Q. Cao, *Characterization and adsorption performance of chitosan/diatomite membranes for Orange G removal.* e-Polymers, 2016. **16**(2): p. 99-109.
25. Hu, C., et al., *Effects of acidity and inorganic ions on the photocatalytic degradation of different azo dyes.* Applied Catalysis B: Environmental, 2003. **46**(1): p. 35-47.
26. Fox, M.A. and M.T. Dulay, *Heterogeneous photocatalysis.* Chemical Reviews, 1993. **93**(1): p. 341-357.
27. Abdellah, M.H., et al., *Photocatalytic decolorization of methylene blue using TiO2/UV system enhanced by air sparging.* Alexandria Engineering Journal, 2018. **57**(4): p. 3727-3735.
28. Saquib, M. and M. Muneer, *TiO2-mediated photocatalytic degradation of a triphenylmethane dye (gentian violet), in aqueous suspensions.* Dyes and Pigments, 2003. **56**(1): p. 37-49.
29. Sivalingam, G., et al., *Photocatalytic degradation of various dyes by combustion synthesized nano anatase TiO2.* Applied Catalysis B: Environmental, 2003. **45**(1): p. 23-38.
30. Daneshvar, N., D. Salari, and A. Khataee, *Photocatalytic Degradation of Azo Dye Acid Red 14 in Water: Investigation of the Effect of Operational Parameters.* Journal of Photochemistry and Photobiology A: Chemistry, 2003. **157**: p. 111-116.

Recovering Zinc from electroplating wastewater by crystallization in a pellet reactor

Nhung Thi-Tuyet Hoang and Anh Thi-Kim Tran*

Faculty of Chemical and Food technology, Ho Chi Minh City University of Technology and Education, 01 Vo Van Ngan, Thu Duc city, Ho Chi Minh city, Vietnam.

*Email: anhttk@hcmute.edu.vn

Abstract. A pellet reactor (PR) was used to investigate the ability for zinc recovery from electroplating wastewater. The pellet reactor is a fluidized bed reactor, in which the nucleated precipitation of heavy metals occurred on the surface of seeding material. The zinc removal efficiency was 75% at molar ratio $[CO_3^{2-}]/[Zn^{2+}]$ of 2.5, the flowrate of 16 L/h, sand's diameter of 0.25 – 0.5 mm, and sand's mass of 50 g. The elemental analysis of zinc carbonate and zinc hydroxide crystallization was analyzed by X-ray energy dispersive spectrometer (EDS) and the surface was characterised by Scanning electron microscope (SEM) to get the morphological observation of the pellets after 38-day operation. This study demonstrated that the fluidized bed reactor can be considered a feasible method for zinc removal efficiency from electroplating wastewater to achieve sustainable development.

Keywords: zinc recovery; pellet reactor; crystallization; electroplating wastewater.

1. Introduction

Environmental protection has been concerned in Vietnam due to rapid industrialization development, in which environmental pollution caused by wastewater is one of the most important issues that needs to be considered. In particular, plating industry wastewater contains certain toxic substances such as chromium, nickel, copper, zinc, cyanide ...which adversely affect harmfully aquatic environment if not being well treated before discharging. This problem not only affects the quality of the water resource, but also damage to the human health such as cancer, cholera, diarrhea and typhoid [1]. So far, there have been many methods of heavy metals removal in wastewater such as: coagulation [2], ion exchange resin, precipitation, biological methods [3], adsorption [4].... In reality, these techniques are affected by the following factors: operation cost, toxic sludge production, removal efficiency, easy or complex operation [5]. Some previous studies mentioned that pellet reactor for metal recovery has these advantages like low sludge formation (dry pellets), a small footprint, less chemical usage, and no sludge handling [6]. The pellet reactor was used to remove hardness [7,8], fluoride [9,10], phosphate [11], heavy metals [12,13]. By using fluidized bed with quartz sand, the calcium hardness could be removed with 86% when the reactor was operated with the optimum conditions such as flowrate of 60 L/h, the seeding height of 0.5 m, pH of 9.5, seeding material size of 0.2 -0.4 mm, and recycle ratio of 60% [8]. Similarly, Tran et al., 2015 showed that by using pellet reactor as the pretreatment method, the removal efficiency of calcium and nickel from the wastewater could be achieved with 74% and 94.4%, respectively to reduce ion exchange membrane scaling potential and enhance the longevity of electrodialysis bipolar membrane system [14]. In addition, Aldaco et al., 2007 optimized the operational parameters to get the highest calcium and heavy metal removal efficiency and used mathematic model to see the kinetic of seeding material size growth in the pellet reactor [10]. Moreover, the reactor network-oriented model was denoted for calcium phosphate precipitation to validate with experimental calcium phosphate precipitation efficiency [10]. In practice, this is a reactor with a height of 5 to 10 m and a diameter of 1.5 to 3 m, containing seeding material such as garnet sand from 0.2 to 0.6 mm. The seeding materials become fluidized when the inlet water a is pumped upward at a suitable superficial velocity. In the pellet reactor, crystallization occurs on the surface of common used silica sand. In this zone, the water is mixed with a caustic reagent (NaOH or $Ca(OH)_2$). After crystallization on the surface, the pellets become heavier (to a maximum pellet diameter of 1 mm), then sink to the bottom of the reactor, from where they are regularly discharged and replaced by fresh seeding material.

In this study, the heavy metals removal and recovery from electroplating wastewater using pellet reactor were investigated with precipitation/ crystallization process using Na_2CO_3 chemicals. By adding Na_2CO_3 together with the inlet water from the bottom of the reactor, more $ZnCO_3$ and $Zn(OH)_2$ are covered on seeding material leading the bigger size, having a higher sedimentation velocity, and will be extracted at the reactor bottom.

$$Na_2CO_3 \rightarrow 2Na^+ + CO_3^{2-}$$

$$Zn^{2+} + CO_3^{2-} \rightarrow ZnCO_3 \downarrow$$

$$Zn^{2+} + 2OH^- \rightarrow Zn(OH)_2 \downarrow$$

$$ZnCO_3 + \text{sand} \rightarrow \text{nucleated precipitation}$$

$$Zn(OH)_2 + \text{sand} \rightarrow \text{nucleated precipitation}$$

Optimizing parameters affecting treatment efficiency such as flow rate, influent pH of wastewater, $(CO_3^{2-})/(Zn^{2+})$ molar ratio, seeding mass were studied. The surface of pellets was examined by SEM microphotography to get the morphological observation. Finally, the zinc removal and recovery efficiency with the real wastewater in long-time operation was investigated to see the technical feasibility in electroplating wastewater treatment.

2. Material and methods

2.1. Materials

Wastewater for experiment was taken from the electroplating company (Binh Duong province) with the influent concentration of Zn^{2+} of 30 - 120 mg/L, pH from 4-5 and COD of 307 – 399 mg/L. The synthetic water in this investigation was prepared by adding $ZnCl_2.2H_2O$ (China) into the water to get the initial zinc concentration of 30 mg/L. Sodium carbonate ($Na_2CO_3.H_2O$) (China) was used as an agent of crystallization reaction to inject to the pellet reactor by peristaltic pumps TRIWIN 8800-2004 and Labor-Schlauchpumpe PLP 66. Sand was washed and sieved with size of 0.25 – 0.5 mm for a seeding material.

2.2. Pellet reactor

The lab-scale pellet reactor is columnar with conical reactors on the top, 150 mm height and a diameter of 16 mm (Figure 1). The wastewater was pumped upwards the fluidized bed through inlet water tube (d =5 mm) without recirculation. At the same time, a mixture of Na_2CO_3 with a flow rate of 1 L/h was injected at a point 3 cm away from the bottom of the pellet reactor (D = 2 mm) to adjust the operational conditions as given in Table 1.

In this study, the synthetic water was used to run in the series of experiments to optimize the operational parameters (Table 1). The real electroplating wastewater was used in the long-term experiment to see the technical feasibility in zinc removal and recovery. Every run of experiment was operated 6 hours continuously. Samples of the effluent were taken at 60 min intervals, and the value of efficiency was the average of 3 times taking sample. These three samples are at 4 h, 5 h, and 6 h running to make sure the stability of the reactor. Zinc concentration was analyzed by with 797 VA Computrace for Voltammetric Trace Analysis (Metrohm AG, Switzerland). The zinc concentration was analyzed with the filtered and non-filtered samples. The non-filtered sample included discrete precipitation of Zn^{2+} and dissolved Zn^{2+} ion which did not crystalize on sand while the filtered sample existed dissolved Zn^{2+} ion.

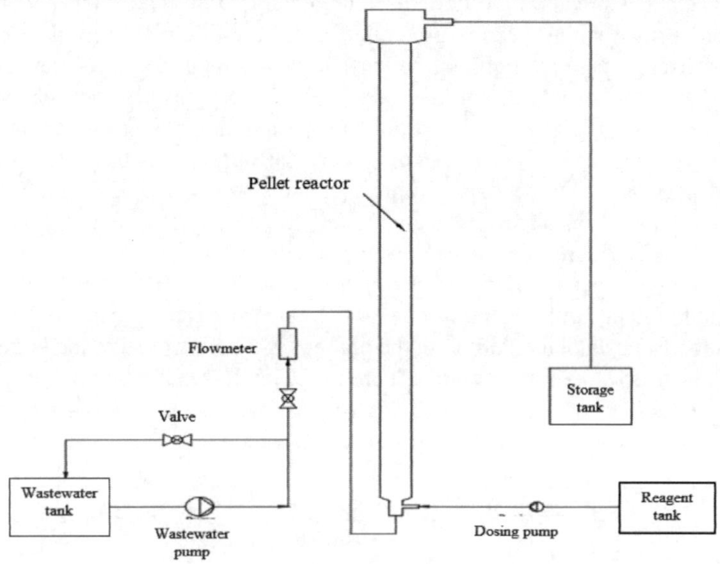

Figure 1. Pellet reactor configuration.

Table 1. Operational conditions

Run	Flowrate, L/h	pH	$[CO_3^{2-}]$: $[Zn^{2+}]$	Sand mass, g	$[Zn^{2+}]$, mg /l
1	12, 14, 16	8	2	40	30
2	Optimum	7, 8, 9, 10	2	40	30
3	Optimum	Optimum	1.5, 2, 2.5, 3	40	30
4	Optimum	Optimum	Optimum	40, 50, 60	30
5	Optimum	Optimum	Optimum	Optimum	30

The surface morphology of pellets was analyzed by scanning electron microscopy (SEM) (FESEM S4800) (Hitachi, Japan) and elemental analysis was performed using an X-ray energy dispersive spectrometer (EDS) (HORIBA H-7593 (ENGLAND)) with 10 kV accelerating voltage.

3. Results and discussion

3.1. Effect of superficial velocity on zinc removal efficiency

The effect of superficial velocity (SV) on zinc removal performances was studied by changing from 60 m/h to 80 m/h. Correspondingly, the flowrate increased from 12 to 16 L/h and the hydraulic retention time decreased from 1.8 to 1.13 min, respectively. These experiments were carried out by keeping pH at 8, the $[CO_3^{2-}]/[Zn^{2+}]$ molar ratio of 2 and sand mass of 40 g. The zinc removal efficiency increased with the increase in superficial velocity from 60 to 80 m/h (Figure 2). Since the superficial velocity increased, the percentage of expanded bed raised from 70 cm to 130 cm, respectively. When the porosity of the expanded bed increased, there were more surface area for the precipitation attached on the sand. This led to an increase in the zinc removal efficiency from 17.2% to 46.7% with the non-filtered samples. However, it can be seen that the increase in filtered samples was not significant due to the occurrence of homogeneous crystallization.

When sodium carbonate was injected into the pellet reactor, the precipitation of heavy metals occurred with two types: nucleated precipitation and discrete precipitation. The nucleated precipitation was coated on the sand surface while discrete precipitation (fine particles) referred to occur in the bulk solution. Since the discrete precipitation as fine particle would be gone out of the reactor, the filtered removal efficiency was not considered significantly [15]. The optimized conditions was chosen based on the non-filtered removal efficiency. In this study, the precipitation of zinc carbonate or zinc hydroxide took place not only on the surface of the seeding material but also in the solution, which are both homogeneous and heterogeneous processes. Figure 2 showed that the zinc removal efficiencies of filtered and non-filtered samples were 17.2% and 56.1% (SV 60 m/h); 34.9% and 63.2% (SV 70 m/h); 46.7% and 60.2% (SV 80 m/h), respectively. These results indicated that when increasing the superficial velocity, the production of fine particles (discrete precipitation) in the effluent decreased, most of $ZnCO_3$ and $Zn(OH)_2$ precipitation occurred on the surface of the sand (nucleated precipitation) under conditions of the increasing superficial velocity. Although the higher superficial velocity, the better the non-filtered removal efficiency, in this study, if the flowrate increased over 16 L/h, the seeding material was washed out of the reactor, the effect of higher superficial velocity was not studied. Therefore, the optimum superficial velocity was chosen at 80 m/h corresponding with the flowrate of 16 L/h.

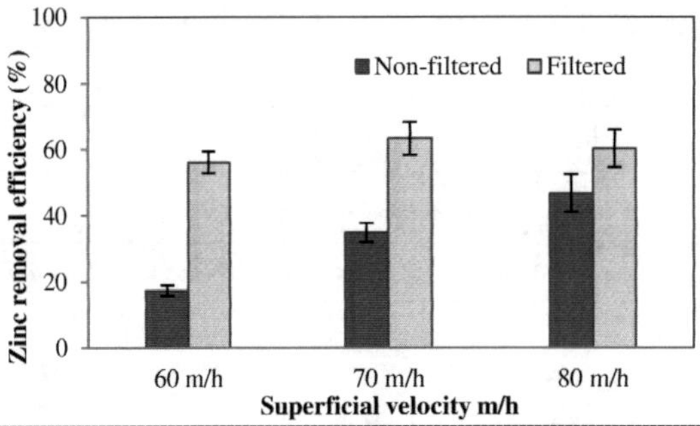

Figure 2. Influence of the superficial velocity on Zinc removal efficiencies.

3.2. Effect of pH of wastewater on Zinc removal efficiency

To investigate the effect of pH on zinc removal efficiency, the pH was adjusted from 7 to 10, other parameters such as flowrate, the molar ratio of carbonate and zinc, mass of sand were kept constant with the value of 16 L/h, 2 and 40 g, respectively.

During pellet reactor treatment, two reactions occurred:

$Zn^{2+} + CO_3^{2-} \rightarrow ZnCO_3$ with $K_{sp} = 3.10^{-17}$ at 25°C [13] (1)

$Zn^{2+} + OH^- \rightarrow Zn(OH)_2$ with $K_{sp} = 1.46.10^{-10}$ at 25°C [13] (2)

Zinc precipitates as $ZnCO_3$ and $Zn(OH)_2$ are formed when the saturation degree exceeds the solubility product constant K_{sp}. According to [13], the K_{sp} of hydroxide was smaller than that of carbonate, therefore, $Zn(OH)_2$ is less soluble than $ZnCO_3$, and so precipitated before $ZnCO_3$. In this study, sodium carbonate was injected into the reactor, zinc was combined with carbonate, forming not only zinc carbonate but also zinc hydroxide. When pH increased from 7 to 10, the zinc removal performance increased from 19.1% to 48.6% for non-filtered samples. However, the efficiencies of filtered samples were decreased from 64.8% to 53.7% with the increase pH from 8 to 10 (Figure 3). This proved that when pH increased, CO_3^{2-} and OH^-concentration increased, that led to higher supersaturation and more precipitation of $ZnCO_3$ and $Zn(OH)_2$ occurred as the nucleated precipitation. This meant that most of $ZnCO_3$ and $Zn(OH)_2$ precipitated on the sand surface as pH increased. Therefore, it can be concluded that pH 10 was the optimum condition to get the high zinc removal efficiency in both filtered and non-filtered samples.

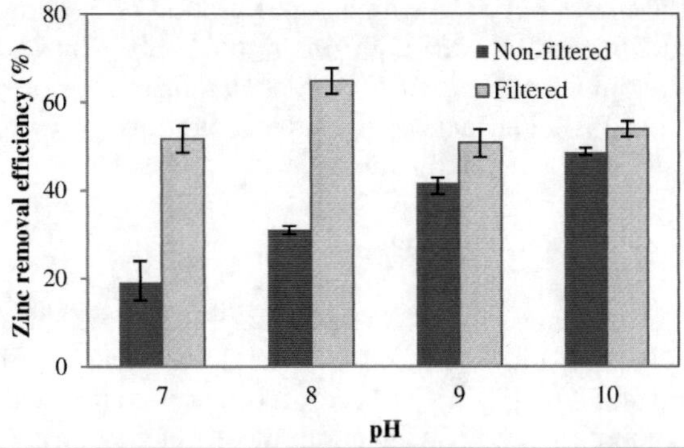

Figure 3. Influence of pH on Zinc removal efficiencies.

3.3. *Effect of [CO_3^{2-}]/[Zn^{2+}] molar ratio on Zinc removal efficiency*

The effect of [CO_3^{2-}]/[Zn^{2+}] molar ratio was carried out to determine the suitable ratio required for optimal zinc removal efficiency by Na_2CO_3 solution Figure 4. The flowrate was 16 L/h, mass of sand was 40 g, and pH was kept at 10. In theory, the precipitation of $ZnCO_3$ was achieved theoretically as the molar ratio of [CO_3^{2-}]/ [Zn^{2+}] = 1. However, when running the pellet reactor with continuous mode, the hydraulic retention time HRT was short with 1.13 minutes (corresponding with flow rate of 16 L/h); therefore, it is difficult to get 100% precipitation at molar ratio of [CO_3^{2-}]/ [Zn^{2+}] =1. With this reason, in this study, the effect of this molar ratio was investigated.

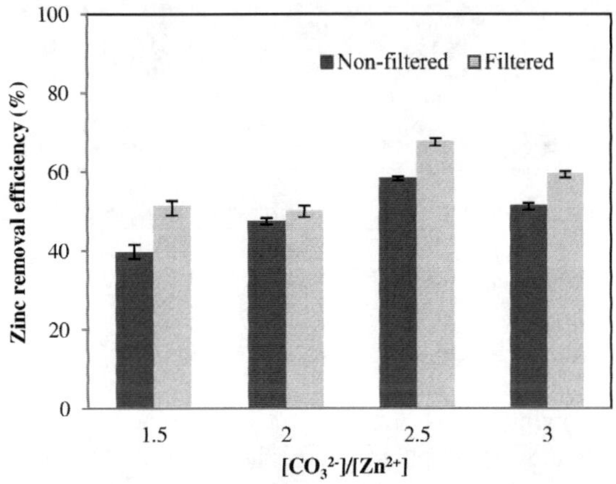

Figure 4. Effect of [CO_3^{2-}]/[Zn^{2+}] ratio on zinc removal efficiencies.

Figure 4 showed that the increase in [CO_3^{2-}]/[Zn^{2+}] led to the increase in the zinc removal efficiency. As the molar ratio of [CO_3^{2-}]/[Zn^{2+}] increased, more carbonate was added, the equation: CO_3^{2-} + H_2O →

HCO_3^- + OH^- shifted to the left; therefore, increased the amounts of carbonate and hydroxide in the solution. With higher CO_3^{2-} and OH^- concentration, higher supersaturation, which in turn achieved fast primary nucleation. Nevertheless, when the ratio of $[CO_3^{2-}]/[Zn^{2+}]$ was higher than 3, the zinc removal efficiency seemed to be lower (Figure 4). According to [9], nucleation took place only heterogeneously on the sand at low supersaturation, while, both homogeneous and heterogeneously nucleation occurred at higher supersaturation, which was proved by the obtained results of 8% of fines formation with a fluoride concentration of 100 mg/L, but 51% with a concentration of 300 mg/L. The optimum ratio of $[CO_3^{2}]/[Zn^{2+}]$ was chosen at 2.5 to achieve the high zinc removal efficiency.
In conclusion, the effects of initial pH and $[CO_3^{2-}]/[Zn^{2+}]$ molar ratio on the precipitation of zinc compound were significant due to their effects on the supersaturation level which is the thermodynamic driving force behind precipitation and crystallization.

3.4. *Effect of sand mass on Zinc removal efficiency*

Figure 5 presents the effect of sand mass in pellet reactor on zinc removal efficiency at a flowrate of 16 L/h corresponding with HRT of 1.13 min. The mass of sand was changed from 40 g to 60 g, pH was kept at 10 and the molar ratio of $[CO_3^{2-}]/ [Zn^{2+}]$ was 2.5. It is observed that when seeding mass increased in the pellet reactor leading more places for nucleation, the efficiency of zinc removal also increased. When sand mass increased from 40 to 50 g, the efficiency increased from 78.9% to 83.6% with filtered samples and 62.3% to 79.6% with non-filtered samples; however, the sand mass increased more to 60g, there was a decrease in zinc removal efficiency from 83.6% to 71.3%. This is due to with same superficial velocity but higher sand mass, the fluidized bed could not be expanded leading to less contacting surface area, decreasing the nucleation potential. From these above results, the sand mass of 50 g was the optimized value to obtain both non-filtered and filtered zinc removal efficiency.

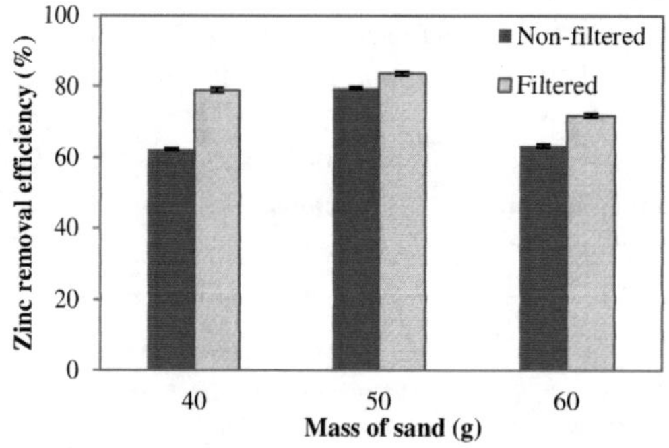

Figure 5. Effect of seeding material mass on zinc removal efficiency.

3.5. *Long-term operation for zinc recovery*

In order to investigate whether the long-tern operation feasible or not, recovery experiment was carried out for more than one month with 2 stages with the optimum parameters: the ratio of chemicals $[CO_3^{2-}]/[Zn^{2+}]$ of 2.5, the flowrate of 16 L/h, pH of 10, the sand mass of 50 grams with the real electroplating wastewater. In the first stage within 30 days, the initial concentration was 30 mg/L, the removal efficiency was around 78%. In the second stage, the zinc concentration was increased to 120 mg/L to test the zinc removal and recovery feasibility of the pellet reactor with higer concentration. The efficiency decreased slightly to 60%. After 38-day operation, the change of sand weight before and after the treatment process (by dissolved in HNO_3 1N solution) was showed in Table 2, in which the increase in mass of sand proved that the precipitation of zinc occurred on the sand.

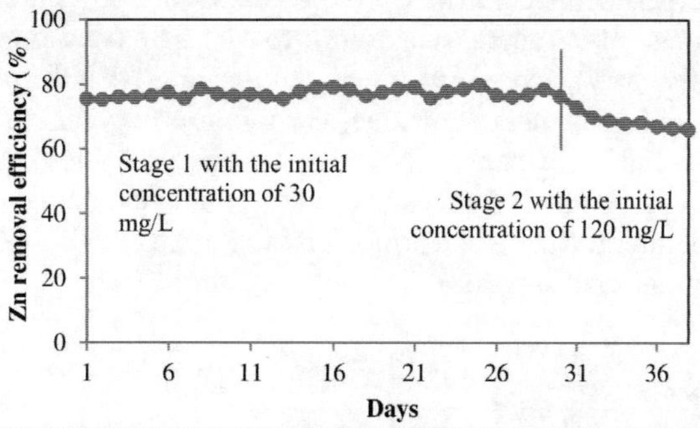

Figure 6. Long-term operation for zinc recovery.

As seen in Figure 7, the sand was larger and the mass increased significantly. Table 2 showed the heavy metal precipitation on the sand proved by the increase in sand's mass. After washing with acid solution, the mass of the origin sand just reduced little from 2 g to 1.96 g, because some components in the sand were dissolved in acid, however, with the treated sand, the increase in mass was 0.61 g after stage 1 and 1.66 g after stage 2, which proves that after the recovery process, the zinc carbonate and zinc hydroxide were crystallized and adhered to the surface of sand grains. With 50g used in the pellet reactor, zinc could be recovered with mass of 26.25 g within 8 days when the initial concentration was 120 mg/L.

Table 2. The change in mass of sand before and after running in the pellet reactor

	Mass before dissolved in acid, g	**Mass after dissolved in acid, g**	**Difference, g**
Initial sand	2	1.96	0.04
Sand after 1st stage	2	1.39	0.61
Sand after 2nd stage	2	0.34	1.66

Figure 7. Sand before (a) and after pellet reactor treatment (b).

3.6. Microscopic analysis

The color and size of pellets after and before heavy metal removal were different, the color of used pellet was whiter (Fig. 7). Therefore, it is clear that $ZnCO_3$ and $Zn(OH)_2$ were precipitated and covered on the surface of the sand. Figure 8 showed SEM and EDS microphotographs of the pellets before and after the crystallization process. With SEM, there were changes in the morphology of the pellets during the experiments: some precipitates appeared on the sand. The difference in elemental percentage of sand before and after recovery experiments confirmed Zn elements were present on the surface of the sand. The relative atomic percentage of Zn increased from 0 to 56.22 % (w/w), while the percentage of Si decreased from 31.15 to 2.7% (w/w). Moreover, it was demonstrated that the recovered zinc had a purity of about 56 wt %. The atomic percentages of zinc, carbon, and oxygen were 23.87%, 12.76%, and 60.45%, respectively, which can easily satisfy the molar ratios (1:1:3) in the zinc carbonate compound ($ZnCO_3$). Moreover, the rest percentage of zinc and oxygen could be counted in the contribution of zinc hydroxide ($Zn(OH)_2$). Therefore, it can be confirmed that the mixture of $ZnCO_3$ and $Zn(OH)_2$ probably attached and crystallized on the sand surface.

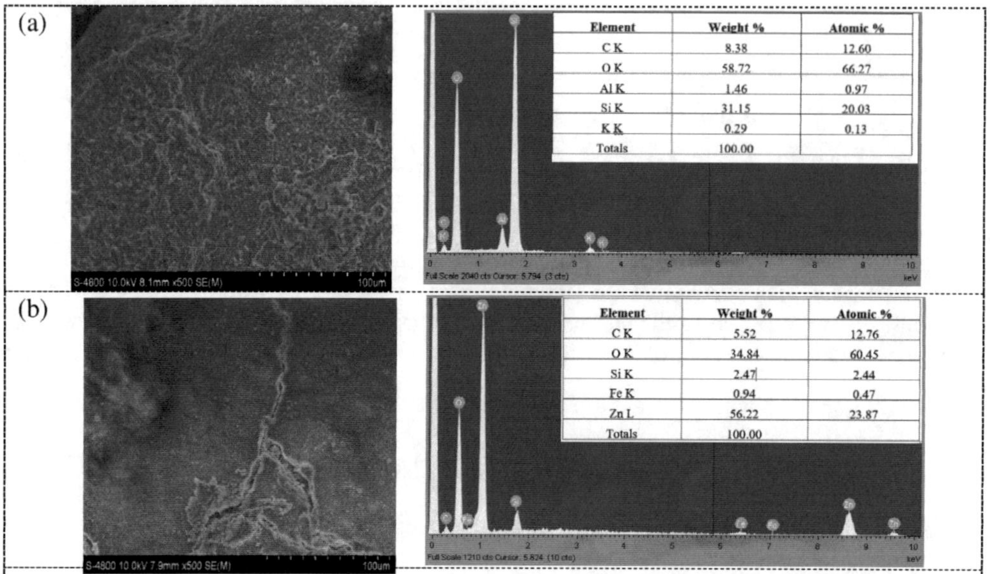

Figure 8. SEM (x 500) and EDS of pellets before (a) and after treatment (b) with sand.

3.7. *The feasibility of pellet reactor in zinc removal from wastewater*

In view of the natural resource scarcity, heavy metal recovery and wastewater treatment by a pellet reactor is very interesting concerning sustainability. The optimized operational conditions were obtained at: the ratio of chemicals $[CO_3^{2-}]$ / $[Zn^{2+}]$ = 2.5; the flowrate of 16 L/h; pH of 10; sand mass of 50 grams, the heavy metal removal efficiency was 75% when the initial concentration of 30 ppm. The high removal efficiency was up to 75% without sludge production shows that the obtained environmental benefits of the pellet reactor were resource recovery and wastewater treatment before being discharged. In this study, the synthetic wastewater was used for optimizing the operational conditions and the real electroplating wastewater was tested for long-term experiment. The results showed that the zinc removal efficiency from real wastewater was decreased slightly compared to that of synthetic wastewater. Therefore, it is believed that even though there were other interfering ions in the real wastewater and maybe with the scale up of the column for full-scale application, the effectiveness of the heterogeneous crystallization process still remains if the optimum conditions are controlled well. By recovering zinc from wastewater, the effluent concentration decreases, contaminant loading is reduced, and the quality of the surrounding aquatic environment is improved. Moreover, zinc is recovered on the sand so it is easy to collect and reuse for other production. Therefore, zinc recovery as zinc carbonate and zinc hydroxide by fluidized bed reactor can be considered a feasible method to achieve sustainable development.

4. Conclusions

A feasible method was proposed to remove and recover zinc from wastewater by a pellet reactor. A pellet reactor was used to precipitate and crystallize the zinc carbonate and zinc hydroxide on the sand. The optimized conditions were obtained at: the ratio of chemicals $[CO_3^{2-}]/[Zn^{2+}] = 2.5/1$; the flowrate of 16 L/h; pH of 10; sand grains' mass of 50 grams, the heavy metal removal efficiency was 75% ± 0.4% when the initial concentration was 30 ppm. After treatment, Zn^{2+} concentration in the effluent water nearly meets Vietnam discharge standard QCVN 40: 2011/ BTNMT for industrial wastewater. The heavy metal recovery mass was 26.25 after 8 days with the initial concentration of 120 mg/L. To summary, the pellet reactor is technically and economically feasible for removing and recovering the heavy metal from wastewater with optimum parameters. This method helps recovering heavy metal from the wastewater, reducing the amount of sludge discharged into the environment and exploiting the natural resource for raw materials supply.

References

1. Islam, M.S., et al., *Heavy metal pollution in surface water and sediment: A preliminary assessment of an urban river in a developing country.* Ecological Indicators, 2015. **48**: p. 282-291.
2. Kalaitzidou, K.; Bakouros, L.; Mitrakas, M. *Techno-economic evaluation of iron and aluminum coagulants on Se(IV) removal.* Water (Switzerland), **2020**, *12*, 672
3. Fu, F. and Q. Wang, *Removal of heavy metal ions from wastewaters: A review.* Journal of Environmental Management, 2011. **92**(3): p. 407-418.
4. Duan, C. et al. *Removal of heavy metals from aqueous solution using carbon-based adsorbents: A review.* Journal of Water Process Engineering, 2020, 37, 101339.
5. Eccles, H., *Treatment of metal-contaminated wastes: Why select a biological process?* Trends in biotechnology, 2000. **17**: p. 462-5.
6. Houwelingen, G., et al., *Experiences with pellet reactor softening as pretreatment for inland desalination in the USA.* Desalination and Water Treatment - DESALIN WATER TREAT, 2010. **13**: p. 259-266.
7. Harms, W. and R. Robinson, *Softening by Fluidized Bed Crystallizers.* Journal of Environmental Engineering-asce - J ENVIRON ENG-ASCE, 1992. **118**.
8. Chen, Y., et al., *Water softening by induced crystallization in fluidized bed.* Journal of Environmental Sciences, 2016. **50**: p. 109-116.
9. Aldaco, R., A. Garea, and A. Irabien, *Fluoride Recovery in a Fluidized Bed: Crystallization of Calcium Fluoride on Silica Sand.* Industrial & Engineering Chemistry Research, 2006. **45**(2): p. 796-802.
10. Aldaco, R., A. Garea, and A. Irabien, *Modeling of particle growth: Application to water treatment in a fluidized bed reactor.* Chemical Engineering Journal, 2007. **134**(1): p. 66-71.
11. Rugaika, A.M., et al., *Phosphorus recovery as calcium phosphate by a pellet reactor pre-treating domestic wastewater before entering a constructed wetland.* International Journal of Environmental Science and Technology, 2019. **16**(7): p. 3851-3860.
12. Park, N.-K., et al., *Optimization of Nickel-Based Catalyst Composition and Reaction Conditions for the Prevention of Carbon Deposition in Toluene Reforming.* Energies, 2019. **12**(7).
13. Tran, A.T.K., et al., *Pellet reactor pretreatment: A feasible method to reduce scaling in bipolar membrane electrodialysis.* Journal of colloid and interface science, 2013. **401**.
14. Tran, A.T.K., et al., *Simultaneous regeneration of inorganic acid and base from a metal washing step wastewater by bipolar membrane electrodialysis after pretreatment by crystallization in a fluidized pellet reactor.* Journal of Membrane Science, 2015. **473**: p. 118-127.

IOP Conf. Series: Earth and Environmental Science **964** (2022) 012029 doi:10.1088/1755-1315/964/1/012029

15. Zhou, P., et al., *Heavy metal removal from wastewater in fluidized bed reactor.* Wat. Res. Vol. 33, No. 8, pp. 1918–1924, 1999

Treatment of C.I Reactive Blue 160 by ozonation system

Pham- Hung Duong[1,2], Ngoc-Han T. Huynh[3,*], Yong-Soo Yoon[2]

[1] Faculty of Technology, Van Lang University, Ho Chi Minh city, Vietnam
[2] Faculty of Chemical engineering, Dankook University, Yongin-si, Korea
[3] Faculty of Environment, Ho Chi Minh City University of Natural Resources and Environment, Ho Chi Minh city, Viet Nam

*Email: han.htn_mt@hcmunre.edu.vn

Abstract. This study was carried out to assess the treatment ability of color, dye, and COD in the dyeing wastewater containing C.I Reactive Blue 160 by ozonation system. Both batch and continuous operating modes with concurrent and counter-current flows were investigated. The effects of the ozone gas flow rate, pH, temperature, Na_2CO_3 concentration, and initial dye concentration were evaluated. The decolorization, dye removal efficiencies, and mineralization ability of COD by ozonation were determined. The results indicated that ozonation had high efficiency in the treatment of dyeing wastewater containing C.I Reactive Blue 160. The treatment performance was affected by the ozone gas flow rate, pH, temperature, Na_2CO_3 concentration, and initial dye concentration. The removal efficiency of color, dye, and COD were 98.04%, 99.84%, and 87.31% for the treatment of 200 mg/L initial dye concentration in batch mode with 30 min ozonation time, respectively. In the continuous operation and counter-current flow, the color, dye, and COD removal efficiencies reached 97.24%, 99.76%, and 86.38% after 30 min HRT, respectively, and higher than concurrent flow. The reaction of ozone and C. I Reactive Blue 160 was the first-order reaction in both batch and continuous operation. The complete mineralization required 90 min ozonation time.

Keywords: C.I. Reactive Blue 160, reactive dye, ozonation, dyeing wastewater

1. Introduction

In the textile industry, dyeing wastewater is the most concerned problems due to it is not only a large volume but also containing wasted dyes. Untreated dyes in such wastewater can cause serious problem for the receiving water environment and surround land area because of the color and toxicity organic compounds including original and degradable intermediate substances. Thus, there were many studies carried out to find the feasible method for the dyeing wastewater treatment. Some methods that may be applied in the dyeing wastewater treatment were found such as adsorption, oxidation by chemical, photocatalyst, ozonation, membrane filtration, coagulation, electrocoagulation, and biological methods, etc [1, 2, 3, 4]. However, the brightly colored types of dye, water soluble reactive and acidic dyes cannot be effectively removed through conventional treatment system [5]. Ozonation is an impressed method in dyeing wastewater treatment. Ozone can decompose the aromatic rings of some textile dyes, azo dyes and other organic pollutants in wastewater [6, 7]. The advantages of this process are the absence of increasing volume of wastewater and no secondary waste solid production (sludge) [6, 7]. Some previous studies showed that ozonation had high reactive dye treatment efficiency such as Reactive dye

Orange 13, Reactive dye Blue 19, Reactive Orange 16 [7, 8]. However, the application ability of ozonation method in the dyeing wastewater treatment depends on types of dye and operating conditions. It needs detailing research before applying it in the practical condition.

C.I Reactive Blue 160 belongs the Reactive dyes group that may become the most usage in the near future. It contains Azo group that is the most popular group of Reactive dye. It is difficult to degrade and cause problems for the health of human and animals. And the blue color is the loveliest color on the world [9, 10]. So, C.I Reactive Blue 160 was concerned in this study.

In case of the treatment of C.I Reactive Blue 160 in the dyeing wastewater, some methods were studied are biodegradation, photodegradation and ozonation. The photocatalytic degradation resulted on the treatment of C.I Reactive Blue 160 but the treated wastewater samples were too small, maximum volume of 180 mL, the initial C.I Reactive Blue 160 concentration was too low about 10 – 25 mg/L, whereas the real dyeing wastewater contains high dye concentration and large volume [11-14, 15-18]. So, this method is too difficult to apply in the practical treatment. The biological treatment was effective in the degradation of C.I Reactive Blue 160 as well. However, it still remains some disadvantages as long treatment time (4 hours – 4 days), selected and uncommon biological agent [19-23]. Ozonation was carried out with too low concentration from 5 mg/L – 15 mg/L, and small volume 1.2 L and 1.5 L, and batch operating mode. The effects of factors such as pH, ozone gas flow rate, temperature, Na_2CO_3, initial dye concentration have not investigated yet [24, 25]. In practical condition, the dyeing wastewater contains high dye concentration. Therefore, the treatment of C.I Reactive Blue 160 in dyeing wastewater by ozonation needs to be studied more with high dye concentration to find the effects of operating factor on treatment efficiency and assess its application ability in practical treatment condition. the feasible treatment method that can be effective, easily applied, and suit the trend of the development of wastewater treatment on the world.

For these reasons, ozonation was selected and carried out for the treatment of the dyeing wastewater containing C.I Reactive Blue 160.

2. Materials and methods

2.1. Materials

The synthetic dyeing wastewater was prepared by adding C.I Reactive Blue 160 powder (Oh Young Inc., South Korea) into the tap water. pH and conductivity of such wastewater were adjusted by Sodium hydroxide 99% or Sulfuric acid 99.9% (Samchun Pure Chemical Co., Ltd., South Korea) and sodium chloride 99.5% (Duksan Pure Chemicals Co., Ltd., South Korea), respectively. Its characteristics were presented in Table 1.

In addition, Sodium Carbonate Anhydrous (Na_2CO_3 99%) was also added to synthetic wastewater in a range of 0.00% - 0.03% to determine its effect on the removal efficiency by ozonation.

Table 1. Characteristics of the synthetic wastewater

Parameters	Units	Range of value
pH	-	3.00 ~ 10.76
Dye concentration	mg/L	100 ~ 400
Color	Pt-Co	490 ~ 1900
COD	mg/L	25 ~ 80
Conductivity	µS/cm	1500 ~ 2500
Sodium chloride	%	0.05
Sodium carbonate	%	0.00 ~ 0.03
Temperature	^oC	15 ~ 35

A bench-scale system was set up as Figure 1, includes:

- ✓ A polyacrylic bubble column reactor (D x H = 100 x 1500 mm) installed with a ceramic diffuser (∅60 mm) at the bottom has a total volume of 11.78 liters and a working volume of 10 liters.
- ✓ A mixing tank made of PVC and PE was used to store the synthetic wastewater with a stirrer to uniform the wastewater during the experimental procedure.

- ✓ An Ozone generator: Model PC-57 provided by Ozonetech Co., Ltd., South Korea. Gaseous ozone flow rate can be adjusted by a flowmeter. Generating Ozone concentration in gas flow was 64.17 mg/L measured by ozone analyzer (Aeroqual series 200, New Zealand).
- ✓ A magnet pump and a water flowmeter were used to supply wastewater into the bubble column reactor with the determined flows.

2.2. Experimental procedure

The experimental setup is presented in Figure. 1. Both batch and continuous operation modes were carried out in this study.

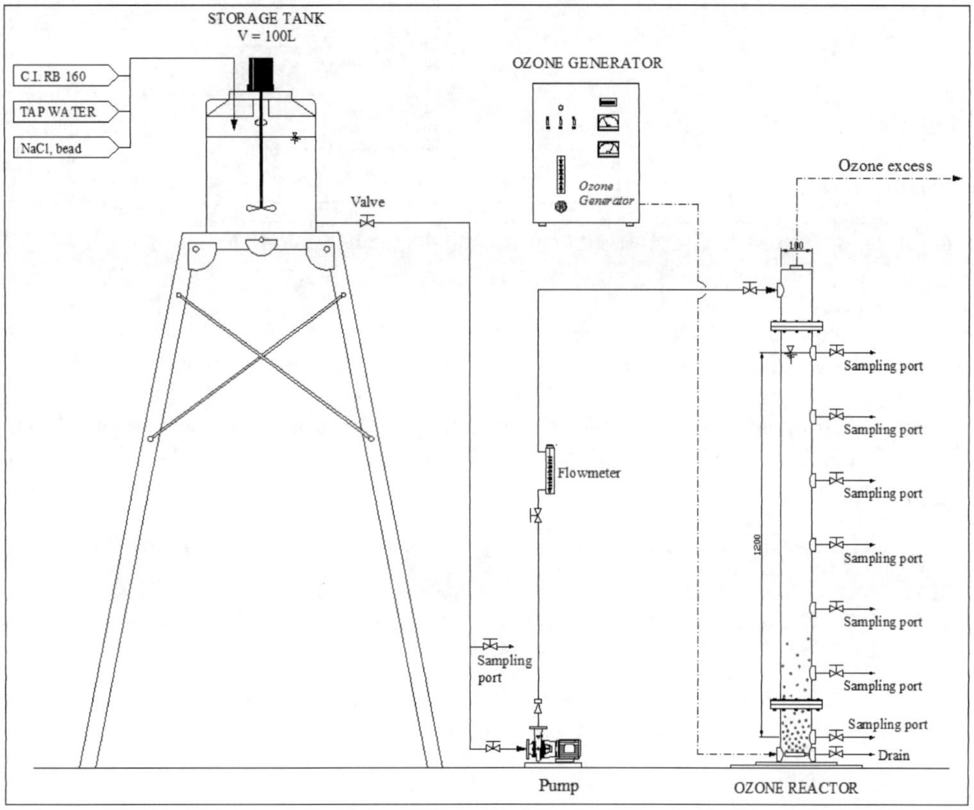

Figure 1. Experimental setup

Determine the effect of operating factors on the treatment efficiency by ozonation system: 10 liters of the synthetic wastewater contained C.I Reactive Blue 160 was filled into the bubble column reactor. The effect of ozone gases flow rate, pH, temperature, sodium carbonate anhydrous concentration, and the initial dye concentration on the treatment performance of ozonation system were evaluated in the batch operation mode by sequent changing of ozone gases flow rate from 0.5 – 1.3 L/min, pH from 3 – 10, temperature from 15°C – 35°C, Na_2CO_3 concentration from 0.01 – 0.03%, and initial dye concentration from 100 – 400 mg/L.

Evaluate the treatment efficiency of dyeing wastewater by ozonation system: Color, CODcr, and dye removal efficiency by ozonation system with both batch and continuous operating modes were determined under the determined optimal operating condition.

All experiments were repeatedly carried out three times to ensure the reliability of the experimental results.

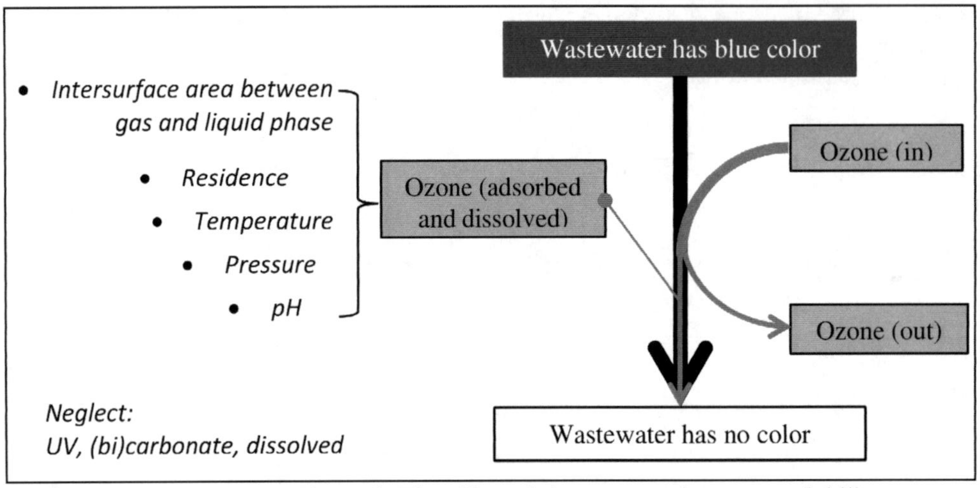

Figure 2. The effect of elements on the ozone solubility

2.3. Analytical Methods

In this study, the maximum absorbance wavelength was carried out by HS 3300 UV-vis. The color of dyes is occurred by absorbing light in the visible spectrum in a range of 400 – 700nm. Namely, dyes have an absorbance wavelength in a range of 580 – 700nm observed in blue to green [26]. In the previous studies, the maximum absorbance wavelength of C.I. Reactive Blue 160 was 630nm, 618nm, 616nm, 610nm, and 604nm, depends on the purity of dye and spectrophotometer equipment [11-13, 18, 20, 23]. In this study, the maximum absorbance wavelength of C.I Blue 160 was carried out by HS 3300 UV-vis and was 611nm

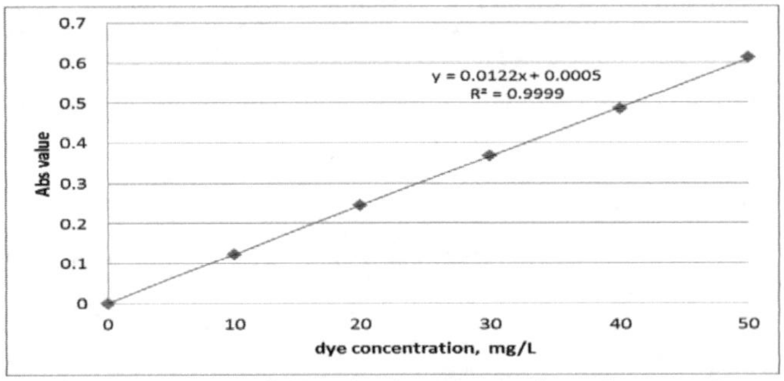

Figure. 3. Calibration curve of dissolved C.I. Reactive Blue 160 dye concentration.

Dissolved C.I Blue 160 dye concentration was determined by the previous built calibration curve, Abs measured at the maximum wavelength, $R^2 = 0.9999$, as in Fig. 3.

There was relationship between dye concentration and color. In this study, with synthetic raw wastewater, the color was found to be linear with dye concentration follow: $y = 4.9964x + 2.5$, $R^2 = 0.9996$. After ozonation, this relationship was changed to approximate the quadratic equation.

During the experiments, all samples were taken from column reactor with fixed reaction time and analyzed as follows:

- ✓ pH, conductivity, and temperature were measured by pH meter (Denver UltraBasic UB-10), conductivity meter (ATI Orion 130) and thermometer (Okaya Handy Thermo Pt-100 OH-IT), respectively.
- ✓ Color was analyzed by Spectrophotometer Hach DR 2800.

- ✓ The COD was analyzed by titrimetric method, which according to the Standard Methods for The Examination of Water and Wastewater [27].

2.4. Calculation methods

2.4.1. Removal efficiency

The removal efficiency was determined as the followed function [28]

$$E_{eff}\% = \frac{(C_0 - C_t)}{C_0} \times 100 \qquad (1)$$

Where: C_0: initial color (Pt-Co), or dye concentration (mg/L), or COD (mg/L).
C_t: color (Pt-Co), or dye concentration (mg/L), or COD (mg/L) respect to reaction time t.

2.4.2. The energy consumption

The energy consumption of the ozonation process was determined as the energy required for the ozone generator to remove a specified quantity of pollutants, namely, COD, color, dye [28].

$$E_{O_3} = \frac{P \times t}{V \times (C_o - C_t) \times 60} \qquad (2)$$

Where: E_{O3} is energy consumption, Wh/mg-COD, or Wh/mg-dye, or Wh/color unit.
P is the power of Ozone generator, W
t is the reaction time (min)
V is the working volume of reactor column (L)
C_0: initial color (Pt-Co), or dye concentration (mg/L), or COD (mg/L).
C_t: color (Pt-Co), or dye concentration (mg/L), or COD (mg/L) at reaction time t.

All experiment was repeated 3 times and collected data was analyzed by using the ANOVA with $\alpha = 0.05$ in Microsoft Excel.

3. Results and discussion

3.1. The effects of operating factors

The treatment efficiency of dyeing wastewater by ozonation system depends on the amount of dissolved ozone in wastewater and dye concentration. The ozone solubility is the function of some elements including the inter-surface area between gas phase and liquid phase, retention time, temperature, pressure, pH, and the appearance of ozone decomposition inhibitor.

All experiments were carried out under atmospheric pressure. Most the wastewater treatment systems are operated under atmospheric pressure in practical conditions, and controlling pressure in the practical conditions is infeasible, especially in the large wastewater treatment systems so the effect of pressure on the removal efficiency was neglected.

In this study, the effects of ozone gas flow rate, pH, temperature, Na_2CO_3 concentration (the inhibitor of ozone decomposition), and the initial dye concentration on the removal efficiency were investigated and alternately presented as follows:

3.1.1. The effect of ozone gas flow rate:

The ozone gas flow rate is one of factors affecting on treatment performance by ozonation because of its effect on the gas holdup and gas bubble velocity (contact retention time between ozone bubble and pollutants). The gas holdup is proportional function of gas flow rate [28].

In this study, the effect of ozone gas flow rate were determined by decolorization and dye treatment ability with the ozone gas flow rate changed in a range of 0.5 L/min - 1.3 L/min, reaction time from 0 –

10 min (2 min interval), initial dye concentration of 200 mg/L. The results of ANOVA analysis showed that both the ozone gas flow rate and reaction time had significant effects on the color and dye removal efficiencies. The interaction between the ozone gas flowrate and reaction time significantly affected on the color and dye treatment ability by ozonation as well (both P-values << 0.05)

As seen in Fig.4, at the gas flow rate of 0.9 L/min, the color and dye treatment efficiency were higher than others. When the gas flow rate increased in ranges of 0.5 L/min – 0.9 L/min, the color and dye concentration efficiencies increased due to increasing the gas holdup led to rise the contact surface area between pollutants and ozone bubbles.

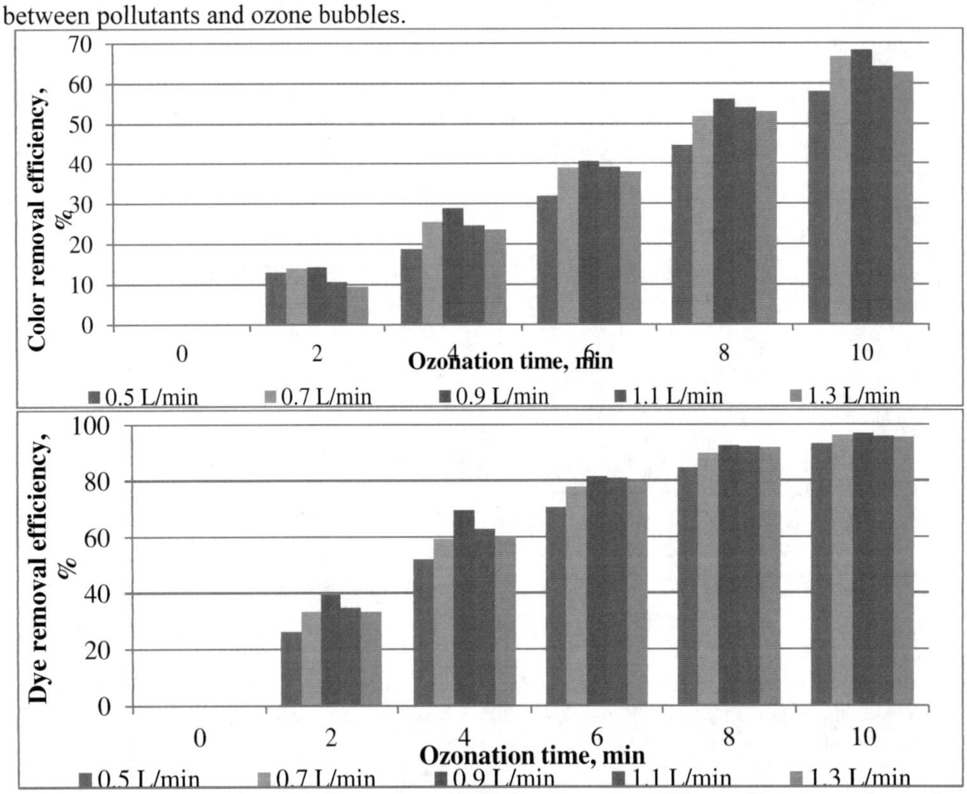

Figure 4. Color and dye treatment efficiency versus different flowrate and reaction time

When ozone gas was diffused into the wastewater, many bubbles were formed and caused the gas holdup. However, increasing the gas flow rate also leads to increasing the bubble size. Because the bubbles can be combined to form the bigger bubbles lead to decrease surface area. This study clearly indicated that the bubbles not only increased the amount and their size but also changed their shape from sphere shape to more oval shape when the gas flow rate increased from 0.5 L/min to 1.3 L/min, respectively, Fig.5.

According to George Gabriel Stokes (1819-1903), the velocity of sphere matter increases following the exponential function of its diameter, expressed by the Stokes equation:

$$v = \frac{d^2 \times (\rho - \rho_1) \times g}{18.\mu} \quad (3)$$

Where: v is velocity of sphere matter
g is gravitational acceleration
d is diameter of sphere matter
ρ is density of environment (gas or liquid)
μ is viscosity of environment (gas or liquid)

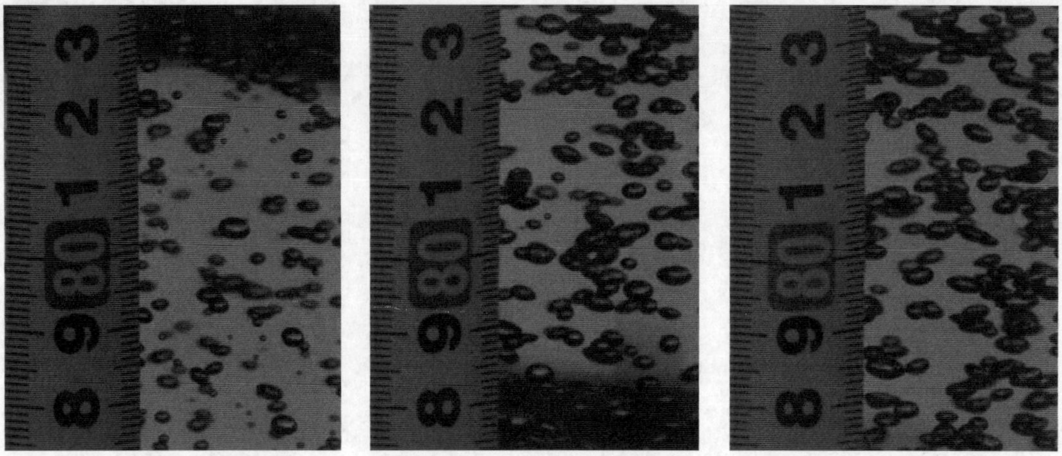

a. Flow rate 0.5 L/min b. Flow rate 0.9 L/min a. Flow rate 1.3 L/min

Figure 5. Morphologic of bubbles in different flow rates

Notes: initial dye concentration is 100 mg/L. Pictures were taken at 0.8m height. Shooting modes: shutter speed 1/800 second, aperture 5.0, and ISO 1600.

Those were all reasons for the gas flow rate increased over 0.9 L/min, the removal efficiency decreased. These results were the same trend in previous study [28]. It may be explained as: Although the gas holdup was higher in 1.1 L/min gas flow rate case, the formed bubbles size was bigger and the ozone bubbles velocity were faster led to reducing the contacting time and contact surface area between the ozone bubbles with pollutants. So the removal efficiency reduced. The gas flow rate of 0.9 L/min was suitable for the ozonation treatment in this system.

3.1.2. The effect of pH:

pH is the important factor in the ozonation process because it decides the reaction mechanism: the direct reaction mechanism occurred in an acidic solution whereas the indirect reaction mechanism occurred in the high alkaline condition.

According to Erdal Kusvuran et. al. (2010 and 2011), pH of direct and indirect mechanisms was 3 and 10, respectively [29, 30]. Especially, both of these mechanisms occurred contemporary when pH at neutral and slightly basic [30, 24]. Ozone (O_3) was dominated reactant in acidic and the radical species ($OH^{\bullet}$, HO_2, O_2^{-}...) were dominated reactants in alkaline. In a neutral and slightly basic solution, both of them occurred together.

Although the oxidation potential of hydroxyl radical (2.8 V) is higher than ozone (2.42 V), the attack of hydroxyl radical with hydroxyl radical is not always stronger than ozone [31, 32]. It depends on type of dye. According to Erdal Kusvuran et al., cationic dyes were destroyed better than anionic dyes by ozone reagent in acidic. Reversely, the anionic dyes were destroyed better than the cationic dyes by hydroxyl radical in basic [30].

In the case of C.I. Reactive Blue 160, the obtained data presented that the removal efficiencies of color and dye were reduced when increasing pH from 3 to 10. The color removal efficiency reached up 79% after 10 min ozonation in pH 3 whereas these were 68% and 61% at pH 7 and pH 10, respectively. The color removal efficiencies at all pH values were gradually converging when increasing reaction time and the color removal efficiency of pH 3 samples were still higher than others (Fig.6).

The dye removal efficiencies had a little different trend compare to the color treatment between different pH values. The dye removal efficiencies at pH 10 were significantly lower than others during the first 15 min reaction time and then they were slowly increased and evenly approaching 99.9% (Fig. 6).

IOP Conf. Series: Earth and Environmental Science **964** (2022) 012030 doi:10.1088/1755-1315/964/1/012030

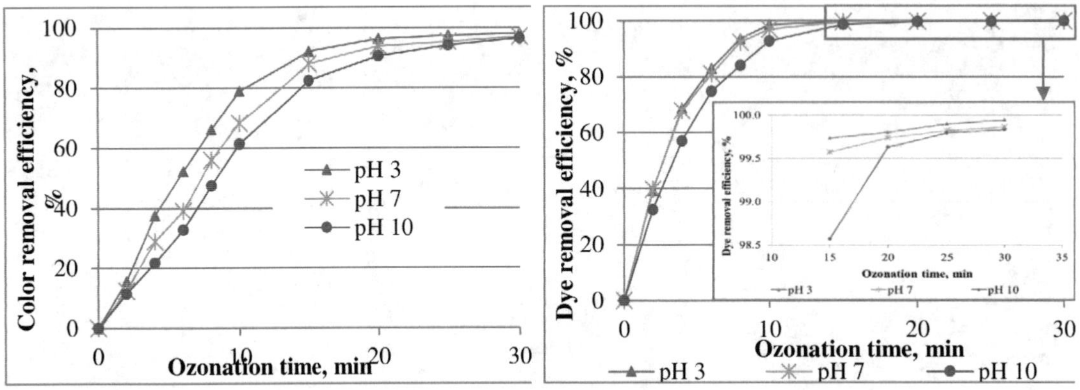

Figure 6. Effect of pH on the color and dye removal efficiency

These results proved that the decomposition reaction of C.I. Reactive Blue 160 molecules by ozone was more selective than by hydroxyl radical. According to Chedly Tizaoui and Naser Grima (2011), the hydroxyl radicals were interested in reaction with the intermediates products rather than parent molecules [7]. Thus, the removal efficiencies of color and dye were decreased at high pH conditions.

The results of ANOVA analysis also showed that pH, the interaction between pH, and reaction time were significant effects on the color and dye removal by ozonation. P values were $3.0 \times 10^{-24} - 5.76 \times 10^{-24}$ and $2.5 \times 10^{-17} - 9.8 \times 10^{-12}$, respectively.

3.1.3. The effect of temperature:

Temperature is a very important factor in a chemical reaction. It accelerates reaction rate lead to shortening reaction time. Besides accelerating reaction rate, the temperature is also the important factor that affect the ozone dissolving from the gaseous phase into the aqueous phase. Unfortunately, dissolved ozone will be decreased when temperature increases because of increasing bubble size. More detailed, high temperature leads to reduce the surface tension of water. Thus, bubble size will be increase. Therefore, the retention time of bubbles in wastewater was shortened [28]. Moreover, the increasing temperature makes to reducing ozone life in solution. The half-life values of ozone in neutral water are 30 min, 15 min, and 8 min respect to 15°C, 25°C, and 35°C.

Thus, temperature may have a positive or a negative effect on the treatment efficiency of ozonation. Therefore, the dependence of treatment efficiency on temperature needs to be concerned, especially in temperate and frigid countries, temperature changes so much between seasons, even in the day.

In this study, the effects of temperature on the removal efficiencies were surveyed at 15°C, 25°C, and 35°C. The surveying results were presented in Fig. 7.

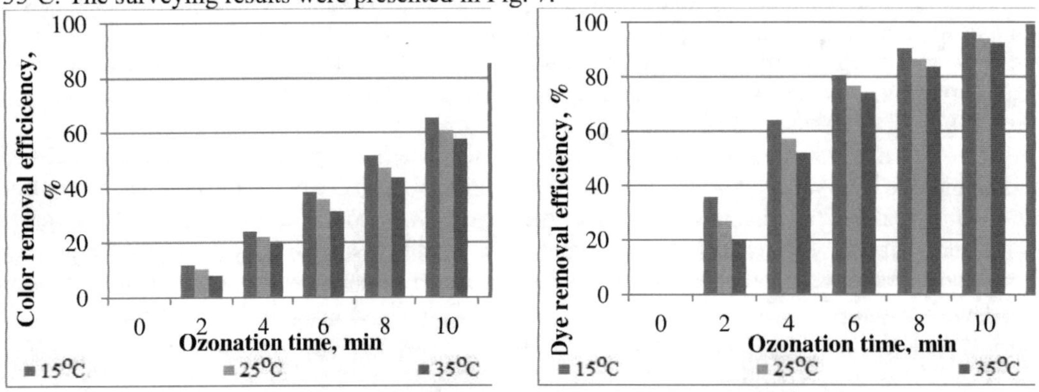

Figure 7. Effect of temperature on the color removal efficiency

Results showed that increasing temperature led to reduce the color and dye treatment ability by ozonation due to the increased reaction rate was not enough to compensate for the decrease of dissolved ozone and ozone decomposition by increasing temperature. Therefore, when the temperature of wastewater rose from 15°C to 35°C, the average color and dye removal efficiencies decreased.

In addition, the results of ANOVA analysis indicated that temperature and reaction time had a significant effect on the treatment efficiency, both of dye and color removal. However, the interaction between temperature and reaction time only had an effect on dye removal.

3.1.4. The effect of Na_2CO_3 concentration:

Sodium carbonates anhydrous is an inhibitor in the decomposition of ozone. So, they help maintain the ozone domination reagents.

In this study, the experiments were carried out with 4 samples: without Na_2CO_3, 0.01% Na_2CO_3, 0.02% Na_2CO_3, and 0.03% Na_2CO_3. The without Na_2CO_3 samples were adjusted pH up to 10, similarly pH conditions of samples containing Na_2CO_3. The results of ANOVA analysis showed that Na_2CO_3 significantly affected the color, dye, and COD removal efficiencies (P-value were 9.03×10^{-6}, 3.95×10^{-6}, and $1.98\times10^{-37} << 0.05$, respectively). The interaction between Na_2CO_3 and reaction time significantly affected on the color, dye, and COD removal efficiencies as well (P-value are 4.54×10^{-37}, 7.15×10^{-28} and $3.31\times10^{-46} << 0.05$, respectively).

As results in the effect of pH, the decomposition reaction between C.I. Reactive Blue 160 and ozone was more selective than hydroxyl radical. Thus, the removal efficiencies of samples containing Na_2CO_3 were better than samples without containing Na_2CO_3. That was proved by the results presented in the Fig.6, the removal efficiencies increased at higher sodium carbonate anhydrous concentration. In particular, the removal efficiencies of sample containing 0.03% Na_2CO_3 were impressive. With only 2 min ozonation, the dye removal efficiency of this sample was 47.3% whereas it was 34.49%, 31.58%, and 30.52% for 0.02% Na_2CO_3 sample, 0.01% Na_2CO_3 sample, and without Na_2CO_3 sample, respectively.

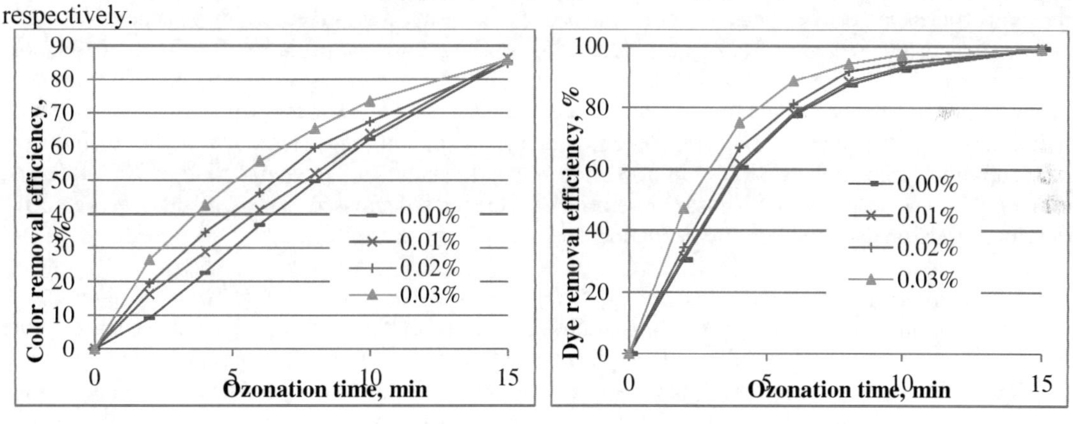

Figure 8. Effect of sodium carbonate anhydrous on the color and dye removal efficiency at the first 15 min ozonation

However, this trends is only true for the first 15 min ozonation. The longer ozonation, the better removal efficiencies in without Na_2CO_3 sample. Reversely, the removal efficiencies of samples containing Na_2CO_3 decreased when increasing Na_2CO_3 concentration. This result may be explained by precipitation of $CaCO_3$ formed by Ca^{2+} in water and CO_3^{2-} released by ozonation of organic components. Because pH of wastewater is around 10, the dominant alkaline agents are CO_3^{2-}. This results was consistent with the previous study of Nelson Saksono et al. (2009), the precipitation of calcium carbonate occurs immediately when initial pH ≥ 8.5 [33]. Thus, maybe the precipitation of calcium carbonate occurs parallel with the decomposition C.I. RB 160.

3.1.5. The effect of initial dye concentration

The initial dye was also an important factor markedly affect the removal efficiencies [11-14, 15-18]. It was proved by the results of ANOVA analysis. The dye concentration had significant effect on the color, and dye removal efficiencies (P-value are 2.88×10^{-87}, and $9.8\times10^{-125} << 0.05$, respectively). The interaction between dye concentration and reaction time also significantly affected the color and dye removal efficiencies (P-value are 1.44×10^{-81}, and $3.31\times10^{-127} << 0.05$, respectively). Namely, the color and dye removal efficiencies decreased when increasing initial dye concentration. Three samples, have initial dye concentrations of 100 mg/L, 200 mg/L and 400 mg/L, were treated under the same conditions: without pH adjustment, ozone gas flow rate 0.9 L/m. The results were showed in Fig. 9.

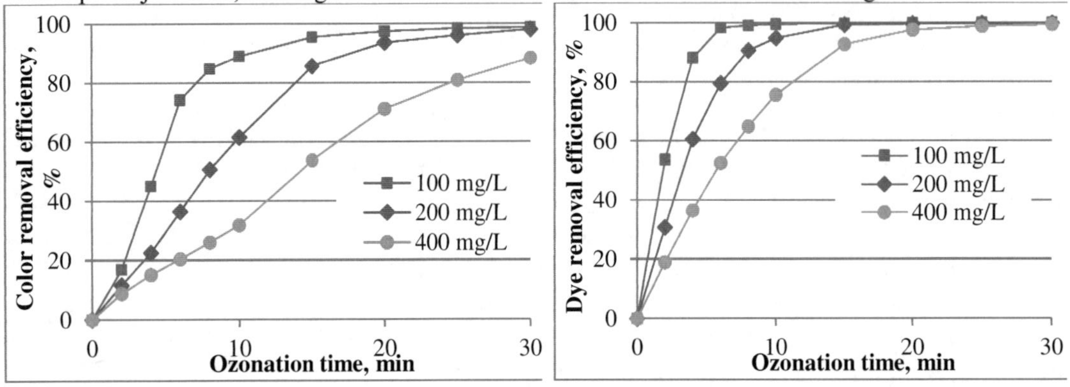

Figure 9. Effect of initial dye concentration on the color and dye removal efficiency

During the first 10 min ozonation, the color removal efficiency of sample 100 mg/L rose very fast and reached up 89% whereas it was only 61.63% and 31.86% for sample containing 200 mg/L and 400 mg/L initial dye concentration. Moreover, the dye removal efficiencies were more impressive than color removal efficiencies. Only during the first 10 min ozonation, the dye removal efficiency of sample 100 mg/L, 200 mg/L, and 400 mg/L are 99.49%, 94.81%, and 75.55%, respectively. Specially, after 30 min ozonation, the dye and color removal of sample 100 mg/L and 200 mg/L were nearly complete. These results occurred due to a larger amount of dye molecules and the generated intermediate products in the solution in higher initial dye concentration sample. So, ozone consumption requirement was higher than others. And supplied ozone in the first 20 min ozonation was not enough for the degradation of 200 mg/L and 400 mg/L dye concentration. Thus, the removal efficiencies decreased. These results agreed to the previous study on dye treatment by ozonation [28].

3.2. The C.I Reactive Blue 160 treatment efficiency by ozonation system with operating batch mode

The experiments were performed with ozone gas flow rate of 0.9L/min, no adjusted pH, temperature of 15-16°C, the obtained data indicated that the effluent of the dyeing wastewater containing 200 mg/L C.I. RB 160 was satisfy the strictest discharge standard after 25 min ozonation. Output parameters were as follows: pH = 6.47, Color = 40 Pt-Co, COD = 24.09 mg/L, and dye residual concentration = 0.53 mg/L. The removal efficiency of color, dye, and COD are 96.05%, 99.62%, and 76.44%, respectively (Fig. 10). The effluents can be satisfying to the discharge standard in Vietnam, Korea.

The reaction between ozone and C.I. RB 160 was found as the first-order reaction and the reaction rate coefficients were 0.1402 min^{-1}, 0.8119 min^{-1}, and 0.0605 min^{-1} corresponding to the color, dye, and COD removals (Fig.11). These results were same with the previous research on treatment Acid Red 114 dye by ozonation [28]. The maximum energy consumption was 1.744 kWh/m^3 treated water.

In fact, the dyeing wastewater usually contains sodium carbonate. According to Dieter Sedlak [34], sodium carbonate anhydrous was usually used with concentration 0.02%. From the results in determination of the effects of sodium carbonate on the removal efficiencies in ozonation, the presence of Na_2CO_3 in the dyeing wastewater enhanced the removal efficiencies of color, dye, and COD. Therefore, in case of the real dyeing wastewater that contains Na_2CO_3, the removal efficiencies will be higher and the energy consumption will be lower than the results obtained in this study.

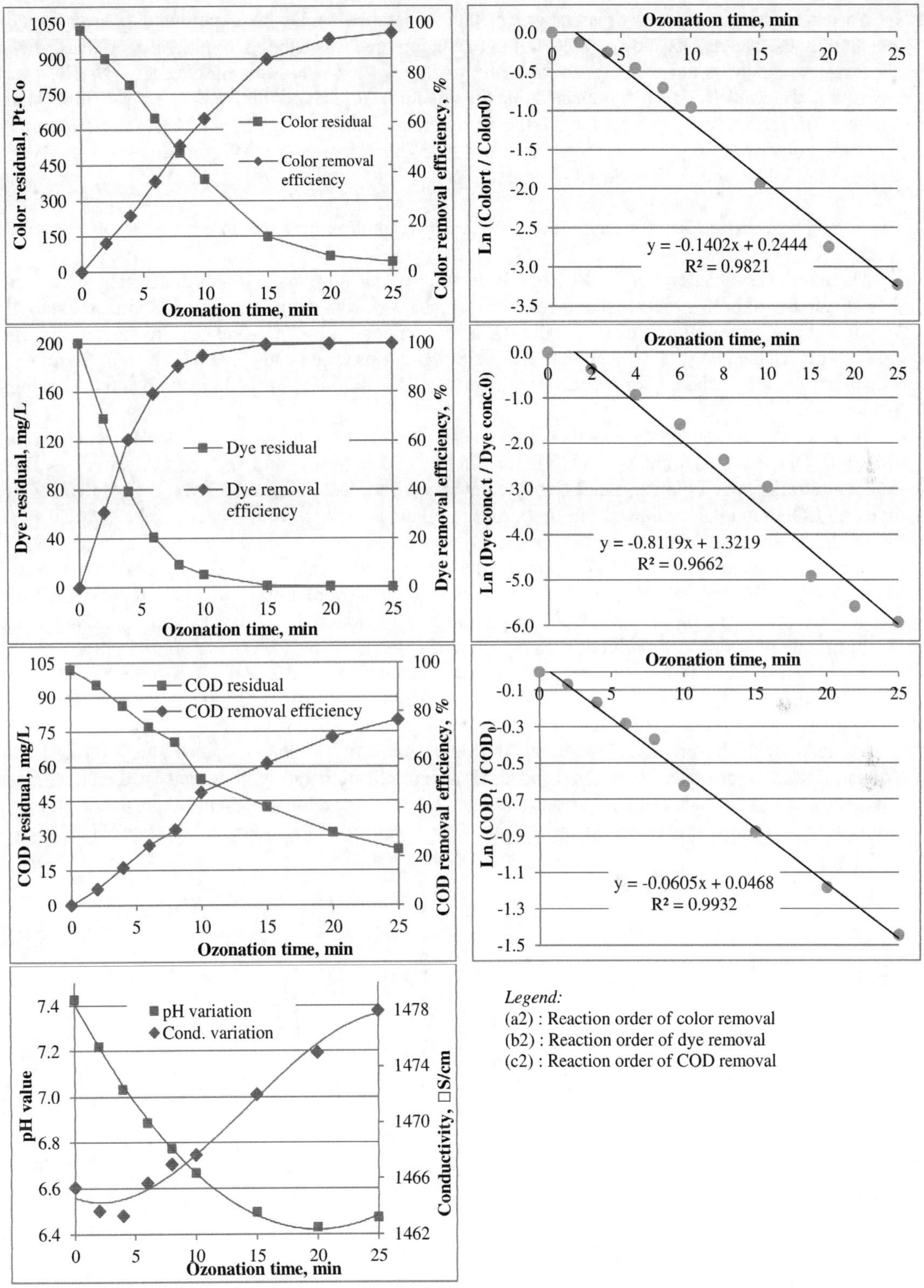

Figure 10. Characteristics of treatment process of sample 200 mg/L without Na_2CO_3

To determine the reaction time for completely mineralize COD, the experiments were extended reaction times. The results showed that 90 min ozonation time was enough to mineralize 100% COD in the wastewater without Na_2CO_3. In this study, the wastewater containing 0.02% Na_2CO_3 was investigated as well. The results indicated that the mineralization was complete with only 70 min. Thus, the ozonation had ability to treat the dyeing wastewater containing C.I Reactive Blue 160 approach to reuse purpose. Energy consumption were 4.8 kWh/m^3 and 3.733 kWh/m^3 for the wastewater containing 0.00% and 0.02% Na_2CO_3, respectively.

3.3. The C.I Reactive Blue 160 treatment efficiency by ozonation system with continuous operating mode

The removal efficiencies of C.I Reactive Blue 160 by ozonation in continuous operating mode were investigated with both concurrent and counter current. The synthetic wastewater containing 200 mg/L initial dye concentration was treated under the operation conditions as ozone gas flow rate 0.9 L/min, without pH adjustment (pH in range of 7.48 – 7.58). HRT was sequentially changed 10 min, 20 min, 25 min and 30 min by adjusting of wastewater flow rate of 1 L/min, 0.5 L/min, 0.4 L/min and 0.33 L/min, respectively.

With counter current flow, ozonation also had effective results in treatment of C.I RB 160. The color, dye and COD residuals after 30 min HRT were 28 Pt-Co, 0.48 mg/L and 13.83 mg/L, respectively. The removal efficiency of color, dye and COD achieved 97.24%, 99.76% and 86.38%, Fig. 11. The color, dye and COD removals followed the first-order rate law as well, with the reaction rates were 0.1224 min^{-1}, 0.2027 min^{-1} and 0.0632 min^{-1}, respectively.

In case of concurrent flow, HRT = 30 min, the removal efficiency of COD, color, and dye reached 85.61%, 97.24%, and 99.76%, respectively. Both of them were lower than counter current flow because in concurrent, there was velocity resonance of wastewater and ozone gas bubbles led to reduce reaction time of ozone and pollutants. Remaining of color, dye and COD after 30 min HRT were 35 Pt-Co, 0.59 mg/L and 14.6 mg/L, respectively. The reaction of C.I RB 160 with ozone in concurrent flow was also the first-order reactions. The reaction rates were 0.1164 min^{-1}, 0.1959 min^{-1} and 0.0618 min^{-1} in the color, dye and COD removals, respectively.

Compare with the previous studies on the degradation of C.I RB 160, this study showed that ozonation was a more effective method and had higher application ability. It not only had high removal efficiency but also could treat a larger amount of wastewater containing higher pollutant concentration with shorter reaction time, easier operation, and without generated secondary solid waste [11-23].

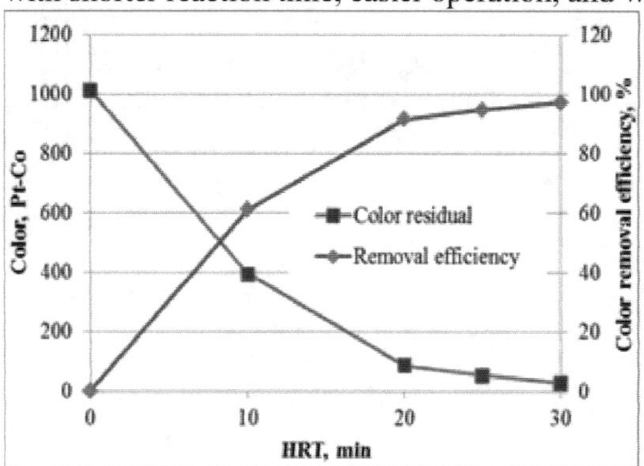

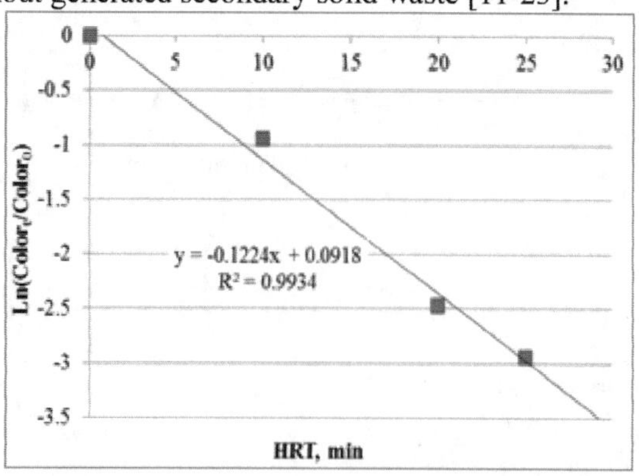

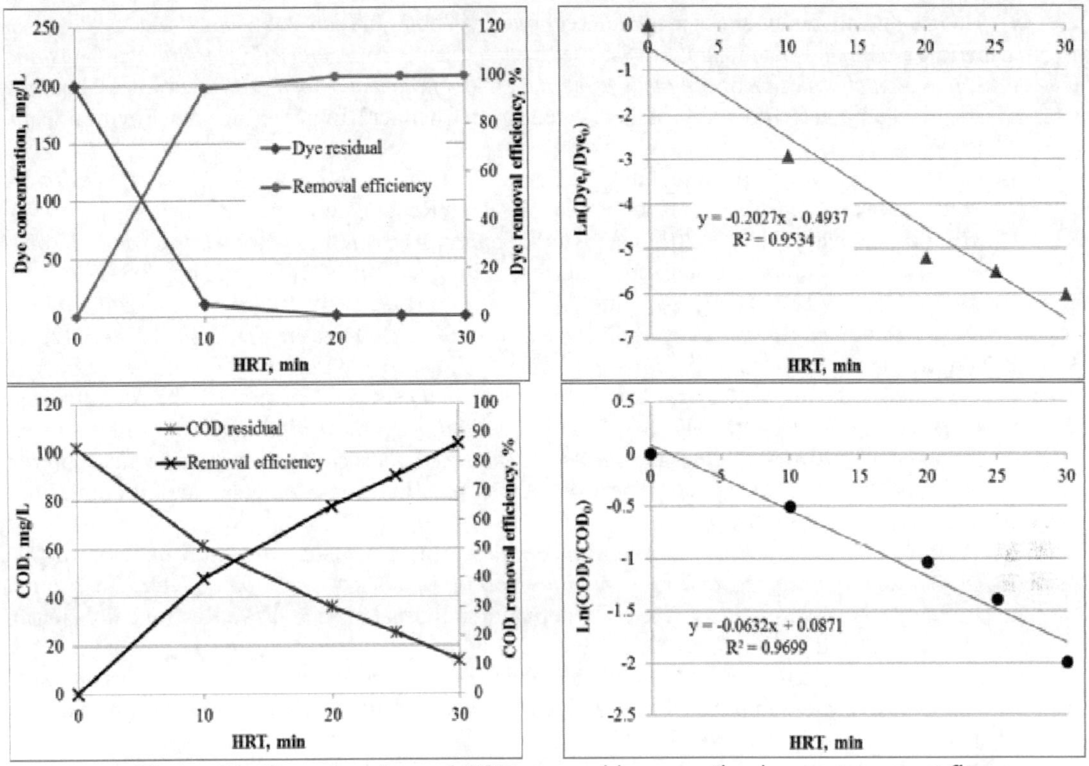

Figure 11. The color, dye and COD removal by ozonation in counter current flow

4. Conclusions

The removal efficiency of color, dye and COD in the dyeing wastewater containing C.I Reactive Blue 160 by ozonation were significantly affected by the ozone gas flow rate, temperature, pH, Na_2CO_3 concentration, and initial dye concentration. Ozonation was performed in the bubble column reactor with both the batch and continuous modes with concurrent and counter current flow. Ozonation had high efficiency in the treatment of dyeing wastewater containing C.I RB 160. The effluent of the dyeing wastewater containing 200 mg/L C.I. RB 160 was satisfy the strictest discharge standard after 25 min ozonation with both operating mode systems. Ozonation time was required longer, approximately 90 min to remove completely COD, mineralized 100%. The presence of Na_2CO_3 in wastewater reduced reaction time and saved energy. The impressive advantages of ozonation were high degradation ability of refractory organic contaminants, short HRT, no sludge production and complete mineralization ability.

References

[1] Nabi Bidhendi, G. R.; Torabian, A.; Ehsani, H.; Razmkhah, N. (2007). *Evaluation of industrial dyeing wastewater treatment with coagulants and polyelectrolyte as a coagulant aid.* Iran. J. Environ. Health. Sci. Eng., **Vol. 4**, No. 1, pp 29 – 36.

[2] Langlais, B.; Reckhow, D.A.; Brink, D.R (1991). *Ozone in water treatment application and engineering*. CRC Press.

[3] Carmen, Z. and Daniela, S. (2012). *Textile organic dyes – characteristics, polluting effects and separation/elimination procedures from industrial effluents – A critical Overview.* 'Gheorghe Asahi' Technical University of Iasi, Faculty of Chemical Engineering and Environmental Protection, InTech Publisher.

[4] Zongping, W.; Miaomiao, X.; Kai, H.; Zizheng, L. (2011). *Texttile dyeing wastewater treatment.*

Huazhong University of Science and Technology China, InTech Publisher.

[5] Zaharia Carmen and Suteu Daniela. *Textile organic dyes – characteristics, polluting effects and separation/elimination procedures from industrial effluents – A critical Overview.* 'Gheorghe Asahi' Technical University of Iasi, Faculty of Chemical Engineering and Environmental Protection.

[6] Bruno Langlais, David A. Reckhow, Deborah R. Brink, 1991. *Ozone in water treatment application and engineering*. Cooperative Research Report, Lewis publishers.

[7] Chedly Tizaoui, Naser Grima (2011). *Kinetics of the ozone oxidation of Reactive Orange 16 azo-dye in aqueous solution.* Chemical Engineering Journal, **Vol. 173**, Issue 2, pp. 463-473.

[8] Chen, T.Y.; Kao, C.M.; Hong, A.; Lin, C.E.; Liang, S.H. (2019). *Application of ozone on the decolorization of reactive dyes — Orange-13 and Blue-19.* Desalination, **249**, 1238-1242.

[9] Kakonged. *Blue most popular colour worldwide, finds Dulux paints survey*. Donna magazine

[10] Joe Hallock 2003. *Coloure Assignment.* Undergraduate thesis, the University of Washington.

[11] Mohammad, H. B.; Maryam, M. (2015). *Co-precipitation synthesis of nano-composites consists of zinc and tin oxides coatings on glass with enhanced photocatalytic activity on degradation of Reactive Blue 160 KE-2B.* Spectrochimica Acta Part A: Molecular and Biomolecular Spectroscopy **137**, 785-789.

[12] Niyaz, A. M.; Khan, A.; Dar, A.A. and Muneer, M. (2014). *Photocatalytic Study of Two Azo Dye Derivatives, Ponceau Bs and Reactive Blue 160 in Aqueous Suspension of TiO2: Adsorption Isotherm and decolorization Kinetics.* International Journal of Innovative Research in Science, Engineering and Technology, **Vol. 3**, Issue 2.

[13] Priti, B.; Dhiraj, S. (2012). *Photodegradation of commercial dye, CI Reactive Blue 160 using ZnO nanopowder: degradation pathway and identification of intermediates by GS/MS.* Separation and Purification Technology, **85**, 112-119

[14] Saraschandra, N.; Thejaswini, T.V.L; Prabhakaran, D.; Sivakumar, A.; Satyanarayana, VSV.; Arun Prasad, A.S (2015). *Enhanced photo-catalytic activity of Sr and Ag co-doped TiO2 nanoparticles for the degradation of Direct Green-6 and Reactive Blue-160 under UV & visible light.* Spectrochimica Acta Part A: Molecular and Biomolecular Spectroscopy, **149**, pp. 571–579.

[15] Mohammad Hossein Babibi, Maryam Mardani (2015). *Co-precipitation synthesis of nano-composites consists of zinc and tin oxides coatings on glass with enhanced photocatalytic activity on degradation of Reactive Blue 160 KE-2B.* Spectrochimica Acta Part A: Molecular and Biomolecular Spectroscopy, **137**, 785-789.

[16] Niyaz A. Mir, A. Khan, A. A. Dar and M. Muneer (2014). *Photocatalytic Study of Two Azo Dye Derivatives, Ponceau Bs and Reactive Blue 160 in Aqueous Suspension of TiO2: Adsorption Isotherm and decolorization Kinetics*. International Journal of Innovative Research in Science, Engineering and Technology, **Vol. 3**, Issue 2.

[17] Priti Bansal, Dhiraj Sud (2012). *Photodegradation of commercial dye, CI Reactive Blue 160 using ZnO nanopowder: degradation pathway and identification of intermediates by GS/MS.* Separation and Purification Technology **85**, 112-119

[18] Saraschandra Naraginti et al. (2015). *Enhanced photo-catalytic activity of Sr and Ag co-doped TiO_2 nanoparticles for the degradation of Direct Green-6 and Reactive Blue-160 under UV & visible light.* Spectrochimica Acta Part A: Molecular and Biomolecular Spectroscopy **149**, pp. 571–579.

[19] Patil, A.V.; Jadhav, J.P., (2013). *Evaluation of phytoremediation potential of Tagetes patula L. for the degradation of textile dye Reactive Blue 160 and assessment of the toxicity ofdegraded etabolites bycytogenotoxicity.* Chemosphere, **92**, pp.225–232.

[20] Chetana, R.; Avinash, K.; Tallika, P. and Shailesh, D. (2016). *Biodegradation of diazo dye, Reactive Blue 160 by Isolate Microbacterium sp. B12 Mutant: Identification of intermediates by LC-MS.* International Journal of Current Microbiology and Applied Sciences, **Vol. 5** Number 3, pp 534-547.

IOP Conf. Series: Earth and Environmental Science **964** (2022) 012030 doi:10.1088/1755-1315/964/1/012030

[21] Selvaraj Barathi, Chinnannan Karthik, Nadanasabapathi S, Indra Arulselvi Padikasan (2020). *Biodegradation of textile dye Reactive Blue 160 by* Bacillus firmus *(Bacillaceae: Bacillales) and non-target toxicity screening of their degraded products*, Toxicology Reports, **vol.7**, pages 16-22.

[22] Selvaraj Barathi, K.N. Aruljothi, Chinnannan Karthik, Indra Arulselvi Padikasan (2020). *Optimization for enhanced ecofriendly decolorization and detoxification of Reactive Blue160 textile dye by Bacillus subtilis*. Biotechnology Reports, **28**, e00522.

[23] Kshama, H. B.; Kunal, J.; Sananda, C.; Nikhil, S.B., (2014). *Co-metabolic degradation of diazo dye—Reactive blue 160 byenriched mixed cultures BDN.* Journal of Hazardous Materials, **279**, pp. 85–95.

[24] Idil Arslan Alaton, Isil Akmehmet Balcioglub, Detlef W. Bahnemann (2002). *Advanced oxidation of a reactive dyebath effluent:comparison of O_3; H_2O_2/UV-C and TiO_2/UV-A processes.* Water Research **36**, pp. 1143-1154.

[25] Burcu Sancar, Onur Balci (2013). *Decolorization of different reactive dye wastewaters by O_3 and O_3/ultrasound alternatives depending on different working parameters.* Textile research journal, **83(6)**, 574-590.

[26] WHO-IARC (2010). *IARC monographs on the evaluation of carcinogenic risks to humans.* **Volume 99**: Some aromatic amines, Organic dyes, and related exposures, Lyon, France.

[27] *Standard Methods for the Examination of Water and Wastewater.* 20th Edition 1999, Method 5220C

[28] Huynh Thi Ngoc Han (2016), *Characteristics of dyeing wastewater treatment by ozonation and electrocoagulation combined system*, Doctoral dissertation, Faculty of Chemical Engineering, Dankook University, South Korea.[30]

[29] Erdal Kusvuran, Osman Gulnaz, Ali Samil, Mehmet Erbil (2010). *Detection of double bond-ozone stoichiometry by an iodimetric method during ozonation processes.* Journal of Hazardous Materials **175**, pp. 410-416.

[30] Erdal Kusvuran, Osman Gulnaz, Ali Samil, Özlem Yildirim (2011). *Decolorization of malachite green, decolorization kinetics and stoichiometry of ozone-malachite green and removal of antibacterial activity with ozonation processes.* Journal of Hazardous Materials **186**, pp. 133-143.

[31] William Kerfoot (2007). *Patent of Superoxidant poiser for groundwater and soil treatment with in-situ oxidation-reduction and acidity-basicity adjustment.* Publication number EP1773723 A2, Application number EP20050793889, USA

[32] Frederick W. Pontius (1990). *Water quality and treatment: A handbook of community water supplies.* fourth edition, American Water Works Association, McGraw-Hill Inc., USA.

[33] Nelson Saksono, Yuliusman, Setijo Bismo, Roekmijati W. Soemantojo, Azwar Manaf (2009). *Effects of pH on calcium carbonate precipitation under magnetic field.* Makara, Teknologi, **Vol. 13**, No. 2, 79-85

[34] Dieter Sedlak (2012). *Chemical Guidance.* Afirm Group.

Potential of Earthworm (Perionyx Excavatus) in food treatment in Ho Chi Minh City

Vo Thanh Hang[1,3,*],Trinh Mai Hoang Anh[2,3], Pham Kim Khanh[2,3]

[1] Faculty of Environment and Natural Resources, Ho Chi Minh City University of Technology (HCMUT), 268 Ly Throng Kiet Street, District 10, Ho Chi Minh City, Vietnam
[2] Faculty of Chemical Engineering, Ho Chi Minh City University of Technology (HCMUT), 268 Ly Thuong Kiet Street, District 10, Ho Chi Minh City, Vietnam
[3] Vietnam National University Ho Chi Minh City, Linh Trung Ward, Thu Duc District, Ho Chi Minh City, Vietnam

*Email: hang_vothanh2003@hcmut.edu.vn

Abstract. In Ho Chi Minh City, Vietnam, there are about 9 thousand tons of domestic waste every day and 73% of them are treated by using landfill sites [1]. With the population growth rate in recent years, the generated waste volume will then increase in the future and is estimated to reach a total of around 13 thousand tons that could cost Ho Chi Minh City about 260 thousand dollars in treating this waste in 2025 [2]. Although this treatment method is not without advantages, the downsides will be far more significant. Moreover, in recent years, the cultivation of Perionyx Excavatus, a species that had been reported to have the ability to treat organic waste in some foreign articles, has become increasingly popular in Vietnam. So, this research was aimed to evaluate the potentiality of Earthworm (Perionyx Excavatus) for food waste management of households in Lab-scale in Ho Chi Minh City, Vietnam. According to this study, the average outdoor humidity and temperature of Ho Chi Minh City are suitable the optimal conditions to raise Earthworm. In addition, our team recommends that to achieve the highest biomass ratio of Earthworm (Perionyx Excavatus), the proportion of food waste and cow dung should be 4:6.

Keywords: Earthworm (Perionyx Excavatus), food waste management, Ho Chi Minh City.

1. Introduction

Vietnam is one of the fastest-growing economies in the world. According to Vietnam's economic statistics in the last ten years (2009 - 2019), gross domestic product (GDP) reached about 266 billion USD, and per capita reached 2,800 USD from a growth rate of 5.40% in 2009 to 7.02% in 2019, the highest in the past ten years [3]. In parallel, Vietnam's population growth rate is also increasing rapidly, especially in Ho Chi Minh City which is estimated to reach a population of 12 million by 2025, including peri-urban areas [4]. As a result of the rapid population growth, there is also a steadily increasing volume of domestic solid waste, especially food waste. According to the National State of Environment Report, in 2011, the total volume of domestic waste generated nationwide was about 44,400 tons/day, which increased to 64,658 tons/day in 2019 (approximately higher 46% compared to 2010). In addition, the volume of domestic waste increased significantly in localities with high urbanization, industrialization, and tourism such as Ho Chi Minh City (9,400 tons/day) and 50-65% of domestic waste are biodegradable organic components (food waste, plant and animal carcasses) [3].

In recent years, the habitations of human beings generate a huge amount of domestic waste, especially food waste, which affects the quality of nature and our living environment. Uncared dumping of leaf litter in public places and the premises of educational institutions might lead to the decomposition of solid wastes particularly during the rainy season resulting in air, water, and soil pollution. It also causes social, ecological, aesthetic, economic problems, human health, and quality of life [5]. Hence, these wastes must be treated carefully. In the face of waste treatment, Ho Chi Minh City has to spend 1,100 - 1,200 billion VND for landfilling (69% by volume), the rest being composted (20%) and burned without energy recovery (11%) [4]. However, waste disposal still faces challenges and a lack of operational experience.

All these problems require a more efficient alternative to reduce both processing costs and the area of the landfill in Vietnam. Especially in terms of food waste, old traditional techniques of composting and organic farming are encouraged because of its environmentally friendly. During composting, the waste organic materials are broken down into different organic products which can be added to the soil with a much beneficial impact on crop growth. The benefits of compost in improving plant growth and yield due to its effects on nutrient availability and improving physic-chemical and physical characteristics of the soil are well known. However, compost is also known to play a role in suppressing various fungal and bacterial plant diseased via the activities of constituent biological, physical, or chemical agents [6].

It is recognized that Earthworms (Perionyx Excavatus) are a potential species based on their high protein and high levels of enzymes [7]. Therefore, many studies have been conducted on the speed of the process of composting (Vermicomposting) and enhance the nutritional quality of compost of Earthworm (Perionyx Excavatus), for example, research about the utilization of Earthworm (Perionyx Excavatus) in organic waste management in India [8] or Evaluation of Earthworm (Perionyx Excavatus) Growth Raised in Different Cultures [9]. In Vietnam, Earthworms (Perionyx Excavatus) are popular species for farming. Although temperature and humidity conditions, especially in Ho Chi Minh City are very suitable for the development of Earthworms, there are very few studies on using Earthworms for waste treatment.

This study was focused on evaluating the potentiality of Earthworm (Perionyx Excavatus) for food waste management of households in Lab-scale in Ho Chi Minh City, Vietnam. Through this results, it was proposed an alternative method in the treatment of organic waste toward smart city in Ho Chi Minh city in 2050 .

2. Research methods

2.1. Method of observation via experiments

The research has implemented 2 experimental models which were implemented from 20th December 2019 to 15th March 2020.

2.1.1. Model 1: Earthworm (Perionyx Excavatus) raising model (100% feed of cow dung, from Earthworm breeding farms) serves for maintaining the breeding stock provided for Model 2.

Designed with dimensions Lx H x D: 100cmx100cmx50cm, respectively. Model 1 was designed outdoors (Figure 2). To increase efficiency, the model of storing Earthworm (Perionyx Excavatus) was designed to have a 1m^2 area and a height of 50 centimeters including 4 layers: clean sand, plastic mesh, biomass, and food. Each layer has its function in raising the Earthworm (Perionyx Excavatus), as below (Figure 1, 2).

- Layer 1: 3 centimeter-thickness clean sand. This layer is the buffer between the ground and the plastic mesh layer which allows maintaining the humidity at an acceptable level. When it rains, water will pass through this layer, seep into the ground helping the model not to be flooded and Earthworms not to suffocate. Moreover, the clean sand layer could prevent other insects in the soil from entering the model.
- Layer 2: 3 millimeter-thickness plastic mesh layer includes 2 plastic meshes sized 32 holes/cm^2. The hole meshes are small enough to help the Earthworm (Perionyx Excavatus) to stay away from the risk of being floated through the mesh and going out of the environment.

- Layer 3: 6 centimeter-thickness biomass layer. This layer is the 'house' of the Earthworm (Perionyx Excavatus) that should keep at an optimal temperature for the Earthworm life (at 20°C-28°C). Besides, the soil must be porous, clean, and nutritious so that the Earthworm (Perionyx Excavatus) can develop well.
- Layer 4: Food layer. This layer is the mixture of cow dung and food after incubating for 2-3 days.

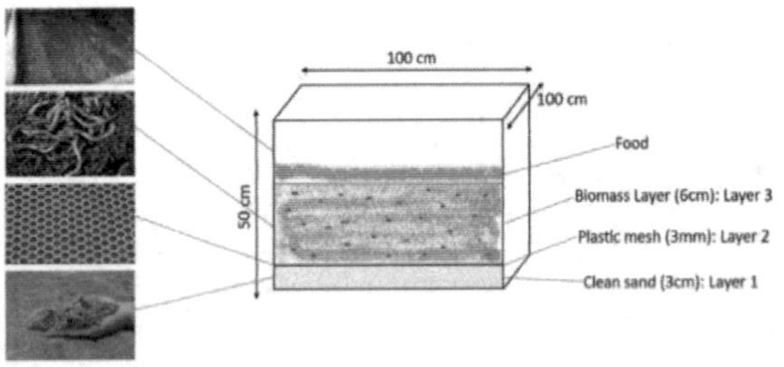

Figure 1. Design of Model 1.

Figure 2. The reality storing model in Cu Chi district.

2.1.2. Model 2: Earthworm (Perionyx Excavatus) raising model by organic waste.

Model 2 for evaluating the efficiency of Earthworm (Perionyx Excavatus) by feeding organic waste with a variety of mixing organic feed rates. This model was designed with dimensions Lx H x D: 40cm x 20cm x 20cm by 5 mica pieces (0.5cm in thickness). The mica material was used because of the waterproof, the transfer limitation, and the transparency, which allows observation of biomass easily (as Figure 3, 4). There were 4 layers like the storing Perionyx Excavatus model:

- Layer 1: 3 centimeter-thickness clean sand.
- Layer 2: 3 millimeter-thickness plastic mesh layer.
- Layer 3: 7 centimeter-thickness biomass layer.
- Layer 4: Food layer.

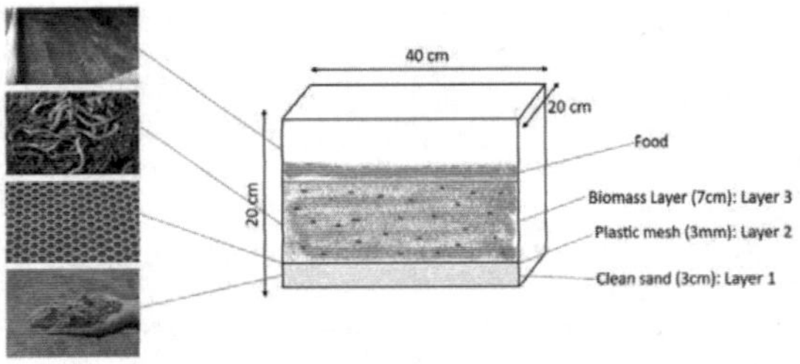

Figure 3. Design of Experimental Model 2.

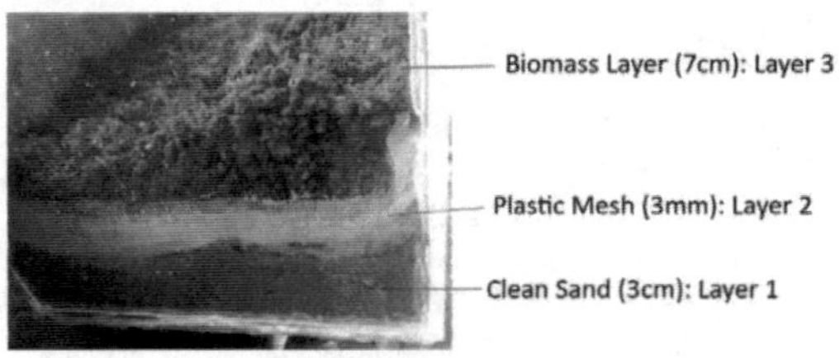

Figure 4. The reality experimental model.

2.2. Method records

The evaluation parameters include temperature (surroundings and in the experimental model), humidity, and biomass were analyzed in the laboratory to assess the factors affecting the culture process and the biomass speed of Earthworms. The experiment was recorded every two weeks because the feed of the Earthworm (Perionyx Excavatus) also changed after two weeks. Moreover, using rule, scale, and handheld digital humidity & temperature meter to determine. Moreover, using Microsoft Excel to take notes during the experimental period.

2.3. Method calculation and conversion temperature and humidity

Temperature and humidity are two crucial factors in raising Earthworms [9, 10]. According to the previous researches, the optimum temperature for Earthworm (Perionyx Excavatus) is between 25°C and 28°C [10], and the temperature difference between outdoors and in the experiment is 10°C – 20°C [9]. In the experiment, it was not possible to directly measure the temperature at the raising model, so it was calculated by the outdoor temperature + 1°C.

2.4. Method of calculation Perionyx Excavatus' biomass

The mass ratio of Earthworm (Perionyx Excavatus) over time was determined by the following formula (according to Mazanraeva):

$$\text{Growth rate (kg/worm/day)} = \frac{W_2 - W_1}{t_2 - t_1} \quad [1]$$

W_1: The initial weight of worms (kilograms)
W_2: The final weight of worms (kilograms)
t_1: Age of the worms at the start of the experiment (in days)
t_2: Age of the worms at the end of the experiment (in days)

2.5. Statistical methods and data analysis via Microsoft Excel

The experimental data were analyzed and evaluated by Microsoft Excel Office 2019 to construct the humidity and temperature graph, the growth rate of Perionyx Excavatus, and the percentage of biomass increased. In addition, this software was used to calculate the mass of Earthworms in Model 1 and Model 2.

3. Results and discussion

3.1. Domestic Solid Waste Management in Ho Chi Minh City

The socio-economic development and the rapid population growth cause pressure on the environment when the amount of solid waste and especially domestic solid waste is arising more and more. Ho Chi Minh City is the core of Vietnam's largest urban area, which is expected to reach 12 million people by 2025, including peri-urban regions. Ho Chi Minh City is facing a steady increase in domestic waste volume, with a generation rate of 10% – 15% which is higher than the rest of Vietnam (8% – 10%) [12]. It is estimated that Ho Chi Minh City has a high rate of domestic waste generation (over 1.0kg/person/day) and the total volume of domestic waste is at about 7,200 –7,800 tons/day (excluding waste sludge) [4].

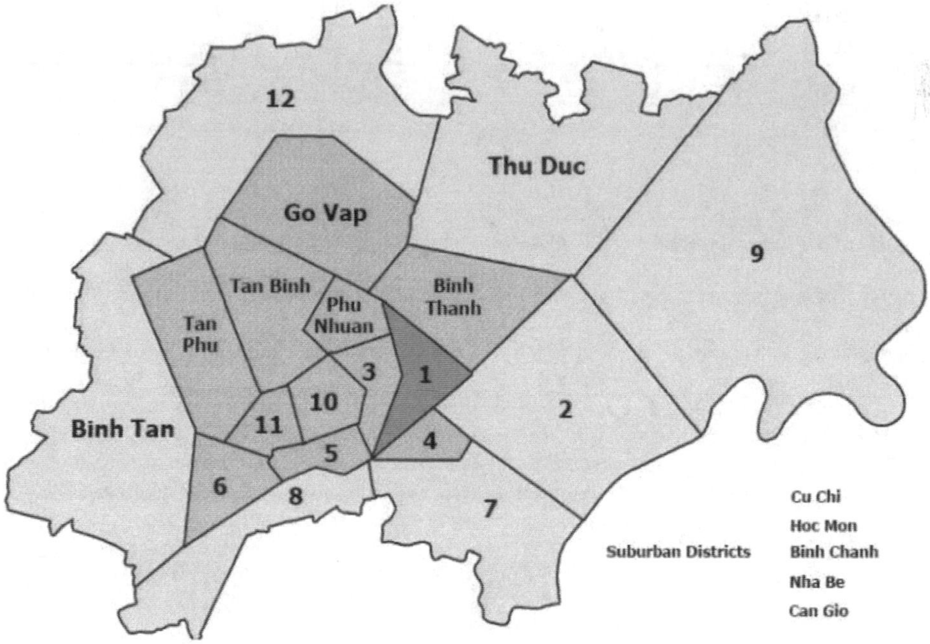

Figure 5. Ho Chi Minh City Map.

Source: [13]

IOP Conf. Series: Earth and Environmental Science **964** (2022) 012031 doi:10.1088/1755-1315/964/1/012031

Sources of domestic waste arise from households, commercial and service areas, offices, public areas, and the daily activities of production facilities. In addition, the main components of domestic waste are organic (food waste, paper, cloth, etc.), inorganic (plastic, rubber, metal, etc.), and other wastes such as e-waste, batteries, waste oil (shown in Table 1).

Table 1. Composition of Domestic Solid Waste from households, schools, markets, sanitary landfills, and compost processing plants in Ho Chi Minh City (%by wet weight)

Waste component	Households			Schools		Markets		Sanitary landfills		Compost plants
	2009	2015	2017	2009	2015	2009	2015	2012	2014	2015
Food waste	74.3	64.8	59.2	28.7	25.5	86.8	87.8	68.9	67.9	53.2
Wood, straw	2.8	0.9	2.4	6.9	-	3,6	1.4	0.7	0.3	1.0
Paper	6.2	5.1	6.4	17.6	35.0	2.5	1.9	3.0	2.5	5.7
Plastics	5.5	11.5	13.9	35.7	36.4	4.7	7.7	16.0	16.4	13.7
Diapers	1.8	8.1	0.6	-	-	-	-	2.3	0.6	10.7
Textiles	1.0	3.2	4.0	1.1	1.0	0.4	-	5.0	7.2	10.7
Leather	0.2	-	0.6	0.1	-	-	-	-	-	-
Rubber	0.9	0.9	2.0	1.4	-	0.4	-	0.7	0.7	0.7
Glass	1.3	1.4	2.6	0.5	1.2	0.2	-	1.2	0.2	1.7
Metal	1.0	0.8	5.5	2.8	-	0.3	0.1	1.6	3.6	0.3
Crockery	0.8	0.5	2.8	0.6	-	0.1	-	-	-	2.4
Ash and dirt	3.2	2.8	-	4.0	-	1.0	1.2	-	-	-
Cinders	0.4	-	-	-	-	-	-	-	-	-
Seashell	0.8	-	-	-	-	0.2	-	0.8	0.6	-
Hazardous waste	0.002	-	-	0.1	-	0.1	-	-	-	0.1

Source: [4]

It is shown that food waste is the main component of domestic waste, which takes up to more than 50% of the wet weight of the domestic waste volume, followed by recycling components such as plastic, paper, and metal. According to statistics, biodegradable organic components (food waste, plant, and animal carcasses) were found in nearly 9,000 tons of domestic waste collected every day in Ho Chi Minh City which accounts for about 50 - 65% of the volume (wet) equivalent to the volume of 4,500 - 5,850 tons of organic matter generated per day [4].

To handle the amount of domestic waste collected, Ho Chi Minh City has to spend 1,100 - 1,200 billion VND for landfilling (69% by volume), the rest being composted (20%) and burned without energy recovery (11%) [4]. However, it has not yet achieved high efficiency and especially landfills are a problem causing environmental pollution in recent years. For instance, Da Phuoc Landfill is reported in air pollution and recommended that an alternative sustainable technology is needed to reduce the rate of landfilling (Figure 6) [14].

Figure 6. Da Phuoc Landfill.

Source: [16]

For another example, Tay Bac Solid Waste Treatment Complex (Figure 7) directly affects groundwater, air, and surface water sources, which is the cause of serious environmental pollution for many years [15].

Figure 7. Tay Bac Solid Waste Treatment Complex.

Source: [17]

3.2. Processing of Experiments

3.2.1. Model 1: Earthworm raising model at the laboratory

The purpose of the experiment in Model 1 is to create a source of Perionyx Excavatus for their experiments and check the ability of the worms to grow when it is completely fed with cow dung. The preparation of the mixture and the incubation process took place within 2-3 days. After incubating, the mixture was spread into lines with 2.5-3.5 kilograms of food and feeding Earthworm (Perionyx Excavatus) 2-3 times a week. The feeding process and the growth of Earthworms are shown in Figure 8.

Figure 8. The feeding process of Model 1.

3.2.2. Model 2: Earthworm raising model by organic waste

The mixture of cow dung and organic waste was incubated for 5 -7 days and then spread into lines in which 0.3-0.5 kilograms each line. The Earthworms were fed 2-3 times per week like model 1. Firstly, it fed Earthworm (Perionyx Excavatus) with a ratio of 9.5:0.5 (cow dung: food waste) in 2 weeks, and then up the rate to 2.b (9:1) feed in the next 2 weeks. Next, the ratio components like 2.c (6:4) were an experiment in the 5th and 6th weeks. Then, Perionyx Excavatus was fed with a ratio of 8:2 in the last 2 weeks. Finally, it is going on to feed 100% of food waste in this model.

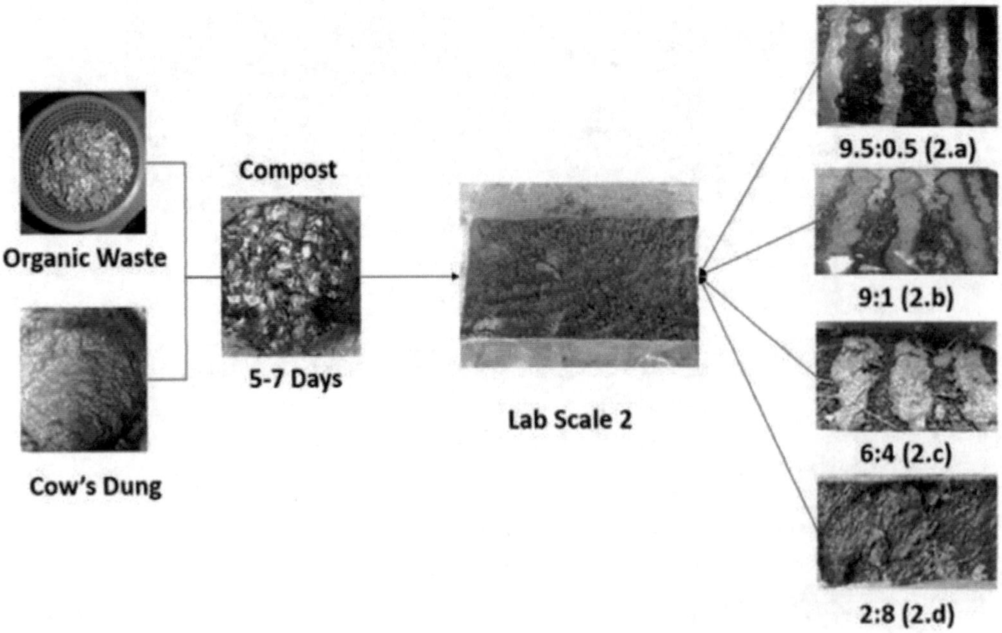

Figure 9. The feeding process of Model 2.

3.3. Results

3.3.1. Preliminary Data Collection and analysis

3.3.1.1. The moisture levels

The interplay of the moisture level and temperature dramatically affects both the growth rate and biomass of Earthworm (Perionyx Excavatus). A suitable level of moisture for Perionyx Excavatus is about 60-70% [18] and the average outdoor humidity of Ho Chi Minh City in the first three months of the year ranges from 71% to 73.8% (Figure 10) [12]. Therefore, during the research, models always are ensured the appropriate moisture level.

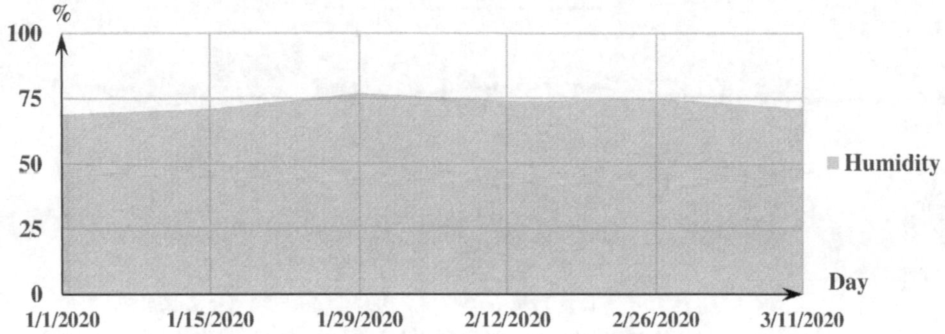

Figure 10. Humidity in Ho Chi Minh City in 3 months (%).

3.3.1.2. The temperature

One of the main factors that affect both the growth of the rate of Earthworm (Perionyx Excavatus) as well biomass and vermicomposting production is temperature, therefore, the too high or too low temperature can lead to reduced worm production. According to statistics, Ho Chi Minh City has two distinct seasons: the rainy season from May to November and the dry season from December to April next year. The average temperature is 27.55°C with high humidity and rainfall. The month with the highest average temperature is in April (28.8°C) and the month with the lowest average temperature is between December and January (25.7°C) (shown in Figure 11). In addition, the optimal temperature of the background for Earthworms is in the range of 25°C to 28°C [10]. Therefore, it can be seen that the temperature conditions of Ho Chi Minh City are suitable for the development of Earthworms.

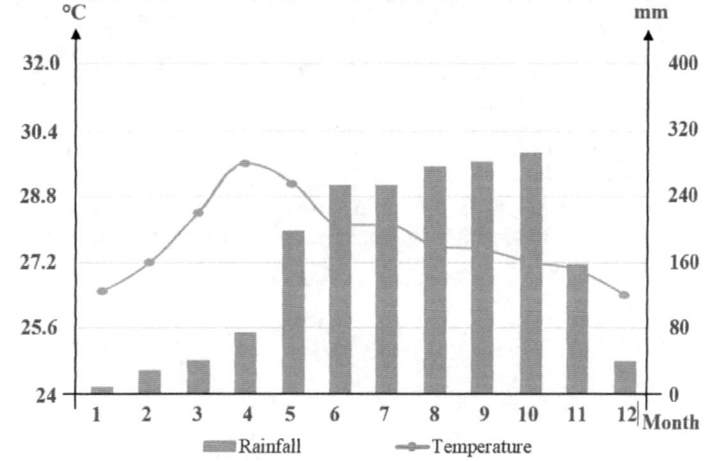

Figure 11. Ho Chi Minh City average temperature and rainfall in 12 months.

The below chart indicates the temperature variation of the experimental model, which is conducted in the period from January to February. In the first month, the temperature is quite stable, fluctuating between 22°C and 33.5°C. Although the environment and substrate temperature in the first month are not optimal for the development of Earthworms, the growth rate of Earthworms remains stable. In the second and third months, the temperature fluctuates from 25°C to 35°C, higher than the first month. However, the application of appropriate methods helped a stable growth rate of Earthworm (Perionyx Excavatus) as Figure 13. The data is from [12].

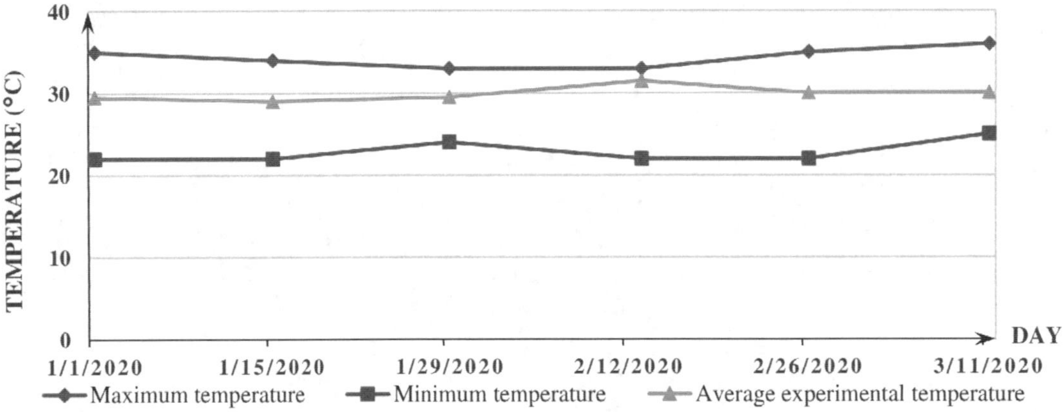

Figure 12. The change in ambient temperature during the experiments.

3.3.1.3. The growth rates

To evaluate the growth of Earthworm (Perionyx Excavatus) as well biomass and vermicomposting production, the mass of model 1 and 2 with different feed rate had investigated and shown in table 2.

Table 2. The mass of Earthworm in Model 1 and Model 2.

	Model 1	**Model 2**			
		2.a	**2.b**	**2.c**	**2.d**
Input (kg)	20	1.7	3.1	5.8	11.7
Output (kg)	35.5	3.1	5.8	11.7	11.7

According to Table 2, the proportion of biomass of Model 2 is higher than that of Model 1. Especially, in terms of increasing the ratio of food waste in the feed instead of cow manure like 2.c, the growth rates of Earthworm (Perionyx Excavatus) in 2.c rose nearly 1.56 times than its Model 1 and 4.42 and 2.21 than the growth rate of Model 2.b and 2.c, respectively. The growth rate of Perionyx Excavatus in Model 2.d is zero (as Figure 13).

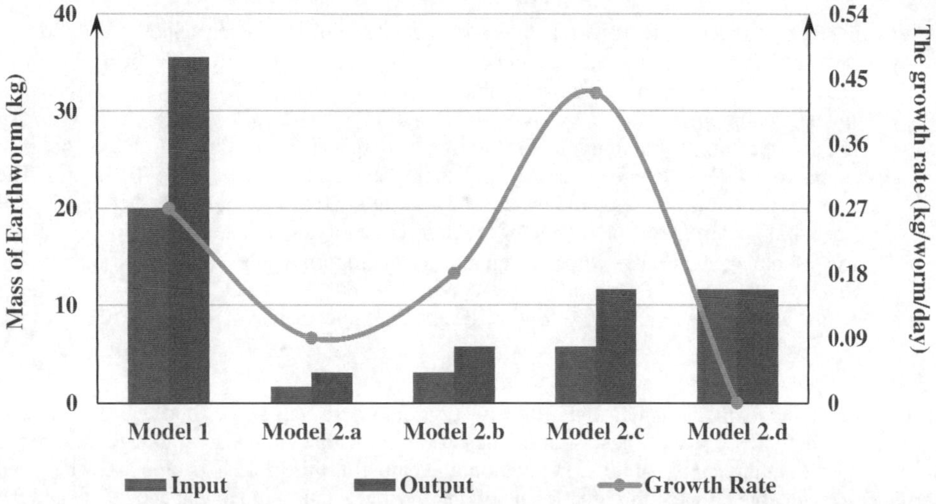

Figure 13. The growth rate of Perionyx Excavatus (kg/worm/day).

In addition, the percentage of biomass in Model 2.c is the highest compared to the remaining experiments (about 50.43%), while that of Model 2.b and 2.a is 46.56% and 45.16%, respectively. The percentage in Model 1 is smaller, with 43.66%. The biomass of Model 2.d for 2 weeks has no change (as Figure 14).

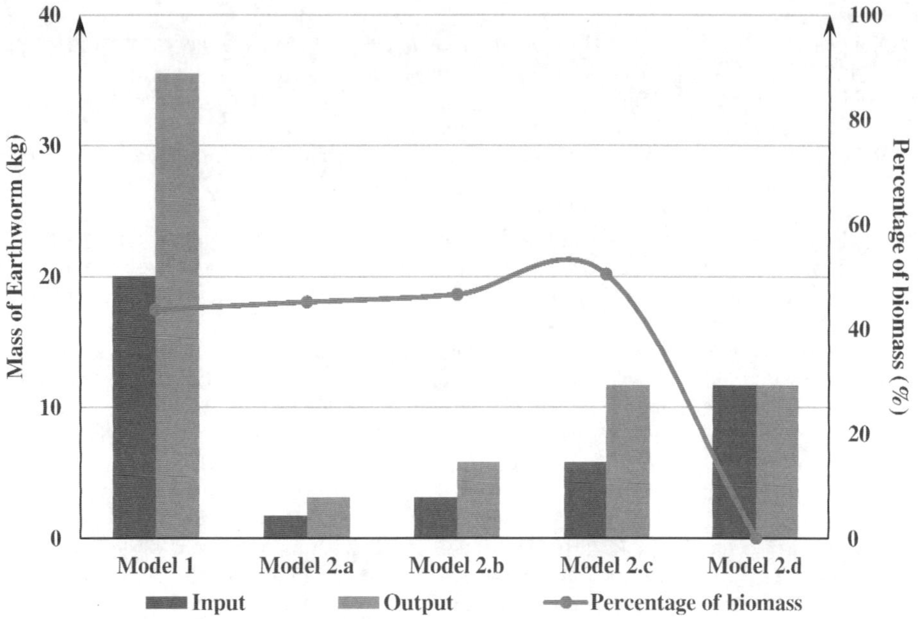

Figure 14. The percentage of biomass increased (%).

3.3.2. Preliminary Findings

After 4 months of project implementation, the experimental results have been achieved as follows:

- Due to experiments placed outdoors, the temperature of the Earthworm (Perionyx Excavatus) raising model (estimated) fluctuated in the range 29°C to 31.5°C, higher than optimal conditions.
- The average outdoor humidity of Ho Chi Minh City ranges from 71% to 73.8%, higher than optimal conditions. The feed directly affects the development of Perionyx Excavatus. The highest growth rate and biomass increase of Perionyx Excavatus are in the feed like Model 2.c (0.42 kg/worm/day) with a ratio of 6:4 (cow dung: organic waste).
- The percentage of biomass witnessed an increasing proportion when rising food waste in the feed to 4:6 (50.43%) with food waste and cow dung, respectively, before declining dramatically in the ratio of 80% food waste and 20% cow manure.

4. Conclusions

This research shows the potential of Earthworm (Perionyx Excavatus) for food waste treatment, which was help not only to reduce the cost of municipal waste treatment but also minimize the amount of waste needed to dumpling at landfills. In addition, the output of this process is biomass, which can be fertilizer to vegetable farms in this area for supplying fresh vegetables in Ho Chi Minh City and the Southern of Vietnam.

Acknowledgments

This research is funded by the Undergraduate Research Initiative (URI) program of Build-it project and Office for International Study Programs (OISP), Ho Chi Minh City University of Technology (HCMUT) for running experiments. Moreover, we would like to thank Ho Chi Minh City University of Technology (HCMUT), VNU-HCM for the support of time and facilities for this study.

References

[1] Nguyen H. 2019. Each day Ho Chi Minh City generates 13,000 tons of waste [Internet]. BAOMOI.COM. Available from: https://baomoi.com/

[2] Tran H. 2019. Ho Chi Minh city may need more than 2000 billion VND per year for waste disposal [Internet]. Zing.vn. Available from: https://news.zing.vn/

[3] P. Schneider, L. Anh, J. Wagner, J. Reichenbach and A. Hebner. 2017. Solid Waste Management in Ho Chi Minh City, Vietnam: Moving towards a Circular Economy?. *Sustainability*. vol. 9. no. 2. p. 286. Available from: https://doi.org/10.3390/su9020286.

[4] Ministry of Natural Resources and Environment. 2019. *Báo cái hiện trạng môi trường quốc gia 2019 – Chuyên đề: Quản lý chất thải rắn sinh hoạt.* Available from: https://moitruong.com.vn/Upload/2020/Ban%20Giay%20phep%20in_Bao%20cao%20hien%20trang%20moi%20truong%20(1).pdf

[5] R McDougall F, R White P, Franke M and Hindle P. 2001. Integrated Waste Management. Integrated Solid Waste Management: A Life-Cycle Inventory, 2nd Edition [Internet]. p. 15-31. Available from: https://doi.org/10.1002/9780470999677.ch2

[6] C G St Martin C and Ramsubhag A. 2015. Potential of Compost for Suppressing Plant Diseases. *Sustainable Crop Disease Management Using Natural Products*. CABI; p. 345-388.

[7] Trang D. 2020. Classification of Earthworm in Viet Nam the world [Internet]. SFARM. Available from: https://sfarm.vn/

[8] Alagesaran P and Dheeba R. 2010. Utilization of Earthworms in organic waste management. *Proceedings of International Forestry and Environment Symposium*;**15**(0).

[9] Binh D and Linh N. Evaluation of Earthworm (Perionyx Excavatus) Growth Raised in Different Cultures.

[10] Bay N. 2016. Technical guide to Earthworm raising [Internet]. SFARM. Available from: http://sfarm.vn

[11] Birundha M, Paul J and Mariappan P. 2013. Growth and reproduction of Perionyx Excavatus in

different organic waste. International Journal of Current Microbiology and Applied Science;2:28-35.

[12] Climate & Weather Averages in Ho Chi Minh, Vietnam [Internet]. AccuWeather. Available from: https://www.accuweather.com/vi/vn/ho-chi-minh-city/353981/february-weather/353981?year=2020&fbclid=IwAR0y5m1klKvVuAE0sDTpevx7LaSBaJrwXlr7rw6a_ihw3oWn2KDzBLUka30.

[13] Ho Chi Minh City Map [Internet]. ITOURVN. Available from: https://mapscatalogonline.blogspot.com/2018/06/ho-chi-minh-city-map.html

[14] Khai Q and Thao L. 2019. Vô khu xử lý rác Đa Phước, sợ cả mùi hôi lẫn hóa chất khử mùi [Internet]. TUOI TRE NEWS. Available from: https://tuoitre.vn/vo-khu-xu-ly-rac-da-phuoc-so-ca-mui-hoi-lan-hoa-chat-khu-mui-20190807122443185.htm

[15] Nguyen N. 2021. Hai công ty xử lý rác thải ở TP HCM vi phạm kéo dài [Internet]. THE LAO DONG (LABOUR) NEWSPAPER Available from: https://nld.com.vn/thoi-su/hai-cong-ty-xu-ly-rac-thai-o-tp-hcm-vi-pham-keo-dai-20210117075819457.htm

[16] Huy Q. 2020. Gần 10.000 tấn rác của TP.HCM đi về đâu mỗi ngày? [Internet]. Electronic Newspaper of Construction. Available form: https://baoxaydung.com.vn/gan-10000-tan-rac-cua-tphcm-di-ve-dau-moi-ngay-284798.html

[17] Hai M. 2021. Đặt hàng xử lý thêm rác thải [Internet]. Sai Gon Giai Phong Newspaper. Available from: https://www.sggp.org.vn/dat-hang-xu-ly-them-rac-thai-750924.html

[18] Vuong P and Hien B. 2017. Raising Earthworm from cattle, poultry manure and agricultural waste.

Synthesis of hydrated ferric oxide on cation exchange resin for phosphate and hardness removal in water

Le Ba Tran[1,2,3,4], Trung Thanh Nguyen[1,2,*], Tri Thich Le[1,2], Quynh Anh Nguyen Thi[1,2], Phuoc Toan Phan[2,3,4], Surapol Padungthon[5] and Nhat Huy Nguyen[2,4,*]

[1] Nanomaterial Laboratory, An Giang University, 18 Ung Van Khiem St., Dong Xuyen Dist., Long Xuyen City, An Giang Province, Vietnam
[2] Vietnam National University Ho Chi Minh City, Linh Trung Ward, Thu Duc City, Ho Chi Minh City, Vietnam
[3] Faculty of Engineering - Technology - Environment, An Giang University, 18 Ung Van Khiem St., Dong Xuyen Dist., Long Xuyen City, An Giang Province, Vietnam
[4] Faculty of Environment and Natural Resources, Ho Chi Minh City University of Technology (HCMUT), 268 Ly Thuong Kiet St., Dist. 10, Ho Chi Minh City, Vietnam, Vietnam
[5] Department of Environmental Engineering, Khon Kaen University, 123 Moo 16 Mittraphap Rd., Nai-Muang, Muang District, Khon Kaen 40002, Thailand

*Email: ntthanh@agu.edu.vn, nnhuy@hcmut.edu.vn

Abstract**.** In this study, a potential adsorbent was synthesized from iron salt and cation exchange resin (FeOOH@CR) and applied for phosphate adsorption in batch experiments. The characteristics of FeOOH@CR materials before and after phosphate adsorption were determined by FTIR, XRD, and SEM. The factors affecting the adsorption process such as reaction time, solution pH, material dosage, concentration, temperature, and competing ions were tested. Kinetic, thermodynamic and isothermal models of the adsorption process were applied to study the nature of the adsorption process. The properties of phosphate adsorption, effect of competitive ions and material reusability were also examined. Results showed that the adsorption time reached equilibrium after 48 h and the suitable adsorption condition was found at solution pH of 6.5, material dosage of 5 g/L. In addition, the durability of the material after 5 times of regeneration was investigated with the remained adsorption ability of about 55% as compared to the original one.

Keywords: phosphate adsorption; ion exchange resins; hydrated iron oxide, hardness

1. Introduction

The anthropogenic nutrients in the aquatic environment should be reduced to protect the water sources for human uses and reduce eutrophication (e.g., algae growth) in the water environment, which decreases the oxygen content in the water and damages aquatic life conditions[1, 2]. The concentration of phosphorus have been significantly increased in water due to excess fertilizer use and also from the discharge of municipal wastewater. Therefore, controlling phosphorus is necessary for the country's ecological survival [3, 4]. Phosphate concentration limit in domestic wastewater in Vietnam is less than or equal to 6 mgP-PO_4^{3-}/L (Column A, QCVN 14:2008/BTNMT). Nevertheless, to meet increasingly stringent requirements, methods such as chemical precipitation, biologic processes, and adsorption by functionalized materials are all common methods for phosphorus removal [5]. Chemical precipitation has been shown to be an uneconomically process due to the high cost of key chemicals and feeding systems in some circumstances, as well as the massive amount of metal salts produced [5, 6]. Because of its cost-effectiveness, biologic treatment has become a popular method for removing phosphorus from wastewater [5]. Nonetheless, advanced biological methods can remove more than 95% of total phosphorus. However, the operational problems in the process can be difficult to control to achieve stability [7]. Besides, this process are unable to meet the mandated level or lower phosphate concentration to near zero or below 10 mg/L [8, 9]. In addition, large reactor volumes, energy, and sludge disposal need result in significant capital and operational costs, ineffective at low phosphate concentrations of less than 10 mg/L. Physical and chemical techniques such as electrodialysis and chemical denitrification have also been reported for nutrient removal [10]. Both methods are successful, but they are non-selective, making them costly and with limited application potential [10].

Among these methods, adsorption is a low-cost, small-footprint, and effective technique for phosphate at low (or trace) quantities [10]. Adsorption techniques have been extensively studied in recent years due to the advantages of simplicity of design, ease of operation, and minimal sludge generation [11, 12]. Adsorbents such as fly ash, sludge, zeolite, aluminum, iron oxides, and even waste items have been used to remove phosphate [5]. The adsorbents' performance in terms of phosphate removal has been described in the literature with varied degrees of success [13]. Some materials have been the focus of recent phosphate adsorption experiments such as anion exchange resin [14, 15], aluminum/iron oxide-hydroxide [16], magnetic chitosan composites [17], fly ash and activated carbon [18] and inorganic adsorbent [19, 20]. Furthermore, prior studies of ferric oxyhydroxide (FeOOH) adsorption onto cation exchange resin have demonstrated good phosphate adsorption activity [5, 13]. Some studies show that hydrated iron(III) oxide is innocuous, inexpensive, readily available, and chemically stable over a wide pH range [21]. As a result, ferric oxyhydroxide nanomaterials can be applied to phosphate treatment and it gives good effect. Co-precipitation was a method applied to synthesize this material onto anion exchange resin by NH_3 solution to treat phosphate, Cr(VI), arsenate, chromate, Pb^{2+}, and Zn^{2+} [5, 22, 23]. Hydrous iron oxide onto anion exchange resin (Purolite FerrIX A33E) was also used to treat phosphate in batch and the reaction time reached equilibrium was 48h [24]. In addition, material hydrated Fe(III) oxide onto anion exchange resin had been successfully synthesized by co-precipitation method with NaOH solution for phosphate treatment and the batch reaction time to reach equilibrium was 72h [10].

On the other hand, the presence of divalent ions like iron, manganese, calcium, and magnesium in water is referred to as hardness. The most common species for water hardness are calcium and magnesium [25]. Water hardness can be a nuisance in some situations. It is an important aesthetic parameter to assess water quality. However, it also depends on the local conditions that can accept different hardness [26]. Consumers can tolerate water supplies with total hardness greater than 200 mg/L, but they are regarded poor resources and values greater than 500 mg/L are not appropriate for most home consumptions [25, 26]. Hard water causes deposits in boilers and domestic appliances, as well as having a variety of effects on detergent cleaning performance [27]. Several human investigations have linked magnesium and calcium to cardiovascular disease and cancer [28-30]. Besides, lime and soda ash are used to remove hardness from water in water purification and treatment plants. The significant volume of liquid sludge produced, as well as the requirement to re-carbonate the softened water, are two major downsides of this procedure [31, 32]. Furthermore, the use of extra chemicals to minimize sludge creation is prohibited, therefore water hardness species cannot be completely removed in most circumstances [26]. Various methods have recently been studied for the removal of a wide variety of ionic and molecular species from various water streams, including electro-deionization [32], electro-membrane processes [33], capacitive deionization [34], membrane and fluidized pellet reactor [35], ion exchange process [36, 37] and adsorption [38, 39], including those responsible for hardness (e.g. Ca^{2+} and Mg^{2+} cationic species). Ion exchange methods have been proposed as an alternate option that is currently in commercial usage around the world [26]. The fact demonstrates that ion exchange resins are frequently used in the treatment of water hardness, lowering costs and improving treatment efficiency.

Recent research have revealed that hardness and phosphate treatment can be done separately, making treatment systems more complex, with a larger footprint and higher costs. In this study, iron oxyhydroxide on cation exchange resins was used to create a new FeOOH@CR material for simultaneous removal of phosphate and cations such as Ca^{2+} and Mg^{2+} in this study. This research demonstrates a vision for the material's practical use in the treatment of phosphate at low concentrations and water with significant hardness.

2. Materials and Methods

2.1. Chemicals

Lab-grade chemicals used in the study such as $FeCl_3.6H_2O$, NH_4Cl, NH_3 (28-34 vol.%), and NaCl were from China, and HCl, KH_2PO_4, KCl, K_2SO_4, $MgCl_2$, KNO_3, and $CaCl_2$ were from Merck. Cation exchange resin H^+ (225H) was from India and deionized water was taken from the laboratory. The solutions contain phosphate (the concentration of $mgPO_4^{3-}$ in the solution of the experiments is calculated by the concentration of mgP in PO_4^{3-} in the solution), calcium, magnesium were prepared by dissolving KH_2PO_4, $CaCl_2$, $MgCl_2$ salts into deionized water.

2.2. Material synthesis and characterizations

FeOOH@CR material was prepared according to a process in the previous publications [5, 22, 23]. The cation-exchange resin hybrid adsorbent was dispersed with ferric oxyhydroxide particles prepared by the following procedure. At first, 2 mL of HCl solution was added into 0.2 L of 12.5% w/w $FeCl_3.6H_2O$ solution and the mixture was stirred for 5 min with a stirrer. After the mixture was completely dissolved, the solution had a low pH of 3. Next, 10 g of dry H^+ ion exchange resins (225H of India) was added to the solution and the solution was stirred at 500 rpm for 20 min. The solids were then separated from the mixture and washed with deionized water 3 times, followed by agitating for 10 min at 500 rpm in 200 mL of deionized water containing 1 g of NH_4Cl and 2 mL of NH_3 solution (28-34 vol. %). Subsequently, the material was separated from the solution and washed with deionized water 3 times. Finally, the material was dried in the open air until it turns red.

Scanning electron microscopy - SEM (JCM-7000, JEOL Ltd, Japan) was used to examine the morphology of the material. Energy dispersive X-ray analysis EDX (JCM-7000, JEOL Ltd, Japan) was used to determine various elemental compositions. Fourier-transform infrared spectroscopy - FTIR (Alpha, Bruker, Germany) was used to determine the surface chemistry. X-ray diffraction –XRD (D2 Phaser Benchtop X-ray Powder Diffraction, AERIS, Malvern Panalytical Ltd, Netherlands) was used to evaluate the crystal phase and crystalline structure of the materials .

2.3. Adsorption tests

Batch adsorption studies were used to investigate the adsorption of phosphate and cations (i.e. Mg and Ca). In the first test, 0.25 g of material was added to 50 mL of phosphate solution (20 $mgPO_4^{3-}/L$) and allowed to adsorb for up to 80 h. The material was then removed from the solution, and a sample of the solution was collected for UV-Visible spectroscopy analysis of the phosphate concentration (V-730 UV, JASCO International Co.Ltd, Japan). The investigated influencing parameters include: contact time (0 - 80 h), adsorbent dosage (2 - 30 g/L), solution pH (2 - 12), and competing ions in solution (sulfate, bicarbonate, chloride, and nitrate), initial phosphate concentration (10 - 50 mg/L) and temperature (5 - 40 °C). The phosphate adsorption capacity (in terms of phosphorous, ($mgPO_4^{3-}/g$ or mgP/g) was calculated as the following equation [5, 22].

$$Q_e = \frac{C_0 - C_t}{m} \times V \qquad (2.1)$$

IOP Conf. Series: Earth and Environmental Science **964** (2022) 012032 doi:10.1088/1755-1315/964/1/012032

Where C_o and C_e are initial and equilibrium concentrations of phosphate in terms of phosphorous ($mgPO_4^{3-}/L$). V is the volume of solution (L) and m was the mass of the material (g).
The simultaneous adsorption of phosphate and cations (Mg and Ca) was achieved by adding 0.25 g of resin into 50 mL of a solution containing 20, 100, and 100 mg/L of phosphate, calcium, and magnesium, respectively. The other steps of the experiment were similar to the previous phosphate adsorption experiment. The solution was then taken for measurement of phosphate content using UV-Visible spectroscopy and calcium and magnesium content using atomic absorption spectroscopy (AAS, Perkin Elmer Aanalyst 400, PerkinElmer, Inc., USA).

3. Results and discussion

3.1. Characterizations of materials

The morphology and surface elemental composition of the materials were examined by SEM and EDX-mapping, respectively (Figure S1, S2, and S3 of Supplementary Data). Accordingly, the composition of the elements on the surface of the materials is summarized in Table 1 with the main components of C, O, Fe, Ca, Mg, and Al. Phosphate was only detected in the structure of the FeOOH@CR material after phosphate adsorption (Figure S2), but not in those of the resin (Figure S1) and fresh FeOOH@CR material (Figure S3). This result showed that the process was successful when iron was present on the surface of the material. Besides, phosphate was also adsorbed on the surface of the material after the experimental process.

Table 1. Elemental composition (wt. %) of material as determined by EDX.

Material/ Elements	Resin	Fresh FeOOH@CR	FeOOH@CR after adsorption
C	42.37 ± 0.12	42.45 ± 0.14	52.41±0.11
O	57.63 ± 0.16	38.48 ± 0.19	18.83±0.15
Fe		19.± 0.35	20.44±0.28
Ca		0.01 ± 0.01	0.01 ± 0.01
Mg		0.07 ± 0.01	0.06 ± 0.01
P			0.92 ± 0.02

Surface functional groups of the nanomaterials are usually determined by FTIR [40, 41]. In this study, the surface chemical structure of FeOOH@CR before and after adsorption was examined to clarify the adsorption mechanism occurring on the surface of the material. The FTIR spectra of FeOOH (i.e., goethite and lepidocrocite) were also provided for reference purposes. As can be observed in Figure 1, all samples have a peak at wavenumber of 3390 cm^{-1}, which is assigned to the H-O-H vibration of the hydrated group. The peaks at the region of 789 and 880 cm^{-1} characterize for Fe-OH-Fe vibration [41-43] while the peaks at 466 and 746 cm^{-1} are attributed to the Fe-O-H vibrations [42, 43]. The absorption bands at 3133 and 3384 cm^{-1} are related to the stretching vibrations of the OH groups. The peak of phosphate ranges from 1000-1100 cm^{-1} [44-46]. In this study, the dose of phosphate was much smaller than iron and the peak of phosphate was also located at the same position as the peak of iron (lepidocrocite and goethite reference [42, 43]). These results are similar to those reported in the previous studies [42, 43].

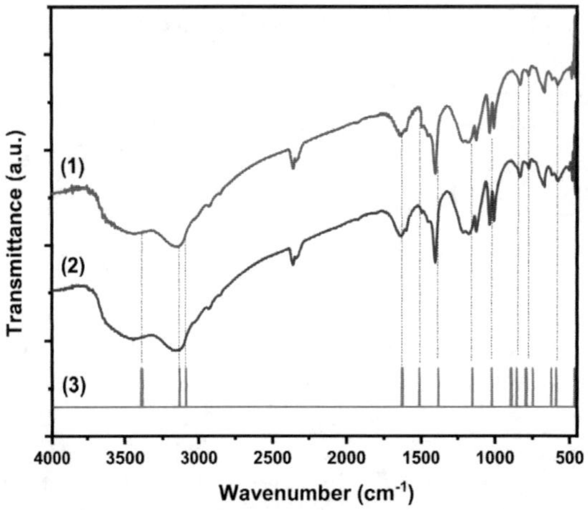

Figure 1. FTIR spectra of (1) FeOOH@CR after phosphate adsorption, (2) fresh FeOOH@CR, and (3) lepidocrocite and goethite reference [42, 43].

Figure 2 shows the XRD patterns of the materials in the 2θ range of 10 - 70°, indicating that the substance is hydrated ferric oxide nanomaterial (FeOOH). The orthorhombic iron oxide-hydroxide structure of the FeOOH@CR material is shown by the distinctive peaks at 2θ of 31.06, 33.10, 38.99, 41.93, 46.19, 48.21, 54.66, and 57.55° (JCPDS card no 18-0639). However, the material is usually considered as amorphous phase because the intensities of these peaks are quite low and sometimes interfered with background noise. This could be attributed to the material's low-temperature production without calcination for the crystallization of iron oxide-hydroxide.

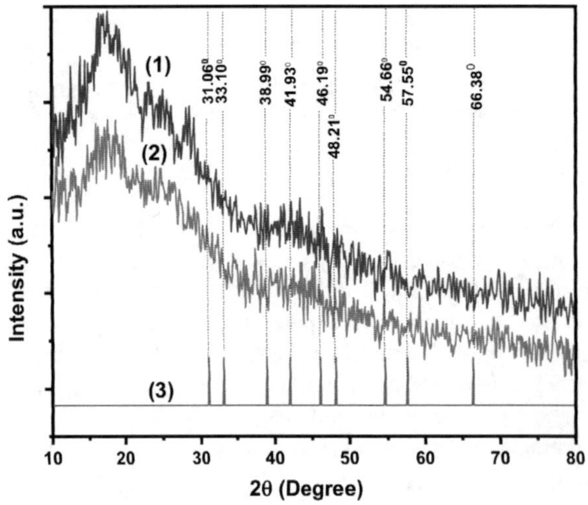

Figure 2. XRD patterns of (1) FeOOH@CR after phosphate adsorption, (2) fresh FeOOH@CR, and (3) iron oxide-hydroxide reference (JCPDS card no. 18-0639) [47].

3.2. Adsorption of phosphate onto FeOOH@CR

During the 80-h experiment, Figure 3 demonstrates that the adsorption capacity of FeOOH@CR increased with increasing adsorption time. With an adsorption capacity of 3.10 mgPO_4^{3-}/g, it reached equilibrium after about 48 h, which is similar to that described in the literature. Using FeOOH@CR material, there are three primary stages in phosphate adsorption, including the first stage of quick adsorption in the first 24 h, the second stage of slow adsorption in the next 24 h, and the last stage of equilibrium after 48 h of adsorption.

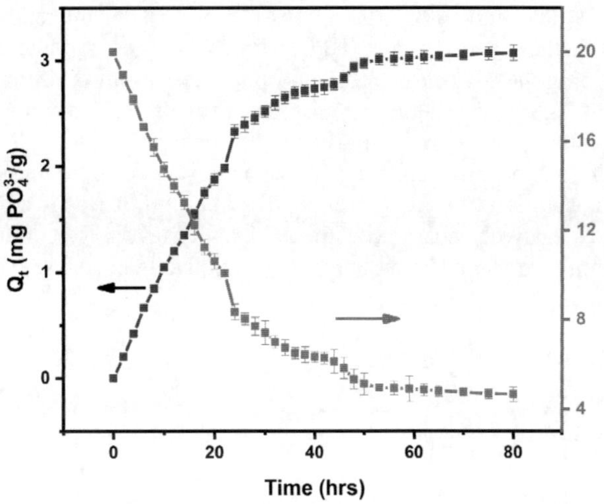

Figure 3. Phosphate adsorption capacity and concentration during 80 h of adsorption.

(Condition: 5 gFeOOH@CR/L, 20 mgPO_4^{3-}/L, pH 6.0, room temperature of ~ 27 °C)

As shown in Table 2, the kinetics of phosphate adsorption onto FeOOH@CR were investigated using three kinetic models of pseudo-first-order, pseudo-second-order, and intra-particle diffusion models. All three models can be utilized to describe phosphate adsorption based on the correlation coefficient ($R^2 > 0.9$). However, because it has the greatest R^2 value of 0.9848 (Figure S4) with similar predicted and tested adsorption capacities, the pseudo-first-order model is the best suitable adsorption model for phosphate adsorption utilizing FeOOH@CR. At a phosphate concentration of 20 mg/L, the K adsorption rate constant was calculated to be 0.0641 (g.mg^{-1}.h^{-1}).

Table 2. Equations and parameters of kinetic models for phosphate adsorption using FeOOH@CR.

Kinetic model	Linear equation	R^2	Capacity (Q_e,cal) mg.g^{-1}	
			Modeling	Experimental
Pseudo-first-order	y = -0.0641x + 1,237	0.9848	3.45	3.07
Pseudo-second-order	y = 0.2599x+4.268	0.9701	3.85	

Intraparticle diffusion	y = 0.5484x + 1.0439	0.9467

The adsorption capacity of materials is affected by solution pH, which is an important parameter. Figure 4 shows the phosphate adsorption capability of the materials from pH 2 to 12. It was also shown that the adsorption capacity increased steadily from pH 2 to pH 6.5, peaking at 2.78 mgPO_4^{3-}/g. In addition, the point of zero charge (pHpzc) of the material was determined to be 6.5 (Figure S5). When the pH is less than 6.5, the material surface has a positive charge, which increases the adsorption of negatively charged phosphates in the forms of HPO_4^{2-} and $H_2PO_4^-$ (Reactions 1, 2, and 3), and are retained by iron oxyhydroxide by forming a complex inside the sphere. Because HPO_4^{2-} forms stronger amphoteric complexes than $H_2PO_4^-$ (Reactions 6 and 7), reducing the pH causes more phosphate to stay in the form of $H_2PO_4^-$, which has a lower negative charge, making the adsorption process more difficult. Furthermore, the solution with low pH reduces the electrostatic attraction between the positively charged iron oxyhydroxide groups and the phosphate ions by reducing the negative charge of phosphate from divalent to monovalent. When the pH was higher than 6.5, the adsorption capacity gradually dropped until it was just 2.12 mgPO_4^{3-}/g at pH 12. FeOOH is deionized and negatively charged at these high pH levels (Reaction 4 and 5). As a result, phosphate adsorption is inhibited by electrostatic repulsion and Donnan co-ion exclusion [48, 49].

$$H_3PO_4 \leftrightarrows H^+ + H_2PO_4^- \quad pKa_1 = 2.15 \quad (3.1)$$

$$H_2PO_4^- \leftrightarrows H^+ + HPO_4^{2-} \quad pKa_2 = 7.20 \quad (3.2)$$

$$HPO_4^- \leftrightarrows H^+ + PO_4^{3-} \quad pKa_3 = 12.35 \quad (3.3)$$

$$Fe\overline{O}H_2^+ \leftrightarrows Fe\overline{O}H + H^+ \quad pKa_1 = 6.5 \quad (3.4)$$

$$Fe\overline{O}H \leftrightarrows FeO^- + H^+ \quad pKa_2 = 8.5 \quad (3.5)$$

$$Fe\overline{O}H_2^+(Cl^-) + H_2PO_4^- \leftrightarrows Fe\overline{O}H_2^+{:}H_2PO_4^- + Cl^- \quad (3.6)$$

$$Fe\overline{O}H_2^+(Cl^-) + HPO_4^{2-} \leftrightarrows 2(Fe\overline{O}H_2^+){:}HPO_4^{2-} + 2Cl^- \quad (3.7)$$

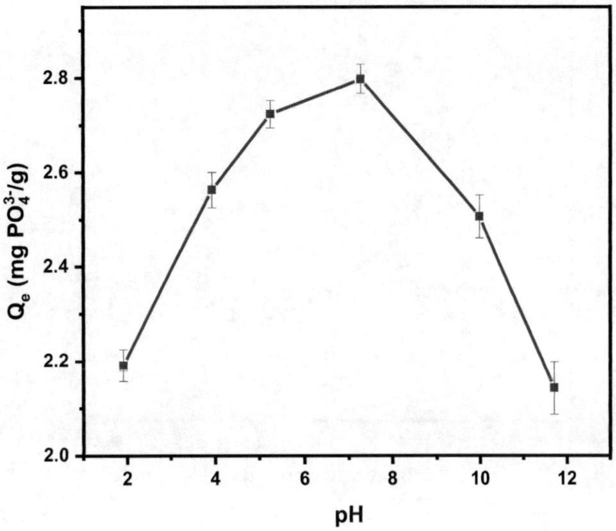

Figure 4. Effect of solution pH on phosphate adsorption.
(Condition: 48 h, 5 gFeOOH@CR/L, 20 mgPO_4^{3-}/L, and room temperature of ~ 27 °C)

The adsorption capacity of phosphate is influenced by the dose of FeOOH@CR adsorbent, as shown in Figure 5. The adsorption capacity declines as the adsorbent dose increases, whereas the adsorption efficiency and the output water quality gradually improves. The phosphate adsorption must meet the QCVN 14:2008/BTNMT (Column A) standard with an output phosphate content of less than 6 mgP/L. When the adsorbent dose was raised in the range of 2 - 10 g/L, the phosphate adsorption capacity declines drastically, then slowly in the range of 10 - 30 g/L. For the following studies, the optimum adsorbent dosage was chosen at 5 g/L.

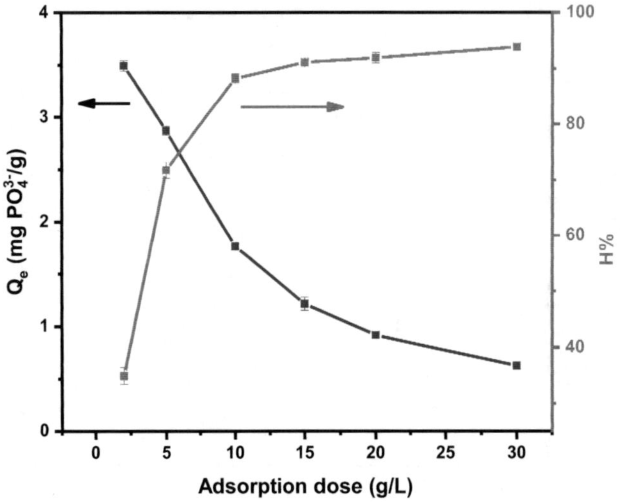

Figure 5. Effect of dosage on the adsorption of phosphate.

(Condition: 48 h, 20 $mgPO_4^{3-}/L$, pH 6.5, and room temperature of ~ 27 °C)

Figure 6 describes the effect of phosphate concentration and temperature on the adsorption capacity, which gradually increases as the concentration rises. This is due to the high phosphate concentration in the solution increasing the density of phosphate ions in the solution, making the adsorption occurs easier. Similarly, as the temperature rises, the mobility of phosphate ions increases, favoring adsorption. This is readily visible in solutions with starting concentrations ranging from 10 to 50 $mgPO_4^{3-}/L$ at temperatures of 5, 20, 30, and 40 °C.

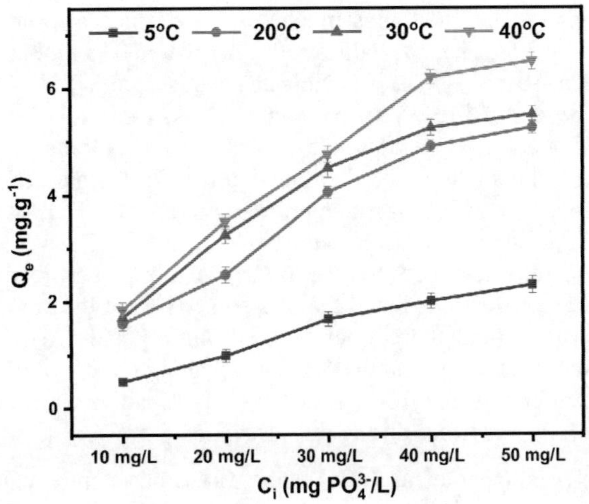

Figure 6. Effect of initial phosphate concentration and temperature on the adsorption of phosphate. (Condition: 48 h, 5 gFeOOH@CR/L, pH 6.5)

The link between the phosphate on the materials surface and its equilibrium concentration in the solution at constant temperature was described using the Langmuir and Freundlich isotherm models [50, 51]. Table 3 shows the adsorption isotherm parameters. Although both the Langmuir and Freundlich models are suitable for describing phosphate adsorption at 30 and 40 ° C, the Langmuir model has a better correlation coefficient with R^2 of 0.9974 at 30 °C, making it more favorable. The K_L value, which is the Langmuir constant that represents the adsorbent's affinity for its surface [52-54], was estimated from the equation to be 0.26552 and 0.3705 L/mg at 30 °C and 40 °C, respectively.

Table 3. The isotherm parameters of phosphate adsorption onto FeOOH@CR material.

Isotherm model	Parameter	Temperature (°C)			
		5	25	30	40
Langmuir	Q_{max} (mg/g)	20.534	7.5245	6.51466	7.5245
	K_L (L/mg)	0.0035	0.1002	0.26552	0.3705
	R^2	0.2013	0.8818	0.9974	0.9903
Freundlich	K_f((mg/g)(L/mg)^n)	13.37	0.95	0.61	0.45
	n	1.04	1.90	2.29	2.41
	R^2	0.9802	0.9288	0.9249	0.9664

The thermodynamic process of adsorption, on the other hand, is stated using equilibrium constants at different temperatures (such as 278, 293, 303, and 313 ^{0}K) utilizing factors such as Gibbs free energy (△G), entropy (△S), and enthalpy (△H). The Gibbs free energy (△G) is a measure of whether a substance is physically or chemically adsorbed. The △G values between -20 and 0 kJ/mol indicate a physical adsorption process while those between -80 and -400 kJ/mol indicate chemisorption [55]. At 278, 293, 303, and 313°K, the △G was estimated to be -0.845, -4.389, -6.751, and -9.114 kJ/mol, showing that the adsorption process was spontaneous at all investigated temperatures [53]. Van der Waals forces, hydrogen bonds, ionic pairs, and other polar and nonpolar interactions should all play a part in the interaction between phosphate ions and the FeOOH@CR surface during the adsorption process [53, 55, 56]. The positive value of △H indicates that this is an endothermic adsorption process with 64.84 (kJ/mol). The value of △S is higher than 0, indicating that the affinity and contact between the adsorbate and the adsorbent of the adsorption process are increasing [56]. This can also be explained that ΔS and ΔH were the key factors for determining the overall sign and magnitude of the binding free energy (ΔG). Only when the free energy change was negative, the adsorption can occur spontaneously. And the positive ΔS along with negative ΔH contributed favorably to the overall binding free energy (ΔG) [53]. Besides, the positive value of ΔS suggests increased randomness at the solid/solution interface during the adsorption [55, 57, 58].

Table 4. Thermodynamic parameters of phosphate adsorption onto FeOOH@CR material.

Thermodynamic parameters	Temperature (K)	Value
ΔG (kJ/mol)	278 K	-0.845
	393 K	-4.389
	303 K	-6.751
	313 K	-9.114
ΔH (kJ/mol)		64.84
ΔS (kJ/(mol.K))		0.24

3.3. Effect of coexisted ions, durability, and comparison with other studies

Figure 7 shows the impacts of anions on the phosphate adsorption capacity of FeOOH@CR material. Anions such as Cl^-, SO_4^{2-}, NO_3^-, and HCO_3^- commonly coexist with PO_4^{3-} in water and wastewater. The anions HCO_3^-, at a concentration of 100 mg/L, had the greatest impact on the phosphate adsorption capacity. Other anions, such as Cl^-, NO_3^-, and SO_4^{2-} had a lower effect than HCO_3^- on the adsorption process, demonstrating the specific interaction and adsorption of FeOOH@CR material with HCO_3^- and PO_4^{3-} anions.

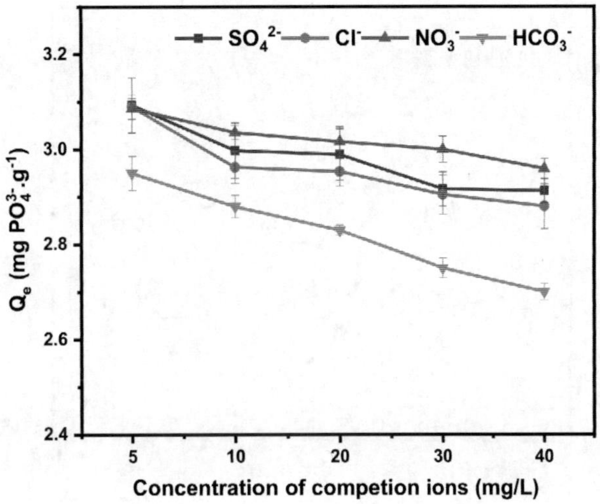

Figure 7. Effect of other anions on the adsorption of phosphate.

(Condition: 48 h, 5 g FeOOH@CR/L, 20 mgPO_4^{3-}/L, pH 6.5, room temperature of ~ 30 °C)

As seen in Figure 8, the presence of cations such as calcium and magnesium in the solution increased the phosphate adsorption. The fundamental reason for this is that the presence of Ca and Mg increases the positive charge of the material surface, which makes negative phosphate anions easier to adsorb. The material was also capable of eliminating Ca^{2+} from 210 to 128 mg/L and Mg^{2+} from 202 to 142 mg/L at the same time. One probable explanation is that after treating the resin with NH_3 solution, iron and other cations dissociate from their ionic bonds with the resin, forming oxides/hydroxides particles inside the pores. As a result, the ion exchange resin is still capable of accepting new Ca^{2+} and Mg^{2+} cations as they arrive in the solution, positively charging the surface to aid phosphate adsorption. In the addition of Ca^{2+} and Mg^{2+}, the phosphate adsorption capacity increased by 1.21 times, reaching 3.652 mgPO_4^{3-}/g in Table 5.

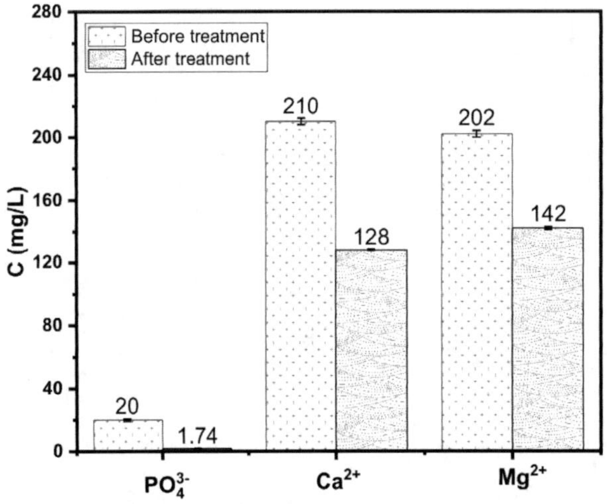

Figure 8. Effect of Ca and Mg cations on the adsorption of phosphate.

(Condition: 48 h, 5 gFeOOH@CR/L, 20 $mgPO_4^{3-}/L$, pH 6.5, room temperature of ~ 30 °C)

Table 5. The comparison of phosphate adsorption capacity.

Parameters	Solution (PO_4^{3-} without Ca^{2+} and Mg^{2+})	Solution (PO_4^{3-}, Ca^{2+}, Mg^{2+})		
	($mgPO_4^{3-}/g$)	($mgPO_4^{3-}/g$)	($mgCa^{2+}/g$)	($mgMg^{2+}/g$)
Adsorption capacity	3.1 ± 0.049	3.625 ± 0.043	16.4 ± 0.057	12 ± 0.096

Besides, this material showed that the total adsorption capacity of phosphate (3.625 mg/g) and hardness (28.4mg/g) was 32.025 mg/g compared to a commercial product (INDION MB-6 SR, Ion Exchange (India) Ltd., India) which had a total adsorption capacity of phosphate (4.4 mg/g) and hardness (8.6 mg/L) was 13 mg/g. This can be seen that the adsorption capacity of the material in this study was 2.46 times higher than that of the INDION MB-6 SR material. INDION MB-6 SR was a commercially available material on the market, it had the ability to remove both cations and anions in water.

3.4. Application to real wastewater, durability, and comparisons with other studies

After that, the FeOOH@CR material was used to treat real wastewater. As shown in Figure S6, the phosphate concentration in synthetic wastewater decreased from 20 to 4.76 $mgPO_4^{3-}/L$, while domestic wastewater decreased from 14.34 to 2.15 $mgPO_4^{3-}/L$, meeting the requirement of QCVN 14: 2008/BTNMT (Column A). Although competing ions may exist in domestic wastewater, they had less effect on the phosphate adsorption process. Figure 9 shows five cycles of adsorption-regeneration that were used to verify the material durability. After five reuse cycles, the adsorption capacity was only 1.75 mgPO_4^{3-}/g, which was approximately 56.68% of the capacity of the fresh material. Table 6 compares the adsorption capacity of various nanomaterials, showing that FeOOH@CR is potential for phosphate application.

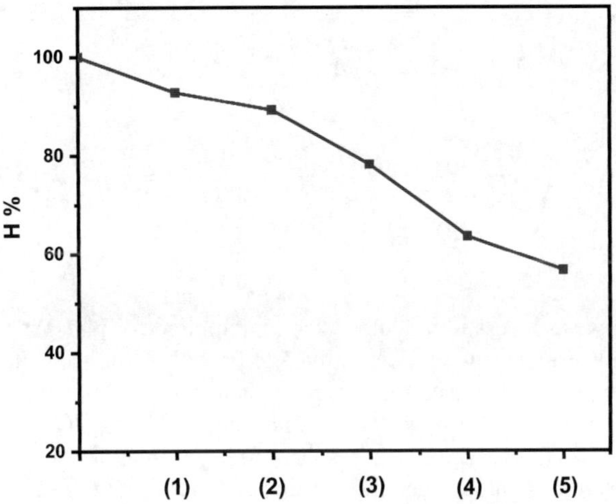

Figure 9. Durability test of FeOOH@CR material for phosphate adsorption.

(Condition: 48 h, 5 gFeOOH@CR/L, 20 mgPO_4^{3-}/L, pH 6.5, room temperature of ~ 30 °C)

Table 6. Phosphate adsorption capacity from various developed adsorbents.

Materials	Experimental conditions	Adsorption capacity (mg P per g of adsorbent)	References
Akagenite	25 °C, pH 7 C_e = 1-250 mgP/L	60	[59]
Akagenite granulated	20 °C, pH 5.5 C_e = 0-4 mgP/L	17-23	[60]
Activated alumina	20 °C, pH 5.5 C_e = 0-4 mgP/L	12-14	[60]
Activated alumina	C_e = 0-2 mgP/L	3.3	[61]
Iron/zirconium binary oxide	25 °C, pH 4 C_e = 0-80 mgP/L	13.7	[12]
MgAlLDH (granular)	25 °C, pH 6.9 C_e = 0-150 mgP/L	47.3	[62]
Activated carbon	C_e = 0-28 mgP/L	3.2	[63]

Anion exchange resin from soybean hulls	25 °C, pH 7 Dose = 10 g C_e = 10-620 mgP/L	20	[64]
Purolite A500P anion exchange resin	Dose = 0.5-10 g/L C_e = 1-15 mgP/L	23.55	[65]
Iron oxide tailings (23% w/w Fe)	20-21 °C, pH 6.6-6.8 C_e = 0-130 mgP/L	7 30.435 (mgPO_4^{3-}/gFe)	[66]
Hydrated Fe(III) oxide nanoparticles on anion exchange resin (35% w/w Fe)	pH 6.5 Dose = 0.2-4.0 g/L	12.24 19.97 (mgPO_4^{3-}/gFe)	[48]
FeOOH@CR (5%)	25-30 °C, pH 6 Dose = 2-10 g/L C_o = 2-20 mgPO_4^{3-}/L	3.1 62 (mg PO_4^{3-}/g Fe)	This study

4. Conclusions

The FeOOH@CR material was successfully synthesized and applied for phosphate and hardness removal in water. The equilibrium adsorption time was determined to be 48 h while the suitable pH was around 6.5. The adsorption capacity increased with the increase of phosphate concentration and adsorption temperature. The maximum adsorption capacity of FeOOH@CR material is 3.1 mgPO_4^{3-}/g (62 mgPO_4^{3-}/gFe).The presence of Ca^{2+} and Mg^{2+} cations enhanced the phosphate adsorption up to 1.21 times, suggesting the use of the material for simultaneous phosphate and hardness removal. In both synthetic and domestic wastewater, the FeOOH@CR was effective for removal phosphate to meet the discharging standards, indicating the practicability of the material in wastewater treatment.

Supplementary Data

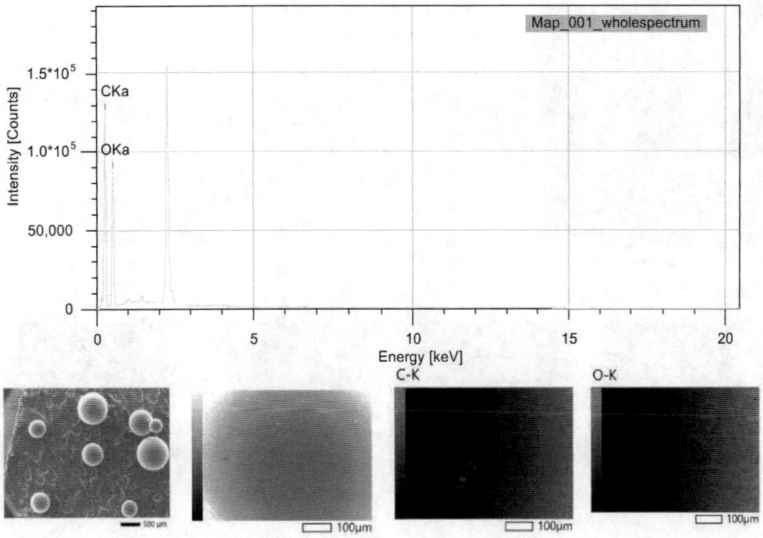

Figure S1. SEM image and EDX-mapping result of CR.

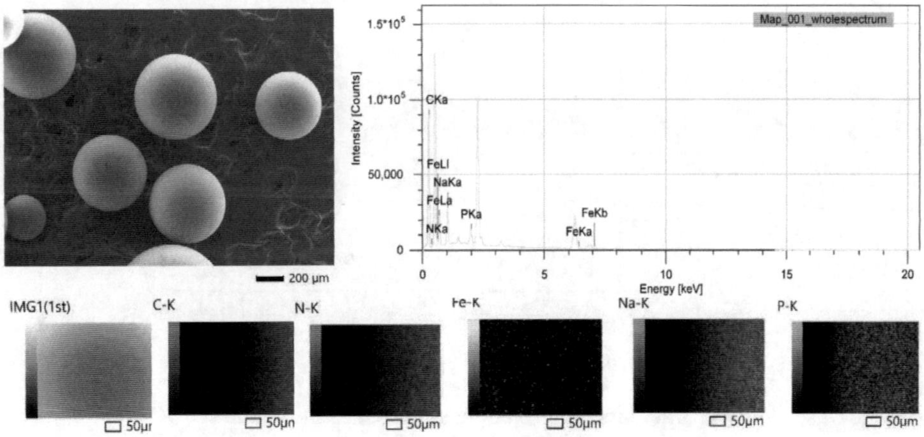

Figure S2. SEM image and EDX-mapping result of FeOOH@CR after phosphate adsorption.

IOP Conf. Series: Earth and Environmental Science **964** (2022) 012032 doi:10.1088/1755-1315/964/1/012032

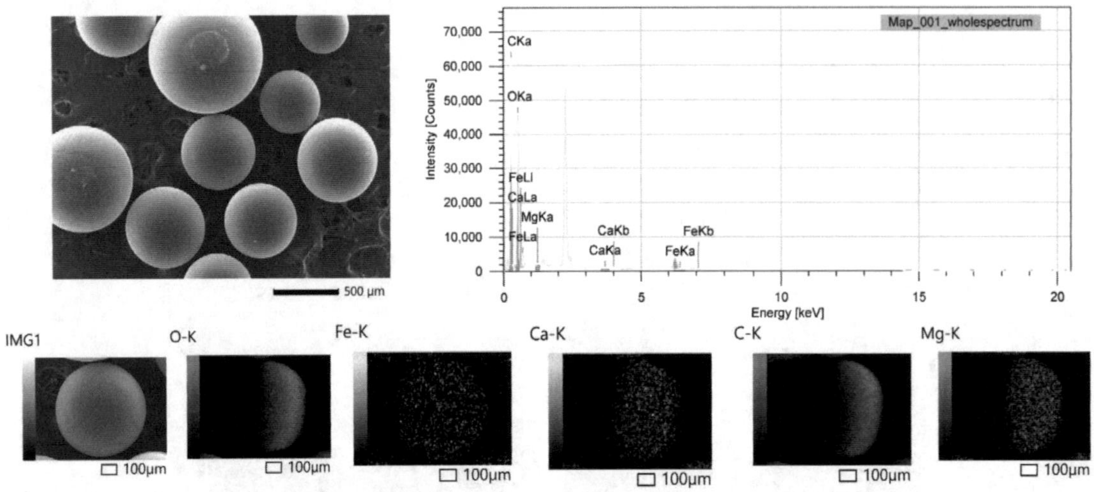

Figure S3. SEM image and EDX-mapping result of fresh FeOOH@CR.

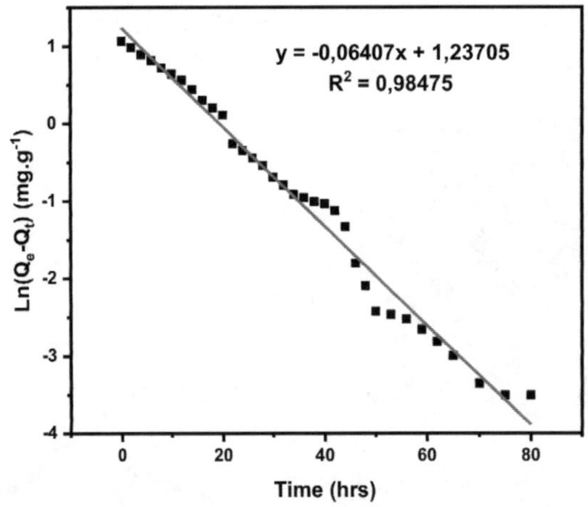

Figure S4. Pseudo- first-order kinetic plot of phosphate adsorption onto FeOOH@CR material.

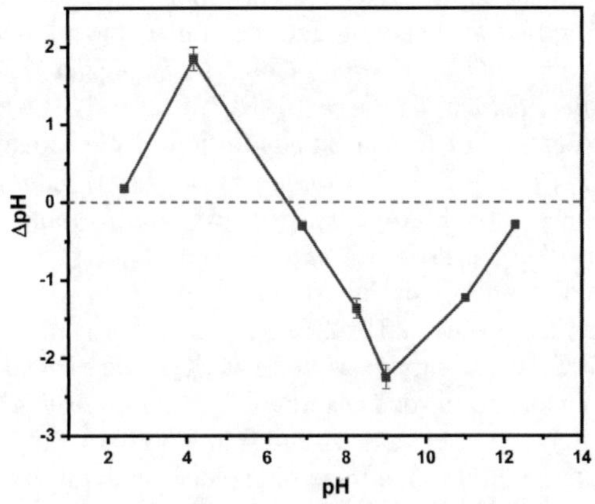

Figure S5. Plot for determining the point of zero charge (pH_{zpc}) of FeOOH@CR material.

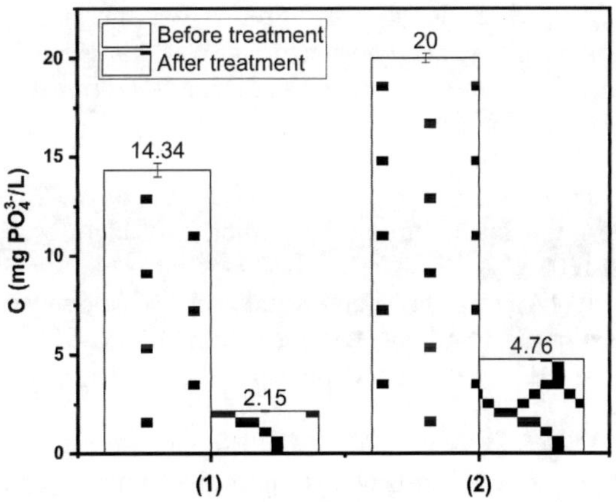

Figure S6. The concentration before and after treatment of synthetic wastewater and domestic wastewater.

References

[1] Sun L, Tong H, Gao C, Liu Y and Feng C 2021 Removal of antibiotic resistance genes from secondary effluent by processes combining nano-iron, ultrasound-activated persulfate, and ultrafiltration *Water Sci Technol* **83** 1578-90
[2] Zhao D and Sengupta A K 1998 Ultimate removal of phosphate from wastewater using a new class of polymeric ion exchangers *Water Research* **32** 1613-25

[3] Conley D J, Paerl H W, Howarth R W, Boesch D F, Seitzinger S P, Havens K E, Lancelot C and Likens G E 2009 Controlling eutrophication: nitrogen and phosphorus *Science* **323** 1014-5
[4] Anzecc A 2000 Australian and New Zealand guidelines for fresh and marine water quality *Australian and New Zealand Environment and Conservation Council and Agriculture and Resource Management Council of Australia and New Zealand, Canberra* 1-103
[5] Ren J, Li N, Zhao L and Ren N 2014 Enhanced adsorption of phosphate by loading nanosized ferric oxyhydroxide on anion resin *Frontiers of Environmental Science & Engineering* **8** 531-8
[6] Zhang T, Ding L and Ren H 2009 Pretreatment of ammonium removal from landfill leachate by chemical precipitation *Journal of Hazardous Materials* **166** 911-5
[7] Oguz E 2005 Thermodynamic and kinetic investigations of PO3− 4 adsorption on blast furnace slag *Journal of Colloid and Interface Science* **281** 62-7
[8] Awual M R, Jyo A, Ihara T, Seko N, Tamada M and Lim K T 2011 Enhanced trace phosphate removal from water by zirconium(IV) loaded fibrous adsorbent *Water Research* **45** 4592-600
[9] Awual M R, Shenashen M, Jyo A, Shiwaku H and Yaita T 2014 Preparing of novel fibrous ligand exchange adsorbent for rapid column-mode trace phosphate removal from water *Journal of Industrial and Engineering Chemistry* **20** 2840-7
[10] Wiriyathamcharoen S, Sarkar S, Jiemvarangkul P, Nguyen T T, Klysubun W and Padungthon S 2020 Synthesis optimization of hybrid anion exchanger containing triethylamine functional groups and hydrated Fe (III) oxide nanoparticles for simultaneous nitrate and phosphate removal *Chemical Engineering Journal* **381** 122671
[11] Xu X, Gao B Y, Yue Q Y and Zhong Q Q 2010 Preparation of agricultural by-product based anion exchanger and its utilization for nitrate and phosphate removal *Bioresour Technol* **101** 8558-64
[12] Long F, Gong J-L, Zeng G-M, Chen L, Wang X-Y, Deng J-H, Niu Q-Y, Zhang H-Y and Zhang X-R 2011 Removal of phosphate from aqueous solution by magnetic Fe–Zr binary oxide *Chemical Engineering Journal* **171** 448-55
[13] Nur T, Johir M A H, Loganathan P, Nguyen T V and Kandasamy J 2014 Phosphate removal from water using an iron oxide impregnated strong base anion exchange resin *Journal of Industrial and Engineering Chemistry* **20** 1301–7
[14] Awual M R, Urata S, Jyo A, Tamada M and Katakai A 2008 Arsenate removal from water by a weak-base anion exchange fibrous adsorbent *Water Research* **42** 689-96
[15] Awual M R and Jyo A 2011 Assessing of phosphorus removal by polymeric anion exchangers *Desalination* **281** 111-7
[16] Chubar N I, Kanibolotskyy V A, Strelko V V, Gallios G G, Samanidou V F, Shaposhnikova T O, Milgrandt V G and Zhuravlev I Z 2005 Adsorption of phosphate ions on novel inorganic ion exchangers *Colloids and Surfaces A: Physicochemical and Engineering Aspects* **255** 55-63
[17] Reddy D H K and Lee S-M 2013 Application of magnetic chitosan composites for the removal of toxic metal and dyes from aqueous solutions *Advances in Colloid and Interface Science* **201-202** 68-93
[18] Chen J, Kong H, Wu D, Chen X, Zhang D and Sun Z 2007 Phosphate immobilization from aqueous solution by fly ashes in relation to their composition *Journal of Hazardous Materials* **139** 293-300
[19] Jellali S, Wahab M A, Hassine R B, Hamzaoui A H and Bousselmi L 2011 Adsorption characteristics of phosphorus from aqueous solutions onto phosphate mine wastes *Chemical Engineering Journal* **169** 157-65
[20] Huang W, Wang S, Zhu Z, Li L, Yao X, Rudolph V and Haghseresht F 2008 Phosphate removal from wastewater using red mud *Journal of Hazardous Materials* **158** 35-42
[21] Nie G, Wang J, Pan B and Lv L 2015 Surface chemistry of polymer-supported nano-hydrated ferric oxide for arsenic removal: effect of host pore structure *Science China Chemistry* **58** 722-30

[22] Ren J, Li N and Zhao L 2012 Adsorptive Removal of Cr(VI) from Water by Anion Exchanger Based Nanosized Ferric Oxyhydroxide Hybrid Adsorbent *Chemical and Biochemical Engineering Quarterly* **26** 111-8
[23] Ding M, de Jong B H W S, Roosendaal S J and Vredenberg A 2000 XPS studies on the electronic structure of bonding between solid and solutes: adsorption of arsenate, chromate, phosphate, Pb2+, and Zn2+ ions on amorphous black ferric oxyhydroxide *Geochimica et Cosmochimica Acta* **64** 1209-19
[24] Nur T, Johir M A H, Loganathan P, Nguyen T, Vigneswaran S and Kandasamy J 2014 Phosphate removal from water using an iron oxide impregnated strong base anion exchange resin *Journal of Industrial and Engineering Chemistry* **20** 1301-7
[25] Yan M, Wang D, Ni J, Qu J, Yan Y and Chow C W 2008 Effect of polyaluminum chloride on enhanced softening for the typical organic-polluted high hardness North-China surface waters *Separation and purification technology* **62** 401-6
[26] Sepehr M N, Zarrabi M, Kazemian H, Amrane A, Yaghmaian K and Ghaffari H R 2013 Removal of hardness agents, calcium and magnesium, by natural and alkaline modified pumice stones in single and binary systems *Applied Surface Science* **274** 295-305
[27] Suzuki K, Tanaka Y, Osada T and Waki M 2002 Removal of phosphate, magnesium and calcium from swine wastewater through crystallization enhanced by aeration *Water research* **36** 2991-8
[28] Rubenowitz E, Axelsson G and Rylander R 1996 Magnesium in drinking water and death from acute myocardial infarction *American journal of epidemiology* **143** 456-62
[29] Rosborg I, Soni V and Kozisek F 2015 pp 79-101
[30] Rapant S, Cvečková V, Fajčíková K, Sedláková D and Stehlíková B 2017 Impact of Calcium and Magnesium in Groundwater and Drinking Water on the Health of Inhabitants of the Slovak Republic *International Journal of Environmental Research and Public Health* **14** 278
[31] Sheikholeslami R 2011 Composite scale formation and assessment by the theoretical Scaling Potential Index (SPI) proposed previously for a single salt *Desalination* **278** 259-67
[32] Fu L, Wang J and Su Y 2009 Removal of low concentrations of hardness ions from aqueous solutions using electrodeionization process *Separation and Purification Technology* **68** 390-6
[33] Park J-S, Song J-H, Yeon K-H and Moon S-H 2007 Removal of hardness ions from tap water using electromembrane processes *Desalination* **202** 1-8
[34] Seo S-J, Jeon H, Lee J K, Kim G-Y, Park D, Nojima H, Lee J and Moon S-H 2010 Investigation on removal of hardness ions by capacitive deionization (CDI) for water softening applications *Water research* **44** 2267-75
[35] Li C-W, Liao J-C and Lin Y-C 2005 Integrating a membrane and a fluidized pellet reactor for removing hardness: effects of NOM and phosphate *Desalination* **175** 279-88
[36] Apell J N and Boyer T H 2010 Combined ion exchange treatment for removal of dissolved organic matter and hardness *Water Research* **44** 2419-30
[37] Kazemian H, Modarres H and Mobtaker H G 2003 Iranian natural clinoptilolite and its synthetic zeolite P for removal of cerium and thorium from nuclear wastewaters *Journal of Radioanalytical and Nuclear Chemistry* **258** 551-6
[38] Torabian A, Kazemian H, Seifi L, Bidhendi G N, Azimi A A and Ghadiri S K 2010 Removal of petroleum aromatic hydrocarbons by surfactant-modified natural zeolite: the effect of surfactant *CLEAN–Soil, Air, Water* **38** 77-83
[39] Seifi L, Torabian A, Kazemian H, Bidhendi G N, Azimi A A, Nazmara S and AliMohammadi M 2011 Adsorption of BTEX on surfactant modified granulated natural zeolite nanoparticles: parameters optimizing by applying Taguchi experimental design method *CLEAN–Soil, Air, Water* **39** 939-48

[40] Hifumi H, Ewing A V and Kazarian S G 2016 ATR-FTIR spectroscopic imaging to study the drying and dissolution of pharmaceutical polymer-based films *International Journal of Pharmaceutics* **515** 57-68

[41] Ruan H D, Frost R L, Kloprogge J T and Duong L 2002 Infrared spectroscopy of goethite dehydroxylation. II. Effect of aluminium substitution on the behaviour of hydroxyl units *Spectrochimica Acta Part A: Molecular and Biomolecular Spectroscopy* **58** 479-91

[42] Liao S, Wang J, Zhu D, Ren L, Lu J, Geng M and Langdon A 2007 Structure and Mn2+ adsorption properties of boron-doped goethite *Applied Clay Science* **38** 43-50

[43] Rahimi S, Moattari R M, Rajabi L, Derakhshan A A and Keyhani M 2015 Iron oxide/hydroxide (α , γ -FeOOH) nanoparticles as high potential adsorbents for lead removal from polluted aquatic media *Journal of Industrial and Engineering Chemistry* **23** 33-43

[44] Lochaiwatana Y, Poolthong S, Hirata I, Okazaki M, Swasdison S and Vongsavan N 2015 The synthesis and characterization of a novel potassium chloride-fluoridated hydroxyapatite varnish for treating dentin hypersensitivity *Dental Materials Journal* **34**

[45] Gheisari H, Karamian E and Abdellahi M 2015 A novel hydroxyapatite –Hardystonite nanocomposite ceramic *Ceramics International* **41** 5967-75

[46] Furuzono T, Yasuda S, Kimura T, Kyotani S, Tanaka J and Kishida A 2004 Nano-scaled hydroxyapatite/polymer composite IV. Fabrication and cell adhesion properties of a three-dimensional scaffold made of composite material with a silk fibroin substrate to develop a percutaneous device *Journal of Artificial Organs* **7** 137-44

[47] Raul P, Devi R, Umlong I, Banerjee S and Purkait M 2012 Removal of Fluoride from Water Using Iron Oxide-Hydroxide Nanoparticles *Journal of nanoscience and nanotechnology* **12** 3922-30

[48] Wiriyathamcharoen S, Sarkar S, Jiemvarangkul P, Nguyen T T, Klysubun W and Padungthon S 2020 Synthesis optimization of hybrid anion exchanger containing triethylamine functional groups and hydrated Fe(III) oxide nanoparticles for simultaneous nitrate and phosphate removal *Chemical Engineering Journal* **381** 122671

[49] Nguyen T T, Le T T, Phan P T and Nguyen N H 2020 Preparation, Characterization, and Application of Novel Ferric Oxide-Amine Material for Removal of Nitrate and Phosphate in Water *Journal of Chemistry* **2020** 8583543

[50] Belhachemi M and Addoun F 2011 Comparative adsorption isotherms and modeling of methylene blue onto activated carbons *Applied Water Science* **1** 111-7

[51] Chabani M, Amrane A and Bensmaili A 2009 Equilibrium sorption isotherms for nitrate on resin Amberlite IRA 400 *Journal of Hazardous Materials* **165** 27-33

[52] Limousin G, Gaudet J P, Charlet L, Szenknect S, Barthès V and Krimissa M 2007 Sorption isotherms: A review on physical bases, modeling and measurement *Applied Geochemistry* **22** 249-75

[53] Venkiteshwaran K, Wells E and Mayer B K 2020 Kinetics, Affinity, Thermodynamics, and Selectivity of Phosphate Removal Using Immobilized Phosphate-Binding Proteins *Environmental Science & Technology* **54** 10885-94

[54] Das R, Das Tuhi S and Zaidi S M J 2018 *Carbon Nanotubes for Clean Water,* ed R Das (Cham: Springer International Publishing) pp 85-106

[55] Bekçi Z, Seki Y and Kadir Yurdakoç M 2007 A study of equilibrium and FTIR, SEM/EDS analysis of trimethoprim adsorption onto K10 *Journal of Molecular Structure* **827** 67-74

[56] Du X, Li Y, Xia Y-L, Ai S-M, Liang J, Sang P, Ji X-L and Liu S-Q 2016 Insights into Protein–Ligand Interactions: Mechanisms, Models, and Methods *International Journal of Molecular Sciences* **17** 144

[57] Manju G N, Raji C and Anirudhan T S 1998 Evaluation of coconut husk carbon for the removal of arsenic from water *Water Research* **32** 3062-70

[58] Namasivayam C and Ranganathan K 1995 Removal of Cd(II) from wastewater by adsorption on "waste" Fe(III)Cr(III) hydroxide *Water Research* **29** 1737-44
[59] Deliyanni E A, Peleka E N and Lazaridis N K 2007 Comparative study of phosphates removal from aqueous solutions by nanocrystalline akaganéite and hybrid surfactant-akaganéite *Separation and Purification Technology* **52** 478-86
[60] Genz A, Kornmüller A and Jekel M 2004 Advanced phosphorus removal from membrane filtrates by adsorption on activated aluminium oxide and granulated ferric hydroxide *Water Research* **38** 3523-30
[61] Wang J, N L I, Meng H and Fang C 2009 Research of LUCC for Industrial Zone Based on the Landsat TM Image. In: *2009 International Conference on Environmental Science and Information Application Technology,* pp 260-2
[62] Kuzawa K, Jung Y-J, Kiso Y, Yamada T, Nagai M and Lee T-G 2006 Phosphate removal and recovery with a synthetic hydrotalcite as an adsorbent *Chemosphere* **62** 45-52
[63] Mahmudov R and Huang C P 2011 Selective adsorption of oxyanions on activated carbon exemplified by Filtrasorb 400 (F400) *Separation and Purification Technology* **77** 294-300
[64] Marshall W E and Wartelle L H 2004 An anion exchange resin from soybean hulls *Journal of Chemical Technology & Biotechnology* **79** 1286-92
[65] Martin B D, Parsons S A and Jefferson B 2009 Removal and recovery of phosphate from municipal wastewaters using a polymeric anion exchanger bound with hydrated ferric oxide nanoparticles *Water Science and Technology* **60** 2637-45
[66] Zeng L, Li X and Liu J 2004 Adsorptive removal of phosphate from aqueous solutions using iron oxide tailings *Water Research* **38** 1318-26

IOP Conf. Series: Earth and Environmental Science **964** (2022) 012033 doi:10.1088/1755-1315/964/1/012033

Fabrication Of Carbon Aerogels From Coir For Oil Adsorption

Hieu M Nguyen[1,2], Khoi A Tran[1,2], Tram T N Nguyen[1,2], Nga N H Do[1,2], Kien A Le[3,*], Phung K Le[1,2,*]

[1] Refinery and Petrochemical Technology Research Centre, Ho Chi Minh City University of Technology (HCMUT), 268 Ly Thuong Kiet Street, District 10, Ho Chi Minh City, Vietnam
[2] Vietnam National University Ho Chi Minh City, Linh Trung Ward, Thu Duc District, Ho Chi Minh City, Vietnam
[3] Institute for Tropical Technology and Enviromental Protection, 57A Truong Quoc Dung Street, Phu Nhuan District, Ho Chi Minh City, Vietnam

*Email: leanhkien@vnn.vn, phungle@hcmut.edu.vn

Abstract. Coir, known as coconut fibers, are an abundant cellulosic source in Vietnam, which are mostly discarded when copra and coconut water are taken, causing environmental pollution and waste of potential biomass. In this research, carbon aerogels from chemically pretreated coir were successfully synthesized via simple sol-gel process with NaOH-urea solution, economical freeze-drying, and carbonization. The samples, including pretreated coir, coir aerogels, and carbon aerogels, are characterized using FTIR spectroscopy, SEM, XRD spectroscopy, and TGA. The carbon aerogels exhibit low density (0.034–0.047 g/cm^3), high porosity (97.63–98.32 %), and comparable motor oil sorption capacity (22.71 g/g). The properties of carbon aerogels are compared with those of coir aerogels, indicating such better values than those of coir aerogels. Coir-derived carbon aerogels is a potential replacement for the hydrophobically-coated cellulose aerogels in term of treating oil spills.

1. Introduction

Coconut is considered the tree of life that nearly all parts of the plant are valuable in industry [1]. The tree of life is grown in over 90 nations, especially in Asian and Pacific regions [1]. In Vietnam, coconuts are among perennial crops with large harvesting area. Nowadays, the total area under coconut is about 175,000 ha, in which Mekong Delta accounts for roughly 80 % of the total area [2]. The coconut husks, which are the source to obtain coir, are mostly wasted [2] when coconut water and copra are already taken. The husks are burnt, left on the sea, or landfill, causing environmental pollution. To address the problem, coir can be extracted from the husks to be further applied in industry. Coir have many applications; they can be used in reinforcing composite materials in construction [3,4], in environmental remediation [3,5], or in biotechnology [3,6]. Its composition is presented in table 1. Among the solutions, the transformation of coir to aerogels to be utilized in adsorption of liquids is a considerable option [3,7–9].

Table 1. Chemical composition of coir [10].

Chemical content	Lignin	Cellulose	Hemicellulose	Pectin	Water soluble	Ash
wt%	40–45	32–43	0.15–0.25	3–4	5	2

Oil spillage which occurs frequently every year is a part of the petroleum exploration, production and transportation business [8,9]. Oil-contaminated water will bring about extremely harsh hazards to the environment and to public health. Among the existing technologies (e.g. containment, skimming, microorganism utilizing [8]), adsorption is considered to be greatly effective for the removal of oils and organic pollutants from water [8,9,11] as other aforementioned methods are reviewed to be ineffective [8]. Traditional sorbents are polyethylene or polypropylene, which are proved to treat the oil well; however, they are controversial of its plastic content and non-biodegradability [8].

Aerogel is a gas-filled porous solid network of bonded particles, fibers, or both, exhibiting very low density, low porosity, and high specific surface area [12]. Since about 40 years after the introduction, it has been massively studied and developed because of its wide ranges of useful applications [13]. High specific surface area helps the aerogel be a perfect sorbents for metal ions, oil, organic liquids etc. [12]. Many reports have shown cellulose aerogels and carbon aerogels are those chosen one [14,15]. As the awareness about green products and sustainability draws such interest, agricultural waste has risen to be a promising candidate to produce the aerogels. A number of reports have shown such optimistic views towards the transformation of the waste into aerogels with good adsorption ability [5,11,15–18]. In 2014, Li *et al.* [16] converted winter melon directly into carbon aerogels for adsorption of oils and organic solvents. The materials had low density of 0.048 g/cm^3, high water contact angle of 135°, and can adsorb organic liquids at 16–50 times of its own weight. Later, Jing *et al.* [11] was able to fabricate biomass carbon aerogel using corn bracts. The product was applied in selective oil/water separation with high specific surface area (675.9 m2/g), superhydrophobicity (water contact angle > 150°), and high sorption capacity (77.7–143.6 g/g) towards toluene, hexane, soybean, engine oil, diesel etc.

In order to make the best use of potential waste coir, this research aims to fabricate carbon aerogels from waste coir via a facile synthesis process involving pretreatment, sol-gel process, freeze-drying and carbonization. The related samples were characterized, evaluated, as well as compared with each other to comprehend the study. Previous studies also attempted the carbonization of coir-based cellulose aerogel; however, the approach to oil adsorption application was not mentioned. Fauziyah *et al.* [5,7] successfully made cellulose aerogels from coir for oil and methylene blue treatment. The group later developed cellulose to carbon aerogels to research in the catalytic field but environmental remediation [19]. This paper studies the adsorption ability of coir-based carbon aerogel towards oil, giving a new solution to address the oil spillage.

2. Experimental section

2.1. Materials

Coir is purchased in Tien Giang Province, Vietnam. Sodium hydroxide (NaOH), urea, and hydrogen peroxide (H_2O_2) are purchased from China. Ethanol (EtOH), and motor oil 5w30 are purchased from Vietnam. All solutions are made with deionized water.

2.2. Experiment

The first stage is the pretreatment of original coir (OC). OC is firstly blended into smaller pieces (1–3 cm). Secondly, the fibers were treated with alkali solutions. The fibers and NaOH 3 % solution are stirred in a chamber at 80°C. After that, delignified coir (DC) is obtained. The fibers are further bleached with 1% NaOH-10% H_2O_2 solution. When bleaching was finished, the coir was washed till pH 7 and dried to get bleached coir (BC). To assist the preparation of aerogels, BC are blended one more time into smaller fibers (below 5 mm).

The second stage comes the synthesis of coir aerogels and carbon aerogels. Foremost, the pretreated coir is taken with desired amount (0.5 g, 0.75 g, 1 g, 1.25 g, and 1.5 g) then dispersed and dissolved in

25 mL of the solution of 7:12:81 wt% of NaOH:urea:water by a magnetic stirrer (2000 rpm). After 30 minutes, the suspension is put in the freezer for a day. The frozen gel was then thawed for 20 min at room temperature before EtOH is gently introduced to the suspension for gelation. The mixture is left for a day. Subsequently, the coir gel was washed with deionized water for several times until pH of the last liquid is 7, then, freeze-dried to obtain coir aerogel (DA). To make carbon aerogel (CA), DA is inserted into a tubular furnace of the carbonization equipment. The carbonization is conducted at 800°C with a heating rate of 5°C/min, and 100 mL/min of N_2 flow.

2.3. Characterization

The density of the aerogels is measured by mass divided by volume with the measurement of the volume of a cylinder. The porosity of the aerogels (φ) is determined by equation (1) based on their density (ρ_a) and density of cellulose fibers (ρ_b = 1.528 g/cm^3 [5]), or density of carbon (ρ_b = 2 g/cm^3 [16])

$$\rho = \left(1 - \frac{\rho_a}{\rho_c}\right) \times 100\% \tag{1}$$

The surface area of carbon aerogels was determined by nitrogen physisorption measurements, performed by BET Sorptometer. Scanning electron microscopy (SEM) was performed by FE-SEM S4800 Hitachi to study the morphology of the aerogels. XRD spectroscopy was performed by D8 Advance. The Cu-Kα radiation source (λ = 0.1541 nm) was applied. The scanning step was 0.02° with the rate of 0.6 °/min in the 2θ range from 5° to 80°. Crystallinity index (CI) is determined by the method of Segal *et al.* [20] via equation (2), where I_{002} is the intensity of the peak at 2θ around 22° (for cellulose I), or the intensity of the peak around 16° (for cellulose II), and I_{am} is the intensity of the peak at 2θ around 18° (for cellulose I), or the weakest intensity of between the peaks 18° and 22° (for cellulose II), both have the same arbitrary unit.

$$CI = \frac{I_{002} - I_{am}}{I_{002}} \times 100\% \tag{2}$$

FTIR spectra of the samples were generated by MIR/NIR Frontier. The spectroscopy was performed in the range from 400 to 4,000 cm^{-1} with a scanning speed of 1 cm^{-1}s^{-1}. TGA was performed by TGA/DSC 3+ thermal analyzer. The aerogels were heated up in from room temperature to 900°C at a rate of 5°C/min in N_2. The maximum oil adsorption capacity of the aerogels is calculated using equation (3).

$$Q_m = \frac{m_f - m_i}{m_i} \tag{3}$$

where Q_m is the maximum oil adsorption capacity of the aerogel (g/g), m_i (g) and m_f (g) are the aerogel weights before and after the oil adsorption test. To model the adsorption kinetics, two common pseudo-first and pseudo-second order models, represented by equation (4) and (5), are used.

$$\ln\left(\frac{Q_m}{Q_m - Q_t}\right) = k_1 t \tag{4}$$

$$\frac{1}{Q_t} = \frac{1}{Q_m^2 k_2} + \frac{t}{Q_m} \tag{5}$$

where Q_t are the oil adsorption capacity of the aerogel (g/g) at time t (s), respectively; k_1, and k_2 is the rate constant of the pseudo-first, and pseudo-second order model, determined from least-squares method for the graph.

3. Results and discussion

3.1. Morphology, density and porosity

DA is well-formed with sol-gel process and freeze-drying method, without any structural collapse. After pyrolysis, the three-dimensional solid structure is still maintained, and the mass reduction in the carbonization step, from coir to carbon, is about 78 %, while the volume reduction is determined to be around 70 %. Figure 1 shows the morphology and structure of DA and CA while figure 2 illustrates the density and porosity of them. The images show that DA have porous structure with randomly arranged fibers, and the fibers are 11.0–15.0 μm in term of diameter. After carbonized, the aerogels still keep the well-built structure but the size of the fibers is reduced to 4.8–8.6 μm. This is because of the loss of chemical elements when cellulose was pyrolyzed. There are also layers at some part of the fibers, as noted in figure 1(b) and 1(f). It seems they are the result of the pretreatment process, in which they outer skin of the fibers were partially destroyed; or the result of the swelling and regeneration of cellulose in sol-gel process with NaOH-urea solution. With increasing coir content, it can be seen that the fibers occupy more space and stay closer to one another in the whole volume of the aerogel. It is discernible in the images of both coir and carbon samples, and corresponds well to the trend of density and porosity of the aerogels.

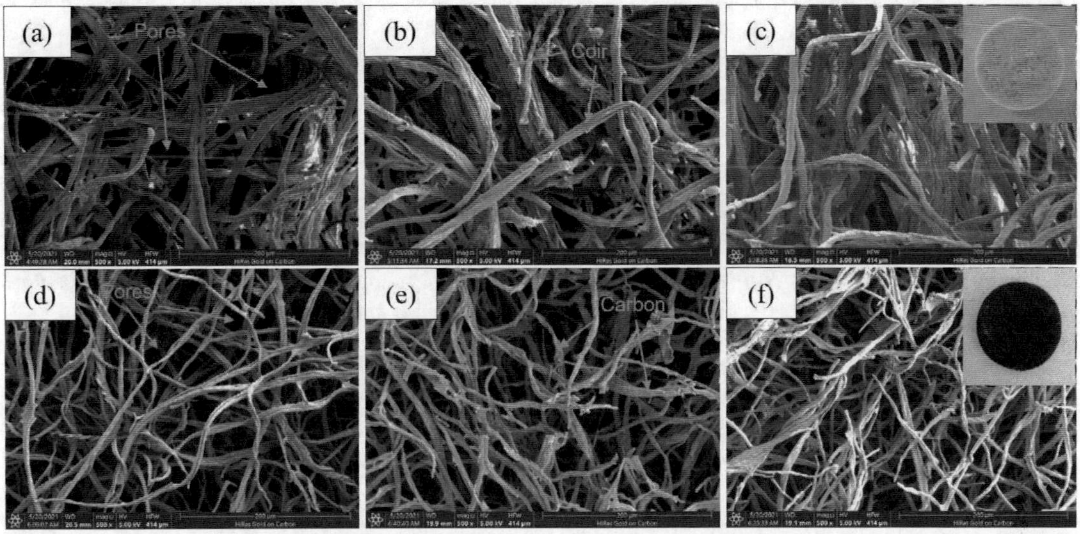

Figure 1. FE-SEM images of the coir aerogels at various fiber contents: (a) 0.5 g, (b) 1 g, and (c) 1.5 g; and the corresponding carbon aerogels: (d) 0.5 g, (e) 1 g, and (f) 1.5 g

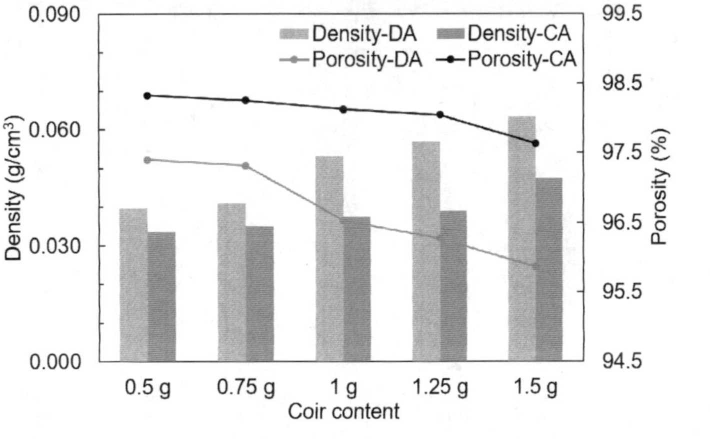

Figure 2. Density and porosity of coir aerogels and carbon aerogels.

When it comes to the density and porosity, DA possesses low density of 0.040–0.063 g/cm^3, and high porosity of 95.86–97.40 % while those values for CA are 0.034–0.047 g/cm^3, and 97.63–98.32 %, respectively. CA has lower density than DA does, which is relevant to the basic difference between the two types of aerogels. Both of them witness the same trend regarding the properties that if more fibers are added, the density will be increased but the porosity will be decreased. The specific surface area of the carbon aerogel is representatively measured for the sample of 0.5 g of fibers. The result shows the value of 289.4 m^2/g, which is higher than coir carbon aerogel (150.85 m^2/g) [19], green-leaf carbon aerogel (47.28 m^2/g) [18].

3.2. XRD and FTIR spectra

Chemical composition and its change in each pretreatment step are analyzed by XRD and FTIR spectroscopy. XRD spectra are illustrated in figure 3. It can be seen that the first three spectra [figure 3(a–c)] show the characteristic peaks of cellulose I, at 16.3°, 21.9°, and 34.0° [5,18,21,22], indicating that the cellulosic structure of the fibers is not changed after the pretreatment [21]. Besides, it is known that hemicellulose and lignin are in amorphous form in nature but cellulose is both crystalline and amorphous [21]. Therefore, changes in crystallinity of the coir can also attribute to changes in their chemical composition. Those characteristic peaks of DC and BC are sharper and clearer than that of OC, especially 15.0° and 21.9°, implying the cellulose content in pretreated coir is improved [21]. Furthermore, the crystallinity indices of the coir are shown in table 2. The increase in CI in the coir after each pretreatment stage shows the progressive elimination of the non-crystalline contents, hemicellulose and lignin [21,22]. The spectrum of DA shows characteristic peaks of cellulose II at 12.0°, 20.6°, and 22.3°, which are distinguished from the those of cellulose I [5,19,23]. The bleached fibers have CI of 57.2, which decreases after dissolution with NaOH-urea to 15.2. This might be caused by the crosslinking of cellulose [5]. So the sol-gel process of pretreated coir and NaOH-urea solution is confirmed to be successful. After carbonization, it can be observed most of the peaks of cellulose are weakened and broadened, or disappeared [figure 3(e)]. The broad bands from 20° to 25° and from 40° to 45° attribute to the existence of amorphous carbon [18,19]. This means the cellulose structure was destroyed and amorphous carbon formed [18].

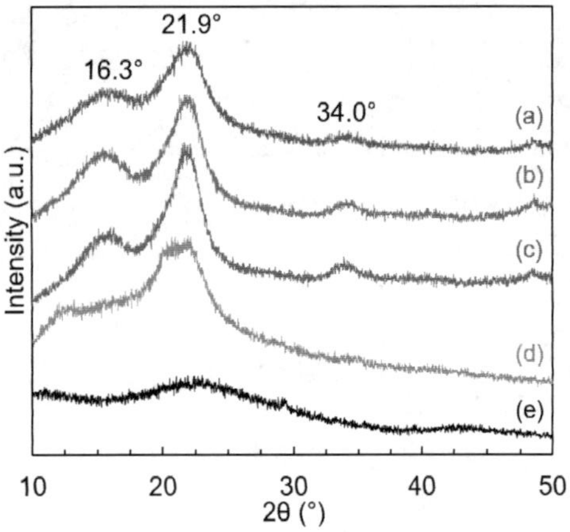

Figure 3. XRD spectra of (a) OC, (b) DC, (c) BC, (d) DA, and (e) CA.

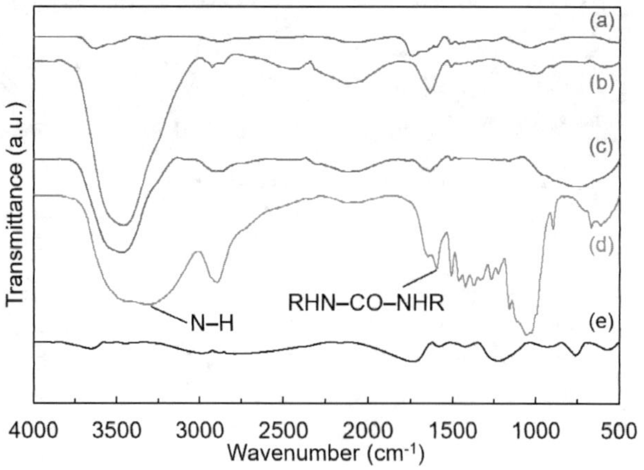

Figure 4. FTIR spectra of (a) OC, (b) DC, (c) BC, (d) DA, and (e) CA.

Table 2. Crystallinity indices of the samples.

Sample	Crystallinity index
(a) Original coir	47.0
(b) Delignified coir	51.1
(c) Bleached coir	57.2
(d) Coir aerogel	15.2

The chemical composition of the aerogels is again confirmed by FTIR spectra (figure 4). It can be observed that the first four spectra [figure 4(a–d)] present the existence of cellulose as typical peaks. The bands at around 3500, 2900, 2080, 1650 and 1450, 1080, and 890 cm^{-1} are those characterize FTIR spectrum of cellulose [18,21,22,24]. In term of the pretreatment of coir, the band at 1750 cm^{-1}, corresponding to lignin [21], and 1510 cm^{-1}, corresponding to hemicellulose [22], of OC are observed to be shortened in that of DC and OC. The characteristic bands of cellulose at roughly 3500, 2900, 2080 cm^{-1} become clearer in the spectra of the treated coir than in that of the untreated one. Also, the spectrum of DC and BC result in the reduction of the intensity of the peaks from about 1610 cm^{-1} to 1100 cm^{-1} and improved the peak at 1726 cm^{-1}, weakening the peak at 1033 cm^{-1} (aromatic C–H of lignin [24]). This may be due to breaking of some α–ether interlinks of lignin [24]. These features confirm the removal of lignin and hemicellulose and the increase of cellulose constituent in the fibers. Besides, figure 4(c) shows the change of the group of peaks 1610 to 1100 cm^{-1} more clearly, thus it demonstrates that lignin was more removed in BC than in DC. Figure 4(d) shows additional peaks at 3314 cm^{-1} (presenting the N–H stretching [19]), and 1644 cm^{-1} (indicating the functional group of urea, RHN–CO–NHR [5]). It can be implied that NaOH-urea solution contributes, to an extent, to the physical and chemical change in the structure of coir. From figure 4(e) of CA, it can be seen that all of the bands of cellulose are significantly weakened or disappear, especially 3500, 2900, 1080, and 890 cm^{-1}. Some characteristic bands of carbon are detected, which are 1725 and 1580 cm^{-1} (C=C in aromatic rings), 1225 cm^{-1} (C–O), 925 and 755 cm^{-1} (C–H) [19]. This attributes to the fact that cellulose decomposed and was transformed into carbon through pyrolysis [22]. Furthermore, the color change from light yellow to black also illustrates coir were completely pyrolyzed to form carbon.

3.3. TGA diagrams

Thermal behaviors of the coir and aerogels are studied with TGA. The thermograms are shown in figure 5(a–d), including OC, DC, BC, DA. All of the samples are observed likely the same trend of decomposition. They lose small quantity of moisture till 110°C, degraded substantially from 250°C to 400°C, in which hemicellulose, cellulose, and lignin are decomposed [5,21]; then, level off gradually. The moisture of the samples is observed to be 5–8 %. It is clearly witnessed that the substantial degradation of OC occurs more slowly and ends at a higher temperature, in which DC, BC, and DA degraded quickly from 250–355°C whereas the temperature range of OC was determined to be 260–400°C. It is because of the higher amount of lignin in the fibers [5]. Lignin has complex structure and needs higher temperature (250–700°C [21]) to be decomposed than hemicellulose (220–300°C [21]) and cellulose (275–400°C [21]). Char content of the samples ranges from 18 % to 24 %. Figure 5(d) also indicates that if coir aerogels are carbonized at 800°C, the mass of resulting carbon aerogels equal to about 20 % of the original coir aerogels, implying a mass reduction of 80 %.

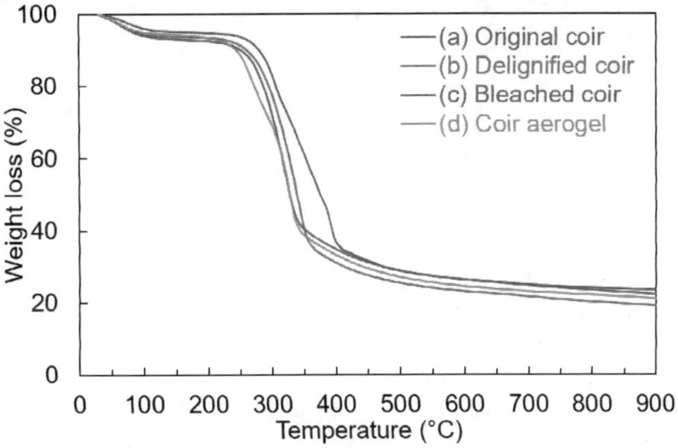

Figure 5. TGA curves of (a) OC, (b) DC, (c) BC, and (d) DA.

3.4. Oil adsorption test

Figure 6 illustrates the oil sorption of the aerogels with respect to time. DA and CA of 0.5 g of fibers were used to test the ability because they have high porosity and low density among the studied range. It can be observed from the curves that both kinds of the aerogels adsorbed the oil very quickly at the first 10 s, then, reached the equilibrium stage of adsorption after about 20–40 s. Owing to the highly porous structure of the aerogels, the oil is quickly transferred into the interior of the fibers via capillary force [25]. The maximum oil capacity of coir aerogel is relatively high, 15.20 g/g. This is due to the abundance of air space inside the aerogel, which are for the oil to be captured within. The maximum oil capacity of carbon aerogel is 22.71 g/g, which is comparative to 16 g/g of coffee-cotton aerogel [17], 20.5 g/g of paper-waste cellulose aerogel [9]. It is larger than that of coir sample, which can be explained by the larger porosity and high specific surface area, as the two factors that significantly contribute to the sorption ability [12]. In practice, cellulose aerogel has to be hydrophobically coated to perform well in oil-water mixture, which was time-consuming (up to a day [17]) and costly (e.g., the price of MTMS, HDMS). But with carbonization, the fabrication time and cost can be saved to an extent.

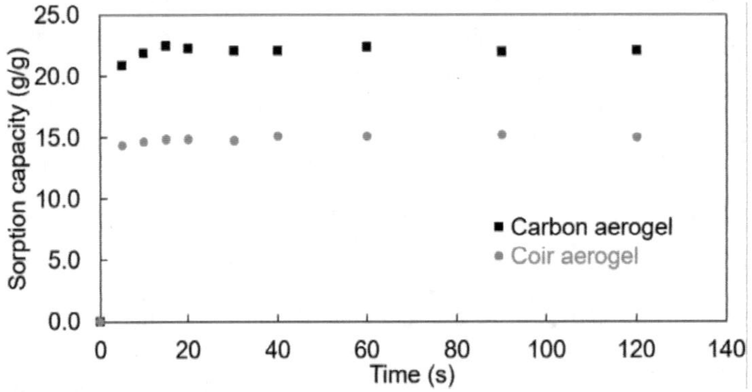

Figure 6. Oil adsorption kinetics of coir aerogel and carbon aerogel.

The analysis of pseudo-first order and pseudo-second order kinetic model is depicted in table 3. The R^2 values for both pseudo-first order and pseudo-second order models are also calculated. Since the R^2 values for the pseudo-second order model is closer to 1 than those of the pseudo-first order, it can be concluded that the oil adsorption kinetics of the both coir aerogels and carbon aerogels follows pseudo-second order model more accurately.

Table 3. Data of pseudo-first order and pseudo-second order kinetic models.

		Coir aerogel	Carbon aerogel
Pseudo-first order model	R^2	0.3607	0.1194
	k_1	0.0002	0.0001
Pseudo-second order model	R^2	0.9999	0.9999
	k_2	1.6554	1.2193

4. Conclusion

The pretreatment of coir is evaluated to be relatively successful with the increase of cellulose constituent in the fibers, displayed in the FTIR, and XRD spectra. Carbon aerogels were successfully fabricated from pretreated coir via sol-gel process, freeze-drying, and carbonization. They can maintain the porous structure of DA after carbonization. CA have low density of 0.034–0.047 g/cm^3, high porosity of 97.63–98.32 %, and high specific surface area of 289.4 m^2/g. The carbon samples show lower density and higher porosity than those of coir aerogel, which contributes to a higher oil sorption capacity (22.71 g/g compared to 15.20 g/g, respectively). The research shows a possibility of transforming coir into carbon aerogels for effective oil adsorption. For the practical adsorption application, instead of time-consuming and costly hydrophobic silane-coating of cellulose aerogel, carbonization can be used as a potential alternative.

Acknowledgements
We also acknowledge the support of time and facilities from Ho Chi Minh City University of Technology (HCMUT), VNU-HCM for this study.

References

[1] Adkins S, Foale M, Bourdeix R, Nguyen Q and Biddle J 2020 *Coconut Biotechnology: Towards the Sustainability of the "Tree of Life"* (Cham: Springer Nature Switzerland AG).
[2] Duc D N and Nananukul N 2018 Supply chain model for renewable energy resource from biomass in Vietnam *5th Int. Conf. on Industrial Engineering and Applications* 26-28 April 2018 Singapore .
[3] Reddy N 2019 *Sustainable Applications of Coir and Other Coconut By-products* (Cham: Springer Nature Switzerland AG).
[4] Murthi P, Poongodi K, Gobinath R and Saravanan R 2020 Evaluation of material performance of coir fibre reinforced quaternary blended concrete *IOP Conf. Ser.: Mater. Sci. Eng.* **872** 012133.
[5] Fauziyah M, Widiyastuti W, Balgis R and Setyawan H 2019 Production of cellulose aerogels from coir fibers via an alkali–urea method for sorption applications *Cellulose* **26** 9583–9598.
[6] Patel S K S, Kalia V C, Joo J B, Kang Y C and Lee J K 2019 Biotransformation of methane into methanol by methanotrophs immobilized on coconut coir *Bioresour. Technol.* **297** 122433
[7] Fauziyah M, Widiyastuti W and Setyawan H 2020 A hydrophobic cellulose aerogel from coir fibers waste for oil spill application *IOP Conf. Ser.: Mater. Sci. Eng.* **778** 012019.
[8] Feng J, Nguyen S T, Fan Z and Duong H M 2015 Advanced fabrication and oil absorption properties of super-hydrophobic recycled cellulose aerogels *Chem. Eng. J.* **270** 168–175.
[9] Nguyen S T, Feng J, Le N T, Nguyen H, Tan V B C and Duong H M 2013 Cellulose Aerogel from Paper Waste for Crude Oil Spill Cleaning *Ind. Eng. Chem. Res.* **52** 18386−18391.
[10] Ayrilmis N, Jarusombuti S, Fueangvivat V, Bauchongkol P and White R H 2011 Coir fiber reinforced polypropylene composite panel for automotive interior applications *Fibers Polym.* **12** 919–926.
[11] Jing Z, Ding J, Zhang T, Yang D, Qiu F, Chen Q and Xu J 2019 Flexible, versatility and superhydrophobic biomass carbon aerogels derived from corn bracts for efficient oil/water separation *Food Bioprod. Process.* **115** 134–142.
[12] Smirnova I and Gurikov P 2017 Aerogels in Chemical Engineering: Strategies Toward Tailor-Made Aerogels *Annu. Rev. Chem. Biomol. Eng.* **8** 307–334.
[13] Duong H M and Nguyen S T 2017 Green Recycled Cellulose Aerogels: Properties and Applications *Green Polymer Composites Technology: Properties and Applications*, ed Inamuddin (Boca Raton: CRC Press) p 237–250.
[14] Gan G, Li X, Fan S, Wang L, Qin M, Yin Z and Chen G 2019 Carbon Aerogels for Environmental Clean-Up *Eur. J. Inorg. Chem.* **2019** 3126–3141.
[15] Zhao S, Malfait W J, Guerrero-Alburquerque N, Koebel M M and Nyström G 2018 Biopolymer Aerogels and Foams: Chemistry, Properties, and Applications *Angew. Chemie Int. Ed.* **57** 7580–7608.
[16] Li Y Q, Samad Y A, Polychronopoulou K, Alhassan S M and Liao K 2014 Carbon Aerogel from Winter Melon for Highly Efficient and Recyclable Oils and Organic Solvents Absorption *ACS Sustain. Chem. Eng.* **2** 1492–1497.
[17] Zhang X, Kwek L P, Le D K, Tan M S and Duong H M 2019 Fabrication and Properties of Hybrid Coffee-Cellulose Aerogels from Spent Coffee Grounds *Polymers (Basel)* **11** 1–12.

[18] Zhanga T, Yuana D, Guoa Q, Qiua F, Yanga D and Ou Z 2019 Preparation of a renewable biomass carbon aerogel reinforced with sisal for oil spillage clean-up: Inspired by green leaves to green Tofu *Food Bioprod. Process.*, vol. **114**, pp. 154–162.

[19] Fauziyah M, Widiyastuti W and Setyawan H 2020 Nitrogen-Doped Carbon Aerogels Prepared by Direct Pyrolysis of Cellulose Aerogels Derived from Coir Fibers Using an Ammonia– Urea System and Their Electrocatalytic Performance toward the Oxygen Reduction Reaction *Ind. Eng. Chem. Res.* **59** 21371–21382.

[20] Segal L, Creely J J, Martin A E and Conrad C M 1959 An Empirical Method for Estimating the Degree of Crystallinity of Native Cellulose Using the X-Ray Diffractometer *Text. Res. J.* **29** 786–794.

[21] Feng Y H, Cheng T Y, Yang W G, Ma P T, He H Z, Yin X C and Yu X X 2018 Characteristics and environmentally friendly extraction of cellulose nanofibrils from sugarcane bagasse *Ind. Crops Prod.* **111** 285–291.

[22] Ditzel F I, Prestes E, Carvalho B M, Demiate I M and Pinheiro L A 2017 Nanocrystalline cellulose extracted from pine wood and corncob *Carbohydr. Polym.* **157** 1577–1585.

[23] Qi H, Yang Q, Zhang L, Liebert T and Heinze T 2011 The dissolution of cellulose in NaOH-based aqueous system by two-step process *Cellulose* **10** 237–245.

[24] Samanta A K, Basu G and Mishra L 2018 Role of major constituents of coconut fibres on absorption of ionic dyes *Ind. Crops Prod.* **117** 20–27.

[25] Mahani A A, Motahari S and Mohebbi A 2018 Sol-gel derived flexible silica aerogel as selective adsorbent for water decontamination from crude oil *Mar. Pollut. Bull.* **129** 438–447.

AUTHOR INDEX